Grundlehren der mathematischen Wissenschaften 333

A Series of Comprehensive Studies in Mathematics

Geoffrey Grimmett

The Random-Cluster Model

With 37 Figures

Geoffrey R. Grimmett
University of Cambridge
Statistical Laboratory
Centre for Mathematical Sciences
Wilberforce Road
Cambridge CB3 0WB
United Kingdom
E-mail: *g.r.grimmett@statslab.cam.ac.uk*

Library of Congress Control Number: 2006925087

Mathematics Subject Classification (2000): 60K35, 82B20, 82B43

ISSN 0072-7830
ISBN-10 3-540-32890-4 Springer Berlin Heidelberg New York
ISBN-13 978-3-540-32890-2 Springer Berlin Heidelberg New York

Springer is a part of Springer Science+Business Media
springer.com

Printed in The Netherlands

Typesetting: by the author
Cover design: *design & production* GmbH, Heidelberg

Printed on acid-free paper SPIN: 11411086 41/SPI Publisher Services 5 4 3 2 1 0

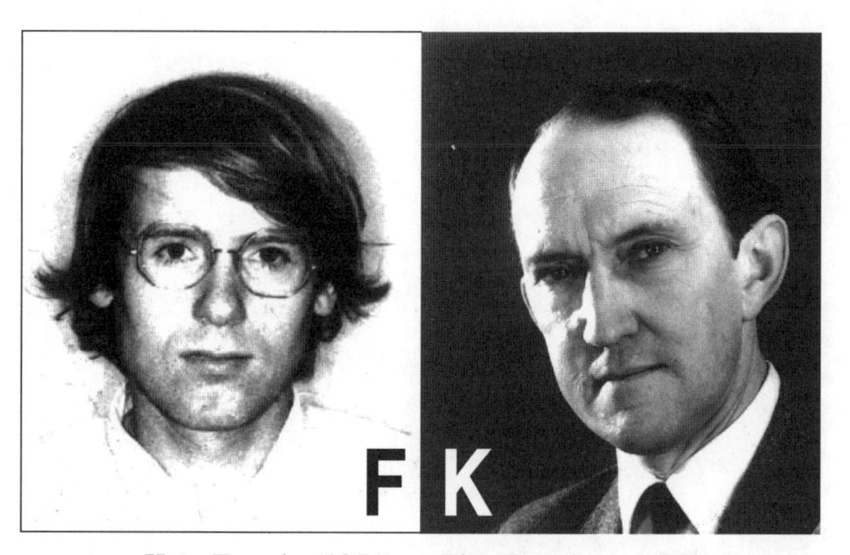

Kees Fortuin (1971)　Piet Kasteleyn (1968)

Preface

The random-cluster model was invented by Cees [Kees] Fortuin and Piet Kasteleyn around 1969 as a unification of percolation, Ising, and Potts models, and as an extrapolation of electrical networks. Their original motivation was to harmonize the series and parallel laws satisfied by such systems. In so doing, they initiated a study in stochastic geometry which has exhibited beautiful structure in its own right, and which has become a central tool in the pursuit of one of the oldest challenges of classical statistical mechanics, namely to model and analyse the ferromagnet and especially its phase transition.

The importance of the model for probability and statistical mechanics was not fully recognized until the late 1980s. There are two reasons for this period of dormancy. Although the early publications of 1969–1972 contained many of the basic properties of the model, the emphasis placed there upon combinatorial aspects may have obscured its potential for applications. In addition, many of the geometrical arguments necessary for studying the model were not known prior to 1980, but were developed during the 'decade of percolation' that began then. In 1980 was published the proof that $p_c = \frac{1}{2}$ for bond percolation on the square lattice, and this was followed soon by Harry Kesten's monograph on two-dimensional percolation. Percolation moved into higher dimensions around 1986, and many of the mathematical issues of the day were resolved by 1989. Interest in the random-cluster model as a tool for studying the Ising/Potts models was rekindled around 1987. Swendsen and Wang utilized the model in proposing an algorithm for the time-evolution of Potts models; Aizenman, Chayes, Chayes, and Newman used it to show discontinuity in long-range one-dimensional Ising/Potts models; Edwards and Sokal showed how to do it with coupling.

One of my main projects since 1992 has been to comprehend the (in)validity of the mantra 'everything worth doing for Ising/Potts is best done via random-cluster'. There is a lot to be said in favour of this assertion, but its unconditionality is its weakness. The random-cluster representation has allowed beautiful proofs of important facts including: the discontinuity of the phase transition for large values of the cluster-factor q, the existence of non-translation-invariant 'Dobrushin' states for large values of the edge-parameter p, the Wulff construction in two and more dimensions, and so on. It has played important roles in the studies of other classical and quantum systems in statistical mechanics, including for example the Widom–Rowlinson two-type lattice gas and the Edwards–Anderson spin-glass model. The last model is especially challenging because it is non-ferromagnetic, and thus gives rise to new problems of importance and difficulty.

The random-cluster model is however only one of the techniques necessary for the mathematical study of ferromagnetism. The principal illustration of its limitations concerns the Ising model. This fundamental model for a ferromagnet has exactly two local states, and certain special features of the number 2 enable a beautiful analysis via the so-called 'random-current representation' which does not appear to be reproducible by random-cluster arguments.

In pursuing the theory of the random-cluster model, I have been motivated not only by its applications to spin systems but also because it is a source of beautiful problems in its own right. Such problems involve the stochastic geometry of interacting lattice systems, and they are close relatives of those treated in my monograph on percolation, published first in 1989 and in its second edition in 1999. There are many new complications and some of the basic questions remain unanswered, at least in part. The current work is primarily an exposition of a fairly mature theory, but prominence is accorded to open problems of significance.

New problems have arrived recently to join the old, and these concern primarily the two-dimensional phase transition and its relation to the theory of stochastic Löwner evolutions. SLE has been much developed for percolation and related topics since the 1999 edition of *Percolation*, mostly through the achievements of Schramm, Smirnov, Lawler, and Werner. We await an extension of the mathematics of SLE to random-cluster and Ising/Potts models.

Here are some remarks on the contents of this book. The setting for the vast majority of the work reported here is the d-dimensional hypercubic lattice $\mathbb{Z}^d$ where $d \geq 2$. This has been chosen for ease of presentation, and may usually be replaced by any other finite-dimensional lattice in two or more dimensions, although an extra complication may arise if the lattice is not vertex-transitive. An exception to this is found in Chapter 6, where the self-duality of the square lattice is exploited.

Following the introductory material of Chapter 1, the fundamental properties of monotonic and random-cluster measures on finite graphs are summarized in Chapters 2 and 3, including accounts of stochastic ordering, positive association, and exponential steepness.

A principal feature of the model is the presence of a phase transition. Since singularities may occur only on infinite graphs, one requires a definition of the random-cluster model on an infinite graph. This may be achieved as for other systems either by passing to an infinite-volume weak limit, or by studying measures which satisfy consistency conditions of Dobrushin–Lanford–Ruelle (DLR) type. Infinite-volume measures in their two forms are studied in Chapter 4.

The percolation probability is introduced in Chapter 5, and this leads to a study of the phase transition and the critical point $p_c(q)$. When $p < p_c(q)$, one expects that the size of the open cluster containing a given vertex of $\mathbb{Z}^d$ is controlled by exponentially-decaying probabilities. This is unproven in general, although exponential decay is proved subject to a further condition on the parameter p.

The supercritical phase, when $p > p_c(q)$, has been the scene of recent major developments for random-cluster and Ising/Potts models. A highlight has been the proof of the so-called 'Wulff construction' for supercritical Ising models. A version of the Wulff construction is valid for the random-cluster model subject to a stronger condition on p, namely that $p > \widehat{p}_c(q)$ where $\widehat{p}_c(q)$ is (for $d \geq 3$) the limit of certain slab critical points. We have no proof that $\widehat{p}_c(q) = p_c(q)$ except when $q = 1, 2$, and to prove this is one of the principal open problems of the day. A second problem is to prove the uniqueness of the infinite-volume limit whenever $p \neq p_c(q)$.

The self-duality of the two-dimensional square lattice $\mathbb{Z}^2$ is complemented by a duality relation for random-cluster measures on planar graphs, and this allows a fuller understanding of the two-dimensional case, as described in Chapter 6. There remain important open problems, of which the principal one is to obtain a clear proof of the 'exact calculation' $p_c(q) = \sqrt{q}/(1 + \sqrt{q})$. This calculation is accepted by probabilists when $q = 1$ (percolation), $q = 2$ (Ising), and when q is large, but the "exact solutions" of theoretical physics seem to have no complete counterpart in rigorous mathematics for general values of q satisfying $q \in [1, \infty)$. There is strong evidence that the phase transition with $d = 2$ and $q \in [1, 4)$ will be susceptible to an analysis using SLE, and this will presumably enable in due course a computation of its critical exponents.

In Chapter 7, we consider duality in three and more dimensions. The dual model amounts to a probability measure on surfaces and certain topological complications arise. Two significant facts are proved. First, it is proved for sufficiently large q that the phase transition is discontinuous. Secondly, it is proved for $q \in [1, \infty)$ and sufficiently large p that there exist non-translation-invariant 'Dobrushin' states.

The model has been assumed so far to be static in time. Time-evolutions may be introduced in several ways, as described in Chapter 8. Glauber dynamics and the Gibbs sampler are discussed, followed by the Propp–Wilson scheme known as 'coupling from the past'. The random-cluster measures for different values of p may be coupled via the equilibrium measure of a suitable Markov process on $[0, 1]^E$, where E denotes the set of edges of the underlying graph.

The so-called 'random-current representation' was remarked above for the Ising model, and a related representation using the 'flow polynomial' is derived in Chapter 9 for the q-state Potts model. It has not so far proved possible to exploit this in a full study of the Potts phase transition. In Chapter 10, we consider the random-cluster model on graphs with a different structure than that of finite-dimensional lattices, namely the complete graph and the binary tree. In each case one may perform exact calculations of mean-field type.

The final Chapter 11 is devoted to applications of the random-cluster representation to spin systems. Five such systems are described, namely the Potts and Ashkin–Teller models, the disordered Potts model, the spin-glass model of Edwards and Anderson, and the lattice gas of Widom and Rowlinson.

There is an extensive literature associated with ferromagnetism, and I have not aspired to a complete account. Salient references are listed throughout this book, but inevitably there are omissions. Amongst earlier papers on random-cluster models, the following include a degree of review material: [8, 44, 136, 149, 156, 169, 240].

I first encountered the random-cluster model one day in late 1971 when John Hammersley handed me Cees Fortuin's thesis. Piet Kasteleyn responded enthusiastically to my 1992 request for information about the history of the model, and his letters are reproduced with his permission in the Appendix. The responses from fellow probabilists to my frequent requests for help and advice have been deeply appreciated, and the support of the community is gratefully acknowledged. I thank Laantje Kasteleyn and Frank den Hollander for the 1968 photograph of Piet, and Cees Fortuin for sending me a copy of the image from his 1971 California driving licence. Raphaël Cerf kindly offered guidance on the Wulff construction, and has supplied some of his beautiful illustrations of Ising and random-cluster models, namely Figures 1.2 and 5.1. A number of colleagues have generously commented on parts of this book, and I am especially grateful to Rob van den Berg, Benjamin Graham, Olle Häggström, Chuck Newman, Russell Lyons, and Senya Shlosman. Jeff Steif has advised me on ergodic theory, and Aernout van Enter has helped me with statistical mechanics. Catriona Byrne has been a source of encouragement and support. I express my thanks to these and to others who have, perhaps unwittingly or anonymously, contributed to this volume.

G. R. G.
Cambridge
January 2006

Contents

Chapter 1

Random-Cluster Measures

Summary. The random-cluster model is introduced, and its relationship to Ising and Potts models is presented via a coupling of probability measures. In the limit as the cluster-weighting factor tends to 0, one arrives at electrical networks and uniform spanning trees and forests.

1.1 Introduction

In 1925 came the Ising model for a ferromagnet, and in 1957 the percolation model for a disordered medium. Each has since been the subject of intense study, and their theories have become elaborate. Each possesses a phase transition marking the onset of long-range order, defined in terms of correlation functions for the Ising model and in terms of the unboundedness of paths for percolation. These two phase transitions have been the scenes of notable exact (and rigorous) calculations which have since inspired many physicists and mathematicians.

It has been known since at least 1847 that electrical networks satisfy so-called 'series/parallel laws'. Piet Kasteleyn noted during the 1960s that the percolation and Ising models also have such properties. This simple observation led in joint work with Cees Fortuin to the formulation of the random-cluster model. This new model has two parameters, an 'edge-weight' p and a 'cluster-weight' q. The (bond) percolation model is retrieved by setting $q = 1$; when $q = 2$, we obtain a representation of the Ising model, and similarly of the Potts model when $q = 2, 3, \dots$. The discovery of the model is described in Kasteleyn's words in the Appendix of the current work.

The mathematics begins with a finite graph $G = (V, E)$, and the associated Ising model[1] thereon. A random variable σ_x taking values -1 and $+1$ is assigned to each vertex x of G, and the probability of the configuration $\sigma = (\sigma_x : x \in V)$ is taken to be proportional to $e^{-\beta H(\sigma)}$, where $\beta > 0$ and the 'energy' $H(\sigma)$ is the

[1] The so-called Ising model [190] was in fact proposed (to Ising) by Lenz. The Potts model [105, 278] originated in a proposal (to Potts) by Domb.

negative of the sum of $\sigma_x \sigma_y$ over all edges $e = \langle x, y \rangle$ of G. As β increases, greater probability is assigned to configurations having a large number of neighbouring pairs of vertices with equal signs. The Ising model has proved extraordinarily successful in generating beautiful mathematics of relevance to the physics, and it has been useful and provocative in the mathematical theory of phase transitions and cooperative phenomena (see, for example, [118]). The proof of the existence of a phase transition in two dimensions was completed by Peierls, [266], by way of his famous "argument".

There are many possible generalizations of the Ising model in which the σ_x may take a general number q of values, rather than $q = 2$ only. One such extension, the so-called 'Potts model', [278], has attracted especial interest amongst physicists, and has displayed a complex and varied structure. For example, when q is large, it possesses a discontinuous phase transition, in contrast to the continuous transition believed to take place for small q. Ising/Potts models are the first of three principal ingredients in the story of random-cluster models. Note that they are 'vertex-models' in the sense that they involve random variables σ_x indexed by the vertices x of the underlying graph. (There is a related extension of the Ising model due to Ashkin and Teller, [21], see Section 11.3.)

The (bond) percolation model was inspired by problems of physical type, and emerged from the mathematics literature[2] of the 1950s, [70]. In this model for a porous medium, each edge of the graph G is declared 'open' (to the passage of fluid) with probability p, and 'closed' otherwise, different edges having independent states. The problem is to determine the typical large-scale properties of connected components of open edges as the parameter p varies. Percolation theory is now a mature part of probability lying at the core of the study of random media and interacting systems, and it is the second ingredient in the story of random-cluster models. Note that bond percolation is an 'edge-model', in that the random variables are indexed by the set of edges of the underlying graph. (There is a variant termed 'site percolation' in which the vertices are open/closed at random rather than the edges, see [154, Section 1.6].)

The theory of electrical networks on the graph G is of course more ancient than that of Ising and percolation models, dating back at least to the 1847 paper, [215], in which Kirchhoff set down a method for calculating macroscopic properties of an electrical network in terms of its local structure. Kirchhoff's work explains in particular the relevance of counts of certain types of spanning trees of the graph. To import current language, an electrical network on a graph G may be studied via the properties of a 'uniformly random spanning tree' (UST) on G (see [31]).

The three ingredients above seemed fairly distinct until Fortuin and Kasteleyn discovered around 1970, [120, 121, 122, 123, 203], that each features within a certain parametric family of models which they termed 'random-cluster models'. They developed the basic theory of such models — correlation inequalities and the like — in this series of papers. The true power of random-cluster models as

[2] See also the historical curiosity [323].

a mechanism for studying Ising/Potts models has emerged progressively over the intervening three decades.

The configuration space of the random-cluster model is the set of all subsets of the edge-set E, which we represent as the set $\Omega = \{0, 1\}^E$ of $0/1$-vectors indexed by E. An edge e is termed *open* in the configuration $\omega \in \Omega$ if $\omega(e) = 1$, and it is termed *closed* if $\omega(e) = 0$. The random-cluster model is thus an *edge*-model, in contrast to the Ising and Potts models which assign spins to the *vertices* of G. The subject of current study is the subgraph of G induced by the set of open edges of a configuration chosen at random from Ω according to a certain probability measure. Of particular importance is the existence (or not) of paths of open edges joining given vertices x and y, and thus the random-cluster model is a model in stochastic geometry.

The model may be viewed as a parametric family of probability measures $\phi_{p,q}$ on Ω, the two parameters being denoted by $p \in [0, 1]$ and $q \in (0, \infty)$. The parameter p amounts to a measure of the density of open edges, and the parameter q is a 'cluster-weighting' factor. When $q = 1$, $\phi_{p,q}$ is a product measure, and the ensuing probability space is usually termed a percolation model or a random graph depending on the context. The integer values $q = 2, 3, \dots$ correspond in a certain way to the Potts model on G with q local states, and thus $q = 2$ corresponds to the Ising model. The nature of these 'correspondences', as described in Section 1.4, is that 'correlation functions' of the Potts model may be expressed as 'connectivity functions' of the random-cluster model. When extended to infinite graphs, it turns out that long-range order in a Potts model corresponds to the existence of infinite clusters in the corresponding random-cluster model. In this sense the Potts and percolation phase transitions are counterparts of one another.

Therein lies a major strength of the random-cluster model. Geometrical methods of some complexity have been derived in the study of percolation, and some of these may be adapted and extended to more general random-cluster models, thereby obtaining results of significance for Ising and Potts models. Such has been the value of the random-cluster model in studying Ising and Potts models that it is sometimes called simply the 'FK representation' of the latter systems, named after Fortuin and Kasteleyn. We shall see in Chapter 11 that several other spin models of statistical mechanics possess FK-type representations.

The random-cluster and Ising/Potts models on the graph $G = (V, E)$ are defined formally in the next two sections. Their relationship is best studied via a certain coupling on the product $\{0, 1\}^E \times \{1, 2, \dots, q\}^V$, and this coupling is described in Section 1.4. The 'uniform spanning-tree' (UST) measure on G is a limiting case of the random-cluster measure, and this and related limits are the topic of Section 1.5. This chapter ends with a section devoted to basic notation.

1.2 Random-cluster model

Let $G = (V, E)$ be a finite graph. The graphs considered here will usually possess neither loops nor multiple edges, but we make no such general assumption. An edge e having endvertices x and y is written as $e = \langle x, y \rangle$. A random-cluster measure on G is a member of a certain class of probability measures on the set of subsets of the edge set E. We take as state space the set $\Omega = \{0, 1\}^E$, members of which are 0/1-vectors $\omega = (\omega(e) : e \in E)$. We speak of the edge e as being *open* (in ω) if $\omega(e) = 1$, and as being *closed* if $\omega(e) = 0$. For $\omega \in \Omega$, let $\eta(\omega) = \{e \in E : \omega(e) = 1\}$ denote the set of open edges. There is a one–one correspondence between vectors $\omega \in \Omega$ and subsets $F \subseteq E$, given by $F = \eta(\omega)$. Let $k(\omega)$ be the number of connected components (or 'open clusters') of the graph $(V, \eta(\omega))$, and note that $k(\omega)$ includes a count of isolated vertices, that is, of vertices incident to no open edge. We associate with Ω the σ-field $\mathcal{F}$ of all its subsets.

A *random-cluster measure* on G has two parameters satisfying $p \in [0, 1]$ and $q \in (0, \infty)$, and is defined as the measure $\phi_{p,q}$ on the measurable pair $(\Omega, \mathcal{F})$ given by

$$\phi_{p,q}(\omega) = \frac{1}{Z_{\mathrm{RC}}} \left\{ \prod_{e \in E} p^{\omega(e)} (1-p)^{1-\omega(e)} \right\} q^{k(\omega)}, \qquad \omega \in \Omega, \tag{1.1}$$

where the 'partition function', or 'normalizing constant', Z_{RC} is given by

$$Z_{\mathrm{RC}} = Z_{\mathrm{RC}}(p, q) = \sum_{\omega \in \Omega} \left\{ \prod_{e \in E} p^{\omega(e)} (1-p)^{1-\omega(e)} \right\} q^{k(\omega)}. \tag{1.2}$$

This measure differs from product measure through the inclusion of the term $q^{k(\omega)}$. Note the difference between the cases $q \le 1$ and $q \ge 1$: the former favours fewer clusters, whereas the latter favours a larger number of clusters. When $q = 1$, edges are open/closed independently of one another. This very special case has been studied in detail under the titles 'percolation' and 'random graphs', see [61, 154, 194]. Perhaps the most important values of q are the integers, since the random-cluster model with $q \in \{2, 3, \dots\}$ corresponds, in a way described in the next two sections, to the Potts model with q local states. The bulk of the work presented in this book is devoted to the theory of random-cluster measures when $q \ge 1$. The case $q < 1$ seems to be harder mathematically and less important physically. There is some interest in the limit as $q \downarrow 0$; see Section 1.5.

We shall sometimes write $\phi_{G,p,q}$ for $\phi_{p,q}$ when the choice of graph G is to be stressed. Computer-generated samples from random-cluster measures on $\mathbb{Z}^2$ are presented in Figures 1.1–1.2. When $q = 1$, the measure $\phi_{p,q}$ is a product measure with density p, and we write $\phi_{G,p}$ or ϕ_p for this special case.

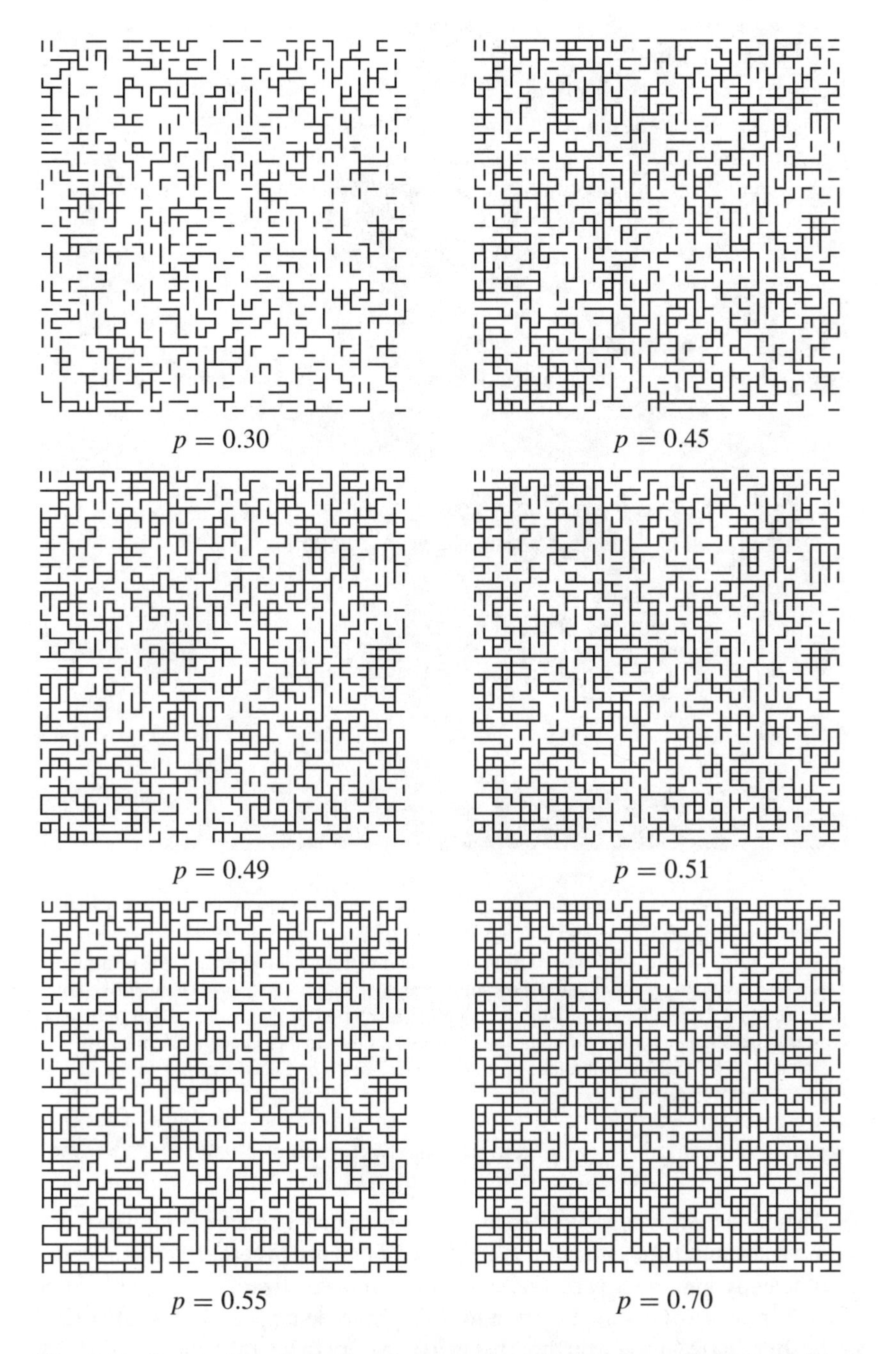

Figure 1.1. Samples from the random-cluster measure with $q = 1$ on a 40×40 box of the square lattice. We have set $q = 1$ for ease of programming, the measure being of product form in this case. The critical value is $p_c(1) = \frac{1}{2}$. Samples with more general values of q may be obtained by the method of 'coupling from the past', as described in Section 8.4.

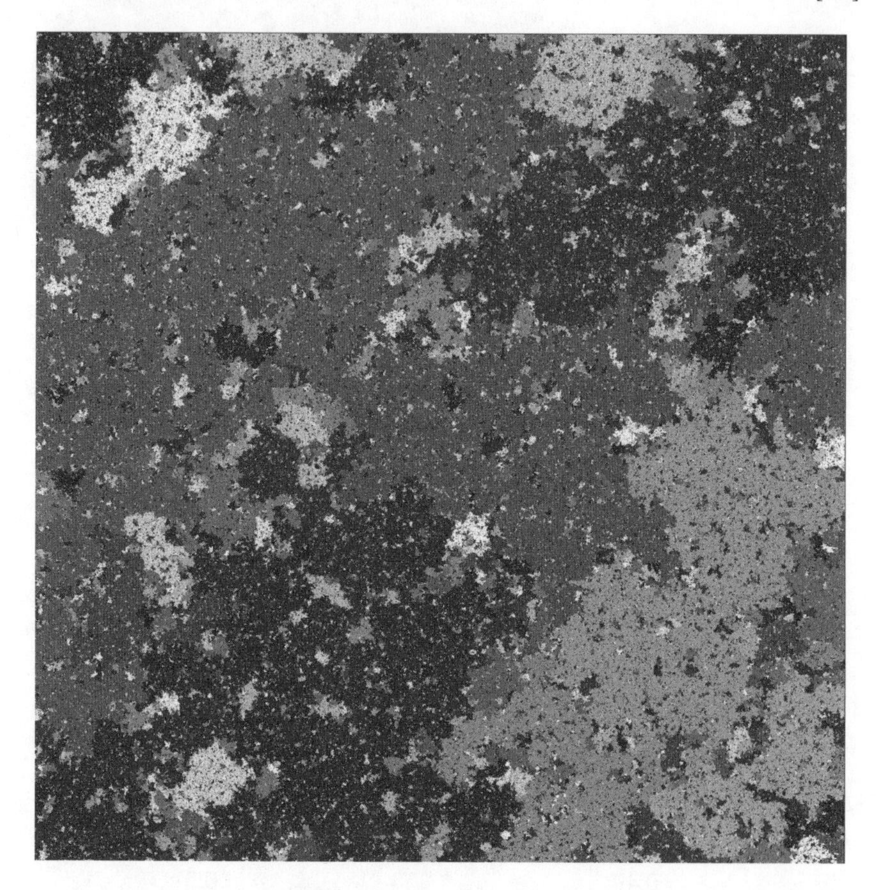

Figure 1.2. A picture of the random-cluster model with free boundary conditions on a 2048×2048 box of $\mathbb{L}^2$, with $p = 0.585816$ and $q = 2$. The critical value of the model with $q = 2$ is $p_c = \sqrt{2}/(1+\sqrt{2}) = 0.585786\ldots$, and therefore the simulation is of a mildly supercritical system. It was obtained by simulating the Ising model using Glauber dynamics (see Section 8.2), and then applying the coupling illustrated in Figure 1.3. Each individual cluster is highlighted with a different tint of gray, and the smaller clusters are not visible in the picture. This and later simulations in Section 5.7 are reproduced by kind permission of Raphaël Cerf.

1.3 Ising and Potts models

In a famous experiment, a piece of iron is exposed to a magnetic field. The field is increased from zero to a maximum, and then diminished to zero. If the temperature is sufficiently low, the iron retains some residual magnetization, otherwise it does not. There is a critical temperature for this phenomenon, often called the *Curie point* after Pierre Curie, who reported this discovery in his 1895 thesis, [98][3]. The

[3] In an example of Stigler's Law, [309], the existence of such a temperature was discovered before 1832 by Pouillet, see [198].

famous (Lenz–)Ising model for such ferromagnetism, [190], may be summarized as follows. One supposes that particles are positioned at the points of some lattice embedded in Euclidean space. Each particle may be in either of two states, representing the physical states of 'spin-up' and 'spin-down'. Spin-values are chosen at random according to a certain probability measure, known as a 'Gibbs state', which is governed by interactions between neighbouring particles. The relevant probability measure is given as follows.

Let $G = (V, E)$ be a finite graph representing part of the lattice. We think of each vertex $x \in V$ as being occupied by a particle having a random spin. Since spins are assumed to come in two basic types, we take as sample space the set $\Sigma = \{-1, +1\}^V$. The appropriate probability mass function $\lambda_{\beta,J,h}$ on Σ has three parameters satisfying $\beta, J \in [0, \infty)$ and $h \in \mathbb{R}$, and is given by

$$\lambda_{\beta,J,h}(\sigma) = \frac{1}{Z_{\mathrm{I}}} e^{-\beta H(\sigma)}, \qquad \sigma \in \Sigma, \tag{1.3}$$

where the partition function Z_{I} and the 'Hamiltonian' $H : \Sigma \to \mathbb{R}$ are given by

$$Z_{\mathrm{I}} = \sum_{\sigma \in \Sigma} e^{-\beta H(\sigma)}, \qquad H(\sigma) = -J \sum_{e=\langle x,y\rangle \in E} \sigma_x \sigma_y - h \sum_{x \in V} \sigma_x. \tag{1.4}$$

The physical interpretation of β is as the reciprocal $1/T$ of temperature, of J as the strength of interaction between neighbours, and of h as the external magnetic field. For reasons of simplicity, we shall consider here only the case of zero external-field, and we assume henceforth that $h = 0$.

Each edge has equal interaction strength J in the above formulation. Since β and J occur only as a product βJ, the measure $\lambda_{\beta,J,0}$ has effectively only a single parameter βJ. In a more complicated measure not studied here, different edges e are permitted to have different interaction strengths J_e, see Chapter 9. In the meantime we shall wrap β and J together by setting $J = 1$, and we write $\lambda_\beta = \lambda_{\beta,1,0}$

As pointed out by Baxter, [26], the Ising model permits an infinity of generalizations. Of these, the extension to so-called 'Potts models' has proved especially fruitful. Whereas the Ising model permits only two possible spin-values at each vertex, the Potts model [278] permits a general number $q \in \{2, 3, \dots\}$, and is governed by a probability measure given as follows.

Let q be an integer satisfying $q \geq 2$, and take as sample space the set of vectors $\Sigma = \{1, 2, \dots, q\}^V$. Thus each vertex of G may be in any of q states. For an edge $e = \langle x, y\rangle$ and a configuration $\sigma = (\sigma_x : x \in V) \in \Sigma$, we write $\delta_e(\sigma) = \delta_{\sigma_x,\sigma_y}$ where $\delta_{i,j}$ is the Kronecker delta. The relevant probability measure is given by

$$\pi_{\beta,q}(\sigma) = \frac{1}{Z_{\mathrm{P}}} e^{-\beta H'(\sigma)}, \qquad \sigma \in \Sigma, \tag{1.5}$$

where $Z_{\mathrm{P}} = Z_{\mathrm{P}}(\beta, q)$ is the appropriate normalizing constant and the Hamiltonian H' is given by

$$H'(\sigma) = - \sum_{e=\langle x,y\rangle \in E} \delta_e(\sigma). \tag{1.6}$$

In the special case $q = 2$, the multiplicative formula

$$\delta_{\sigma_x,\sigma_y} = \tfrac{1}{2}(1 + \sigma_x\sigma_y), \qquad \sigma_x, \sigma_y \in \{-1, +1\}, \tag{1.7}$$

is valid. It is now easy to see in this case that the ensuing Potts model is simply the Ising model with an adjusted value of β, in that $\pi_{\beta,2}$ is the measure obtained from $\lambda_{\beta/2}$ by re-labelling the local states.

Here is a brief mention of one further generalization of the Ising model, namely the so-called n-vector or $O(n)$ model. Let $n \in \{1, 2, \dots\}$ and let $\mathbb{I}$ be the set of vectors of $\mathbb{R}^n$ with unit length. The n-vector model on $G = (V, E)$ has configuration space $\mathbb{I}^V$ and Hamiltonian

$$H_n(\mathbf{s}) = -\sum_{e=\langle x,y\rangle \in E} \mathbf{s}_x \cdot \mathbf{s}_y, \qquad \mathbf{s} = (\mathbf{s}_v : v \in V) \in \mathbb{I}^V,$$

where $\mathbf{s}_x \cdot \mathbf{s}_y$ denotes the dot product. When $n = 1$, this is the Ising model. It is called the X/Y model when $n = 2$, and the Heisenberg model when $n = 3$.

1.4 Random-cluster and Ising/Potts models coupled

Fortuin and Kasteleyn discovered that Potts models may be re-cast as random-cluster models, and furthermore that the relationship between the two systems facilitates an extended study of phase transitions in Potts models, see [121, 122, 123, 203]. Their methods were elementary in nature. In a more modern approach, we construct the two systems on a common probability space. There may in principle be many ways to do this, but the standard coupling of Edwards and Sokal, [108], is of special value.

Let $q \in \{2, 3, \dots\}$, $p \in [0, 1]$, and let $G = (V, E)$ be a finite graph. We consider the product sample space $\Sigma \times \Omega$ where $\Sigma = \{1, 2, \dots, q\}^V$ and $\Omega = \{0, 1\}^E$ as above. We define a probability mass function μ on $\Sigma \times \Omega$ by

$$\mu(\sigma, \omega) \propto \prod_{e \in E} \bigl\{(1 - p)\delta_{\omega(e),0} + p\delta_{\omega(e),1}\delta_e(\sigma)\bigr\}, \qquad (\sigma, \omega) \in \Sigma \times \Omega, \tag{1.8}$$

where, as before, $\delta_e(\sigma) = \delta_{\sigma_x,\sigma_y}$ for $e = \langle x, y\rangle \in E$. The constant of proportionality is exactly that which ensures the normalization

$$\sum_{(\sigma,\omega) \in \Sigma \times \Omega} \mu(\sigma, \omega) = 1.$$

By an expansion of (1.8),

$$\mu(\sigma, \omega) \propto \psi(\sigma)\phi_p(\omega)1_F(\sigma, \omega), \qquad (\sigma, \omega) \in \Sigma \times \Omega,$$

where ψ is the uniform probability measure on Σ, ϕ_p is product measure on Ω with density p, and 1_F is the indicator function of the event

$$F = \bigl\{(\sigma, \omega) : \delta_e(\sigma) = 1 \text{ for any } e \text{ satisfying } \omega(e) = 1\bigr\} \subseteq \Sigma \times \Omega. \tag{1.9}$$

Therefore, μ may be viewed as the product measure $\psi \times \phi_p$ conditioned on F.

Elementary calculations reveal the following facts.

(1.10) Theorem (Marginal measures of μ) [108]. *Let $q \in \{2, 3, \dots\}$, $p \in [0, 1)$, and suppose that $p = 1 - e^{-\beta}$.*

(a) Marginal on Σ. *The marginal measure $\mu_1(\sigma) = \sum_{\omega \in \Omega} \mu(\sigma, \omega)$ on Σ is the Potts measure*

$$\mu_1(\sigma) = \frac{1}{Z_{\mathrm{P}}} \exp\left\{\beta \sum_{e \in E} \delta_e(\sigma)\right\}, \qquad \sigma \in \Sigma.$$

(b) Marginal on Ω. *The marginal measure $\mu_2(\omega) = \sum_{\sigma \in \Sigma} \mu(\sigma, \omega)$ on Ω is the random-cluster measure*

$$\mu_2(\omega) = \frac{1}{Z_{\mathrm{RC}}} \left\{\prod_{e \in E} p^{\omega(e)} (1-p)^{1-\omega(e)}\right\} q^{k(\omega)}, \qquad \omega \in \Omega.$$

(c) Partition functions. *We have that*

$$\sum_{\omega \in \Omega} \left\{\prod_{e \in E} p^{\omega(e)} (1-p)^{1-\omega(e)}\right\} q^{k(\omega)} = \sum_{\sigma \in \Sigma} \prod_{e \in E} \exp[\beta(\delta_e(\sigma) - 1)], \quad (1.11)$$

which is to say that

$$Z_{\mathrm{RC}}(p, q) = e^{-\beta|E|} Z_{\mathrm{P}}(\beta, q). \qquad (1.12)$$

The conditional measures of μ are given in the following theorem[4], and illustrated in Figure 1.3.

(1.13) Theorem (Conditional measures of μ) [108]. *Let $q \in \{2, 3, \dots\}$, $p \in [0, 1)$, and suppose that $p = 1 - e^{-\beta}$.*

(a) *For $\omega \in \Omega$, the conditional measure $\mu(\cdot \mid \omega)$ on Σ is obtained by putting random spins on entire clusters of ω (of which there are $k(\omega)$). These spins are constant on given clusters, are independent between clusters, and each is uniformly distributed on the set $\{1, 2, \dots, q\}$.*

(b) *For $\sigma \in \Sigma$, the conditional measure $\mu(\cdot \mid \sigma)$ on Ω is obtained as follows. If $e = \langle x, y \rangle$ is such that $\sigma_x \neq \sigma_y$, we set $\omega(e) = 0$. If $\sigma_x = \sigma_y$, we set*

$$\omega(e) = \begin{cases} 1 & \text{with probability } p, \\ 0 & \text{otherwise,} \end{cases}$$

the values of different $\omega(e)$ being (conditionally) independent random variables.

---

[4]The corresponding facts for the infinite lattice are given in Theorem 4.91.

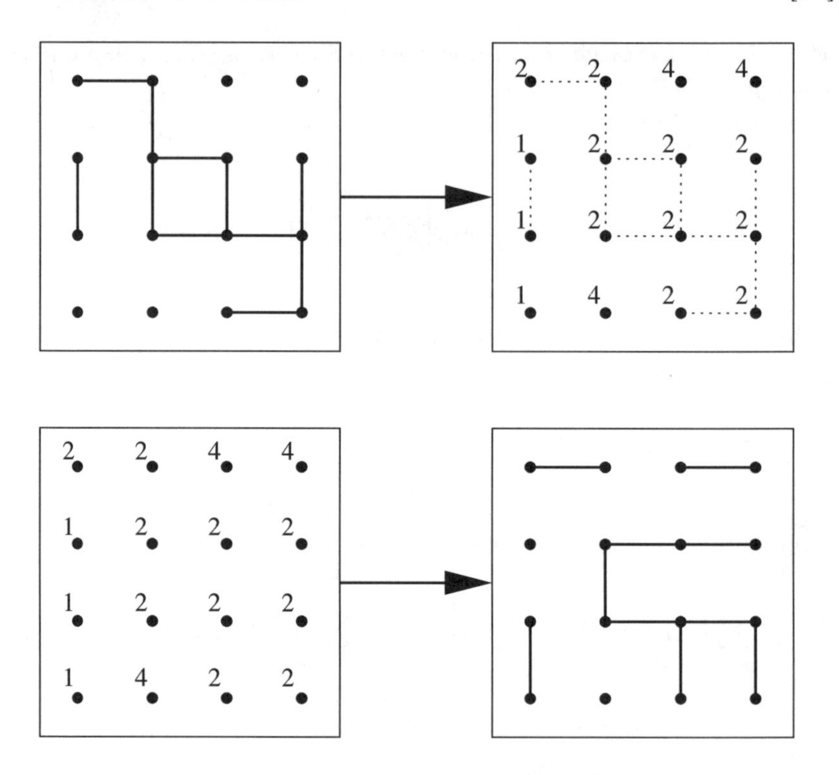

Figure 1.3. The upper diagram is an illustration of the conditional measure of μ on Σ given ω, with $q = 4$. To each open cluster of ω is allocated a spin-value chosen uniformly from $\{1, 2, 3, 4\}$. Different clusters are allocated independent values. In the lower diagram, we begin with a configuration σ. An edge is placed between vertices x, y with probability p (respectively, 0) if $\sigma_x = \sigma_y$ (respectively, $\sigma_x \neq \sigma_y$), and the outcome has as law the conditional measure of μ on Ω given σ.

In conclusion, the measure μ is a coupling of a Potts measure $\pi_{\beta,q}$ on V, together with the random-cluster measure $\phi_{p,q}$ on Ω. The parameters of these measures are related by the equation $p = 1 - e^{-\beta}$. Since $0 \leq p < 1$, we have that $0 \leq \beta < \infty$.

This special coupling may be used in a particularly simple way to show that correlations in Potts models correspond to open connections in random-cluster models. When extended to infinite graphs, this will imply that the phase transition of a Potts model corresponds to the creation of an infinite open cluster in the random-cluster model. Thus, arguments of stochastic geometry, and particularly those developed for the percolation model, may be harnessed directly in order to understand the correlation structure of the Potts system. The basic step is as follows.

Let $\{x \leftrightarrow y\}$ denote the set of all $\omega \in \Omega$ for which there exists an open path joining vertex x to vertex y. The complement of the event $\{x \leftrightarrow y\}$ is denoted by $\{x \nleftrightarrow y\}$.

The 'two-point correlation function' of the Potts measure $\pi_{\beta,q}$ on the finite graph $G = (V, E)$ is defined to be the function $\tau_{\beta,q}$ given by

$$\tau_{\beta,q}(x, y) = \pi_{\beta,q}(\sigma_x = \sigma_y) - \frac{1}{q}, \qquad x, y \in V. \tag{1.14}$$

The term q^{-1} is the probability that two independent and uniformly distributed spins are equal. Thus[5],

$$\tau_{\beta,q}(x, y) = \frac{1}{q}\pi_{\beta,q}(q\delta_{\sigma_x,\sigma_y} - 1). \tag{1.15}$$

The 'two-point connectivity function' of the random-cluster measure $\phi_{p,q}$ is defined as the function $\phi_{p,q}(x \leftrightarrow y)$ for $x, y \in V$, that is, the probability that x and y are joined by a path of open edges. It turns out that these 'two-point functions' are (except for a constant factor) the same.

(1.16) Theorem (Correlation/connection) [203]. *Let $q \in \{2, 3, \dots\}$, $p \in [0, 1)$, and suppose that $p = 1 - e^{-\beta}$. Then*

$$\tau_{\beta,q}(x, y) = (1 - q^{-1})\phi_{p,q}(x \leftrightarrow y), \qquad x, y \in V.$$

The theorem may be generalized as follows. Suppose we are studying the Potts model, and are interested in some 'observable' $f : \Sigma \to \mathbb{R}$. The mean value of $f(\sigma)$ satisfies

$$\begin{aligned}\pi_{\beta,q}(f) &= \sum_{\sigma} f(\sigma)\pi_{\beta,q}(\sigma) = \sum_{\sigma,\omega} f(\sigma)\mu(\sigma, \omega)\\ &= \sum_{\omega} F(\omega)\phi_{p,q}(\omega) = \phi_{p,q}(F)\end{aligned}$$

where $F : \Omega \to \mathbb{R}$ is given by

$$F(\omega) = \mu(f \mid \omega) = \sum_{\sigma} f(\sigma)\mu(\sigma \mid \omega).$$

Theorem 1.16 is obtained by setting $f(\sigma) = \delta_{\sigma_x,\sigma_y} - q^{-1}$.

The Potts models considered above have zero external-field. Some complications arise when an external field is added; see the discussions in [15, 44].

Proof of Theorem 1.10. (a) Let $\sigma \in \Sigma$ be given. Then

$$\begin{aligned}\sum_{\omega\in\Omega} \mu(\sigma, \omega) &\propto \sum_{\omega\in\Omega} \prod_{e\in E} \bigl\{(1 - p)\delta_{\omega(e),0} + p\delta_{\omega(e),1}\delta_e(\sigma)\bigr\}\\ &= \prod_{e\in E} [1 - p + p\delta_e(\sigma)].\end{aligned}$$

[5] If μ is a probability measure and X a random variable, the expectation of X with respect to μ is written $\mu(X)$.

Now $p = 1 - e^{-\beta}$ and

$$1 - p + p\delta = e^{\beta(\delta-1)}, \qquad \delta \in \{0, 1\},$$

whence

$$(1.17) \quad \sum_{\omega\in\Omega} \prod_{e\in E} \{(1-p)\delta_{\omega(e),0} + p\delta_{\omega(e),1}\delta_e(\sigma)\} = \prod_{e\in E} \exp[\beta(\delta_e(\sigma) - 1)].$$

Viewed as a set of weights on Σ, the latter expression generates the Potts measure.
(b) Let $\omega \in \Omega$ be given. We have that

$$(1.18) \quad \prod_{e\in E} \{(1-p)\delta_{\omega(e),0} + p\delta_{\omega(e),1}\delta_e(\sigma)\} = p^{|\eta(\omega)|}(1-p)^{|E\setminus\eta(\omega)|} 1_F(\sigma, \omega),$$

where $1_F(\sigma, \omega)$ is the indicator function that $\delta_e(\sigma) = 1$ whenever $\omega(e) = 1$, see (1.9). Now, $1_F(\sigma, \omega) = 1$ if and only if σ is constant on every open cluster of ω. There are $k(\omega)$ such clusters, and therefore $q^{k(\omega)}$ qualifying spin-vectors σ. Thus,

$$(1.19) \quad \sum_{\sigma\in\Sigma} \prod_{e\in E} \{(1-p)\delta_{\omega(e),0} + p\delta_{\omega(e),1}\delta_e(\sigma)\} = p^{|\eta(\omega)|}(1-p)^{|E\setminus\eta(\omega)|} q^{k(\omega)}.$$

This set of weights on Ω generates the random-cluster measure.
(c) We obtain the same answer if we sum (1.17) over all σ, or we sum (1.19) over all ω. □

Proof of Theorem 1.13. (a) Let $\omega \in \Omega$ be given. From (1.18)–(1.19),

$$\mu(\sigma \mid \omega) = \frac{1_F(\sigma, \omega)}{q^{k(\omega)}}, \qquad \sigma \in \Sigma,$$

whence the conditional measure is uniform on those σ with $1_F(\sigma, \omega) = 1$.
(b) Let $\sigma \in \Sigma$ be given. By (1.8),

$$\mu(\omega \mid \sigma) = K_\sigma \prod_{e\in E:\, \delta_e(\sigma)=0} \delta_{\omega(e),0} \prod_{e\in E:\, \delta_e(\sigma)=1} \{(1-p)\delta_{\omega(e),0} + p\delta_{\omega(e),1}\},$$

where $K_\sigma = K_\sigma(p, q)$. Therefore, $\mu(\omega \mid \sigma)$ is a product measure on Ω with

$$\omega(e) = 1 \quad \text{with probability} \begin{cases} 0 & \text{if } \delta_e(\sigma) = 0, \\ p & \text{if } \delta_e(\sigma) = 1. \end{cases}$$ □

Proof of Theorem 1.16. By Theorem 1.13(a),

$$\begin{aligned} \tau_{\beta,q}(x, y) &= \sum_{\sigma,\omega} \{1_{\{\sigma_x=\sigma_y\}}(\sigma) - q^{-1}\}\mu(\sigma, \omega) \\ &= \sum_{\omega} \phi_{p,q}(\omega) \sum_{\sigma} \mu(\sigma \mid \omega)\{1_{\{\sigma_x=\sigma_y\}}(\sigma) - q^{-1}\} \\ &= \sum_{\omega} \phi_{p,q}(\omega)\{(1 - q^{-1})1_{\{x\leftrightarrow y\}}(\omega) + 0 \cdot 1_{\{x\not\leftrightarrow y\}}(\omega)\} \\ &= (1 - q^{-1})\phi_{p,q}(x \leftrightarrow y), \end{aligned}$$

where μ is the above coupling of the Potts and random-cluster measures. □

Here is a final note. The random-cluster measure $\phi_{p,q}$ has two parameters p, q. In a more general version, we replace p by a vector $\mathbf{p} = (p_e : e \in E)$ of reals each of which satisfies $p_e \in [0, 1]$. The corresponding random-cluster measure $\phi_{\mathbf{p},q}$ on $(\Omega, \mathcal{F})$ is given by

$$\phi_{\mathbf{p},q}(\omega) = \frac{1}{Z}\left\{\prod_{e \in E} p_e^{\omega(e)}(1-p_e)^{1-\omega(e)}\right\}q^{k(\omega)}, \qquad \omega \in \Omega, \tag{1.20}$$

where Z is the appropriate normalizing factor. The measure $\phi_{p,q}$ is retrieved by setting $p_e = p$ for all $e \in E$.

1.5 The limit as $q \downarrow 0$

Let $G = (V, E)$ be a finite connected graph, and let $\phi_{p,q}$ be the random-cluster measure on G with parameters $p \in (0, 1)$, $q \in (0, \infty)$. We consider in this section the set of weak limits which may arise as $q \downarrow 0$. In preparation, we introduce three graph-theoretic terms.

A subset F of the edge-set E is called:

- a *forest* of G if the graph (V, F) contains no circuit,
- a *spanning tree* of G if (V, F) is connected and contains no circuit,
- a *connected subgraph* of G if (V, F) is connected.

In each case we consider the graph (V, F) containing every vertex of V; in this regard, sets F of edges satisfying one of the above conditions are sometimes termed *spanning*. Note that F is a spanning tree if and only if it is both a forest and a connected subgraph. For $\Omega = \{0, 1\}^E$ and $\omega \in \Omega$, we call ω a forest (respectively, spanning tree, connected subgraph) if $\eta(\omega)$ is a forest (respectively, spanning tree, connected subgraph). Write Ω_{for}, Ω_{st}, Ω_{cs} for the subsets of Ω containing all forests, spanning trees, and connected subgraphs, respectively, and write USF, UST, UCS for the uniform probability measures[6] on the respective sets Ω_{for}, Ω_{st}, Ω_{cs}.

We consider first the weak limit of $\phi_{p,q}$ as $q \downarrow 0$ for fixed $p \in (0, 1)$. This limit may be ascertained by observing that the dominant terms in the partition function

$$Z_{\text{RC}}(p, q) = \sum_{\omega \in \Omega} p^{|\eta(\omega)|}(1-p)^{|E \setminus \eta(\omega)|}q^{k(\omega)}$$

are those for which $k(\omega)$ is a minimum, that is, those with $k(\omega) = 1$. It follows that $\lim_{q \downarrow 0} \phi_{p,q}$ is precisely the product measure $\phi_p = \phi_{p,1}$ (that is, percolation

[6]This usage of the term 'uniform spanning forest' differs from that of [31].

with intensity p) *conditioned on the resulting graph* $(V, \eta(\omega))$ *being connected*. That is, $\phi_{p,q} \Rightarrow \phi_r^{\text{cs}}$ as $q \downarrow 0$, where $r = p/(1-p)$,

$$\phi_r^{\text{cs}}(\omega) = \begin{cases} \dfrac{1}{Z_{\text{cs}}} r^{|\eta(\omega)|} & \text{if } \omega \in \Omega_{\text{cs}}, \\ 0 & \text{otherwise,} \end{cases} \tag{1.21}$$

and $Z_{\text{cs}} = Z_{\text{cs}}(r)$ is the appropriate normalizing constant. In the special case $p = \frac{1}{2}$, we have that $\phi_{p,q} \Rightarrow$ UCS as $q \downarrow 0$.

Further limits arise if we allow both p and q to converge to 0. Suppose $p = p_q$ is related to q in such a way that $p \to 0$ and $q/p \to 0$ as $q \downarrow 0$; thus, p approaches zero slower than does q. We may write Z_{RC} in the form

$$Z_{\text{RC}}(p, q) = (1-p)^{|E|} \sum_{\omega \in \Omega} \left(\frac{p}{1-p} \right)^{|\eta(\omega)|+k(\omega)} \left(\frac{q(1-p)}{p} \right)^{k(\omega)}.$$

Note that $p/(1-p) \to 0$ and $q(1-p)/p \to 0$ as $q \downarrow 0$. Now, $k(\omega) \geq 1$ and $|\eta(\omega)| + k(\omega) \geq |V|$ for $\omega \in \Omega$; these two inequalities are satisfied simultaneously with equality if and only if $\omega \in \Omega_{\text{st}}$. Therefore, in the limit as $q \downarrow 0$, the 'mass' is concentrated on spanning trees, and it is easily seen that the limit mass is uniformly distributed. That is, $\phi_{p,q} \Rightarrow$ UST.

Another limit emerges if p approaches 0 at the same rate as does q. Take $p = \alpha q$ where $\alpha \in (0, \infty)$ is constant, and consider the limit as $q \downarrow 0$. This time we write

$$Z_{\text{RC}}(p, q) = (1-\alpha q)^{|E|} \sum_{\omega \in \Omega} \left(\frac{\alpha}{1-\alpha q} \right)^{|\eta(\omega)|} q^{|\eta(\omega)|+k(\omega)}.$$

We have that $|\eta(\omega)| + k(\omega) \geq |V|$ with equality if and only if $\omega \in \Omega_{\text{for}}$, and it follows that $\phi_{p,q} \Rightarrow \phi_\alpha^{\text{for}}$, where

$$\phi_\alpha^{\text{for}}(\omega) = \begin{cases} \dfrac{1}{Z_{\text{for}}} \alpha^{|\eta(\omega)|} & \text{if } \omega \in \Omega_{\text{for}}, \\ 0 & \text{otherwise,} \end{cases} \tag{1.22}$$

and $Z_{\text{for}} = Z_{\text{for}}(\alpha)$ is the appropriate normalizing constant. In the special case $\alpha = 1$, we find that $\phi_{p,q} \Rightarrow$ USF.

Finally, if p approaches 0 faster than does q, in that $p/q \to 0$ as $p, q \to 0$, it is easily seen that the limit measure is concentrated on the empty set of edges. We summarize the three special cases above in a theorem.

(1.23) Theorem. *We have in the limit as $q \downarrow 0$ that:*

$$\phi_{p,q} \Rightarrow \begin{cases} \text{UCS} & \text{if } p = \frac{1}{2}, \\ \text{UST} & \text{if } p \to 0 \text{ and } q/p \to 0, \\ \text{USF} & \text{if } p = q. \end{cases}$$

The spanning-tree limit is especially interesting for historical and mathematical reasons. As explained in the Appendix, the random-cluster model originated in a systematic study by Fortuin and Kasteleyn of systems of a certain type which satisfy certain parallel and series laws (see Section 3.8). Electrical networks are the best known such systems: two resistors of resistances r_1 and r_2 in parallel (respectively, in series) may be replaced by a single resistor with resistance $(r_1^{-1} + r_2^{-1})^{-1}$ (respectively, $r_1 + r_2$). Fortuin and Kasteleyn [123] realized that the electrical-network theory of a graph G is related to the limit as $q \downarrow 0$ of the random-cluster model on G, where p is given[7] by $p = \sqrt{q}/(1+\sqrt{q})$. It has been known since Kirchhoff's theorem, [215], that the electrical currents which flow in a network may be expressed in terms of counts of spanning trees. We return to this discussion of UST in Section 3.9.

The theory of the uniform-spanning-tree measure UST is beautiful in its own right (see [31]), and is linked in an important way to the emerging field of stochastic growth processes of 'stochastic Löwner evolution' (SLE) type (see [231, 284]), to which we return in Section 6.7. Further discussions of USF and UCS may be found in [165, 268].

1.6 Basic notation

We present some of the basic notation necessary for a study of random-cluster measures. Let $G = (V, E)$ be a graph, with finite or countably infinite vertex-set V and edge-set E. If two vertices x and y are joined by an edge e, we write $x \sim y$, and $e = \langle x, y \rangle$, and we say that x is *adjacent* to y. The (graph-theoretic) distance $\delta(x, y)$ from x to y is defined to be the number of edges in a shortest path of G from x to y.

The configuration space of the random-cluster model on G is the set $\Omega = \{0, 1\}^E$, points of which are represented as vectors $\omega = (\omega(e) : e \in E)$ and called *configurations*. For $\omega \in \Omega$, we call an edge e *open* (or ω-open, when the role of ω is to be emphasized) if $\omega(e) = 1$, and *closed* (or ω-closed) if $\omega(e) = 0$. We speak of a set F of edges as being 'open' (respectively, 'closed') in the configuration ω if $\omega(f) = 1$ (respectively, $\omega(f) = 0$) for all $f \in F$.

The indicator function of a subset A of Ω is the function $1_A : \Omega \to \{0, 1\}$ given by

$$1_A(\omega) = \begin{cases} 0 & \text{if } \omega \notin A, \\ 1 & \text{if } \omega \in A. \end{cases}$$

For $e \in E$, we write $J_e = \{\omega \in \Omega : \omega(e) = 1\}$, the event that the edge e is open. We use J_e to denote also the indicator function of this event, so that $J_e(\omega) = \omega(e)$. A function $X : \Omega \to \mathbb{R}$ is called a *cylinder* function if there exists a finite subset F of E such that $X(\omega) = X(\omega')$ whenever $\omega(e) = \omega'(e)$ for $e \in F$. A subset A of Ω is called a *cylinder event* if its indicator function is a cylinder function. We

[7]This choice of p is convenient, but actually one requires only that $q/p \to 0$, see [166].

take $\mathcal{F}$ to be the σ-field of subsets of Ω generated by the cylinder events, and we shall consider certain probability measures on the measurable pair $(\Omega, \mathcal{F})$. If G is finite, then $\mathcal{F}$ is the set of all subsets of Ω; all events are cylinder events, and all functions are cylinder functions. The complement of an event A is written A^c or $\overline{A}$.

For $W \subseteq V$, let E_W denote the set of edges of G having both endvertices in W. We write $\mathcal{F}_W$ (respectively, $\mathcal{T}_W$) for the smallest σ-field of $\mathcal{F}$ with respect to which each of the random variables $\omega(e)$, $e \in E_W$ (respectively, $e \notin E_W$), is measurable. The notation $\mathcal{F}_F$, $\mathcal{T}_F$ is to be interpreted similarly for $F \subseteq E$. The intersection of the $\mathcal{T}_F$ over all finite sets F is called the *tail* σ-field and is denoted by $\mathcal{T}$. Sets in $\mathcal{T}$ are called *tail events*.

There is a natural partial order on the set Ω of configurations given by: $\omega_1 \le \omega_2$ if and only if $\omega_1(e) \le \omega_2(e)$ for all $e \in E$. Rather than working always with the vector $\omega \in \Omega$, we shall sometimes work with its set of open edges, given by

$$\eta(\omega) = \{e \in E : \omega(e) = 1\}. \tag{1.24}$$

Clearly,

$$\omega_1 \le \omega_2 \qquad \text{if and only if} \qquad \eta(\omega_1) \subseteq \eta(\omega_2).$$

The smallest (respectively, largest) configuration is that with $\omega(e) = 0$ (respectively, $\omega(e) = 1$) for all e, and this is denoted by 0 (respectively, 1). A function $X : \Omega \to \mathbb{R}$ is called *increasing* if $X(\omega_1) \le X(\omega_2)$ whenever $\omega_1 \le \omega_2$. Similarly, X is *decreasing* if $-X$ is increasing. Note that every increasing function $X : \Omega \to \mathbb{R}$ is necessarily bounded since $X(0) \le X(\omega) \le X(1)$ for all $\omega \in \Omega$. A subset A of Ω is called *increasing* (respectively, *decreasing*) if it has increasing (respectively, decreasing) indicator function.

For $\omega \in \Omega$ and $e \in E$, let ω^e and ω_e be the configurations obtained from ω by 'switching on' and 'switching off' the edge e, respectively. That is,

$$\begin{aligned} \omega^e(f) &= \begin{cases} \omega(f) & \text{if } f \ne e, \\ 1 & \text{if } f = e, \end{cases} \qquad \text{for } f \in E, \\ \omega_e(f) &= \begin{cases} \omega(f) & \text{if } f \ne e, \\ 0 & \text{if } f = e, \end{cases} \qquad \text{for } f \in E. \end{aligned} \tag{1.25}$$

More generally, for $J \subseteq E$ and $K \subseteq E \setminus J$, we denote by ω_K^J the configuration that equals 1 on J, equals 0 on K, and agrees with ω on $E \setminus (J \cup K)$. When J and/or K contain only one or two edges, we may omit the necessary parentheses. The *Hamming distance* between two configurations is given by

$$H(\omega_1, \omega_2) = \sum_{e \in E} |\omega_1(e) - \omega_2(e)|, \qquad \omega_1, \omega_2 \in \Omega. \tag{1.26}$$

A *path* of G is defined as an alternating sequence $x_0, e_0, x_1, e_1, \dots, e_{n-1}, x_n$ of distinct vertices x_i and edges $e_i = \langle x_i, x_{i+1} \rangle$. Such a path has *length* n and is said to connect x_0 to x_n. A *circuit* or *cycle* of G is an alternating sequence $x_0, e_0, x_1, \dots, e_{n-1}, x_n, e_n, x_0$ of vertices and edges such that $x_0, e_0, \dots, e_{n-1}, x_n$ is a path and $e_n = \langle x_n, x_0 \rangle$; such a circuit has length $n + 1$. For $\omega \in \Omega$, we call a path or circuit *open* if all its edges are open, and *closed* if all its edges are closed. Two subgraphs of G are called *edge-disjoint* if they have no edges in common, and *disjoint* if they have neither edges nor vertices in common.

Let $\omega \in \Omega$. Consider the random subgraph of G containing the vertex set V and the open edges only, that is, the edges in $\eta(\omega)$. The connected components of this graph are called *open clusters*. We write $C_x = C_x(\omega)$ for the open cluster containing the vertex x, and we call C_x the *open cluster at* x. The vertex-set of C_x is the set of all vertices of G that are connected to x by open paths, and the edges of C_x are those edges of $\eta(\omega)$ that join pairs of such vertices. We shall occasionally use the term C_x to represent the *set of vertices* joined to x by open paths, rather than the graph of this open cluster. We shall be interested in the size of C_x, and we denote by $|C_x|$ the number of vertices in C_x. Note that $C_x = \{x\}$ whenever x is an isolated vertex, which is to say that x is incident to no open edge. We denote by $k(\omega)$ the number of open clusters in the configuration ω, that is, $k(\omega)$ is the number of components of the graph $(V, \eta(\omega))$. The random variable k plays an important role in the definition of a random-cluster measure, and the reader is warned of the importance of including in k a count of the number of isolated vertices of the graph.

Let $\omega \in \Omega$. If A and B are sets of vertices of G, we write '$A \leftrightarrow B$' if there exists an open path joining some vertex in A to some vertex in B; if $A \cap B \neq \varnothing$ then $A \leftrightarrow B$ trivially. Thus, for example, $C_x = \{y \in V : x \leftrightarrow y\}$. We write '$A \nleftrightarrow B$' if there exists no open path from any vertex of A to any vertex of B, and '$A \leftrightarrow B$ off D' if there exists an open path joining some vertex in A to some vertex in B that uses no vertex in the set D.

If W is a set of vertices of the graph, we write ∂W for the *boundary* of A, being the set of vertices in A that are adjacent to some vertex not in A,

$$\partial W = \{x \in W : \text{there exists } y \notin W \text{ such that } x \sim y\}.$$

We write $\Delta_e W$ for the set of edges of G having exactly one endvertex in W, and we call $\Delta_e W$ the *edge-boundary* of W.

We shall be mostly interested in the case when G is a subgraph of a d-dimensional lattice with $d \geq 2$. Rather than embarking on a debate of just what constitutes a 'lattice-graph', we shall, almost without exception, consider only the case of the (hyper)cubic lattice. This restriction enables a clear exposition of the theory and open problems without suffering the complications which arise through allowing greater generality.

Let d be a positive integer. We write $\mathbb{Z} = \{\dots, -1, 0, 1, \dots\}$ for the set of all integers, and $\mathbb{Z}^d$ for the set of all d-vectors $x = (x_1, x_2, \dots, x_d)$ with integral coordinates. For $x \in \mathbb{Z}^d$, we generally write x_i for the ith coordinate of x, and we define

$$\delta(x, y) = \sum_{i=1}^{d} |x_i - y_i|.$$

The *origin* of $\mathbb{Z}^d$ is denoted by 0. The set $\{1, 2, \dots\}$ of natural numbers is denoted by $\mathbb{N}$, and $\mathbb{Z}_+ = \mathbb{N} \cup \{0\}$. The real line is denoted by $\mathbb{R}$.

We turn $\mathbb{Z}^d$ into a graph, called the *d-dimensional cubic lattice*, by adding edges between all pairs x, y of points of $\mathbb{Z}^d$ with $\delta(x, y) = 1$. We denote this lattice by $\mathbb{L}^d$, and we write $\mathbb{Z}^d$ for the set of vertices of $\mathbb{L}^d$, and $\mathbb{E}^d$ for the set of its edges. Thus, $\mathbb{L}^d = (\mathbb{Z}^d, \mathbb{E}^d)$. We shall often think of $\mathbb{L}^d$ as a graph embedded in $\mathbb{R}^d$, the edges being straight line-segments between their endvertices. The *edge-set* $\mathbb{E}_V$ of $V \subseteq \mathbb{Z}^d$ is the set of all edges of $\mathbb{L}^d$ both of whose endvertices lie in V.

Let x, y be vertices of $\mathbb{L}^d$. The (graph-theoretic) distance from x to y is simply $\delta(x, y)$, and we write $|x|$ for the distance $\delta(0, x)$ from the origin to x. We shall make occasional use of another distance function on $\mathbb{Z}^d$, namely

$$\|x\| = \max\big\{|x_i| : i = 1, 2, \dots, d\big\}, \qquad x \in \mathbb{Z}^d,$$

and we note that

$$\|x\| \le |x| \le d\|x\|, \qquad x \in \mathbb{Z}^d.$$

For $\omega \in \Omega = \{0, 1\}^{\mathbb{E}^d}$, we abbreviate to C the open cluster C_0 at the origin.

A *box* of $\mathbb{L}^d$ is a subset of $\mathbb{Z}^d$ of the form

$$\Lambda_{a,b} = \big\{x \in \mathbb{Z}^d : a_i \le x_i \le b_i \text{ for } i = 1, 2, \dots, d\big\}, \qquad a, b \in \mathbb{Z}^d,$$

and we sometimes write

$$\Lambda_{a,b} = \prod_{i=1}^{d} [a_i, b_i]$$

as a convenient shorthand. The expression $\Lambda_{a,b}$ is used also to denote the graph with vertex-set $\Lambda_{a,b}$ together with those edges of $\mathbb{L}^d$ joining two vertices in $\Lambda_{a,b}$. For $x \in \mathbb{Z}^d$, we write $x + \Lambda_{a,b}$ for the translate by x of the box $\Lambda_{a,b}$. The expression Λ_n denotes the box with side-length $2n$ and centre at the origin,

$$\Lambda_n = [-n, n]^d = \{x \in \mathbb{Z}^d : \|x\| \le n\}. \tag{1.27}$$

Note that $\partial\Lambda_n = \Lambda_n \setminus \Lambda_{n-1}$.

In taking what is called a 'thermodynamic limit', one works often on a finite box Λ of $\mathbb{Z}^d$, and then takes the limit as $\Lambda \uparrow \mathbb{Z}^d$. Such a limit is to be interpreted along a sequence $\mathbf{\Lambda} = (\Lambda_n : n = 1, 2, \dots)$ of boxes such that: Λ_n is non-decreasing in n and, for all m, $\Lambda_n \supseteq [-m, m]^d$ for all large n.

For any random variable X and appropriate probability measure μ, we write $\mu(X)$ for the expectation of X,

$$\mu(X) = \int X \, d\mu.$$

Let $\lfloor a \rfloor$ and $\lceil a \rceil$ denote the integer part of the real number a, and the least integer not less than a, respectively. Finally, $a \wedge b = \min\{a, b\}$ and $a \vee b = \max\{a, b\}$.

Chapter 2

Monotonic Measures

Summary. The property of monotonicity of measures leads naturally to positive association and the FKG inequality. A monotonic measure may be used as the seed for a parametric family of measures satisfying probabilistic inequalities including influence, sharp-threshold, and exponential-steepness inequalities.

2.1 Stochastic ordering of measures

The stochastic ordering of probability measures provides a technique which is fundamental to the study of random-cluster measures. Let E be a finite or countably infinite set, let $\Omega = \{0, 1\}^E$, and let $\mathcal{F}$ be the σ-field generated by the cylinder events of Ω. In applications of the arguments of this section, E will be the edge-set of a graph, and thus we refer to members of E as 'edges', although the graphical structure is not itself relevant at this stage.

The configuration space Ω is a partially ordered set with partial order given by: $\omega_1 \le \omega_2$ if $\omega_1(e) \le \omega_2(e)$ for all $e \in E$. A random variable $X : \Omega \to \mathbb{R}$ is called *increasing* if $X(\omega_1) \le X(\omega_2)$ whenever $\omega_1 \le \omega_2$. An event $A \in \mathcal{F}$ is called *increasing* (respectively, *decreasing*) if its indicator function 1_A is increasing (respectively, decreasing). The set Ω, equipped with the topology of open sets generated by the cylinder events, is a metric space, and we speak of a random variable $X : \Omega \to \mathbb{R}$ as being 'continuous' if it is a continuous function on this metric space. Since Ω is compact, any continuous function on Ω is necessarily bounded. In addition, any increasing function $X : \Omega \to \mathbb{R}$ is bounded since $X(0) \le X(\omega) \le X(1)$ for $\omega \in \Omega$.

Given two probability measures μ_1, μ_2 on $(\Omega, \mathcal{F})$, we write $\mu_1 \le_{\text{st}} \mu_2$ (or $\mu_2 \ge_{\text{st}} \mu_1$), and we say that μ_1 is stochastically smaller than μ_2, if[1] $\mu_1(X) \le \mu_2(X)$ for all increasing continuous random variables X on Ω.

---

[1]Recall that $\mu(X)$ denotes the expectation of X under μ, that is, $\mu(X) = \int X \, d\mu$.

For two probability measures ϕ_1, ϕ_2 on $(\Omega, \mathcal{F})$, a *coupling* of ϕ_1 and ϕ_1 is a probability measure κ on $(\Omega, \mathcal{F}) \times (\Omega, \mathcal{F})$ whose first (respectively, second) marginal is ϕ_1 (respectively, ϕ_2). There exist many couplings of any given pair ϕ_1, ϕ_2, and the art of coupling lies in finding one which is useful. Let μ_1, μ_2 be probability measures on $(\Omega, \mathcal{F})$. The theorem known sometimes as 'Strassen's theorem' states that $\mu_1 \le_{\text{st}} \mu_2$ if and only if there exists a coupling κ satisfying $\kappa(S) = 1$, where $S = \{(\omega_1, \omega_2) \in \Omega^2 : \omega_1 \le \omega_2\}$ is the 'sub-diagonal' of the product space Ω^2. A useful account of coupling and its applications may be found in [237].

We call a probability measure μ on $(\Omega, \mathcal{F})$ *strictly positive* if $\mu(\omega) > 0$ for all $\omega \in \Omega$. For $\omega_1, \omega_2 \in \Omega$, we denote by $\omega_1 \vee \omega_2$ and $\omega_1 \wedge \omega_2$ the 'maximum' and 'minimum' configurations given by

$$\begin{aligned} \omega_1 \vee \omega_2(e) &= \max\{\omega_1(e), \omega_2(e)\}, \qquad e \in E, \\ \omega_1 \wedge \omega_2(e) &= \min\{\omega_1(e), \omega_2(e)\}, \qquad e \in E. \end{aligned}$$

We suppose for the remainder of this section that E is finite. There is a useful sufficient condition for the stochastic inequality $\mu_1 \le_{\text{st}} \mu_2$, as follows.

(2.1) Theorem (Holley inequality) [185]. *Let μ_1 and μ_2 be strictly positive probability measures on the finite space $(\Omega, \mathcal{F})$ such that*

$$\mu_2(\omega_1 \vee \omega_2)\mu_1(\omega_1 \wedge \omega_2) \ge \mu_1(\omega_1)\mu_2(\omega_2), \qquad \omega_1, \omega_2 \in \Omega. \tag{2.2}$$

Then

$$\mu_1(X) \le \mu_2(X) \qquad \textit{for increasing functions } X : \Omega \to \mathbb{R},$$

which is to say that $\mu_1 \le_{\text{st}} \mu_2$.

This may be extended in (at least) two ways. Firstly, a similar claim[2] is valid in the more general setting where $\Omega = T^E$ and T is a finite subset of $\mathbb{R}$. Secondly, one may relax the condition that the measures be *strictly* positive. See, for example, [136, Section 4].

Let $S \subseteq \Omega^2$ $(= \Omega \times \Omega)$ be the set of all ordered pairs (π, ω) of configurations satisfying $\pi \le \omega$, as above. In the proof of Theorem 2.1, we shall construct a coupling κ of μ_1 and μ_2 such that $\kappa(S) = 1$. It is an immediate consequence that $\mu_1 \le_{\text{st}} \mu_2$. There is a variety of couplings of measures which play roles in the theory of random-cluster measures. Another may be found in the proof of Theorem 3.45.

One needs to verify condition (2.2) in order to apply Theorem 2.1. It is not necessary to check (2.2) for *all* pairs ω_1, ω_2 but only for those pairs that disagree at two or fewer edges. Recall the notation ω_K^J introduced after (1.25).

---

[2] An application of such a claim may be found in the analysis of the Ashkin–Teller model at Theorem 11.12.

(2.3) Theorem. *A pair μ_1, μ_2 of strictly positive probability measures on $(\Omega, \mathcal{F})$ satisfies* (2.2) *if and only it satisfies the two inequalities*:

$$\mu_2(\omega^e)\mu_1(\omega_e) \geq \mu_1(\omega^e)\mu_2(\omega_e), \qquad \omega \in \Omega,\ e \in E, \tag{2.4}$$

$$\mu_2(\omega^{ef})\mu_1(\omega_{ef}) \geq \mu_1(\omega_f^e)\mu_2(\omega_e^f), \qquad \omega \in \Omega,\ e, f \in E. \tag{2.5}$$

Before moving to the proofs, we point out that inequality (2.2) is equivalent to a condition of monotonicity on the one-point conditional distributions.

(2.6) Theorem. *A pair μ_1, μ_2 of strictly positive probability measures on $(\Omega, \mathcal{F})$ satisfies* (2.2) *if and only if the one-point conditional probabilities satisfy*:

$$\begin{aligned} \mu_2\big(\omega(e) = 1 \,\big|\, \omega(f) = \zeta(f) \text{ for all } f \in E \setminus \{e\}\big) \\ \geq \mu_1\big(\omega(e) = 1 \,\big|\, \omega(f) = \xi(f) \text{ for all } f \in E \setminus \{e\}\big) \end{aligned} \tag{2.7}$$

for all $e \in E$, and all pairs $\xi, \zeta \in \Omega$ satisfying $\xi \leq \zeta$.

Proof of Theorem 2.1. The theorem amounts to a 'mere' numerical inequality involving a finite number of positive reals. It may in principle be proved in a totally elementary manner, using essentially no general mechanism. The proof given here proceeds by constructing certain reversible Markov chains. There is some extra mechanism required, but the method is beautiful, and in addition yields a structure which finds applications elsewhere.

The main step of the proof is designed to show that, under condition (2.2), μ_1 and μ_2 may be 'coupled' in such a way that the sub-diagonal S has full measure. This is achieved by constructing a certain Markov chain with the coupled measure as invariant measure.

Here is a preliminary calculation. Let μ be a strictly positive probability measure on $(\Omega, \mathcal{F})$. We may construct a reversible Markov chain with state space Ω and unique invariant measure μ by choosing a suitable generator (or 'Q-matrix') satisfying the detailed balance equations. The dynamics of the chain involve the 'switching on or off' of components of the current state. Let $G : \Omega^2 \to \mathbb{R}$ be given by

$$G(\omega_e, \omega^e) = 1, \quad G(\omega^e, \omega_e) = \frac{\mu(\omega_e)}{\mu(\omega^e)}, \qquad \omega \in \Omega,\ e \in E. \tag{2.8}$$

We let $G(\omega, \omega') = 0$ for all other pairs ω, ω' with $\omega \neq \omega'$. The diagonal elements $G(\omega, \omega)$ are chosen in such a way that

$$\sum_{\omega' \in \Omega} G(\omega, \omega') = 0, \qquad \omega \in \Omega.$$

It is elementary that

$$\mu(\omega)G(\omega, \omega') = \mu(\omega')G(\omega', \omega), \qquad \omega, \omega' \in \Omega,$$

and therefore G generates a Markov chain on the state space Ω which is reversible with respect to μ. The chain is irreducible, for the following reason. For $\omega, \omega' \in \Omega$, one may add edges one by one to $\eta(\omega)$ thus arriving at the unit vector 1, and then one may remove edges one by one thus arriving at ω'. By (2.8), each such transition has a strictly positive intensity, whence the chain is irreducible. It follows that the chain has unique invariant measure μ. Similar constructions are explored in Chapter 8. An account of the general theory of reversible Markov chains may be found in [164, Section 6.5].

We follow next a similar route for *pairs* of configurations. Let μ_1 and μ_2 satisfy the hypotheses of the theorem, and let S be the set of all ordered pairs (π, ω) of configurations in Ω satisfying $\pi \le \omega$. We define $H : S \times S \to \mathbb{R}$ by

$$(2.9) \qquad H(\pi_e, \omega; \pi^e, \omega^e) = 1,$$

$$(2.10) \qquad H(\pi, \omega^e; \pi_e, \omega_e) = \frac{\mu_2(\omega_e)}{\mu_2(\omega^e)},$$

$$(2.11) \qquad H(\pi^e, \omega^e; \pi_e, \omega^e) = \frac{\mu_1(\pi_e)}{\mu_1(\pi^e)} - \frac{\mu_2(\omega_e)}{\mu_2(\omega^e)},$$

for all $(\pi, \omega) \in S$ and $e \in E$; all other off-diagonal values of H are set to 0. The diagonal terms $H(\pi, \omega; \pi, \omega)$ are chosen in such a way that

$$\sum_{(\pi', \omega') \in S} H(\pi, \omega; \pi', \omega') = 0, \qquad (\pi, \omega) \in S.$$

Equation (2.9) specifies that, for $\pi \in \Omega$ and $e \in E$, the edge e is acquired by π (if it does not already contain it) at rate 1; any edge so acquired is added also to ω if it does not already contain it. (Here, we speak of a configuration ψ 'containing the edge e' if $\psi(e) = 1$.) Equation (2.10) specifies that, for $\omega \in \Omega$ and $e \in E$ with $\omega(e) = 1$, the edge e is removed from ω (and also from π if $\pi(e) = 1$) at the rate given in (2.10). For e with $\pi(e) = 1$, there is an additional rate given in (2.11) at which e is removed from π but not from ω. This additional rate is indeed non-negative, since the required inequality

$$(2.12) \qquad \mu_2(\omega^e)\mu_1(\pi_e) \ge \mu_1(\pi^e)\mu_2(\omega_e) \qquad \text{whenever } \pi \le \omega,$$

follows from (2.2) with $\omega_1 = \pi^e$ and $\omega_2 = \omega_e$.

Let $(Y_t, Z_t)_{t \ge 0}$ be a Markov chain on S with generator H, and set $(Y_0, Z_0) = (0, 1)$, where 0 (respectively, 1) is the state of all zeros (respectively, ones). We write $\mathbb{P}$ for the appropriate probability measure. Since all transitions retain the ordering of the two components of the state, we may assume that the chain satisfies $\mathbb{P}(Y_t \le Z_t \text{ for all } t) = 1$. By examination of (2.9)–(2.11) we see that $Y = (Y_t : t \ge 0)$ is a Markov chain with generator given by (2.8) with $\mu = \mu_1$, and that $Z = (Z_t : t \ge 0)$ arises similarly with $\mu = \mu_2$. Here is a brief explanation of this elementary step in the case of Y, a similar argument holds for Z. For $\pi \in \Omega$ and $e \in E$,

$$\begin{aligned}\mathbb{P}(Y_{t+h} = \pi^e \mid Y_t = \pi_e)&\\ &= \sum_{\omega\in\Omega} \mathbb{P}\big(Y_{t+h} = \pi^e \,\big|\, (Y_t, Z_t) = (\pi_e, \omega)\big)\mathbb{P}(Z_t = \omega \mid Y_t = \pi_e)\\ &= \sum_{\omega\in\Omega} [h + \mathrm{o}(h)]\mathbb{P}(Z_t = \omega \mid Y_t = \pi_e) \qquad \text{by (2.9)}\\ &= h + \mathrm{o}(h).\end{aligned}$$

Similarly, with J_e the event that e is open,

$$\begin{aligned}&\mathbb{P}(Y_{t+h} = \pi_e \mid Y_t = \pi^e)\\ &\quad = \sum_{\omega\in J_e,\ \omega'\in\Omega} \mathbb{P}\big((Y_{t+h}, Z_{t+h}) = (\pi_e, \omega') \,\big|\, (Y_t, Z_t) = (\pi^e, \omega^e)\big)\\ &\qquad\qquad\qquad\qquad \times\mathbb{P}(Z_t = \omega^e \mid Y_t = \pi^e)\\ &\quad = \sum_{\omega\in J_e} \big[\big\{H(\pi^e, \omega^e; \pi_e, \omega_e) + H(\pi^e, \omega^e; \pi_e, \omega^e)\big\}h + \mathrm{o}(h)\big]\\ &\qquad\qquad\qquad\qquad \times\mathbb{P}(Z_t = \omega^e \mid Y_t = \pi^e)\\ &\quad = \sum_{\omega\in J_e} \left[\frac{\mu_1(\pi_e)}{\mu_1(\pi^e)}h + \mathrm{o}(h)\right]\mathbb{P}(Z_t = \omega^e \mid Y_t = \pi^e) \qquad \text{by (2.10) and (2.11)}\\ &\quad = \frac{\mu_1(\pi_e)}{\mu_1(\pi^e)}h + \mathrm{o}(h).\end{aligned}$$

Let κ be an invariant measure for the paired chain $(Y_t, Z_t)_{t\ge0}$. Since Y and Z have (respective) unique invariant measures μ_1 and μ_2, the marginals of κ are μ_1 and μ_2. Since $\mathbb{P}(Y_t \le Z_t \text{ for all } t) = 1$,

$$\kappa(S) = \kappa\big(\{(\pi, \omega) : \pi \le \omega\}\big) = 1,$$

and κ is the required 'coupling' of μ_1 and μ_2.

Let $(\pi, \omega) \in S$ be chosen according to the measure κ. Then

$$\mu_1(X) = \kappa(X(\pi)) \le \kappa(X(\omega)) = \mu_2(X),$$

for any increasing function X. Therefore $\mu_1 \le_{\mathrm{st}} \mu_2$. □

Proof of Theorem 2.3[3]. The implication in one direction is obvious. Suppose conversely that (2.4)–(2.5) hold. By (2.4), the inequality of (2.2) holds for all pairs ω_1, ω_2 with Hamming distance satisfying $H(\omega_1, \omega_2) = 1$. Suppose that $H(\omega_1, \omega_2) = 2$. There are only two non-trivial cases for (2.2). There exist $\omega \in \Omega$

[3] See also [257, Lemma 6.5].

and distinct $e, f \in E$ such that: either $\omega_1 = \omega_f^e, \omega_2 = \omega_e^f$, or $\omega_1 = \omega^{ef}, \omega_2 = \omega$. The first case is handled by (2.5), and the second by (2.4) on noting that

$$\begin{aligned}\frac{\mu_2(\omega^{ef})}{\mu_2(\omega)} &= \frac{\mu_2(\omega^{ef})}{\mu_2(\omega^e)} \cdot \frac{\mu_2(\omega^e)}{\mu_2(\omega)} \\ &\geq \frac{\mu_1(\omega^{ef})}{\mu_1(\omega^e)} \cdot \frac{\mu_1(\omega^e)}{\mu_1(\omega)} = \frac{\mu_1(\omega^{ef})}{\mu_1(\omega)}.\end{aligned} \tag{2.13}$$

We suppose now that $H(\omega_1, \omega_2) \geq 3$, and we shall proceed by induction on $H(\omega_1, \omega_2)$. Let $h \geq 3$ and suppose that the inequality of (2.2) holds for all pairs ω_1, ω_2 satisfying $H(\omega_1, \omega_2) < h$. Let $\omega_1, \omega_2 \in \Omega$ be such that $H(\omega_1, \omega_2) = h$, and furthermore such that neither $\omega_1 \leq \omega_2$ nor $\omega_1 \geq \omega_2$ (the claim holds as in (2.13) otherwise). There exist integers a, b such that $a, b \geq 1$ and $a + b = h$, and disjoint subsets $A, B \subseteq E$ with cardinalities a and b respectively, such that:

$$\begin{aligned}\text{if } e \in A, &\quad (\omega_1(e), \omega_2(e)) = (1, 0), \\ \text{if } e \in B, &\quad (\omega_1(e), \omega_2(e)) = (0, 1), \\ \text{if } e \in E \setminus (A \cup B), &\quad \omega_1(e) = \omega_2(e).\end{aligned}$$

We fix an ordering $(e_i : i = 1, 2, \ldots, |E|)$ of the set E in which edges in A are indexed $1, 2, \ldots, a$, and edges in B are indexed $a + 1, a + 2, \ldots, a + b$. A configuration ω may be written as a 'word' $\omega(e_1) \cdot \omega(e_2) \cdot \cdots \cdot \omega(e_{|E|})$; we write 0^x for a sub-word of length x every entry of which is 0, with a similar meaning for 1^y. Since the entries of the configurations $\omega_1, \omega_2, \omega_1 \vee \omega_2, \omega_1 \wedge \omega_2$ are constant off $A \cup B$, we shall omit explicit reference to these values. Thus, for example, $\omega_1 = 1^a \cdot 0^b$ and $\omega_2 = 0^a \cdot 1^b$.

We may assume that $a \leq b$, whence $b \geq 2$, since an analagous argument holds when $a > b$. By the induction hypothesis,

$$\begin{aligned}\mu_2(1^a \cdot 1^b)\mu_1(0^a \cdot 1 \cdot 0^{b-1}) &\geq \mu_1(0^a \cdot 1^b)\mu_2(1^a \cdot 1 \cdot 0^{b-1}) \\ &\qquad \text{since } H(0^a \cdot 1^b, 1^a \cdot 1 \cdot 0^{b-1}) = h - 1, \\ \mu_2(1^a \cdot 1 \cdot 0^{b-1})\mu_1(0^{a+b}) &\geq \mu_1(0^a \cdot 1 \cdot 0^{b-1})\mu_2(1^a \cdot 0^b) \\ &\qquad \text{since } H(0^a \cdot 1 \cdot 0^{b-1}, 1^a \cdot 0^b) = a + 1 < h,\end{aligned}$$

whence

$$\begin{aligned}&\mu_2(1^a \cdot 1^b)\mu_1(0^a \cdot 1 \cdot 0^{b-1})\mu_1(0^{a+b}) \\ &\qquad \geq \mu_1(0^a \cdot 1^b)\mu_2(1^a \cdot 1 \cdot 0^{b-1})\mu_1(0^{a+b}) \\ &\qquad \geq \mu_1(0^a \cdot 1^b)\mu_1(0^a \cdot 1 \cdot 0^{b-1})\mu_2(1^a \cdot 0^b).\end{aligned}$$

The claim follows on dividing both sides by the strictly positive quantity $\mu_1(0^a \cdot 1 \cdot 0^{b-1})$. □

Proof of Theorem 2.6. Inequality (2.7) is equivalent to

$$\mu_2(\zeta^e)[\mu_1(\xi^e) + \mu_1(\xi_e)] \geq \mu_1(\xi^e)[\mu_2(\zeta^e) + \mu_2(\zeta_e)],$$

or, equivalently,

$$\mu_2(\zeta^e)\mu_1(\xi_e) \geq \mu_1(\xi^e)\mu_2(\zeta_e). \tag{2.14}$$

Assume (2.7) holds. Let $e, f \in E$ be distinct, and let $\omega \in \Omega$. We apply (2.14) with $\xi = \zeta_e = \omega$ to deduce (2.4). As for (2.5), we apply (2.14) to the pair $\xi = \omega_f$, $\zeta = \omega^f$. The required (2.2) holds by Theorem 2.3. Conversely, if (2.2) holds, then so does (2.14) for $\xi \leq \zeta$. □

2.2 Positive association

Let E be a finite set as in the last section, and let $\Omega = \{0, 1\}^E$. A probability measure μ on Ω is said to have the *FKG lattice property* if it satisfies the so-called *FKG lattice condition*:

$$\mu(\omega_1 \vee \omega_2)\mu(\omega_1 \wedge \omega_2) \geq \mu(\omega_1)\mu(\omega_2), \qquad \omega_1, \omega_2 \in \Omega. \tag{2.15}$$

It is a consequence of the Holley inequality (Theorem 2.1), as follows, that any strictly positive probability measure with the FKG lattice property satisfies the so-called FKG inequality. A stronger result will appear at Theorem 2.24.

(2.16) Theorem (FKG inequality) [124, 185]. *Let μ be a strictly positive probability measure on Ω satisfying the FKG lattice condition. Then*

$$\mu(XY) \geq \mu(X)\mu(Y) \qquad \textit{for increasing functions } X, Y : \Omega \to \mathbb{R}. \tag{2.17}$$

There is an extensive literature on the FKG inequality[4] and its extensions. See, for example, [2, 25, 184]. One may extend the inequality to probability measures on sample spaces of the form T^E with T a finite subset of $\mathbb{R}$. In addition, some of the results of this section are valid for measures that are not strictly positive. Any probability measure μ satisfying (2.17) is said to have the property of 'positive association' or, more concisely, to be 'positively associated'. We consider in Section 4.1 the positive association of measures on $\Omega = \{0, 1\}^E$ when E is countably infinite.

Correlation-type inequalities play an important role in mathematical physics. For example, the FKG inequality is a fundamental tool in the study of the Ising and random-cluster models, see Chapter 3. There are many other correlation

[4] The history and origins of the FKG inequality are described in the Appendix.

inequalities in statistical physics (see [118]), but these do not generally have a random-cluster equivalent and are omitted from the current work.

Proof. Since inequality (2.17) involves a finite set of real numbers only, it may in principle be proved in a totally elementary manner, [280]. We follow here the more interesting route via the Holley inequality, Theorem 2.1. Assume that μ satisfies the FKG lattice condition (2.15), and let X and Y be increasing functions. Let $a > 0$ and $Y' = Y + a$. Since

$$\mu(XY') - \mu(X)\mu(Y') = \mu(XY) - \mu(X)\mu(Y),$$

it suffices to prove (2.17) with Y replaced by Y'. We may pick a sufficiently large that $Y(\omega) > 0$ for all $\omega \in \Omega$. Thus, it suffices to prove (2.17) under the additional hypothesis that Y is strictly positive, and we assume henceforth that this holds. Define the strictly positive probability measures μ_1 and μ_2 on $(\Omega, \mathcal{F})$ by $\mu_1 = \mu$ and

$$\mu_2(\omega) = \frac{Y(\omega)\mu(\omega)}{\sum_{\omega' \in \Omega} Y(\omega')\mu(\omega')}, \qquad \omega \in \Omega.$$

Since Y is increasing, inequality (2.2) follows from (2.15). By the Holley inequality, $\mu_2(X) \geq \mu_1(X)$, which is to say that

$$\frac{\sum_{\omega \in \Omega} X(\omega)Y(\omega)\mu(\omega)}{\sum_{\omega' \in \Omega} Y(\omega')\mu(\omega')} \geq \sum_{\omega \in \Omega} X(\omega)\mu(\omega). \qquad \square$$

If X is increasing and Y is decreasing, we may apply (2.17) to X and $-Y$ to find, under the conditions of the theorem, that $\mu(XY) \leq \mu(X)\mu(Y)$. In the special case when $X = 1_A$, $Y = 1_B$, the indicator functions of events A and B, we obtain similarly that

$$\mu(A \cap B) \geq \mu(A)\mu(B) \qquad \text{for increasing events } A, B. \tag{2.18}$$

Let $X = (X_1, X_2, \dots, X_r)$ be a vector of random variables taking values in $\{0, 1\}^r$. We speak of X as being positively associated if its law on $\{0, 1\}^r$ is itself positively associated. Let $Y = h(X)$ where $h : \{0, 1\}^r \to \{0, 1\}^s$ is a non-decreasing function. It is standard that the vector Y is positively associated whenever X is positively associated. The proof is straightforward, as follows. Let A, B be increasing subsets of $\{0, 1\}^s$. Then

$$\begin{aligned}
\mathbb{P}(Y \in A \cap B) &= \mathbb{P}\big(X \in \{h^{-1}A\} \cap \{h^{-1}B\}\big) \\
&\geq \mathbb{P}(X \in h^{-1}A)\mathbb{P}(X \in h^{-1}B) \\
&= \mathbb{P}(Y \in A)\mathbb{P}(Y \in B),
\end{aligned}$$

since $h^{-1}A$ and $h^{-1}B$ are increasing subsets of $\{0, 1\}^r$.

We turn now to a consideration of the FKG lattice condition. Recall the Hamming distance between configurations defined in (1.26). A pair $\omega_1, \omega_2 \in \Omega$ is called *comparable* if either $\omega_1 \leq \omega_2$ or $\omega_1 \geq \omega_2$, and *incomparable* otherwise.

(2.19) Theorem. *A strictly positive probability measure μ on $(\Omega, \mathcal{F})$ satisfies the FKG lattice condition if and only the inequality of* (2.15) *holds for all incomparable pairs $\omega_1, \omega_2 \in \Omega$ with $H(\omega_1, \omega_2) = 2$.*

For pairs ω_1, ω_2 that differ on exactly two edges e and f, the inequality of (2.15) is equivalent to the statement that, conditional on the states of all other edges, the states of e and f are positively associated.

Proof. The inequality of (2.15) is a triviality when $H(\omega_1, \omega_2) = 1$, and the claim now follows by Theorem 2.3. □

The FKG lattice condition is sufficient but not necessary for positive association. It is equivalent for strictly positive measures to a stronger property termed 'strong positive-association' (or, sometimes, 'strong FKG'). For $F \subseteq E$ and $\xi \in \Omega$, we write $\Omega_F = \{0, 1\}^F$ and

$$\Omega_F^\xi = \{\omega \in \Omega : \omega(e) = \xi(e) \text{ for all } e \in E \setminus F\}, \tag{2.20}$$

the set of configurations that agree with ξ on the complement of F. Let μ be a probability measure on $(\Omega, \mathcal{F})$, and let F, ξ be such that $\mu(\Omega_F^\xi) > 0$. We define the conditional probability measure μ_F^ξ on Ω_F by

$$\mu_F^\xi(\omega_F) = \mu(\omega_F \mid \Omega_F^\xi) = \frac{\mu(\omega_F \times \xi)}{\mu(\Omega_F^\xi)}, \qquad \omega_F \in \Omega_F, \tag{2.21}$$

where $\omega_F \times \xi$ denotes the configuration that agrees with ω_F on F and with ξ on its complement. We say that μ is *strongly positively-associated* if: for all $F \subseteq E$ and all $\xi \in \Omega$ such that $\mu(\Omega_F^\xi) > 0$, the measure μ_F^ξ is positively associated.

We call μ *monotonic* if: for all $F \subseteq E$, all increasing subsets A of Ω_F, and all $\xi, \zeta \in \Omega$ such that $\mu(\Omega_F^\xi), \mu(\Omega_F^\zeta) > 0$,

$$\mu_F^\xi(A) \le \mu_F^\zeta(A) \qquad \text{whenever } \xi \le \zeta. \tag{2.22}$$

That is, μ is monotonic if, for all $F \subseteq E$,

$$\mu_F^\xi \le_{\text{st}} \mu_F^\zeta \qquad \text{whenever } \xi \le \zeta. \tag{2.23}$$

We call μ 1-*monotonic* if (2.23) holds for all singleton sets F. That is, μ is 1-monotonic if and only if, for all $f \in E$, $\mu(J_f \mid \Omega_f^\xi)$ is a non-decreasing function of ξ. Here, J_f denotes the event that f is open.

(2.24) Theorem[5]**.** *Let μ be a strictly positive probability measure on $(\Omega, \mathcal{F})$. The following are equivalent.*

(a) *μ is strongly positively-associated.*

(b) *μ satisfies the FKG lattice condition.*

(c) *μ is monotonic.*

(d) *μ is 1-monotonic.*

It is a near triviality to check that any product measure on Ω satisfies the FKG lattice condition, and thus product measures are strongly positively-associated. This is the $q = 1$ case of Theorem 3.8, and is usually referred to as Harris's inequality, [181]. We give two examples of probability measures that are positively associated but do not satisfy the statements of the above theorem.

(2.25) Example[6]**.** Let $\epsilon, \delta \in (0, 1)$, and let μ_0, μ_1 be the probability measures on $\{0, 1\}^3$ given by

$$\mu_0(010) = \mu_0(001) = \delta, \quad \mu_0(000) = 1 - 2\delta,$$
$$\mu_1(111) = \mu_1(100) = \tfrac{1}{2}.$$

Let $\epsilon \in [0, 1]$ and set $\mu = \epsilon\mu_0 + (1 - \epsilon)\mu_1$. Note that

$$\mu(011) = \mu(101) = \mu(110) = 0.$$

It may be checked that μ does not satisfy the FKG lattice condition whereas, for sufficiently small positive values of the constants ϵ, δ, the measure μ is positively associated. Note from the above that μ is not strictly positive. However, a strictly positive example may be arranged by replacing μ by the probability measure $\mu' = (1 - \eta)\mu + \eta\mu_2$ where

$$\mu_2(011) = \mu_2(101) = \mu_2(110) = \tfrac{1}{3}$$

and η is small and positive.

(2.26) Example[7]**.** Let X and Y be independent Bernoulli random variables with parameter $\frac{1}{2}$, so that

$$\mathbb{P}(X = 0) = \mathbb{P}(X = 1) = \tfrac{1}{2},$$

and similarly for Y. Let $Z = XY$. It is clear that

$$\mathbb{P}(X = 1 \mid Z = 1) > \mathbb{P}(X = 1), \quad \mathbb{P}(X = 1 \mid Y = Z = 1) = \mathbb{P}(X = 1).$$

[5] Closely related material is discussed in [204]. The equivalence of (a) and (b) is attributed in [8] to J. van den Berg and R. M. Burton (1987). See [136] for a further discussion of monotonic measures.

[6] Proposed by J. Steif.

[7] Proposed by J. van den Berg.

It is easy to deduce that the law μ of the triple (X, Y, Z) is not monotonic. It is however positively associated since the triple (X, Y, Z) is an increasing function of the independent pair X, Y.

As in the previous example, μ is not strictly positive, a weakness which we remedy differently than before. Let X', Y', Z' (respectively, X'', Y'', Z'') be Bernoulli random variables with parameter δ (respectively, $1-\delta$), and assume the maximal amount of independence. The triple

$$(A, B, C) = \big((X \vee X') \wedge X'', (Y \vee Y') \wedge Y'', (Z \vee Z') \wedge Z''\big)$$

is an increasing function of positively associated random variables, and is therefore positively associated. However, for small positive δ, it is only a small (stochastic) perturbation of the original triple (X, Y, Z), and one may check that (A, B, C) is not monotonic. It is easily verified that $\mathbb{P}((A, B, C) = \omega) > 0$ for all $\omega \in \{0, 1\}^3$.

Proof of Theorem 2.24. Throughout, μ is assumed strictly positive.
(a) $\Longleftrightarrow$ (b). We prove first that (a) implies (b). By Theorem 2.19, it suffices to prove (2.15) for two incomparable configurations ω_1, ω_2 that disagree on exactly two edges. Let e, f be distinct members of E, and take e and f to be first two edges in a given ordering of E. We shall adopt the notation in the proof of Theorem 2.3. Thus we write $\omega_1 = 0 \cdot 1 \cdot w$ and $\omega_2 = 1 \cdot 0 \cdot w$ for some word w of length $|E| - 2$. By strong positive-association, $\alpha(xy) = \mu(x \cdot y \cdot w)$ satisfies

$$\alpha(11)\big[\alpha(00) + \alpha(01) + \alpha(10) + \alpha(11)\big] \geq \big[\alpha(01) + \alpha(11)\big]\big[\alpha(10) + \alpha(11)\big],$$

which may be simplified to obtain as required that

$$\alpha(11)\alpha(00) \geq \alpha(01)\alpha(10).$$

We prove next that (b) implies (a). Suppose (b) holds, and let $F \subseteq E$ and $\xi \in \Omega$. It is immediate from (2.21) that

$$\mu_F^\xi(\omega_1 \vee \omega_2)\mu_F^\xi(\omega_1 \wedge \omega_2) \geq \mu_F^\xi(\omega_1)\mu_F^\xi(\omega_2), \qquad \omega_1, \omega_2 \in \Omega_F.$$

By Theorem 2.16, μ_F^ξ is positively associated.
(b) $\Longrightarrow$ (c). By the Holley inequality, Theorem 2.1, it suffices to prove for $\omega_F, \rho_F \in \Omega_F$ that

$$\mu_F^\zeta(\omega_F \vee \rho_F)\mu_F^\xi(\omega_F \wedge \rho_F) \geq \mu_F^\zeta(\omega_F)\mu_F^\xi(\rho_F) \qquad \text{whenever } \xi \leq \zeta.$$

This is, by (2.21), an immediate consequence of the FKG lattice property applied to the pair $\omega_F \times \zeta$, $\rho_F \times \xi$.
(c) $\Longrightarrow$ (d). This is trivial.
(d) $\Longrightarrow$ (b). Let μ be 1-monotonic. By Theorem 2.6, the pair μ, μ satisfies (2.2), which is to say that μ satisfies the FKG lattice condition. □

2.3 Influence for monotonic measures

Let $N \geq 1$, and let E be an arbitary finite set with $|E| = N$. We write $\Omega = \{0, 1\}^E$ as usual, and $\mathcal{F}$ for the set of all subsets of Ω. Let μ be a probability measure on $(\Omega, \mathcal{F})$, and A an increasing event. The *(conditional) influence* on A of the edge $e \in E$ is defined by

$$I_A(e) = \mu(A \mid J_e = 1) - \mu(A \mid J_e = 0), \tag{2.27}$$

where $J = (J_e : e \in E)$ denotes[8] the identity function on Ω. There has been an extensive study of the largest influence, $\max_e I_A(e)$, when μ is a product measure, and this has been used to obtain concentration theorems for $\phi_p(A)$ viewed as a function of p, where ϕ_p denotes product measure with density p on Ω. Such results have applications to several topics including random graphs, random walks, and percolation. Theorems concerning influence were first proved for product measures, but they may be extended in a natural way to monotonic measures.

(2.28) Theorem (Influence) [141]. *There exists a constant c satisfying* $c \in (0, \infty)$ *such that the following holds. Let* $N \geq 1$, *let E be a finite set with* $|E| = N$, *and let A be an increasing subset of* $\Omega = \{0, 1\}^E$. *Let* μ *be a strictly positive probability measure on* $(\Omega, \mathcal{F})$ *that is monotonic. There exists* $e \in E$ *such that*

$$I_A(e) \geq c \min\{\mu(A), 1 - \mu(A)\} \frac{\log N}{N}.$$

There are several useful references concerning influence for product measures, see [125, 126, 200, 201] and their bibliographies. The order of magnitude $N^{-1} \log N$ is the best possible, see [34].

Proof. Let μ be strictly positive and monotonic. The idea is to encode μ in terms of Lebesgue measure λ on the Euclidean cube $[0, 1]^E$, and then to apply the influence theorem[9] of [67]. This will be done via a certain function $f : [0, 1]^E \to \{0, 1\}^E$ constructed next. A similar argument will be used to prove Theorem 3.45.

We may suppose without loss of generality that $E = \{1, 2, \dots, N\}$. Let $\mathbf{x} = (x_i : i = 1, 2, \dots, N) \in [0, 1]^E$, and let $f(\mathbf{x}) = (f_i(\mathbf{x}) : i = 1, 2, \dots, N) \in \mathbb{R}^E$ be given recursively as follows. The first coordinate $f_1(\mathbf{x})$ is defined by:

$$\text{with} \quad a_1 = \mu(J_1 = 1), \quad \text{let} \quad f_1(\mathbf{x}) = \begin{cases} 1 & \text{if } x_1 > 1 - a_1, \\ 0 & \text{otherwise.} \end{cases} \tag{2.29}$$

Suppose we know the values $f_i(\mathbf{x})$ for $i = 1, 2, \dots, k-1$. Let

$$a_k = \mu\big(J_k = 1 \,\big|\, J_i = f_i(\mathbf{x}) \text{ for } i = 1, 2, \dots, k-1\big), \tag{2.30}$$

[8] Thus, J_e denotes both the event $\{\omega \in \Omega : \omega(e) = 1\}$ and its indicator function.

[9] An interesting aspect of the proof of this theorem is the use of discrete Fourier transforms and hypercontractivity.

and define

$$f_k(\mathbf{x}) = \begin{cases} 1 & \text{if } x_k > 1 - a_k, \\ 0 & \text{otherwise.} \end{cases} \tag{2.31}$$

It may be shown as follows that the function $f : [0,1]^E \to \{0,1\}^E$ is non-decreasing. Let $\mathbf{x} \le \mathbf{x}'$, and write $a_k = a_k(\mathbf{x})$ and $a'_k = a_k(\mathbf{x}')$ for the values in (2.29)–(2.30) corresponding to the vectors $\mathbf{x}$ and $\mathbf{x}'$. Clearly $a_1 = a'_1$, so that $f_1(\mathbf{x}) \le f_1(\mathbf{x}')$. Since μ is monotonic, $a_2 \le a'_2$, implying that $f_2(\mathbf{x}) \le f_2(\mathbf{x}')$. Continuing inductively, we find that $f_k(\mathbf{x}) \le f_k(\mathbf{x}')$ for all k, which is to say that $f(\mathbf{x}) \le f(\mathbf{x}')$.

Let $A \in \mathcal{F}$ be an increasing event, and let B be the increasing subset of $[0,1]^E$ given by $B = f^{-1}(A)$. We make four notes concerning the definition of f.

(a) For given $\mathbf{x}$, each a_k depends only on $x_1, x_2, \dots, x_{k-1}$.

(b) Since μ is strictly positive, the a_k satisfy $0 < a_k < 1$ for all $\mathbf{x} \in [0,1]^N$ and $k \in E$.

(c) For any $\mathbf{x} \in [0,1]^N$ and $k \in E$, the values $f_k(\mathbf{x}), f_{k+1}(\mathbf{x}), \dots, f_N(\mathbf{x})$ depend on $x_1, x_2, \dots, x_{k-1}$ only through the values $f_1(\mathbf{x}), f_2(\mathbf{x}), \dots, f_{k-1}(\mathbf{x})$.

(d) The function f and the event B depend on the ordering of the set E.

Let $U = (U_i : i = 1, 2, \dots, N)$ be the identity function on $[0,1]^E$, so that U has law λ. By the definition of f, $f(U)$ has law μ. Hence,

$$\mu(A) = \lambda(f(U) \in A) = \lambda(U \in f^{-1}(A)) = \lambda(B). \tag{2.32}$$

Let

$$K_B(i) = \lambda(B \mid U_i = 1) - \lambda(B \mid U_i = 0),$$

where the conditional probabilities are interpreted as

$$\lambda(B \mid U_i = u) = \lim_{\epsilon \downarrow 0} \lambda\big(B \bigm| U_i \in (u - \epsilon, u + \epsilon)\big).$$

By [67, Thm 1], there exists a constant $c \in (0, \infty)$, independent of the choice of N and A, such that: there exists $e \in E$ with

$$K_B(e) \ge c \min\{\lambda(B), 1 - \lambda(B)\} \frac{\log N}{N}. \tag{2.33}$$

We choose e accordingly. We claim that

$$I_A(j) \ge K_B(j) \qquad \text{for } j \in E. \tag{2.34}$$

By (2.32) and (2.33), it suffices to prove (2.34). We prove first that

$$I_A(1) \ge K_B(1), \tag{2.35}$$

which is stronger than (2.34) with $j = 1$. By (b) and (c) above,

$$
\begin{aligned}
(2.36) \qquad I_A(1) &= \mu(A \mid J_1 = 1) - \mu(A \mid J_1 = 0) \\
&= \lambda(B \mid f_1(U) = 1) - \lambda(B \mid f_1(U) = 0) \\
&= \lambda(B \mid U_1 > 1 - a_1) - \lambda(B \mid U_1 \le 1 - a_1) \\
&= \lambda(B \mid U_1 = 1) - \lambda(B \mid U_1 = 0) \\
&= K_B(1).
\end{aligned}
$$

We turn to (2.34) with $j \ge 2$. We re-order the set E to bring the index j to the front. That is, we let F be the re-ordered index set $F = (k_1, k_2, \dots, k_N) = (j, 1, 2, \dots, j-1, j+1, \dots, N)$. Let $g = (g_{k_r} : r = 1, 2, \dots, N)$ denote the associated function given by (2.29)–(2.31) subject to the new ordering, and let $C = g^{-1}(A)$. We claim that

$$(2.37) \qquad K_C(k_1) \ge K_B(j).$$

By (2.36) with E replaced by F, $K_C(k_1) = I_A(j)$, and (2.34) follows. It remains to prove (2.37), and we use monotonicity again for this. It suffices to prove that

$$(2.38) \qquad \lambda(C \mid U_j = 1) \ge \lambda(B \mid U_j = 1),$$

together with the reversed inequality given $U_j = 0$. Let

$$(2.39) \qquad \overline{U} = (U_1, U_2, \dots, U_{j-1}, 1, U_{j+1}, \dots, U_N).$$

The $0/1$-vector $f(\overline{U}) = (f_i(\overline{U}) : i = 1, 2, \dots, N)$ is constructed sequentially (as above) by considering the indices $1, 2, \dots, N$ in turn. At stage k, we declare $f_k(\overline{U})$ equal to 1 if U_k exceeds a certain function a_k of the variables $f_i(\overline{U})$, $1 \le i < k$. By the monotonicity of μ, this function is non-increasing in these variables. The index j plays a special role in that: (i) $f_j(\overline{U}) = 1$, and (ii) given this fact, it is more likely than before that the variables $f_k(\overline{U})$, $j < k \le N$, will take the value 1. The values $f_k(\overline{U})$, $1 \le k < j$ are unaffected by the value of U_j.

Consider now the $0/1$-vector $g(\overline{U}) = (g_{k_r}(\overline{U}) : r = 1, 2, \dots, N)$, constructed in the same manner as above but with the new ordering F of the index set E. First we examine index k_1 $(= j)$, and we automatically declare $g_{k_1}(\overline{U}) = 1$ (since $U_j = 1$). We then construct $g_{k_r}(\overline{U})$, $r = 2, 3, \dots, N$, in sequence. Since the a_k are non-decreasing in the variables constructed so far,

$$(2.40) \qquad g_{k_r}(\overline{U}) \ge f_{k_r}(\overline{U}), \qquad r = 2, 3, \dots, N.$$

Therefore, $g(\overline{U}) \ge f(\overline{U})$, and hence

$$(2.41) \qquad \lambda(C \mid U_j = 1) = \lambda(g(\overline{U}) \in A) \ge \lambda(f(\overline{U}) \in A) = \lambda(B \mid U_j = 1).$$

Inequality (2.38) has been proved. The same argument implies the reversed inequality obtained from (2.38) by changing the conditioning to $U_j = 0$. Inequality (2.37) follows, and the proof is complete. □

2.4 Sharp thresholds for increasing events

We consider next certain families of probability measures μ_p indexed by a parameter $p \in (0, 1)$, and we prove a sharp-threshold theorem subject to a hypothesis of monotonicity. The idea is as follows. Let A be a non-empty increasing event in $\Omega = \{0, 1\}^N$. Subject to a certain hypothesis on the μ_p, the function $f(p) = \mu_p(A)$ is non-decreasing with $f(0) = 0$ and $f(1) = 1$. If A has a certain property of symmetry, the sharp-threshold theorem asserts that $f(p)$ increases steeply from 0 to 1 over a short interval of p-values with length of order $1/\log N$.

We use the notation of the previous section. Let μ be a probability measure on $(\Omega, \mathcal{F})$. For $p \in (0, 1)$, let μ_p be the probability measure given by

$$\mu_p(\omega) = \frac{1}{Z_p}\mu(\omega)\left\{\prod_{e\in E} p^{\omega(e)}(1-p)^{1-\omega(e)}\right\}, \qquad \omega \in \Omega, \tag{2.42}$$

where Z_p is the normalizing constant

$$Z_p = \sum_{\omega\in\Omega} \mu(\omega)\left\{\prod_{e\in E} p^{\omega(e)}(1-p)^{1-\omega(e)}\right\}.$$

It is elementary that $\mu = \mu_{\frac{1}{2}}$, and that (each) μ_p is strictly positive if and only if μ is strictly positive. It is easy to check that (each) μ_p satisfies the FKG lattice condition (2.15) if and only if μ satisfies this condition, and it follows by Theorem 2.24 that, for strictly positive μ, μ is monotonic if and only if (each) μ_p is monotonic. In order to prove a sharp-threshold theorem for the family μ_p, we present first a differential formula of the type referred to as Russo's formula, [154, Section 2.4].

(2.43) Theorem [39]. *For a random variable $X : \Omega \to \mathbb{R}$,*

$$\frac{d}{dp}\mu_p(X) = \frac{1}{p(1-p)}\mathrm{cov}_p(|\eta|, X), \qquad p \in (0, 1), \tag{2.44}$$

where cov_p denotes covariance with respect to the probability measure μ_p, and $\eta(\omega)$ is the set of ω-open edges.

We note for later use that

$$\mathrm{cov}_p(|\eta|, X) = \sum_{e\in E} \mathrm{cov}_p(J_e, X). \tag{2.45}$$

Proof. We follow [39, Prop. 4] and [156, Section 2.4]. Write

$$\nu_p(\omega) = p^{|\eta(\omega)|}(1-p)^{N-|\eta(\omega)|}\mu(\omega), \qquad \omega \in \Omega,$$

so that

$$\mu_p(X) = \frac{1}{Z_p} \sum_{\omega \in \Omega} X(\omega) \nu_p(\omega). \tag{2.46}$$

It is elementary that
(2.47)

$$\frac{d}{dp} \mu_p(X) = \frac{1}{Z_p} \sum_{\omega \in \Omega} \left(\frac{|\eta(\omega)|}{p} - \frac{N - |\eta(\omega)|}{1-p} \right) X(\omega) \nu_p(\omega) - \frac{Z'_p}{Z_p} \mu_p(X),$$

where $Z'_p = dZ_p/dp$. Setting $X = 1$, we find that

$$0 = \frac{1}{p(1-p)} \mu_p(|\eta| - pN) - \frac{Z'_p}{Z_p},$$

whence

$$\begin{aligned} p(1-p) \frac{d}{dp} \mu_p(X) &= \mu_p\big([|\eta| - pN]X\big) - \mu_p(|\eta| - pN)\mu_p(X) \\ &= \mu_p(|\eta| X) - \mu_p(|\eta|)\mu_p(X) \\ &= \operatorname{cov}_p(|\eta|, X). \end{aligned}$$

□

Let Π be the group of permutations of E. Any $\pi \in \Pi$ acts[10] on Ω by $\pi\omega = (\omega(\pi_e) : e \in E)$. We say that a subgroup $\mathcal{A}$ of Π *acts transitively* on E if, for all pairs $j, k \in E$, there exists $\alpha \in \mathcal{A}$ with $\alpha_j = k$.

Let $\mathcal{A}$ be a subgroup of Π. A probability measure ϕ on $(\Omega, \mathcal{F})$ is called *$\mathcal{A}$-invariant* if $\phi(\omega) = \phi(\alpha\omega)$ for all $\alpha \in \mathcal{A}$. An event $A \in \mathcal{F}$ is called *$\mathcal{A}$-invariant* if $A = \alpha A$ for all $\alpha \in \mathcal{A}$. It is easily seen that, for any subgroup $\mathcal{A}$, μ is $\mathcal{A}$-invariant if and only if (each) μ_p is $\mathcal{A}$-invariant.

(2.48) Theorem (Sharp threshold) [141]. *There exists a constant c satisfying $c \in (0, \infty)$ such that the following holds. Let $N = |E| \geq 1$ and let $A \in \mathcal{F}$ be an increasing event. Let μ be a strictly positive probability measure on $(\Omega, \mathcal{F})$ that is monotonic. Suppose there exists a subgroup $\mathcal{A}$ of Π acting transitively on E such that μ and A are $\mathcal{A}$-invariant. Then*

$$\frac{d}{dp} \mu_p(A) \geq \frac{c m_p}{p(1-p)} \min\big\{\mu_p(A), 1 - \mu_p(A)\big\} \log N, \qquad p \in (0, 1), \tag{2.49}$$

where $m_p = \mu_p(J_e)(1 - \mu_p(J_e))$.

Let $\epsilon \in (0, \frac{1}{2})$ and let A be non-empty and increasing. Under the conditions of the theorem, $\mu_p(A)$ increases from ϵ to $1 - \epsilon$ over an interval of values of p having length of order $1/\log N$. This amounts to a quantification of the so-called S-shape results described and cited in [154, Section 2.5]. Note that m_p does not depend on the choice of edge e.

The proof is preceded by an easy lemma. Let

$$I_{p,A}(e) = \mu_p(A \mid J_e = 1) - \mu_p(A \mid J_e = 0), \qquad e \in E.$$

[10] This differs slightly from the definition of Section 4.3, for reasons of local convenience.

(2.50) Lemma. *Let $A \in \mathcal{F}$. Suppose there exists a subgroup $\mathcal{A}$ of Π acting transitively on E such that μ and A are $\mathcal{A}$-invariant. Then $I_{p,A}(e) = I_{p,A}(f)$ for all $e, f \in E$ and all $p \in (0,1)$.*

Proof. Since μ is $\mathcal{A}$-invariant, so is μ_p for every p. Let $e, f \in E$, and find $\alpha \in \mathcal{A}$ such that $\alpha_e = f$. Under the given conditions,

$$\begin{aligned} \mu_p(A,\ J_f = 1) &= \sum_{\omega \in A} \mu_p(\omega) J_f(\omega) = \sum_{\omega \in A} \mu_p(\alpha\omega) J_e(\alpha\omega) \\ &= \sum_{\omega' \in A} \mu_p(\omega') J_e(\omega') = \mu_p(A,\ J_e = 1). \end{aligned}$$

We deduce with $A = \Omega$ that $\mu_p(J_f = 1) = \mu_p(J_e = 1)$. On dividing, we obtain that $\mu_p(A \mid J_f = 1) = \mu_p(A \mid J_e = 1)$. A similar equality holds with 1 replaced by 0, and the claim of the lemma follows. □

Proof of Theorem 2.48. By Lemma 2.50, $I_{p,A}(e) = I_{p,A}(f)$ for all $e, f \in E$. Since A is increasing and μ_p is monotonic, each $I_{p,A}(e)$ is non-negative, and therefore

$$\begin{aligned} \mathrm{cov}_p(J_e, 1_A) &= \mu_p(J_e 1_A) - \mu_p(J_e)\mu_p(A) \\ &= \mu_p(J_e)(1 - \mu_p(J_e)) I_{p,A}(e) \\ &\geq m_p I_{p,A}(e), \qquad e \in E. \end{aligned}$$

Summing over the index set E as in (2.44)–(2.45), we deduce (2.49) by Theorem 2.28 applied to the monotonic measure μ_p. □

2.5 Exponential steepness

This chapter closes with a further differential inequality for the probability of a monotonic event. Let $A \in \mathcal{F}$ and $\omega \in \Omega$. We define $H_A(\omega)$ to be the Hamming distance from ω to A, that is,

$$H_A(\omega) = \inf\{H(\omega', \omega) : \omega' \in A\}, \tag{2.51}$$

where $H(\omega', \omega)$ is given in (1.26). Note that

(2.52)

$$H_A(\omega) = \begin{cases} \inf\left\{\sum_e [\omega'(e) - \omega(e)] : \omega' \geq \omega,\ \omega' \in A\right\} & \text{if } A \text{ is increasing,} \\ \inf\left\{\sum_e [\omega(e) - \omega'(e)] : \omega' \leq \omega,\ \omega' \in A\right\} & \text{if } A \text{ is decreasing.} \end{cases}$$

Suppose now that A is increasing (respectively, decreasing). Here are three useful facts concerning H_A.

(i) H_A is a decreasing (respectively, increasing) random variable.

(ii) The function $|\eta| + H_A$ (respectively, $|\eta| - H_A$) is increasing, since the addition of a single open edge to a configuration ω causes $|\eta(\omega)|$ to increase by 1, and $H_A(\omega)$ to decrease (respectively, increase) by at most 1.

(iii) We have that $H_A(\omega)1_A(\omega) = 0$ for $\omega \in \Omega$.

Given a probability measure μ on $(\Omega, \mathcal{F})$, the associated measures μ_p, $p \in (0, 1)$, are given by (2.42).

(2.53) Theorem [153, 163]. *Let μ be a strictly positive probability measure on $(\Omega, \mathcal{F})$ that is monotonic. For a non-empty event $A \in \mathcal{F}$, and $p \in (0, 1)$,*

$$\frac{d}{dp} \log \mu_p(A) \geq \frac{\mu_p(H_A)}{p(1-p)}, \qquad \text{if } A \text{ is increasing,} \tag{2.54}$$

$$\frac{d}{dp} \log \mu_p(A) \leq -\frac{\mu_p(H_A)}{p(1-p)}, \qquad \text{if } A \text{ is decreasing.} \tag{2.55}$$

Inequality (2.54) bears a resemblance to a formula valid for percolation that may be written as

$$\frac{d}{dp} \log \phi_p(A) = \frac{1}{p} \phi_p(N_A \mid A),$$

where N_A is the number of pivotal edges for the increasing event A, and ϕ_p denotes product measure with density p on $(\Omega, \mathcal{F})$. See [154, p. 44] for further details.

Proof. Since μ is assumed strictly positive and monotonic, it satisfies the FKG lattice property. Therefore, every μ_p satisfies the FKG lattice property, and hence is positively associated. Let $A \in \mathcal{F}$ be non-empty and increasing. By (2.44), (ii)–(iii) above, and positive association,

$$\begin{aligned}
\frac{d}{dp} \mu_p(A) &= \frac{1}{p(1-p)} \operatorname{cov}_p(|\eta|, 1_A) \\
&= \frac{1}{p(1-p)} \big[\operatorname{cov}_p(|\eta| + H_A, 1_A) - \operatorname{cov}_p(H_A, 1_A)\big] \\
&\geq -\frac{1}{p(1-p)} \operatorname{cov}_p(H_A, 1_A) \\
&= \frac{\mu_p(H_A)\mu_p(A)}{p(1-p)},
\end{aligned}$$

and (2.54) follows. The argument is easily adapted for decreasing A. □

Let $A \in \mathcal{F}$ be non-empty and increasing. Inequality (2.54) is usually used in integrated form. Integrating over the interval $[r, s]$, and using the facts that $p(1-p) \leq \frac{1}{4}$ and that H_A is decreasing, we obtain that

$$\begin{aligned}
\mu_r(A) &\leq \mu_s(A) \exp\left\{-4 \int_r^s \mu_p(H_A)\, dp\right\} \\
&\leq \mu_s(A) \exp\{-4(s-r)\mu_s(H_A)\}, \qquad 0 < r \leq s < 1.
\end{aligned} \tag{2.56}$$

This may sometimes be combined with a complementary inequality derived by a consideration of 'finite energy', see Theorem 3.45.

Chapter 3

Fundamental Properties

Summary. The basic properties of random-cluster measures are presented in a manner suitable for future applications. Accounts of conditional random-cluster measures and positive association are followed by differential formulae, a sharp-threshold theorem, and exponential steepness. There are several useful inequalities involving partition functions. The series/parallel laws are formulated, and the chapter ends with a discussion of negative correlation.

3.1 Conditional probabilities

Throughout this chapter, $G = (V, E)$ will be assumed to be a finite graph. Let $\phi_{G,p,q}$ be the random-cluster measure on G. Whether or not a given edge e is open depends on the configuration on the remainder of the graph. The relevant conditional probabilities may be described in the following useful manner.

For $e = \langle x, y \rangle \in E$, the expression $G \setminus e$ (respectively, $G.e$) denotes the graph obtained from G by deleting (respectively, contracting) the edge e. We write $\Omega_{\langle e \rangle} = \{0, 1\}^{E \setminus \{e\}}$ and, for $\omega \in \Omega$, we define $\omega_{\langle e \rangle} \in \Omega_{\langle e \rangle}$ by

$$\omega_{\langle e \rangle}(f) = \omega(f), \qquad f \in E, \ f \neq e.$$

Let K_e denote the event that x and y are joined by an open path not using e.

(3.1) Theorem (Conditional probabilities) [122]. *Let $p \in (0, 1)$, $q \in (0, \infty)$.*

(a) *We have for $e \in E$ that*

$$\phi_{G,p,q}(\omega \mid \omega(e) = j) = \begin{cases} \phi_{G \setminus e,p,q}(\omega_{\langle e \rangle}) & \text{if } j = 0, \\ \phi_{G.e,p,q}(\omega_{\langle e \rangle}) & \text{if } j = 1, \end{cases} \tag{3.2}$$

and

$$\phi_{G,p,q}(\omega(e) = 1 \mid \omega_{\langle e \rangle}) = \begin{cases} p & \text{if } \omega_{\langle e \rangle} \in K_e, \\ \dfrac{p}{p + q(1-p)} & \text{if } \omega_{\langle e \rangle} \notin K_e. \end{cases} \tag{3.3}$$

(b) *Conversely, if ϕ is a probability measure on $(\Omega, \mathcal{F})$ satisfying* (3.3) *for all $\omega \in \Omega$ and $e \in E$, then $\phi = \phi_{G,p,q}$.*

The effect of conditioning on the absence or presence of an edge e is to replace the measure $\phi_{G,p,q}$ by the random-cluster measure on the respective graph $G \backslash e$ or $G.e$. In addition, the conditional probability that e is open, given the configuration elsewhere, depends only on whether or not K_e occurs, and is then given by the stated formula. By (3.3),

$$(3.4) \quad 0 < \phi_{G,p,q}(\omega(e) = 1 \mid \omega_{\langle e \rangle}) < 1, \qquad e \in E, \ p \in (0,1), \ q \in (0, \infty).$$

Thus, given $\omega_{\langle e \rangle}$, each of the two possible states of e occurs with a strictly positive probability. This useful fact is known as the 'finite-energy property', and is related to the property of so-called 'insertion tolerance' (see Section 10.12).

We shall sometimes need to condition on the states of more than one edge. Towards this end, we state next a more general property than (3.2), beginning with a brief discussion of boundary conditions; more on the latter topic may be found in Section 4.2. Let $\xi \in \Omega$, $F \subseteq E$, and let Ω_F^ξ be the subset of Ω containing all configurations ψ satisfying $\psi(e) = \xi(e)$ for all $e \notin F$. We define the random-cluster measure $\phi_{F,p,q}^\xi$ on $(\Omega, \mathcal{F})$ by

(3.5)

$$\phi_{F,p,q}^\xi(\omega) = \begin{cases} \dfrac{1}{Z_F^\xi(p,q)} \left\{ \displaystyle\prod_{e \in F} p^{\omega(e)}(1-p)^{1-\omega(e)} \right\} q^{k(\omega,F)} & \text{if } \omega \in \Omega_F^\xi, \\ 0 & \text{otherwise,} \end{cases}$$

where $k(\omega, F)$ is the number of components of the graph $(G, \eta(\omega))$ that intersect the set of endvertices of F, and

$$(3.6) \qquad Z_F^\xi(p,q) = \sum_{\omega \in \Omega_F^\xi} \left\{ \prod_{e \in F} p^{\omega(e)}(1-p)^{1-\omega(e)} \right\} q^{k(\omega,F)}.$$

Note that $\phi_{F,p,q}^\xi(\Omega_F^\xi) = 1$.

(3.7) Theorem. *Let $p \in [0,1]$, $q \in (0,\infty)$, and $F \subseteq E$. Let X be a random variable that is $\mathcal{F}_F$-measurable. Then*

$$\phi_{G,p,q}(X \mid \mathcal{T}_F)(\xi) = \phi_{F,p,q}^\xi(X), \qquad \xi \in \Omega.$$

In other words, given the states of edges not belonging to F, the conditional measure on F is a random-cluster measure subject to the retention of open connections of ξ using edges not belonging to F.

Here is a final note. Let $p \in (0,1)$ and $q \neq 1$. It is easily seen that the states of two distinct edges e, f are independent if and only if the pair e, f lies in no circuit of G. This may be proved either directly or via the simulation methods of Sections 3.4 and 8.2.

Proof of Theorem 3.1. (a) This is easily seen by an expansion of the conditional probability,

$$\phi_{G,p,q}(\omega \mid \omega(e) = j) = \begin{cases} \phi_{G,p,q}(\omega_e)/\phi_{G,p,q}(\overline{J_e}) & \text{if } j = 0, \\ \phi_{G,p,q}(\omega^e)/\phi_{G,p,q}(J_e) & \text{if } j = 1, \end{cases} \qquad \omega \in \Omega,$$

where $J_e = \{\omega \in \Omega : \omega(e) = 1\}$, and ω_e, ω^e are given by (1.25).

Similarly,

$$\begin{aligned} \phi_{G,p,q}(\omega(e) = 1 \mid \omega_{\langle e \rangle}) &= \frac{\phi_{G,p,q}(\omega^e)}{\phi_{G,p,q}(\omega^e) + \phi_{G,p,q}(\omega_e)} \\ &= \frac{[p/(1-p)]^{|\eta(\omega^e)|} q^{k(\omega^e)}}{[p/(1-p)]^{|\eta(\omega^e)|} q^{k(\omega^e)} + [p/(1-p)]^{|\eta(\omega_e)|} q^{k(\omega_e)}} \\ &= \begin{cases} \dfrac{p/(1-p)}{[p/(1-p)] + 1} & \text{if } \omega_e \in K_e, \\ \dfrac{p/(1-p)}{[p/(1-p)] + q} & \text{if } \omega_e \notin K_e, \end{cases} \end{aligned}$$

where $\eta(\omega)$ is, as usual, the set of open edges in Ω.

(b) The claim is immediate by the fact, easily proved, that a strictly positive probability measure ϕ is specified uniquely by the conditional probabilities $\phi(\omega(e) = 1 \mid \omega_{\langle e \rangle})$, $\omega \in \Omega$, $e \in E$. □

Proof of Theorem 3.7. This holds by repeated application of (3.2), with one application for each edge not belonging in F. □

3.2 Positive association

Let $\phi_{p,q}$ denote the random-cluster measure on G with parameters p and q. We shall see that $\phi_{p,q}$ satisfies the FKG lattice condition (2.15) whenever $q \geq 1$, and we arrive thus at the following conclusion.

(3.8) Theorem (Positive association) [122]. *Let $p \in (0, 1)$ and $q \in [1, \infty)$.*

(a) *The random-cluster measure $\phi_{p,q}$ is strictly positive and satisfies the FKG lattice condition.*

(b) *The random-cluster measure $\phi_{p,q}$ is strongly positively-associated, and in particular*

$$\begin{aligned} \phi_{p,q}(XY) &\geq \phi_{p,q}(X)\phi_{p,q}(Y) && \text{for increasing } X, Y : \Omega \to \mathbb{R}, \\ \phi_{p,q}(A \cap B) &\geq \phi_{p,q}(A)\phi_{p,q}(B) && \text{for increasing } A, B \in \mathcal{F}. \end{aligned}$$

It is not difficult to see that $\phi_{p,q}$ is not (in general) positively associated when $q \in (0, 1)$, as illustrated in the example following. Let G be the graph containing just two vertices and having exactly two parallel edges e and f joining these vertices. It is an easy computation that

$$\phi_{p,q}(J_e \cap J_f) - \phi_{p,q}(J_e)\phi_{p,q}(J_f) = \frac{p^2q^2(q-1)(1-p)^2}{Z(p,q)^2}, \tag{3.9}$$

where J_g is the event that g is open. This is strictly negative if $0 < p, q < 1$.

Proof of Theorem 3.8. Let $p \in (0,1)$ and $q \in [1,\infty)$. Part (b) follows from (a) and Theorem 2.24. It is elementary that $\phi_{p,q}$ is strictly positive. We now check as required that $\phi_{p,q}$ satisfies the FKG lattice condition (2.15). Since the set $\eta(\omega)$ of open edges in a configuration ω satisfies

$$|\eta(\omega_1 \vee \omega_2)| + |\eta(\omega_1 \wedge \omega_2)| = |\eta(\omega_1)| + |\eta(\omega_2)|, \qquad \omega_1, \omega_2 \in \Omega, \tag{3.10}$$

it suffices, on taking logarithms, to prove that

$$k(\omega_1 \vee \omega_2) + k(\omega_1 \wedge \omega_2) \geq k(\omega_1) + k(\omega_2), \qquad \omega_1, \omega_2 \in \Omega. \tag{3.11}$$

By Theorem 2.19, we may restrict our attention to incomparable pairs ω_1, ω_2 that differ on exactly two edges. There must then exist distinct edges $e, f \in E$ and a configuration $\omega \in \Omega$ such that $\omega_1 = \omega_f^e$, $\omega_2 = \omega_e^f$. As in the proof of Theorem 2.24, we omit reference to the states of edges other than e and f, and we write $\omega_1 = 10$ and $\omega_2 = 01$. Let D_f be the indicator function of the event that the endvertices of f are connected by no open path of $E \setminus \{f\}$. Since D_f is a decreasing random variable, we have that $D_f(10) \leq D_f(00)$. Therefore,

$$k(10) - k(11) = D_f(10) \leq D_f(00) = k(00) - k(01),$$

which implies (3.11). □

Theorem 3.8 applies only to *finite* graphs G, whereas many potential applications concern *infinite* graphs. We shall see in Sections 4.3 and 4.4 how to derive the required extension.

3.3 Differential formulae and sharp thresholds

One way of estimating the probability of an event A is via an estimate of its derivative $d\phi_{p,q}(A)/dp$. When $q = 1$, there is a formula for this derivative which has proved very useful in reliability theory, percolation, and elsewhere, see [22, 126, 154, 287]. This formula has been extended to random-cluster measures. For $\omega \in \Omega$, let $|\eta| = |\eta(\omega)| = \sum_{e \in E} \omega(e)$ be the number of open edges of ω as usual, and $k = k(\omega)$ the number of open clusters.

(3.12) Theorem [39]. *Let $p \in (0, 1)$, $q \in (0, \infty)$, and let $\phi_{p,q}$ be the corresponding random-cluster measure on a finite graph $G = (V, E)$. We have that*

$$\frac{d}{dp}\phi_{p,q}(X) = \frac{1}{p(1-p)}\mathrm{cov}_{p,q}(|\eta|, X), \tag{3.13}$$

$$\frac{d}{dq}\phi_{p,q}(X) = \frac{1}{q}\mathrm{cov}_{p,q}(k, X), \tag{3.14}$$

for any random variable $X : \Omega \to \mathbb{R}$, where $\mathrm{cov}_{p,q}$ denotes covariance with respect to $\phi_{p,q}$.

In most applications, we set $X = 1_A$, the indicator function of some given event A, and we obtain that

$$\frac{d}{dp}\phi_{p,q}(A) = \frac{\phi_{p,q}(1_A|\eta|) - \phi_{p,q}(A)\phi_{p,q}(|\eta|)}{p(1-p)}, \tag{3.15}$$

with a similar formula for the derivative with respect to q.

Here are two simple examples of Theorem 3.12 which result in monotonicities valid for all $q \in (0, \infty)$. Let $h : \mathbb{R} \to \mathbb{R}$ be non-decreasing. On setting $X = h(|\eta|)$, we have from (3.13) that

$$\frac{d}{dp}\phi_{p,q}(X) = \frac{1}{p(1-p)}\mathrm{cov}_{p,q}(|\eta|, h(|\eta|)) \geq 0.$$

In the special case $h(x) = x$, we deduce that the mean number of open edges is a non-decreasing function of p, for all $q \in (0, \infty)$. Similarly, by (3.14), for non-decreasing h,

$$\frac{d}{dq}\phi_{p,q}(h(k)) = \frac{1}{q}\mathrm{cov}_{p,q}(k, h(k)) \geq 0.$$

This time we take $h = -1_{(-\infty,1]}$, so that $-h$ is the indicator function of the event that the open graph $(V, \eta(\omega))$ is connected. We deduce that the probability of connectedness is a decreasing function of q on the interval $(0, \infty)$. These examples are curiosities, given the failure of stochastic monotonicity when $q < 1$.

Let $q \in [1, \infty)$. Since $\phi_{p,q}$ satisfies the FKG lattice condition (2.15), it is monotonic. Let $\mathcal{A}$ be a subgroup of the automorphism group[1] $\mathrm{Aut}(G)$ of the graph $G = (V, E)$. We call E *$\mathcal{A}$-transitive* if $\mathcal{A}$ acts transitively on E.

[1]The automorphism group $\mathrm{Aut}(G)$ is discussed further in Sections 4.3 and 10.12.

(3.16) Theorem (Sharp threshold) [141]. *There exists an absolute constant $c \in (0, \infty)$ such that the following holds. Let $A \in \mathcal{F}$ be an increasing event, and suppose there exists a subgroup $\mathcal{A}$ of* Aut(G) *such that E is $\mathcal{A}$-transitive and A is $\mathcal{A}$-invariant. Then, for $p \in (0, 1)$ and $q \in [1, \infty)$,*

$$\frac{d}{dp}\phi_{p,q}(A) \geq C \min\{\phi_{p,q}(A), 1 - \phi_{p,q}(A)\} \log |E|, \tag{3.17}$$

where

$$C = c \min\left\{1, \frac{q}{\{p + q(1-p)\}^2}\right\}.$$

Since $q \geq 1$, (3.17) implies that

$$\frac{d}{dp}\phi_{p,q}(A) \geq \frac{c}{q} \min\{\phi_{p,q}(A), 1 - \phi_{p,q}(A)\} \log |E|, \tag{3.18}$$

an inequality that may be integrated directly. Let $p_1 = p_1(A, q) \in (0, 1)$ be chosen such that $\phi_{p_1,q}(A) \geq \frac{1}{2}$. Then

$$-\frac{d}{dp} \log[1 - \phi_{p,q}(A)] \geq \frac{c}{q} \log |E|, \qquad p \in (p_1, 1),$$

and hence, by integration,

$$\phi_{p,q}(A) \geq 1 - \tfrac{1}{2}|E|^{-c(p-p_1)/q}, \qquad p \in (p_1, 1),\ q \in [1, \infty), \tag{3.19}$$

whenever the conditions of Theorem 3.16 are satisfied. If in addition $p_1 \geq \sqrt{q}/(1 + \sqrt{q})$, then $C = c$, and hence

$$\phi_{p,q}(A) \geq 1 - \tfrac{1}{2}|E|^{-c(p-p_1)}, \qquad p \in (p_1, 1). \tag{3.20}$$

An application to box crossings in two dimensions may be found in [141].

Proof of Theorem 3.12. The first formula was proved for Theorem 2.43, and the second is obtained in a similar fashion. □

Proof of Theorem 3.16. With $\mathcal{A}$ as in the theorem, $\phi_{p,q}$ is $\mathcal{A}$-invariant since $\mathcal{A} \subseteq$ Aut(G). The claim is a consequence of Theorem 2.48 on noting from (3.3) that

$$\frac{\phi_{p,q}(J_e)\phi_{p,q}(\overline{J_e})}{p(1-p)} \geq \min\left\{1, \frac{q}{[p + q(1-p)]^2}\right\}, \qquad e \in E. \qquad \square$$

3.4 Comparison inequalities

The comparison inequalities of this section are an important tool in the study of random-cluster measures. As usual, we write $\phi_{p,q}$ for the random-cluster measure on the finite graph $G = (V, E)$.

(3.21) Theorem (Comparison inequalities) [122]. *It is the case that*:

$$(3.22)\quad \phi_{p_1,q_1} \le_{\text{st}} \phi_{p_2,q_2} \quad \textit{if } q_1 \ge q_2,\ q_1 \ge 1,\ \textit{and } p_1 \le p_2,$$

$$(3.23)\quad \phi_{p_1,q_1} \ge_{\text{st}} \phi_{p_2,q_2} \quad \textit{if } q_1 \ge q_2,\ q_1 \ge 1,\ \textit{and } \frac{p_1}{q_1(1-p_1)} \ge \frac{p_2}{q_2(1-p_2)}.$$

The first of these inequalities may be strengthened as in the next theorem. A subset W of the vertex set V is called *spanning* if every edge of E is incident to at least one vertex of W. The degree $\deg(W)$ of a spanning set W is defined to be the maximum degree of its members, that is, the maximum number of edges of G incident to any one vertex in W.

(3.24) Theorem [151]. *For $\Delta \in \{1, 2, \dots\}$, there exists a continuous function $\gamma(p,q) = \gamma_\Delta(p,q)$, which is strictly increasing in p on $(0,1)$, and strictly decreasing in q on $[1,\infty)$, such that the following holds. Let G be a finite graph, and suppose there exists a spanning set W such that* $\deg(W) \le \Delta$. *Then*

$$(3.25)\qquad \phi_{p_1,q_1} \le_{\text{st}} \phi_{p_2,q_2} \quad \textit{if} \quad 1 \le q_2 \le q_1 \ \textit{and } \gamma(p_1,q_1) \le \gamma(p_2,q_2).$$

An application is to be found in Section 5.1, where it is proved that the critical point $p_c(q)$ of an infinite-volume random-cluster model on a lattice is *strictly* increasing in q.

Proof of Theorem 3.21. We may assume that $p_1, p_2 \in (0,1)$, since the other cases are straightforward. We may either apply the Holley inequality (Theorem 2.1) or use the positive association of random-cluster measures (Theorem 3.8) as follows. Let $X : \Omega \to \mathbb{R}$ be increasing. Then

$$\begin{aligned}
\phi_{p_2,q_2}(X) &= \frac{1}{Z(p_2,q_2)} \sum_{\omega\in\Omega} X(\omega) p_2^{|\eta(\omega)|}(1-p_2)^{|E\setminus\eta(\omega)|} q_2^{k(\omega)} \\
&= \left(\frac{1-p_2}{1-p_1}\right)^{|E|} \frac{1}{Z(p_2,q_2)} \sum_{\omega\in\Omega} X(\omega)Y(\omega) p_1^{|\eta(\omega)|}(1-p_1)^{|E\setminus\eta(\omega)|} q_1^{k(\omega)} \\
&= \left(\frac{1-p_2}{1-p_1}\right)^{|E|} \frac{Z(p_1,q_1)}{Z(p_2,q_2)} \phi_{p_1,q_1}(XY)
\end{aligned}$$

where

$$Y(\omega) = \left(\frac{q_2}{q_1}\right)^{k(\omega)} \left(\frac{p_2/(1-p_2)}{p_1/(1-p_1)}\right)^{|\eta(\omega)|}.$$

Setting $X = 1$, we obtain

$$\phi_{p_2,q_2}(1) = 1 = \left(\frac{1-p_2}{1-p_1}\right)^{|E|} \frac{Z(p_1,q_1)}{Z(p_2,q_2)} \phi_{p_1,q_1}(Y),$$

whence, on dividing,

$$\phi_{p_2,q_2}(X) = \frac{\phi_{p_1,q_1}(XY)}{\phi_{p_1,q_1}(Y)}. \tag{3.26}$$

Assume now that the conditions of (3.22) hold. Since $k(\omega)$ is a decreasing function and $|\eta(\omega)|$ is increasing, we have that Y is increasing. Since $q_1 \geq 1$, ϕ_{p_1,q_1} is positively associated, whence

$$\phi_{p_1,q_1}(XY) \geq \phi_{p_1,q_1}(X)\phi_{p_1,q_1}(Y), \tag{3.27}$$

and (3.26) yields $\phi_{p_2,q_2}(X) \geq \phi_{p_1,q_1}(X)$. Claim (3.22) follows.

Assume now that the conditions of (3.23) hold. We write $Y(\omega)$ in the form

$$Y(\omega) = \left(\frac{q_2}{q_1}\right)^{k(\omega)+|\eta(\omega)|} \left(\frac{p_2/[q_2(1-p_2)]}{p_1/[q_1(1-p_1)]}\right)^{|\eta(\omega)|}.$$

Note that $k(\omega) + |\eta(\omega)|$ is an increasing function of ω, since the addition of an extra open edge to ω causes $|\eta(\omega)|$ to increase by 1 and $k(\omega)$ to decrease by at most 1. In addition, $|\eta(\omega)|$ is increasing. Since $q_2 \leq q_1$ and $p_2/[q_2(1-p_2)] \leq p_1/[q_1(1-p_1)]$ by assumption, we have that Y is decreasing. By the positive association of ϕ_{p_1,q_1} as above,

$$\phi_{p_1,q_1}(XY) \leq \phi_{p_1,q_1}(X)\phi_{p_1,q_1}(Y),$$

and (3.26) now implies $\phi_{p_2,q_2}(X) \leq \phi_{p_1,q_1}(X)$. Claim (3.23) follows. □

The proof of Theorem 3.24 begins with a subsidiary result. This contains two inequalities, only the first of which will be used in that which follows.

(3.28) Proposition [151]. *Let $p \in (0,1)$, $q \in [1,\infty)$ and $\Delta \in \{1,2,\dots\}$. There exists a strictly positive and continuous function $\alpha(p,q) = \alpha_\Delta(p,q)$ such that the following holds. Let G be a finite graph, and suppose there exists a spanning set W such that* $\deg(W) \leq \Delta$. *Then*

$$\alpha(p,q)\frac{\partial}{\partial p}\phi_{p,q}(A) \leq -q\frac{\partial}{\partial q}\phi_{p,q}(A) \leq p(1-p)\frac{\partial}{\partial p}\phi_{p,q}(A) \tag{3.29}$$

for all increasing events A.

Proof. Let A be an increasing event, and write $\theta(p,q) = \phi_{p,q}(A)$. As in the proof of Theorem 2.1 we shall construct a Markov chain $Z_t = (X_t, Y_t)$ taking values in the product space Ω^2.

Let $\omega \in \Omega$ and $e \in E$, and let ω_e, ω^e be the configurations given at (1.25). Let $D_e(\omega)$ be the indicator function of the event that the endvertices of e are connected by no open path of $E \setminus \{e\}$. We define the functions $H, H^A : \Omega^2 \to \mathbb{R}$ as follows. First,

$$H(\omega_e, \omega^e) = 1, \tag{3.30}$$

$$H(\omega^e, \omega_e) = \frac{1-p}{p} q^{D_e(\omega)}, \tag{3.31}$$

for $\omega \in \Omega$ and $e \in E$. Secondly, $H(\omega, \omega') = 0$ for other pairs ω, ω' with $\omega \neq \omega'$. Next, we define H^A by

$$H^A(\omega, \omega') = H(\omega, \omega') 1_A(\omega \wedge \omega') \qquad \text{if } \omega \neq \omega'. \tag{3.32}$$

The diagonal terms $H(\omega, \omega)$ and $H^A(\omega, \omega)$ are chosen in such a way that

$$\sum_{\omega' \in \Omega} H(\omega, \omega') = \sum_{\omega' \in \Omega} H^A(\omega, \omega') = 0, \qquad \omega \in \Omega.$$

Let $S = \{(\pi, \omega) \in \Omega^2 : \pi \leq \omega\}$, the set of all ordered pairs of configurations, and let $J : S \times S \to \mathbb{R}$ be given by

$$J(\pi_e, \omega; \pi^e, \omega^e) = 1, \tag{3.33}$$

$$J(\pi, \omega^e; \pi_e, \omega_e) = H^A(\omega^e, \omega_e), \tag{3.34}$$

$$J(\pi^e, \omega^e; \pi_e, \omega^e) = H(\pi^e, \pi_e) - H^A(\omega^e, \omega_e), \tag{3.35}$$

for $e \in E$. All other off-diagonal values of J are set to 0, and the diagonal elements are chosen such that

$$\sum_{(\pi', \omega') \in S} J(\pi, \omega; \pi', \omega') = 0, \qquad (\pi, \omega) \in S.$$

The function J will be used as the generator of a Markov chain $Z = (Z_t : t \geq 0)$ on the state space $S \subseteq \Omega^2$. With J viewed in this way, equation (3.33) specifies that, for $\pi \in \Omega$ and $e \in E$, the edge e is acquired by π (if it does not already contain it) at rate 1; any edge thus acquired is added also to ω if it does not already contain it. Equation (3.34) specifies that, for $\omega \in \Omega$ and $e \in \eta(\omega)$, the edge e is removed from ω (and also from π if $e \in \eta(\pi)$) at rate $H^A(\omega^e, \omega_e)$. For $e \in \eta(\pi)$ $(\subseteq \eta(\omega))$, there is an additional rate at which e is removed from π but not from ω. This additional rate is indeed non-negative, since

$$H(\pi^e, \pi_e) - H^A(\omega^e, \omega_e) = \frac{1-p}{p} \left[q^{D_e(\pi)} - q^{D_e(\omega)} 1_A(\omega_e) \right] \geq 0,$$

by (3.31) and (3.32). We have used the facts that $q \geq 1$, and $D_e(\omega) \leq D_e(\pi)$ for $\pi \leq \omega$. The ensuing Markov chain has no possible transition that can exit the set S. That is, if the chain starts in S, then we may assume it remains in S for all time.

It is easily seen as in Section 2.1 and [39] that there exists a Markov chain $Z_t = (X_t, Y_t)$ on the state space S such that:

(i) Z_t has generator J, that is, for $(\pi, \omega) \neq (\pi', \omega')$,

$$\mathbb{P}\big(Z_{t+h} = (\pi', \omega') \,\big|\, Z_t = (\pi, \omega)\big) = J(\pi, \omega; \pi', \omega')h + o(h),$$

(ii) $X_t \Rightarrow \phi_{p,q}(\cdot)$ as $t \to \infty$,

(iii) $Y_t \Rightarrow \phi_{p,q}(\cdot \mid A)$ as $t \to \infty$,

(iv) $X_t \leq Y_t$ for all t.

See [164, Chapter 6] for an account of the theory of Markov chains.

Differentiating $\theta = \theta(p, q) = \phi_{p,q}(A)$ with respect to p, one obtains as in Theorem 3.12 that

$$\begin{aligned}
(3.36)\qquad \frac{\partial\theta}{\partial p} &= \frac{1}{p(1-p)} \mathrm{cov}_{p,q}(|\eta|, 1_A) \\
&= \frac{1}{p(1-p)} \{\phi_{p,q}(|\eta| 1_A) - \phi_{p,q}(|\eta|)\phi_{p,q}(A)\} \\
&= \frac{\theta(p,q)}{p(1-p)} \left\{ \lim_{t\to\infty} \mathbb{P}\big(|\eta(Y_t)| - |\eta(X_t)|\big) \right\} \\
&= \frac{\theta(p,q)}{p(1-p)} \sum_{e \in E} \lim_{t\to\infty} \mathbb{P}\big(X_t(e) = 0,\ Y_t(e) = 1\big),
\end{aligned}$$

where $|\eta| = |\eta(\omega)|$ is the number of open edges, and $\mathbb{P}$ is the appropriate probability measure for the chain Z. A similar calculation using (3.14) yields that

$$(3.37)\qquad \frac{\partial\theta}{\partial q} = \frac{1}{q} \mathrm{cov}_{p,q}(k, 1_A) = -\frac{1}{q}\theta(p,q) \left\{ \lim_{t\to\infty} \mathbb{P}\big(k(X_t) - k(Y_t)\big) \right\},$$

where $k = k(\omega)$ is the number of open components.

By an elementary graph-theoretic argument,

$$k(X_t) - k(Y_t) \leq |\eta(Y_t)| - |\eta(X_t)|,$$

whence, by (3.36)–(3.37),

$$-q\frac{\partial\theta}{\partial q} \leq p(1-p)\frac{\partial\theta}{\partial p},$$

which is the right-hand inequality of (3.29).

Let Δ be a positive integer, and let W be a spanning set of vertices satisfying $\deg(W) \le \Delta$. For $x \in V$, let I_x be the indicator function of the event that x is an isolated vertex. Clearly,

$$\mathbb{P}\big(k(X_t) - k(Y_t)\big) \ge \sum_{x \in W} \mathbb{P}\big(I_x(X_t) = 1,\ I_x(Y_t) = 0\big), \tag{3.38}$$

since the right-hand side counts the number of vertices of W that are isolated in X_t but not in Y_t. Let $x \in W$, and let e_x be any edge of E that is incident to x. We claim that

$$\nu \mathbb{P}\big(X_t(e_x) = 0,\ Y_t(e_x) = 1\big) \le \mathbb{P}\big(I_x(X_{t+1}) = 1,\ I_x(Y_{t+1}) = 0\big) \tag{3.39}$$

for some $\nu = \nu_\Delta(p, q)$ which is continuous, and is strictly positive on $(0, 1) \times [1, \infty)$. Here, ν is allowed to depend on the value of Δ but not further upon x, e_x, W, G, or the choice of event A. Once (3.39) is proved, the left-hand inequality of (3.29) follows with $\alpha = \nu p(1-p)/\Delta$ by summing (3.39) over x and using (3.36)–(3.38) as follows:

$$\begin{aligned}
-q\frac{\partial \theta}{\partial q} &\ge \theta \left\{ \lim_{t\to\infty} \sum_{x \in W} \mathbb{P}\big(I_x(X_{t+1}) = 1,\ I_x(Y_{t+1}) = 0\big) \right\} \\
&\ge \theta\nu \left\{ \lim_{t\to\infty} \sum_{x \in W} \frac{1}{\Delta} \sum_{e:\, e\sim x} \mathbb{P}\big(X_t(e) = 0,\ Y_t(e) = 1\big) \right\} \\
&\ge \frac{\theta\nu}{\Delta} \left\{ \lim_{t\to\infty} \sum_{e \in E} \mathbb{P}\big(X_t(e) = 0,\ Y_t(e) = 1\big) \right\} \\
&= \frac{\nu p(1-p)}{\Delta} \frac{\partial \theta}{\partial p},
\end{aligned}$$

where $\sum_{e:\, e\sim x}$ denotes summation over all edges e incident to the vertex x.

Finally we prove (3.39). Let E_x be the set of edges of E that are incident to x. Suppose that the event $F_t = \{X_t(e_x) = 0,\ Y_t(e_x) = 1\}$ occurs. Let:

(a) T be the event that, during the time-interval $(t, t+1)$, every edge e of $E_x \setminus \{e_x\}$ with $X_t(e) = 1$ changes its X-state from 1 to 0; the removal of such edges from X may or may not entail their removal from Y,

(b) U be the event that no edge e of $E_x \setminus \{e_x\}$ with $X_t(e) = 0$ changes its state $(X_u(e), Y_u(e))$ in the time-interval $(t, t+1)$,

(c) V be the event that the state $(X_u(e_x), Y_u(e_x))$ of the edge e_x remains unchanged during the time-interval $(t, t+1)$.

By elementary computations using the generator of the chain $Z_t = (X_t, Y_t)$, there exists a strictly positive and continuous function $\nu_W = \nu_W(p, q)$ on $(0, 1) \times [1, \infty)$, which is allowed to depend on G and W only through the quantity $\deg(W)$, such that

$$\mathbb{P}(T \cap U \cap V \mid F_t) \ge \nu_W, \qquad t \ge 0,$$

uniformly in x, e_x, and G. This inequality remains true if we replace ν_W by the strictly positive and continuous function $\nu = \nu_\Delta(p, q)$ defined by

$$\nu_\Delta(p, q) = \min\big\{\nu_W(p, q) : W \text{ a spanning set such that } 0 \le \deg(W) \le \Delta\big\}.$$

If $F_t \cap T \cap U \cap V$ occurs, then x is isolated in X_{t+1} but not in Y_{t+1} (since $Y_{t+1}(e_x) = 1$). Therefore, (3.39) is valid, and the proof of the proposition is complete. A function ν of the required form may be written down explicitly. □

Proof of Theorem 3.24. Let α be as in Proposition 3.28, and let A be an increasing event. Inequality (3.29) may be stated in the form

$$(3.40) \qquad (\alpha, q).\nabla\phi_{p,q}(A) \le 0 \le (p(1-p), q).\nabla\phi_{p,q}(A),$$

where

$$\nabla f = \left(\frac{\partial f}{\partial p}, \frac{\partial f}{\partial q}\right), \qquad f : (0,1) \times [1, \infty) \to \mathbb{R}.$$

In addition, by Theorem 3.21,

$$\frac{\partial}{\partial q}\phi_{p,q}(A) \le 0 \le \frac{\partial}{\partial p}\phi_{p,q}(A).$$

The right-hand inequality of (3.40) may be used to obtain (3.23), but our current interest lies with the left-hand inequality. Let $\gamma : (0,1) \times [1, \infty)$ be a solution of the differential equation $(\alpha, q).\nabla\gamma = 0$ subject to

$$(3.41) \qquad \frac{\partial \gamma}{\partial q} < 0 < \frac{\partial \gamma}{\partial p}, \qquad p \in (0,1),\ q \in (1, \infty).$$

See Figure 3.1 for a sketch of the contours of γ, that is, the curves on which γ is constant. The contour of γ passing through the point (p, q) has tangent (α, q). The directional derivative of $\phi_{p,q}(A)$ in this direction satisfies, by (3.40),

$$(\alpha, q).\nabla\phi_{p,q}(A) = \alpha\frac{\partial}{\partial p}\phi_{p,q}(A) + q\frac{\partial}{\partial q}\phi_{p,q}(A) \le 0,$$

whence $\phi_{p,q}(A)$ is decreasing as (p, q) moves along the contour of γ in the direction of increasing q. Therefore,

$$\phi_{p_1,q_1}(A) \le \phi_{p_2,q_2}(A) \qquad \text{if } \gamma(p_1, q_1) = \gamma(p_2, q_2) \text{ and } 1 \le q_2 \le q_1,$$

and (3.25) follows. □

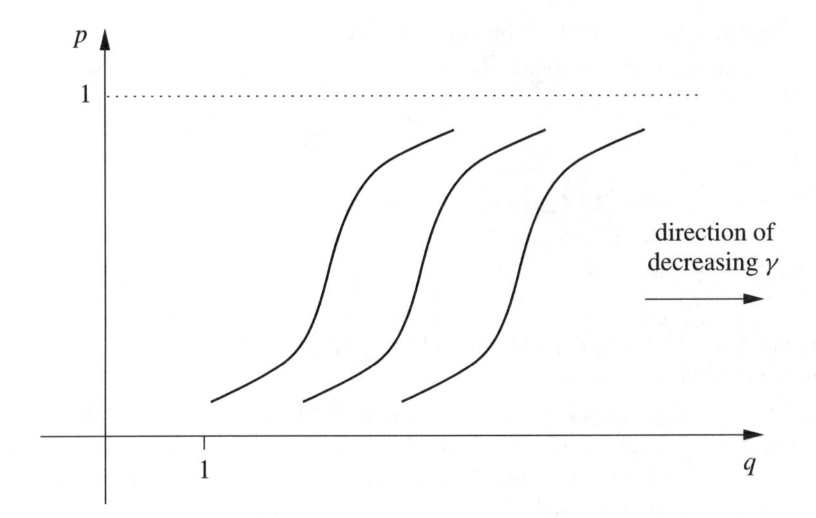

Figure 3.1. A sketch of the contours of the function $\gamma = \gamma(p, q)$.

3.5 Exponential steepness

Let $\phi_{p,q}$ be the random-cluster measure on the graph G, with q assumed to satisfy $q \ge 1$. Let $A \in \mathcal{F}$ and let $H_A(\omega)$ denote the Hamming distance from ω to A. We may apply Theorem 2.53 to obtain the following. A similar inequality holds for decreasing A.

(3.42) Theorem [153, 163]. *Let $p \in (0, 1)$ and $q \in [1, \infty)$. For any non-empty, increasing event $A \in \mathcal{F}$,*

(3.43)
$$\frac{d}{dp} \log \phi_{p,q}(A) \ge \frac{\phi_{p,q}(H_A)}{p(1-p)}.$$

As in (2.56), for increasing A,

(3.44)
$$\phi_{r,q}(A) \le \phi_{s,q}(A) \exp\{-4(s-r)\phi_{s,q}(H_A)\}, \qquad 0 < r \le s < 1.$$

Applications of this inequality are aided by a further relation between $\phi_{p,q}(A)$ and $\phi_{p,q}(H_A)$.

(3.45) Theorem [153, 163]. *Let $q \in [1, \infty)$ and $0 < r < s < 1$. For any non-empty, increasing event $A \in \mathcal{F}$,*

(3.46)
$$\phi_{r,q}(H_A \le k) \le C^k \phi_{s,q}(A), \qquad k = 0, 1, 2, \dots,$$

where

$$C = \frac{q^2(1-r)}{(s-r)[r + q(1-r)]}.$$

This may be used in the following way. By (3.46)

$$\phi_{r,q}(H_A) = \sum_{k=0}^{\infty} \phi_{r,q}(H_A > k) \geq \sum_{k=0}^{K} [1 - C^k \phi_{s,q}(A)],$$

where $K = \max\{k : C^k \phi_{s,q}(A) \leq 1\}$. Since $C > 1$,

$$(3.47) \qquad \phi_{r,q}(H_A) \geq \frac{-\log \phi_{s,q}(A)}{\log C} - \frac{C - \phi_{s,q}(A)}{C - 1}, \qquad 0 < r < s < 1.$$

Inequalities (3.43) and (3.47) provide a mechanism for bounding below the gradient of $\log \phi_{p,q}(A)$.

One area of potential application is the study of connection probabilities. Let S and T be disjoint sets of vertices of G, and let $A = \{S \leftrightarrow T\}$ be the event that there exists an open path joining some $s \in S$ to some $t \in T$. Then H_A is the minimum number of closed edges amongst the family Π of all paths from S to T, which is to say that

$$H_A(\omega) = \min\left\{ \sum_{e \in \pi} [1 - \omega(e)] : \pi \in \Pi \right\}.$$

Before proceeding to the proofs, we note that Theorem 3.45 is closely related to the 'sprinkling lemma' of [6], a version of which is valid for random-cluster models; see also [154]. The argument used to prove Theorem 3.45 may be used also to prove the following, the proof of which is omitted.

(3.48) Theorem. *Let* $q \in [1, \infty)$ *and* $0 < r < s < 1$. *For any non-empty, decreasing event* $A \in \mathcal{F}$,

$$(3.49) \qquad \phi_{r,q}(A) \geq \left(\frac{s - r}{qs} \right)^k \phi_{s,q}(H_A \leq k), \qquad k = 0, 1, 2, \ldots.$$

Proof of Theorem 3.45. Let $q \in [1, \infty)$ and $0 < r < s < 1$. We shall employ a suitable coupling of the measures $\phi_{r,q}$ and $\phi_{s,q}$. Let $E = \{e_1, e_2, \ldots, e_m\}$ be the edges of the graph G, and let $U_1, U_2, \ldots, U_m$ be independent random variables having the uniform distribution on $[0, 1]$. We write $\mathbb{P}$ for the probability measure associated with the U_j. We shall examine the edges in turn, to determine whether they are open or closed for the respective parameters r and s. The outcome will be a pair (π, ω) of configurations each lying in $\Omega = \{0, 1\}^E$ and such that $\pi \leq \omega$. The configurations π, ω are random in the sense that they are functions of the U_j. A similar coupling was used in the proof of Theorem 2.28.

First, we declare

$$\pi(e_1) = 1 \quad \text{if and only if} \quad U_1 < \phi_{r,q}(J_1),$$
$$\omega(e_1) = 1 \quad \text{if and only if} \quad U_1 < \phi_{s,q}(J_1),$$

where J_i is the event that e_i is open. By the comparison inequality (3.22), $\phi_{r,q}(J_1) \le \phi_{s,q}(J_1)$, and therefore $\pi(e_1) \le \omega(e_1)$.

Let M be an integer satisfying $1 \le M < m$. Having defined $\pi(e_i)$, $\omega(e_i)$ for $1 \le i \le M$ such that $\pi(e_i) \le \omega(e_i)$, we define $\pi(e_{M+1})$ and $\omega(e_{M+1})$ as follows. We declare

$$\begin{aligned} \pi(e_{M+1}) = 1 &\quad \text{if and only if} \quad U_{M+1} < \phi_{r,q}(J_{M+1} \mid \Omega_{M,\pi}), \\ \omega(e_{M+1}) = 1 &\quad \text{if and only if} \quad U_{M+1} < \phi_{s,q}(J_{M+1} \mid \Omega_{M,\omega}), \end{aligned}$$

where $\Omega_{M,\gamma}$ is the set of configurations $\nu \in \Omega$ satisfying $\nu(e_i) = \gamma(e_i)$ for $1 \le i \le M$. By the comparison inequalities (Theorem 3.21) and strong positive association (Theorem 3.8),

$$\phi_{r,q}(J_{M+1} \mid \Omega_{M,\pi}) \le \phi_{s,q}(J_{M+1} \mid \Omega_{M,\omega})$$

since $r < s$ and $\pi(e_i) \le \omega(e_i)$ for $1 \le i \le M$. Therefore, $\pi(e_{M+1}) \le \omega(e_{M+1})$. Continuing likewise, we obtain a pair (π, ω) of configurations satisfying $\pi \le \omega$, and such that π has law $\phi_{r,q}$, and ω has law $\phi_{s,q}$.

By Theorem 3.1,

$$\phi_{p,q}(J_i \mid \overline{K_i}) = \frac{p}{p+q(1-p)}, \qquad \phi_{p,q}(J_i \mid K_i) = p,$$

where K_i is the event that there exists an open path of $E \setminus \{e_i\}$ joining the endvertices of e_i. Using conditional expectations and the assumption $q \ge 1$,

$$\frac{p}{p+q(1-p)} \le \phi_{p,q}(J_i \mid D) \le p \tag{3.50}$$

for any event D defined in terms of the states of edges in $E \setminus \{e_i\}$. Therefore[2], by the definition of the $\pi(e_i)$ and $\omega(e_i)$,

$$\begin{aligned} \mathbb{P}\big(\pi(e_{M+1}) = 0 \,\big|\, U_1, U_2, \dots, U_M\big) &= 1 - \phi_{r,q}(J_{M+1} \mid \Omega_{M,\pi}) \\ &\le \frac{q(1-r)}{r+q(1-r)}. \end{aligned}$$

We claim that

$$\begin{aligned} \mathbb{P}\big(\omega(e_{M+1}) = 1, \pi(e_{M+1}) = 0 \,\big|\, U_1, U_2, \dots, U_M\big) \\ &= \phi_{s,q}(J_{M+1} \mid \Omega_{M,\omega}) - \phi_{r,q}(J_{M+1} \mid \Omega_{M,\pi}) \\ &\ge \frac{s-r}{q}, \end{aligned} \tag{3.51}$$

[2] Subject to the correct interpretation of the conditional measure in question.

and the proof of this follows. By Theorem 3.42 with $A = J_i$ (so that $H_{J_i} = 1 - 1_{J_i}$) together with (3.50),

$$\frac{d}{dp}\phi_{p,q}(J_i) \geq \frac{\phi_{p,q}(J_i)(1-\phi_{p,q}(J_i))}{p(1-p)} \geq \frac{1}{p+q(1-p)} \geq \frac{1}{q}.$$

We integrate over the interval $[r, s]$ to obtain that

$$\phi_{s,q}(J_i) - \phi_{r,q}(J_i) \geq \frac{s-r}{q}. \tag{3.52}$$

Finally,

$$\begin{aligned}\phi_{s,q}(J_{M+1} \mid \Omega_{M,\omega}) &- \phi_{r,q}(J_{M+1} \mid \Omega_{M,\pi}) \\ &\geq \phi_{s,q}(J_{M+1} \mid \Omega_{M,\omega}) - \phi_{r,q}(J_{M+1} \mid \Omega_{M,\omega}),\end{aligned}$$

and (3.51) follows by applying (3.52) with $i = M+1$ to the graph obtained from G by contracting (respectively, deleting) any edge e_i (for $1 \leq i \leq M$) with $\omega(e_i) = 1$ (respectively, $\omega(e_i) = 0$). See [152, Theorem 2.3].

By the above,

$$\mathbb{P}\big(\omega(e_{M+1}) = 1 \,\big|\, \pi(e_{M+1}) = 0,\ U_1, U_2, \dots, U_M\big) \geq \frac{s-r}{q} \cdot \frac{r+q(1-r)}{q(1-r)}. \tag{3.53}$$

Let $\xi \in \Omega$, and let B be a set of edges satisfying $\xi(e) = 0$ for $e \in B$. We claim that

$$\mathbb{P}(\pi = \xi,\ \omega(e) = 1 \text{ for } e \in B) \geq \left(\frac{s-r}{q} \cdot \frac{r+q(1-r)}{q(1-r)}\right)^{|B|} \mathbb{P}(\pi = \xi). \tag{3.54}$$

This follows by the recursive construction of π and ω in terms of the family $U_1, U_2, \dots, U_m$, in the light of the bound (3.53).

Inequality (3.54) implies the claim of the theorem, as follows. Let A be an increasing event and let ξ be a configuration satisfying $H_A(\xi) \leq k$. There exists a set $B = B_\xi$ of edges such that:

(a) $|B| \leq k$,

(b) $\xi(e) = 0$ for $e \in B$,

(c) $\xi^B \in A$, where ξ^B is obtained from ξ by allocating state 1 to all edges in B.

If more than one such set B exists, we pick the earliest in some deterministic ordering of all subsets of E. By (3.54),

$$\begin{aligned}\phi_{s,q}(A) &\geq \mathbb{P}\big(H_A(\pi) \leq k,\ \omega(e) = 1 \text{ for } e \in B_\pi\big) \\ &= \sum_{\xi:\, H_A(\xi) \leq k} \mathbb{P}\big(\pi = \xi,\ \omega(e) = 1 \text{ for } e \in B_\xi\big) \\ &\geq \left(\frac{s-r}{q} \cdot \frac{r+q(1-r)}{q(1-r)}\right)^k \phi_{r,q}(H_A \leq k).\end{aligned}$$

□

3.6 Partition functions

The partition function associated with the finite graph $G = (V, E)$ is given by

$$Z_G(p, q) = \sum_{\omega \in \Omega} p^{|\eta(\omega)|}(1-p)^{|E \setminus \eta(\omega)|} q^{k(\omega)}. \tag{3.55}$$

In the usual approach of classical statistical mechanics, one studies phase transitions via the partition function and its derivatives. We prefer in this work to follow a more probabilistic approach, but shall nevertheless have recourse to various arguments based on the behaviour of the partition function, of which we note some basic properties.

The (Whitney) *rank-generating function* of $G = (V, E)$ is the function

$$W_G(u, v) = \sum_{E' \subseteq E} u^{r(G')} v^{c(G')}, \qquad u, v \in \mathbb{R}, \tag{3.56}$$

where $r(G') = |V| - k(G')$ is the *rank* of the subgraph $G' = (V, E')$, and $c(G') = |E'| - |V| + k(G')$ is its *co-rank*. Here, $k(G')$ denotes the number of components of the graph G'. The rank-generating function has various useful properties, and it crops up in several contexts in graph theory, see [40, 313]. It occurs in other forms also. For example, the function

$$T_G(u, v) = (u-1)^{|V|-1} W_G\big((u-1)^{-1}, v-1\big) \tag{3.57}$$

is known as the *dichromatic* (or *Tutte*) *polynomial*, [313]. The partition function Z_G of the graph G is easily seen to satisfy

$$Z_G(p, q) = q^{|V|}(1-p)^{|E|} W_G\left(\frac{p}{q(1-p)}, \frac{p}{1-p}\right), \tag{3.58}$$

a relationship which provides a link with other classical quantities associated with a graph. See [40, 41, 121, 157, 308, 315] and Chapter 9.

Another way of viewing Z_G is as the moment generating function of the number of clusters in a random graph, that is,

$$Z_G(p, q) = \phi_p(q^{k(\omega)}), \tag{3.59}$$

where ϕ_p denotes product measure. This indicates a link to percolation on G, and to the large-deviation theory of the number of clusters in the percolation model. See [62, 298] and Section 10.8.

The partition function Z_G does not change a great deal if an edge is removed from G. Let $F \subseteq E$, and write $G \setminus F$ for the graph G with the edges in F removed. If F is the singleton $\{e\}$, we write $G \setminus e$ for $G \setminus \{e\}$.

(3.60) Theorem. *Let $p \in [0, 1]$ and $q \in (0, \infty)$. Then*

$$(3.61) \qquad (1 \wedge q)^{|F|} \le \frac{Z_{G \setminus F}(p, q)}{Z_G(p, q)} \le (1 \vee q)^{|F|}, \qquad F \subseteq E.$$

We give next an application of these inequalities to be used later. Let $G_i = (V_i, E_i)$, $i = 1, 2$, be finite graphs on disjoint vertex sets V_1, V_2, and write $G_1 \cup G_2$ for the graph $(V_1 \cup V_2, E_1 \cup E_2)$. It is immediate from (3.55) that

$$(3.62) \qquad Z_{G_1 \cup G_2} = Z_{G_1} Z_{G_2},$$

where for clarity we have removed explicit mention of p, q. Taken in conjunction with (3.61), this leads easily to a pair of inequalities which we state as a theorem.

(3.63) Theorem. *Let $G = (V, E)$ be a finite graph, and let F be a set of edges whose removal breaks G into two disjoint graphs $G_1 = (V_1, E_1)$, $G_2 = (V_2, E_2)$. Thus, $V = V_1 \cup V_2$ and $E = E_1 \cup E_2 \cup F$. For $p \in [0, 1]$ and $q \in (0, \infty)$,*

$$Z_{G_1} Z_{G_2} (1 \vee q)^{-|F|} \le Z_G \le Z_{G_1} Z_{G_2} (1 \wedge q)^{-|F|}.$$

Proof of Theorem 3.60. It suffices to prove (3.61) with F a singleton set, that is, $F = \{e\}$. The claim for general F will follow by iteration. For $\omega \in \Omega$, we write $\omega_{\langle e \rangle}$ for the configuration in $\Omega_{\langle e \rangle} = \{0, 1\}^{E \setminus \{e\}}$ that agrees with ω off e. Clearly,

$$k(\omega) \le k(\omega_{\langle e \rangle}) \le k(\omega) + 1,$$

whence

$$(3.64) \qquad (1 \wedge q) q^{k(\omega)} \le q^{k(\omega_{\langle e \rangle})} \le (1 \vee q) q^{k(\omega)}.$$

Now, since $p + (1 - p) = 1$,

$$(3.65) \qquad Z_{G \setminus e}(p, q) = \sum_{\omega_{\langle e \rangle} \in \Omega_{\langle e \rangle}} p^{|\eta(\omega_{\langle e \rangle})|} (1 - p)^{|E \setminus \eta(\omega_{\langle e \rangle})| - 1} q^{k(\omega_{\langle e \rangle})}$$
$$= \sum_{\omega \in \Omega} p^{|\eta(\omega)|} (1 - p)^{|E \setminus \eta(\omega)|} q^{k(\omega_{\langle e \rangle})}.$$

Equations (3.64) and (3.65) imply (3.61) with $F = \{e\}$. □

We develop next an inequality related to (3.61) concerning the addition of a vertex, and which will be useful later. Let $G = (V, E)$ be a finite graph as usual, and let $v \notin V$ and $W \subseteq V$. We augment G by adding the vertex v together with edges $\langle v, w \rangle$ for $w \in W$. Let us write $G + v$ for the resulting graph.

(3.66) Theorem. *Let $p \in [0,1]$ and $q \in [1,\infty)$. In the above notation,*

$$\frac{Z_{G+v}(p,q)}{Z_G(p,q)} \ge q(1-p+pq^{-1})^{|W|}.$$

Proof. Let $\Omega_E = \{0,1\}^E$ and $\Omega_v = \{0,1\}^W$. We identify $\nu \in \Omega_v$ with the edge-configuration on the edge-neighbourhood $\{\langle v,w\rangle : w \in W\}$ of v given by $\nu(\langle v,w\rangle) = \nu(w)$. Now,

$$\begin{aligned}(3.67)\qquad Z_{G+v}(p,q) &= \sum_{\omega\in\Omega_E,\ \nu\in\Omega_v} p^{|\eta(\omega)|}(1-p)^{|E\setminus\eta(\omega)|}q^{k(\omega)} \\ &\qquad \times \left[\left\{\prod_{w\in W} p^{\nu(w)}(1-p)^{1-\nu(w)}\right\}q^{1-k(\omega,\nu)}\right] \\ &= Z_G(p,q)\phi_{G,p,q}\big[q\phi_p(q^{-k(\omega,\nu)})\big],\end{aligned}$$

where ϕ_p is product measure on Ω_v with density p, and $k(\omega,\nu)$ is the number of open clusters of ω containing some $w \in W$ with $\nu(w) = 1$. Let $n_1, n_2, \dots, n_r$ be the sizes of the equivalence classes of W under the equivalence relation $w_1 \sim w_2$ if $w_1 \leftrightarrow w_2$ in ω. For $\omega \in \Omega_E$,

$$\begin{aligned}(3.68)\qquad \phi_p(q^{-k(\omega,\nu)}) &= \prod_{i=1}^r \left[(1-p)^{n_i} + \frac{1}{q}[1-(1-p)^{n_i}]\right] \\ &\ge \prod_{i=1}^r \left[\frac{1}{q} + \left(1-\frac{1}{q}\right)(1-p)\right]^{n_i} \\ &= \left[\frac{1}{q} + \left(1-\frac{1}{q}\right)(1-p)\right]^{|W|},\end{aligned}$$

where we have used the elementary (convexity) inequality

$$\alpha + (1-\alpha)y^n \ge [\alpha + (1-\alpha)y]^n, \qquad \alpha, y \in [0,1],\ n \in \{1,2,\dots\}.$$

We substitute (3.68) into (3.67) to obtain the claim. □

So far in this section we have considered the effect on the partition function of removing edges or adding vertices. There is a related result in which, instead, we identify certain vertices. Let $G = (V,E)$ be a finite graph, and let C be a subset of V separating the vertex-sets A_1 and A_2. That is, V is partitioned as $V = A_1 \cup A_2 \cup C$ and, for all $a_1 \in A_1$, $a_2 \in A_2$, every path from a_1 to a_2 passes through at least one vertex in C. We write c for a composite vertex formed by identifying all vertices in C, and $G_1 = (A_1 \cup \{c\}, E_1)$ (respectively, $G_2 = (A_2 \cup \{c\}, E_2)$) for the graph on the vertex set $A_1 \cup \{c\}$ (respectively, $A_2 \cup \{c\}$) and with the edges derived from G. For example, if $x, y \in A_1$, then $\langle x,y\rangle$ is an edge of G_1 if and only if it is an edge of G; for $a \in A_1$, the number of edges of G_1 between a and c is exactly the number of edges in G between a and members of C.

(3.69) Lemma. *For $p \in [0,1]$ and $q \in [1,\infty)$, $Z_G \ge q^{-1} Z_{G_1} Z_{G_2}$.*

Since $Z_G \ge q$ for all G when $q \ge 1$,

$$Z_G \ge Z_{G_1}. \tag{3.70}$$

Proof. Let $\omega \in \{0,1\}^E$, and let ω_1 and ω_2 be the induced configurations in $\Omega_1 = \{0,1\}^{E_1}$ and $\Omega_2 = \{0,1\}^{E_2}$, respectively. It is easily seen that

$$k(\omega) \ge k(\omega_1) + k(\omega_2) - 1,$$

and the claim follows from the definition (3.55) □

The partition function has a property of convexity which will be useful when studying random-cluster measures on infinite graphs. Rather than working with Z_G, we work for convenience with the function $Y_G : \mathbb{R}^2 \to \mathbb{R}$ given by

$$Y_G(\pi,\kappa) = \sum_{\omega \in \Omega} \exp\bigl\{\pi|\eta(\omega)| + \kappa k(\omega)\bigr\}, \tag{3.71}$$

a function which is related to Z_G as follows. We set $\pi = \pi(p)$ and $\kappa = \kappa(q)$ where

$$\pi(p) = \log\left(\frac{p}{1-p}\right), \quad \kappa(q) = \log q, \tag{3.72}$$

and then

$$Z_G(p,q) = (1-p)^{|E|} Y_G(\pi(p), \kappa(q)).$$

We write ∇X for the gradient vector of a function $X : \mathbb{R}^2 \to \mathbb{R}$.

(3.73) Theorem. *Let the vectors (π,κ) and (p,q) be related by* (3.72).

(a) *The gradient vector of the function* $\log Y_G(\pi,\kappa)$ *is given by*

$$\nabla\{\log Y_G(\pi,\kappa)\} = \bigl(\phi_{p,q}(|\eta|), \phi_{p,q}(k)\bigr). \tag{3.74}$$

(b) *Let $\mathbf{i} = (i_1, i_2)$ be a unit vector in $\mathbb{R}^2$. We have that*

$$\frac{d^2}{d\alpha^2}\Bigl\{\log Y_G\bigl((\pi,\kappa) + \alpha\mathbf{i}\bigr)\Bigr\}\Bigg|_{\alpha=0} = \mathrm{var}_{p,q}(i_1|\eta| + i_2 k) \tag{3.75}$$

where $\mathrm{var}_{p,q}$ denotes variance with respect to $\phi_{p,q}$. In particular, $\log Y_G$ *is a convex function on* $\mathbb{R}^2$.

By (3.71),

$$Y_G\bigl((\pi,\kappa) + \alpha\mathbf{i}\bigr) = Y_G(\pi,\kappa)\phi_{p,q}(e^{\alpha L(\mathbf{i})}),$$

where $L(\mathbf{i}) = i_1|\eta| + i_2 k$. Therefore, the jth derivative as in (3.75) equals the jth cumulant (or semi-invariant[3]) of $L(\mathbf{i})$.

Proof. (a) It is elementary that

$$\begin{aligned}\frac{\partial}{\partial \pi} \log Y_G(\pi, \kappa) &= \frac{1}{Y_G(\pi, \kappa)} \sum_{\omega \in \Omega} |\eta(\omega)| \exp\{\pi |\eta(\omega)| + \kappa k(\omega)\} \\ &= \phi_{p,q}(|\eta|),\end{aligned}$$

with a similar relation for the other derivative.

(b) We have that

$$Y_G\big((\pi, \kappa) + \alpha \mathbf{i}\big) = \sum_{\omega \in \Omega} \exp\big\{\alpha\big(i_1|\eta(\omega)| + i_2 k(\omega)\big)\big\} \exp\big\{\pi |\eta(\omega)| + \kappa k(\omega)\big\},$$

and (3.75) follows as in part (a). The convexity is a consequence of the fact that variances are non-negative. □

3.7 Domination by the Ising model

Stochastic domination is an invaluable tool in the study of random-cluster measures. Since the random-cluster model is an 'edge-model', it is usual to make comparisons with other edge-models. The relationship when $q \in \{2, 3, \dots\}$ to Potts models suggest the possibility of comparison with a 'vertex-model', and a hint of how to achieve this is provided by the case of integral q.

Consider the random-cluster model with parameters p and q on the finite graph $G = (V, E)$. If $q \in \{2, 3, \dots\}$, we may generate a Potts model by assigning a uniformly chosen spin-value to each open cluster. The spin configuration thus obtained is governed by the Potts measure with inverse-temperature β satisfying $p = 1 - e^{-\beta}$. Evidently, this can work only if q is an integer. A weaker conclusion may be obtained if q is not an integer, namely the following. Suppose $p \in [0, 1]$ and $q \in [1, \infty)$. We examine each open cluster of the random-cluster model in turn, and we declare it to be *red* with probability $1/q$ and *white* otherwise, different clusters receiving independent colours[4]. Let R be the set of vertices lying in red clusters. If $q \in \{2, 3, \dots\}$, then R has the same distribution as the set of vertices of the corresponding Potts model that have a pre-determined spin-value. Write $\mathbb{P}_{p,q}$ for an appropriate probability measure. One has for general $q \in (1, \infty)$ that,

[3] See [164, p. 185] and [255, p. 266].

[4] This construction is related to the so-called fuzzy Potts model, see [35, 170, 172, 245].

for $A \subseteq V$,

(3.76)
$$\begin{aligned}\mathbb{P}_{p,q}(R = A) &= \frac{1}{Z_G(p,q)}(1-p)^{|\Delta_e A|} \\ &\quad\times \left\{ \sum_{\omega \in \Omega_A} p^{|\eta(\omega)|}(1-p)^{|E_A \setminus \eta(\omega)|} q^{k(\omega)} \left(\frac{1}{q}\right)^{k(\omega)} \right\} \\ &\quad\times \left\{ \sum_{\omega' \in \Omega_{\overline{A}}} p^{|\eta(\omega')|}(1-p)^{|E_{\overline{A}} \setminus \eta(\omega')|} q^{k(\omega')} \left(1 - \frac{1}{q}\right)^{k(\omega')} \right\} \\ &= \frac{1}{Z_G(p,q)}(1-p)^{|\Delta_e A|} Z_{\overline{A}}(p, q-1)\end{aligned}$$

where $\overline{A} = V \setminus A$, $\Omega_A = \{0,1\}^{E_A}$ with E_A the subset of E containing all edges with both endvertices in A, the Z_G, $Z_{\overline{A}}$ are the appropriate partition functions, and $\Delta_e A$ is the set of edges of G with exactly one endvertex in A. When q is an integer, (3.76) reduces to the usual Potts law for the set of vertices with a given spin-value.

The random set R, with law given in (3.76), is the first element in the proposed stochastic comparison. The second element is the set of $+$ spins of an Ising model with external field, and we recall next from Section 1.3 the definition of an Ising model on the graph G. Let $\Sigma = \{-1, +1\}^V$, and let $\beta \in (0, \infty)$ and $h \in \mathbb{R}$. The Hamiltonian is the function $H : \Sigma \to \mathbb{R}$ given by

(3.77)
$$H(\sigma) = -\sum_{e=\langle u,v\rangle \in E} \sigma_u \sigma_v - h \sum_{v \in V} \sigma_v, \qquad \sigma = (\sigma_u : u \in V) \in \Sigma,$$

and the (Ising) probability measure is given by

(3.78)
$$\pi_{\beta,h}(\sigma) = \frac{1}{Z_{\mathrm{I}}} e^{-\frac{1}{2}\beta H(\sigma)}, \qquad \sigma \in \Sigma,$$

where $Z_{\mathrm{I}} = Z_{\mathrm{I}}(\beta, h)$ is the required normalizing constant[5]. We shall be concerned here with the random set $S = S(\sigma) = \{u \in V : \sigma_u = 1\}$, containing all vertices with spin $+1$.

Let $\deg(u)$ denote the degree of the vertex u in the graph G, and let

$$\Delta = \max\{\deg(u) : u \in V\}.$$

[5]The fraction $\frac{1}{2}$ in the exponent is that appearing in (1.7).

(3.79) Theorem [15]. *Let $\beta \in (0, \infty)$, $p = 1 - e^{-\beta}$, $q \in [2, \infty)$, and let R be the random 'red' set of the random-cluster model, governed by the law given in (3.76). Let $\beta' \in (0, \infty)$ and $h' \in (-\infty, \infty)$ be given by*

$$e^{2\beta'} = e^{\beta}\left(\frac{q-2+e^{\beta}}{q-1}\right), \qquad e^{\beta'(\Delta+h')} = \frac{1}{q-1}e^{\beta\Delta}, \tag{3.80}$$

and let S be the set of vertices with spin $+1$ under the Ising measure $\pi_{\beta',h'}$. Then

$$R \leq_{\text{st}} S. \tag{3.81}$$

Inequality (3.81) is to be interpreted as

$$\mathbb{P}_{p,q}(f(R)) \leq \pi_{\beta',h'}(f(S))$$

for all increasing functions $f : \{0,1\}^V \to \mathbb{R}$. Its importance lies in the deduction that R is small whenever S is small. The Ising model allows a deeper analysis than do general Potts and random-cluster models (see, for example, the results of Chapter 9). Particularly relevant facts are known for the set of $+$ spins in the Ising model when the external field h' is negative, and thus it becomes important to obtain conditions under which $h' < 0$.

Let $q > 2$ and assume that G is such that $\Delta \geq 3$. Setting $h' = 0$ and eliminating β' in (3.80), we find that $\beta = \beta_\Delta$ where

$$e^{\beta_\Delta} = \frac{q-2}{(q-1)^{1-(2/\Delta)} - 1}. \tag{3.82}$$

By (3.80) and an elementary argument using monotonicity,

$$h' < 0 \quad \text{if and only if} \quad \beta < \beta_\Delta. \tag{3.83}$$

We make one further note in advance of proving the theorem. By (3.82), $\beta_\Delta \to 0$ as $\Delta \to \infty$; if the maximum vertex degree is large, the field of application of the theorem is small. In an important application of the theorem, we shall take G to be a box Λ of the lattice $\mathbb{L}^d$ with so-called 'wired boundary conditions' (see Section 4.2). This amounts to identifying all vertices in the boundary $\partial\Lambda$, and thus to the introduction of a single vertex, w say, having large degree. The method of proof of Theorem 3.79 is valid in this slightly more general setting with

$$\Delta = \max\{\deg(u) : u \in V \setminus \{w\}\},$$

under the assumption that the open cluster containing w is automatically designated red. That is, we let R be the union of the cluster at w together with all other clusters declared red under the above randomization, and we let S be the set of $+$ spins in the Ising model with parameters β', h' and with $\sigma_w = +1$. The conclusion

(3.81) is then valid in this setting, with Δ given as above. An application to the exponential decay of connectivity in two dimensions will be found in Section 6.3.

Further work on stochastic domination inequalities for the set S of $+$ spins of the Ising model may be found in [236].

Proof. We present a direct proof based on the Holley inequality, Theorem 2.1. For $A \subseteq V$ and $u, v \in V$ with $u \neq v$, we write

$$A^u = A \cup \{u\}, \quad A_v = A \setminus \{v\}, \quad A_v^u = (A^u)_v, \quad \text{and so on.}$$

Let μ_1 (respectively, μ_2) denote the law of R (respectively, S), so that

$$\mu_1(A) = \mathbb{P}_{p,q}(R = A), \quad \mu_2(A) = \pi_{\beta',h'}(S = A), \qquad A \subseteq V.$$

The measures μ_i are strictly positive, and it therefore suffices by Theorem 2.3 to prove that

$$\mu_2(A \cup B)\mu_1(A \cap B) \geq \mu_1(A)\mu_2(B), \tag{3.84}$$

for incomparable sets A, B with $|A \triangle B| = 1, 2$. The interesting case is when $|A \triangle B| = 2$, and the case $|A \triangle B| = 1$ may be treated similarly.

Suppose that

$$A = C^u, \quad B = C^v, \quad \text{where } C \subseteq V,\ u, v \in V \setminus C,\ u \neq v. \tag{3.85}$$

We shall assume for the moment that $u \nsim v$. Let $r = |\{c \in C : a \sim u\}|$, the number of neighbours of u in C. By (3.77)–(3.78),

$$\frac{\mu_2(C^{uv})}{\mu_2(C^v)} = \exp\big(\beta'(2r - \delta) + \beta' h'\big) \tag{3.86}$$

where $\delta = \deg(u)$. Also, by (3.76) and Theorem 3.66,

$$\begin{aligned} \frac{\mu_1(C)}{\mu_1(C^u)} &= (1-p)^{2r-\delta} \frac{Z_{\overline{C}}(p, q-1)}{Z_{\overline{C^u}}(p, q-1)} \\ &\geq (1-p)^{2r-\delta}(q-1)\big[1 - p + p(q-1)^{-1}\big]^{\delta - r}. \end{aligned} \tag{3.87}$$

Substituting $p = 1 - e^{-\beta}$ and setting $x = e^{\beta}$, we obtain by multiplying (3.86) and (3.87) that

$$\begin{aligned} \frac{\mu_2(A \cup B)\mu_1(A \cap B)}{\mu_1(A)\mu_2(B)} &\geq \exp\big(\beta'(2r - \delta) + \beta' h' - \beta(2r - \delta)\big) \\ &\qquad \times (q-1)\big[e^{-\beta} + (1 - e^{-\beta})(q-1)^{-1}\big]^{\delta - r} \\ &= x^{r - \delta/2} \left(\frac{q - 2 + x}{q-1}\right)^{r - \delta/2} \left(\frac{q-2+x}{q-1}\right)^{-\Delta/2} \frac{x^{\Delta/2}}{q-1} \\ &\qquad \times x^{\delta - 2r}(q-1)\left[\frac{q-2+x}{x(q-1)}\right]^{\delta - r} \\ &= \left(\frac{x(q-1)}{q-2+x}\right)^{(\Delta - \delta)/2}, \end{aligned}$$

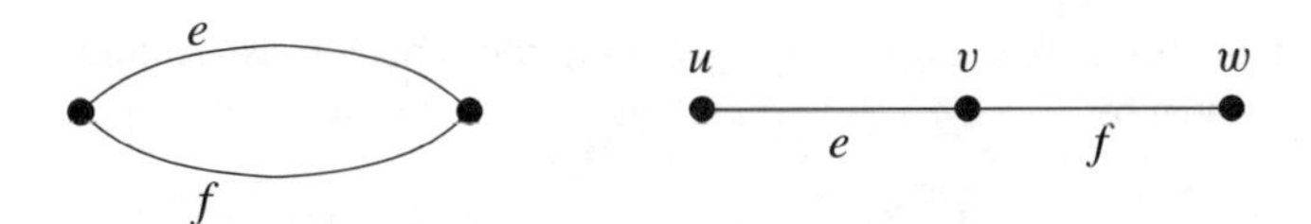

Figure 3.2. Two edges e and f in parallel and in series.

by (3.80). Now $\delta \le \Delta$, and $x(q-1)/(q-2+x) \ge 1$ for $x \ge 1$, whence

$$\frac{\mu_2(A \cup B)\mu_1(A \cap B)}{\mu_1(A)\mu_2(B)} \ge 1 \tag{3.88}$$

as required at (3.84).

In the case when $u \sim v$, the quantity r in (3.86) is replaced by $r+1$, and (3.88) remains valid. We deduce by Theorem 2.3 that $R \le_{\text{st}} S$ as claimed. □

3.8 Series and parallel laws

Kasteleyn observed[6] in the 1960s that electrical networks, percolation, and the Ising and Potts models satisfy the series/parallel laws, and this gave inspiration for the random-cluster model. The series/parallel laws will be used later, and they are described briefly here in the context of the random-cluster model.

Let $G = (V, E)$ be a finite graph (possibly with parallel edges). Two distinct edges $e, f \in E$ are said to be *in parallel* if they have the same endvertices. They are said to be *in series* if they share exactly one endvertex, v say, and in addition v is incident to no further edge of E. See Figure 3.2.

Let e, f be in either parallel or series. In either case, we may define another graph G' as follows. If e, f are in parallel, let $G' = (V', E')$ be obtained from G by replacing e and f by a single edge g with the same endvertices. If e and f are in series, say $e = \langle u, v\rangle$, $f = \langle v, w\rangle$, we obtain $G' = (V', E')$ by deleting both e and f (together with the vertex v) and inserting a new edge $g = \langle u, w\rangle$. We have in either case that $E' = (E \setminus \{e, f\}) \cup \{g\}$.

Let $\pi : [0,1]^2 \to [0,1]$ and $\sigma : [0,1]^2 \times (0,\infty) \to [0,1]$ be given by

$$\begin{aligned} \pi(x,y) &= 1-(1-x)(1-y), \\ \sigma(x,y,q) &= \frac{xy}{1+(q-1)(1-x)(1-y)}. \end{aligned} \tag{3.89}$$

Let $\mathbf{p} = (p_h : h \in E) \in [0,1]^E$ and $q \in (0,\infty)$. We write $\phi_{\mathbf{p},q}$ for the random-cluster measure on $\Omega = \{0,1\}^E$ in which each edge h has an associated parameter-value p_h, see (1.20). We shall see that $\phi_{\mathbf{p},q}$ is invariant (in a manner to made specific soon) when two edges e, f in parallel (respectively, series) are

[6] See the Appendix.

replaced as above by a single edge g having the 'correct' associated parameter-value p_g given by

$$p_g = \begin{cases} \pi(p_e, p_f) & \text{if } e, f \text{ are in parallel,} \\ \sigma(p_e, p_f, q) & \text{if } e, f \text{ are in series.} \end{cases} \tag{3.90}$$

Let $\Omega' = \{0, 1\}^{E'}$ be the configuration space associated with the graph G' given above. We define a mapping $\tau : \Omega \to \Omega'$ by $\tau_\omega(h) = \omega(h)$ for $h \neq g$, and

$$\tau_\omega(g) = \begin{cases} 1 - (1 - \omega(e))(1 - \omega(f)) & \text{if } e, f \text{ are in parallel,} \\ \omega(e)\omega(f) & \text{if } e, f \text{ are in series.} \end{cases}$$

When e, f are in parallel (respectively, series), g is open in τ_ω if and only if either e or f is open (respectively, both e and f are open) in ω. The mapping τ maps open connections to open connections; in particular, for $x, y \in V'$, $x \leftrightarrow y$ in τ_ω if and only if $x \leftrightarrow y$ in ω.

The measure $\phi_{\mathbf{p},q}$ on Ω induces a measure $\phi'_{\mathbf{p},q}$ on Ω' defined by

$$\phi'_{\mathbf{p},q}(\omega') = \phi_{\mathbf{p},q}(\tau^{-1}\omega'), \qquad \omega' \in \Omega',$$

and it turns out that this new measure is simply a random-cluster measure with an adapted parameter-value for the new edge g, as in (3.90).

(3.91) Theorem. *Let e, f be distinct edges of the finite graph G.*

(a) Parallel law. *Let e, f be in parallel. The measure $\phi'_{\mathbf{p},q}$ is the random-cluster measure on G' with parameters p_h for $h \neq g$, $p_g = \pi(p_e, p_f)$.*

(b) Series law. *Let e, f be in series. The measure $\phi'_{\mathbf{p},q}$ is the random-cluster measure on G' with parameters p_h for $h \neq g$, $p_g = \sigma(p_e, p_f, q)$.*

There is a third transformation of value when calculating effective resistances of electrical networks, namely the 'star–triangle' (or 'star–delta') transformation. This plays a part for random-cluster models also, see Section 6.6 and the discussion leading to Lemma 6.64.

Proof. (a) The edge g is open in τ_ω if and only if either or both of e, f is open in ω. Therefore, the numbers of open clusters in ω and τ_ω satisfy $k(\omega) = k(\tau_\omega)$. It is a straightforward calculation to check that, for $\omega' \in \Omega'$,

$$\phi'_{\mathbf{p},q}(\omega') \propto \sum_{\omega:\, \tau_\omega = \omega'} \left\{ \prod_{h:\, h \neq g} p_h^{\omega'(h)}(1 - p_h)^{1-\omega'(h)} \right\} \left[\pi^{\omega'(g)}(1 - \pi)^{1-\omega'(h)}\right] q^{k(\omega')},$$

where $\pi = p_e p_f + p_e(1 - p_f) + p_f(1 - p_e) = \pi(p_e, p_f)$.

(b) Write $e = \langle u, v\rangle$, $f = \langle v, w\rangle$, so that $g = \langle u, w\rangle$. Recall that ω and τ_ω agree off the edges e, f, g, and hence the partial configurations $(\omega(h) : h \neq e, f)$ and $(\tau_\omega(h) : h \neq g)$ have the same law. Let K be the set of all $\omega \in \Omega$ such that there exists an open path from u to w not using e, f; let K' be the corresponding event in Ω' with e, f replaced by g. Note that $K' = \tau K$.

By the remarks above and Theorem 3.1(b), it suffices to show that

$$\phi'_{\mathbf{p},q}(\omega'(g) = 1 \mid K') = \sigma, \tag{3.92}$$

$$\phi'_{\mathbf{p},q}(\omega'(g) = 1 \mid \overline{K'}) = \frac{\sigma}{\sigma + q(1-\sigma)}, \tag{3.93}$$

where $\sigma = \sigma(p_e, p_f, q)$. The edge g is open in τ_ω if and only if both e and f are open in ω. Therefore,

$$\phi'_{\mathbf{p},q}(\omega'(g) = 1 \mid K') = \phi_{\mathbf{p},q}\big(\omega(e) = \omega(f) = 1 \,\big|\, K\big),$$

which is easily seen to equal

$$\frac{p_e p_f}{p_e p_f + p_e(1-p_e) + p_f(1-p_f) + q(1-p_e)(1-p_f)},$$

in agreement with (3.92). Similarly,

$$\phi'_{\mathbf{p},q}(\omega'(g) = 1 \mid \overline{K'}) = \phi_{\mathbf{p},q}\big(\omega(e) = \omega(f) = 1 \,\big|\, \overline{K}\big),$$

which in turn equals

$$\frac{p_e p_f}{p_e p_f + q p_e(1-p_f) + q p_f(1-p_e) + q^2(1-p_e)(1-p_f)},$$

in agreement with (3.93). □

3.9 Negative association

This chapter closes with a short discussion of negative association when $q \le 1$. Let E be a finite set, and let μ be a probability measure on the sample space $\Omega = \{0,1\}^E$. There are four relevant concepts of negative association, of which we start at the 'lowest'. The measure μ is said to be *edge-negatively-associated* if

$$\mu(J_e \cap J_f) \le \mu(J_e)\mu(J_f), \qquad e, f \in E,\ e \ne f. \tag{3.94}$$

Recall that $J_e = \{\omega \in \Omega : \omega(e) = 1\}$.

There is a more general notion of negative association, as follows. For $\omega \in \Omega$ and $F \subseteq E$ we define the cylinder event $\Omega_{F,\omega}$ generated by ω on F by

$$\Omega_{F,\omega} = \{\omega' \in \Omega : \omega'(e) = \omega(e) \text{ for } e \in F\}.$$

For $E' \subseteq E$ and an event $A \subseteq \Omega$, we say that A is *defined on* E' if, for all $\omega \in \Omega$, we have that $\omega \in A$ if and only if $\Omega_{F,\omega} \subseteq A$. We call μ *negatively associated* if

$$\mu(A \cap B) \le \mu(A)\mu(B)$$

for all pairs (A, B) of increasing events with the property that there exists $E' \subseteq E$ such that A is defined on E' and B is defined on its complement $E \setminus E'$. An account of negative association and its inherent problems may be found in [268].

Our third and fourth concepts of negative association involve so-called 'disjoint occurrence' (see [37, 154]). Let A and B be events in Ω. We define $A \,\Box\, B$ to be the set of all vectors $\omega \in \Omega$ for which there exists a set $F \subseteq E$ such that $\Omega_{F,\omega} \subseteq A$ and $\Omega_{\overline{F},\omega} \subseteq B$, where $\overline{F} = E \setminus F$. Note that the choice of F is allowed to depend on the vector ω. We say that μ has the *disjoint-occurrence property* if

$$\mu(A \,\Box\, B) \le \mu(A)\mu(B), \qquad A, B \subseteq \Omega, \tag{3.95}$$

and has the *disjoint-occurrence property on increasing events* if (3.95) holds under the additional assumption that A and B are increasing events.

It is evident that:

μ has the disjoint-occurrence property

$\Rightarrow$ μ has the disjoint-occurrence property on increasing events

$\Rightarrow$ μ is negatively associated

$\Rightarrow$ μ is edge-negatively-associated.

It was proved by van den Berg and Kesten [37] that the product measures ϕ_p on Ω have the disjoint-occurrence property on increasing events, and further by Reimer [283] that ϕ_p has the more general disjoint-occurrence property. It is easily seen[7] that the random-cluster measure $\phi_{p,q}$ cannot in general be edge-negatively-associated when $q > 1$. It may however be conjectured that $\phi_{p,q}$ satisfies some form of negative association when $q < 1$. Such a property would be useful in studying random-cluster measures, particularly in the thermodynamic limit (see Chapter 4), but no such property has yet been proved.

In the absence of a satisfactory approach to the general case of random-cluster measures with $q < 1$, we turn next to the issue of negative association of weak limits of $\phi_{p,q}$ as $q \downarrow 0$; see Section 1.5 and especially Theorem 1.23. Here is a mild conjecture, as yet unproven.

(3.96) Conjecture [156, 165, 199, 268]. *For any finite graph $G = (V, E)$, the uniform-spanning-forest measure* USF *and the uniform-connected-subgraph measure* UCS *are edge-negatively-associated.*

A stronger version of this conjecture is that USF and UCS are negatively associated in one or more of the senses described above.

---

[7] Consider the two events J_e, J_f in the graph G comprising exactly two edges e, f in parallel.

Since USF and UCS are uniform measures, Conjecture 3.96 may be rewritten in the form of two questions concerning subgraph counts. For simplicity we shall consider only graphs with neither loops nor multiple edges. Let $V = \{1, 2, \dots, n\}$, and let K be the set of $N = \binom{n}{2}$ edges of the complete graph on the vertex set V. We think of subsets of K as being graphs on V. Let $E \subseteq K$. For $X \subseteq E$, let $M^X = M^X(E)$ be the number of subsets E' of E with $E' \supseteq X$ such that the graph (V, E') is connected. Edge-negative-association for connected subgraphs amounts to the inequality

$$M^{\{e,f\}} M^{\varnothing} \leq M^e M^f, \qquad e, f \in E,\ e \neq f. \tag{3.97}$$

Here and later in this context, singleton sets are denoted without their braces, and any empty set is suppressed.

In the second such question, we ask if the same inequality is valid with M^X re-defined as the number of subsets $E' \subseteq E$ containing X such that (V, E') is a forest. See [199, 268].

With E fixed as above, and with $X, Y \subseteq E$, let $M^X_Y = M^X_Y(E)$ denote the number of subsets $E' \subseteq E$ of the required type such that $E' \supseteq X$ and $E' \cap Y = \varnothing$. Inequality (3.97) is easily seen to be equivalent to the inequality

$$M^{\{e,f\}} M_{\{e,f\}} \leq M^e_f M^f_e, \qquad e, f \in E,\ e \neq f. \tag{3.98}$$

The corresponding statement for the uniform spanning tree is known.

(3.99) Theorem. *The uniform-spanning-tree measure* UST *is edge-negatively-associated.*

The stronger property of negative association has been proved for UST, see [116], but we do not discuss this here. See also the discussions in [31, 241]. The strongest such conclusion known currently for USF appears to be the following, the proof is computer-aided and is omitted.

(3.100) Theorem [165]. *If $G = (V, E)$ has eight or fewer vertices, or has nine vertices and eighteen or fewer edges, then the associated uniform-spanning-forest measure* USF *has the edge-negative-association property.*

Since forests are dual to connected subgraphs for planar graphs, this implies a property of edge-negative-association for the UCS measure on certain planar graphs having fewer than ten faces.

The conjectures of this section have been expressed in terms of inequalities involving counts of connected subgraphs and forests, see the discussion around (3.97). Such inequalities may be formulated in the following more general way. Let $\mathbf{p} = (p_e : e \in E)$ be a collection of non-negative numbers indexed by E. For $E' \subseteq E$, let

$$f_{\mathbf{p}}(E') = \prod_{e \in E'} p_e.$$

We now ask whether (3.97) holds with $M^X = M^X(\mathbf{p})$ defined by

$$M^X(\mathbf{p}) = \sum_{\substack{E' : X \subseteq E' \subseteq E \\ (V,E') \text{ has property } \Pi}} f_{\mathbf{p}}(E'), \tag{3.101}$$

where Π is either the property of being connected or the property of containing no circuits. Note that (3.97) becomes a polynomial inequality in $|E|$ real variables. Such a formulation is natural when the problem is cast in the context of the Tutte polynomial, see Section 3.6 and [308].

Proof[8] *of Theorem 3.99.* Consider an electrical network on the connected graph G in which each edge corresponds to a unit resistor. The relevant fact from the theory of electrical networks is that, if a unit current flows from a *source* vertex s to a *sink* vertex t ($\neq s$), then the current flowing along the edge $e = \langle x, y\rangle$ in the direction xy equals $N(s, x, y, t)/N$, where N is the number of spanning trees of G and $N(s, x, y, t)$ is the number of spanning trees whose unique path from s to t passes along the edge $\langle x, y\rangle$ in the direction xy.

Let $e = \langle x, y\rangle$, and let μ be the UST measure on G. By the above, $\mu(J_e)$ equals the current flowing along e when a unit current flows through G from source x to sink y. By Ohm's Law, this equals the potential difference between x and y, which in turn equals the effective resistance $R_G(x, y)$ of the network between x and y.

Let $f \in E$, $f \neq e$, and denote by $G.f$ the graph obtained from G by contracting the edge f. There is a one–one correspondence between spanning trees of $G.f$ and spanning trees of G containing f. Therefore, $\mu(J_e \mid J_f)$ equals the effective resistance $R_{G.f}(x, y)$ of the network $G.f$ between x and y.

The so-called Rayleigh principle states that the effective resistance of a network is a non-decreasing function of the individual edge-resistances. It follows that $R_{G.f}(x, y) \leq R_G(x, y)$, and hence $\mu(J_e \mid J_f) \leq \mu(J_e)$. □

The usual proof of the Rayleigh principle makes use of the Thomson/Dirichlet variational principle, which in turn asserts that, amongst all unit flows from source to sink, the true flow of unit size is that which minimizes the dissipated energy. A good account of the Kirchhoff theorem on electrical networks and spanning trees may be found in [59]. Further accounts of the mathematics of electrical networks include [106] and [241], the latter containing also much material about the uniform spanning tree.

---

[8] When re-stated in terms of counts of spanning trees with certain properties, this is a consequence of the 1847 work of Kirchhoff [215] on electrical networks, as elaborated by Brooks, Smith, Stone, and Tutte in their famous paper [71] on the dissection of rectangles. Indeed, the difference $\mu(J_e \cap J_f) - \mu(J_e)\mu(J_f)$ may be expressed in terms of a certain 'transfer current matrix'. See [74] for an extension to more than two edges, and [31, 241] for related discussion.

Chapter 4

Infinite-Volume Measures

Summary. Random-cluster measures on infinite graphs may be defined either by passing to infinite-volume limits or by using the approach of Dobrushin, Lanford, and Ruelle. The problem of the uniqueness of infinite-volume measures is answered in part by way of an argument using the convexity of 'pressure'. The random-cluster and Potts measures in infinite volume may be coupled, thereby permitting a study of the Potts model on the lattice $\mathbb{L}^d$.

4.1 Infinite graphs

Although there is interesting theory associated with random-cluster measures on *finite* graphs, the real action, seen from the point of view of statistical mechanics, takes place in the context of *infinite* graphs. On a finite graph, all probabilities are polynomials in p and q, and are therefore smooth functions, whereas singularities and 'phase transitions' occur when the graph is infinite. These singularities provide most of the mathematical and physical motivation for the study of the random-cluster model.

While one may define random-cluster measures on a broad class of infinite graphs using the methods of this chapter, we shall concentrate here on finite-dimensional lattice-graphs. We shall, almost without exception, consider the (hyper)cubic lattice $\mathbb{L}^d = (\mathbb{Z}^d, \mathbb{E}^d)$ in some number d of dimensions satisfying $d \geq 2$. This restriction enables a clear exposition of the theory and open problems without suffering the complications that arise through allowing greater generality. We note however that many of the basic properties of random-cluster measures on lattices are valid on a much larger class of graphs. Interesting further questions arise in the non-finite-dimensional setting of non-amenable graphs, to which we return in Section 10.12.

There are two ways of defining random-cluster measures on an infinite graph $G = (V, E)$. The first is to consider weak limits of measures on finite subgraphs Λ, in the limit as $\Lambda \uparrow V$. This will be discussed in Section 4.3, following the introduction in Section 4.2 of the notion of boundary conditions. The second way is to restrict oneself to infinite-volume measures whose conditional marginal on any given finite sub-domain Λ is the finite-volume random-cluster measure on Λ with the correct boundary condition. This latter route is inspired by work of Dobrushin [102] and Lanford–Ruelle [226] for Gibbs states, and will be discussed in Section 4.4. In preparation for the required arguments, we summarize next the stochastic ordering and positive association of probability measures on $\mathbb{L}^d$.

Let $\Omega = \{0, 1\}^{\mathbb{E}^d}$, and let $\mathcal{F}$ be the σ-field generated by the cylinder subsets of Ω. Since Ω is a partially ordered set, we may speak of 'increasing' events and random variables. Given two probability measures μ_1, μ_2 on $(\Omega, \mathcal{F})$, we write $\mu_1 \le_{\text{st}} \mu_2$ if

$$\mu_1(X) \le \mu_2(X) \quad \text{for all increasing continuous } X : \Omega \to \mathbb{R}. \tag{4.1}$$

See Section 2.1. Note that any increasing random variable X with range $\mathbb{R}$ satisfies $X(0) \le X(\omega) \le X(1)$ for all $\omega \in \Omega$, and is therefore bounded.

One sometimes wishes to apply (4.1) to increasing random variables X that are semicontinuous rather than continuous[1]. This may be done as follows. For $\omega, \xi \in \Omega$ and a box Λ, we write ω_Λ^ξ for the configuration given by

$$\omega_\Lambda^\xi(e) = \begin{cases} \omega(e) & \text{if } e \in \mathbb{E}_\Lambda, \\ \xi(e) & \text{otherwise.} \end{cases} \tag{4.2}$$

For $X : \Omega \to \mathbb{R}$, we define X_Λ^0 and X_Λ^1 by

$$X_\Lambda^b(\omega) = X(\omega_\Lambda^b), \qquad \omega \in \Omega,\ b = 0, 1. \tag{4.3}$$

Assume that X is increasing. It is easily checked that, as $\Lambda \uparrow \mathbb{Z}^d$,

$$\begin{aligned} X_\Lambda^0 &\uparrow X \quad \text{if and only if } X \text{ is lower-semicontinuous,} \\ X_\Lambda^1 &\downarrow X \quad \text{if and only if } X \text{ is upper-semicontinuous,} \end{aligned} \tag{4.4}$$

where the convergence is pointwise on Ω. The functions X_Λ^0, X_Λ^1 are continuous. Therefore, by the monotone convergence theorem, $\mu_1 \le_{\text{st}} \mu_2$ if and only if

$$\mu_1(X) \le \mu_2(X) \quad \text{for all increasing semicontinuous } X. \tag{4.5}$$

It is a useful fact that, when $\mu_1 \le_{\text{st}} \mu_2$, then $\mu_1 = \mu_2$ whenever their marginals are equal. We state this as a theorem for future use, see also [235, Section II.2]. Recall that J_e is the event that e is open.

---

[1] An important example of an upper-semicontinuous function is the indicator function $X = 1_A$ of an increasing closed event A.

(4.6) Proposition. *Let E be a countable set, let $\Omega = \{0,1\}^E$, and let $\mathcal{F}$ be the σ-field generated by the cylinder subsets of Ω. Let μ_1, μ_2 be probability measures on $(\Omega, \mathcal{F})$ such that $\mu_1 \le_{\rm st} \mu_2$. Then $\mu_1 = \mu_2$ if and only if*

$$\mu_1(J_e) = \mu_2(J_e) \qquad \textit{for all } e \in E. \tag{4.7}$$

We say that a probability measure μ on $(\Omega, \mathcal{F})$ is *positively associated* if

$$\mu(XY) \ge \mu(X)\mu(Y) \quad \text{for all increasing continuous } X, Y. \tag{4.8}$$

Note from the arguments above that μ is positively associated if and only if

$$\mu(XY) \ge \mu(X)\mu(Y) \quad \text{for all increasing semicontinuous } X, Y. \tag{4.9}$$

Stochastic inequalities and positive association are conserved by weak convergence, in the following sense.

(4.10) Proposition. *Let E be a countable set, let $\Omega = \{0,1\}^E$, and let $\mathcal{F}$ be the σ-field generated by the cylinder subsets of Ω.*

(a) *Let $(\mu_{n,i} : n = 1, 2, \dots)$, $i = 1, 2$, be two sequences of probability measures on $(\Omega, \mathcal{F})$ satisfying: $\mu_{n,i} \Rightarrow \mu_i$ as $n \to \infty$, for $i = 1, 2$, and $\mu_{n,1} \le_{\rm st} \mu_{n,2}$ for all n. Then $\mu_1 \le_{\rm st} \mu_2$.*

(b) *Let $(\mu_n : n = 1, 2, \dots)$ be a sequence of probability measures on $(\Omega, \mathcal{F})$ satisfying $\mu_n \Rightarrow \mu$ as $n \to \infty$. If each μ_n is positively associated, then so is μ.*

Proof of Proposition 4.6. If $\mu_1 = \mu_2$ then (4.7) holds. Suppose conversely that (4.7) holds. By [235, Thm 2.4] or [237, Thm II.2.4], there exists a 'coupled' measure μ on $(\Omega, \mathcal{F}) \times (\Omega, \mathcal{F})$ with marginals μ_1 and μ_2, and such that

$$\mu\big(\{(\pi, \omega) \in \Omega^2 : \pi \le \omega\}\big) = 1.$$

For any increasing cylinder event A,

$$\begin{aligned}
\mu_2(A) - \mu_1(A) &= \mu\big(\{(\pi, \omega) : \pi \notin A,\ \omega \in A\}\big) \\
&\le \sum_{e \in E} \mu\big(\pi(e) = 0,\ \omega(e) = 1\big) \\
&= \sum_{e \in E} \big[\mu(\omega(e) = 1) - \mu(\pi(e) = 1)\big] \\
&= \sum_{e \in E} \big[\mu_2(J_e) - \mu_1(J_e)\big] = 0.
\end{aligned}$$

Since $\mathcal{F}$ is generated by the increasing cylinders A, the claim is proved. □

Proof of Proposition 4.10. (a) We have that $\mu_{n,1}(X) \le \mu_{n,2}(X)$ for any increasing continuous random variable X, and the conclusion follows by letting $n \to \infty$. Part (b) is proved similarly. □

4.2 Boundary conditions

An important part of statistical mechanics is directed at understanding the way in which assumptions on the boundary of a region affect what happens in its interior. In order to make precise such a discussion for random-cluster models, we introduce next the concept of a 'boundary condition'.

Let Λ be a finite subset of $\mathbb{Z}^d$. We shall later take Λ to be a box, but we retain the extra generality at this point. For $\xi \in \Omega$, let Ω_Λ^ξ denote the (finite) subset of Ω containing all configurations ω satisfying $\omega(e) = \xi(e)$ for $e \in \mathbb{E}^d \setminus \mathbb{E}_\Lambda$; these are the configurations that 'agree with ξ off Λ'. For $\xi \in \Omega$ and $p \in [0, 1], q \in (0, \infty)$, we shall write $\phi_{\Lambda,p,q}^\xi$ for the random-cluster measure on the finite graph $(\Lambda, \mathbb{E}_\Lambda)$ 'with boundary condition ξ'; this is the equivalent of a 'specification' for Gibbs states, see [134]. More precisely, let $\phi_{\Lambda,p,q}^\xi$ be the probability measure on the pair $(\Omega, \mathcal{F})$ given by

(4.11)
$$\phi_{\Lambda,p,q}^\xi(\omega) = \begin{cases} \dfrac{1}{Z_\Lambda^\xi(p,q)} \left\{ \displaystyle\prod_{e \in \mathbb{E}_\Lambda} p^{\omega(e)}(1-p)^{1-\omega(e)} \right\} q^{k(\omega,\Lambda)} & \text{if } \omega \in \Omega_\Lambda^\xi, \\ 0 & \text{otherwise,} \end{cases}$$

where $k(\omega, \Lambda)$ is the number of components of the graph $(\mathbb{Z}^d, \eta(\omega))$ that intersect Λ, and $Z_\Lambda^\xi(p, q)$ is the appropriate normalizing constant,

$$Z_\Lambda^\xi(p,q) = \sum_{\omega \in \Omega_\Lambda^\xi} \left\{ \prod_{e \in \mathbb{E}_\Lambda} p^{\omega(e)}(1-p)^{1-\omega(e)} \right\} q^{k(\omega,\Lambda)}. \tag{4.12}$$

Note that $\phi_{\Lambda,p,q}^\xi(\Omega_\Lambda^\xi) = 1$.

The boundary condition ξ influences the measure $\phi_{\Lambda,p,q}^\xi$ through the way in which the term $k(\omega, \Lambda)$ in (4.11) counts the number of ω-open clusters of Λ intersecting the boundary $\partial\Lambda$. Let $x, y \in \partial\Lambda$, and suppose there exists a path of ξ-open edges of $\mathbb{E}^d \setminus \mathbb{E}_\Lambda$ from x to y. Then any two ω-open clusters of Λ containing x and y, respectively, will contribute only 1 to the count $k(\omega, \Lambda)$.

Random-cluster measures have an important 'nesting' property which is best expressed in terms of conditional probabilities. For any finite subset Λ of $\mathbb{Z}^d$, we write as usual $\mathcal{F}_\Lambda$ (respectively, $\mathcal{T}_\Lambda$) for the σ-field generated by the states of edges in $\mathbb{E}_\Lambda$ (respectively, $\mathbb{E}^d \setminus \mathbb{E}_\Lambda$).

(4.13) Lemma. *Let $p \in [0, 1]$ and $q \in (0, \infty)$. If Λ, Δ are finite sets of vertices with $\Lambda \subseteq \Delta$, then for every $\xi \in \Omega$ and every event $A \in \mathcal{F}$,*

$$\phi_{\Delta,p,q}^\xi(A \mid \mathcal{T}_\Lambda)(\omega) = \phi_{\Lambda,p,q}^\omega(A), \qquad \omega \in \Omega_\Delta^\xi.$$

Two extremal boundary conditions of special importance are the configurations 0 and 1, comprising 'all edges closed' and 'all edges open' respectively.

One speaks of configurations in Ω_Λ^0 as having 'free' boundary conditions, and configurations in Ω_Λ^1 as having 'wired' boundary conditions. The word 'wired' refers to the fact that, with boundary condition 1, the set of open clusters of $\omega \in \Omega_\Lambda^1$ that intersect $\partial\Lambda$ are 'wired together' and contribute only 1 in all to the count $k(\omega, \Lambda)$ of clusters[2]. This terminology originated in the study of electrical networks. 'Free' is understood as the converse: such clusters are counted in their actual number when the boundary condition is 0.

The free and wired boundary conditions provide random-cluster measures which are extremal (for $q \geq 1$) in the sense of stochastic ordering.

(4.14) Lemma. *Let $p \in [0, 1]$ and $q \in [1, \infty)$, and let $\Lambda \subseteq \mathbb{Z}^d$ be a finite set.*

(a) *For every $\xi \in \Omega$, the probability measure $\phi_{\Lambda,p,q}^\xi$ is positively associated.*

(b) *For $\psi, \xi \in \Omega$, we have that $\phi_{\Lambda,p,q}^\psi \leq_{\text{st}} \phi_{\Lambda,p,q}^\xi$ whenever $\psi \leq \xi$. In particular,*

$$\phi_{\Lambda,p,q}^0 \leq_{\text{st}} \phi_{\Lambda,p,q}^\xi \leq_{\text{st}} \phi_{\Lambda,p,q}^1, \qquad \xi \in \Omega.$$

Proof of Lemma 4.13. We apply Theorem 3.1(a) repeatedly, once for each edge in $\mathbb{E}_\Delta \setminus \mathbb{E}_\Lambda$. □

Proof of Lemma 4.14. The key to the proof is positive association, which is valid by Theorem 3.8 when $q \in [1, \infty)$. The proof is straightforward, if slightly tedious when written out in detail. Since p and q will be held constant, we omit them from future subscripts. Let $q \in [1, \infty)$ and let Λ be a finite subset of $\mathbb{Z}^d$. For $\xi \in \Omega$ and for any increasing continuous function $X : \Omega \to \mathbb{R}$, we define the increasing random variable $X_\Lambda^\xi : \Omega \to \mathbb{R}$ by

$$X_\Lambda^\xi(\omega) = X(\omega_\Lambda^\xi)$$

where ω_Λ^ξ is given in (4.2). We may view X_Λ^ξ as an increasing function on $\{0, 1\}^{\mathbb{E}_\Lambda}$.

We augment the graph $(\Lambda, \mathbb{E}_\Lambda)$ by adding some extra edges as follows around the boundary $\partial\Lambda$. For every distinct unordered pair $x, y \in \partial\Lambda$, we add a new edge, denoted $[x, y]$, between x and y. If the edge $\langle x, y\rangle$ exists already in Λ, we simply add another in parallel. We write $\mathbb{F}$ for the set of new edges, $\overline{\Omega}_\Lambda = \{0, 1\}^{\mathbb{E}_\Lambda \cup \mathbb{F}}$ for the augmented configuration space, and let $\overline{\phi}_\Lambda$ be the random-cluster measure on the augmented graph $(\Lambda, \mathbb{E}_\Lambda \cup \mathbb{F})$. The key point is that $\overline{\phi}_\Lambda$ satisfies the statements in Theorem 2.24.

For $\xi \in \Omega$, let $\overset{\xi}{\sim}$ be the equivalence relation on $\partial\Lambda$ given by: $x \overset{\xi}{\sim} y$ if and only if there exists a ξ-open path of $\mathbb{E}^d \setminus \mathbb{E}_\Lambda$ joining x to y. Let $\mathbb{F}^\xi$ be the set of all edges $[x, y] \in \mathbb{F}$ such that $x \overset{\xi}{\sim} y$.

---

[2]Alternatively, one may omit from the cluster-count all clusters that intersect $\partial\Lambda$. This undercuts $k(\omega, \Lambda)$ by 1 for the wired measure $\phi_{\Lambda,p,q}^1$, and the difference, being constant, has no effect on the measure. See also Section 10.9.

(a) Let X, Y be increasing and continuous on Ω. Then

$$\begin{aligned}\phi_\Lambda^\xi(XY) &= \phi_\Lambda^\xi(X_\Lambda^\xi Y_\Lambda^\xi)\\ &= \overline{\phi}_\Lambda(X_\Lambda^\xi Y_\Lambda^\xi \mid \mathbb{F}^\xi \text{ open}, \mathbb{F}\setminus\mathbb{F}^\xi \text{ closed})\\ &\ge \overline{\phi}_\Lambda(X_\Lambda^\xi \mid \mathbb{F}^\xi \text{ open}, \mathbb{F}\setminus\mathbb{F}^\xi \text{ closed})\overline{\phi}_\Lambda(Y_\Lambda^\xi \mid \mathbb{F}^\xi \text{ open}, \mathbb{F}\setminus\mathbb{F}^\xi \text{ closed})\\ &\qquad\qquad \text{by strong positive-association}\\ &= \phi_\Lambda^\xi(X_\Lambda^\xi)\phi_\Lambda^\xi(Y_\Lambda^\xi) = \phi_\Lambda^\xi(X)\phi_\Lambda^\xi(Y),\end{aligned}$$

whence ϕ_Λ^ξ is positively associated.

(b) In broad terms, the 'greater' the connections off Λ, the larger is the induced measure within Λ. Let $\psi \le \xi$, whence $\mathbb{F}^\psi \subseteq \mathbb{F}^\xi$, and let X be an increasing random variable. Then

$$\begin{aligned}\phi_\Lambda^\psi(X) &= \phi_\Lambda^\psi(X_\Lambda^\psi)\\ &= \overline{\phi}_\Lambda(X_\Lambda^\psi \mid \mathbb{F}^\psi \text{ open}, \mathbb{F}\setminus\mathbb{F}^\psi \text{ closed})\\ &\le \overline{\phi}_\Lambda(X_\Lambda^\psi \mid \mathbb{F}^\xi \text{ open}, \mathbb{F}\setminus\mathbb{F}^\xi \text{ closed}) \qquad \text{by monotonicity}\\ &\le \overline{\phi}_\Lambda(X_\Lambda^\xi \mid \mathbb{F}^\xi \text{ open}, \mathbb{F}\setminus\mathbb{F}^\xi \text{ closed}) \qquad \text{since } X_\Lambda^\psi \le X_\Lambda^\xi\\ &= \phi_\Lambda^\xi(X_\Lambda^\xi) = \phi_\Lambda^\xi(X),\end{aligned}$$

and the claim follows. □

4.3 Infinite-volume weak limits

We begin with a definition of a 'weak-limit' random-cluster measure on $\mathbb{L}^d$. The use of the letter Λ is restricted throughout this section to boxes of $\mathbb{Z}^d$.

(4.15) Definition. Let $p \in [0, 1]$ and $q \in (0, \infty)$. A probability measure ϕ on $(\Omega, \mathcal{F})$ is called a *limit-random-cluster measure* with parameters p and q if, for some $\xi \in \Omega$, ϕ is an accumulation point of the family $\{\phi_{\Lambda,p,q}^\xi : \Lambda \subseteq \mathbb{Z}^d\}$, that is, there exists a sequence $\mathbf{\Lambda} = (\Lambda_n : n = 1, 2, \dots)$ of boxes satisfying $\Lambda_n \uparrow \mathbb{Z}^d$ as $n \to \infty$ such that

$$\phi_{\Lambda_n,p,q}^\xi \Rightarrow \phi \qquad \text{as } n \to \infty.$$

The set of all such measures ϕ is denoted by $\mathcal{W}_{p,q}$, and the closed convex hull of $\mathcal{W}_{p,q}$ is denoted by $\overline{\text{co}\, \mathcal{W}_{p,q}}$.

One might at first sight consider instead the class of all weak limits of the form

$$\phi = \lim_{n\to\infty} \phi_{\Lambda_n,p,q}^{\xi_n} \tag{4.16}$$

for sequences $\mathbf{\Lambda} = (\Lambda_n)$ of boxes and (ξ_n) of configurations. This provides no extra generality over Definition 4.15, as we explain next in two paragraphs which the reader may choose to omit, [152].

The measure $\phi^{\xi}_{\Lambda,p,q}$ is influenced by ξ through the connections it provides between vertices in the boundary $\partial\Lambda$. By arranging for the same connections (and no others) to be provided in a manner which is 'more economical in the use of space' one discovers the following. Let Λ be a box and $\xi \in \Omega$. There exists a box $\Lambda' \supseteq \Lambda$ and a configuration ζ such that: $\phi^{\zeta}_{\Lambda,p,q}(A) = \phi^{\psi}_{\Lambda,p,q}(A)$ for any event $A \in \mathcal{F}_\Lambda$ and any configuration ψ that agrees with ζ on $\mathbb{E}_{\Lambda'} \setminus \mathbb{E}_\Lambda$.

Assume now that (4.16) holds for some $\mathbf{\Lambda}$, ξ. Let A be a cylinder event, and assume that Λ_1 is such that $A \in \mathcal{F}_{\Lambda_1}$. Define the increasing subsequence $(\Delta_n : n = 1, 2, \dots)$ of $\mathbf{\Lambda}$ and the configuration ξ as follows. We set $\Delta_1 = \Lambda_1$ and $\xi(e) = \xi_1(e)$ for $e \in \mathbb{E}_{\Delta_1}$. Having constructed $\Delta_r = \Lambda_{n_r}$ and the partial configuration $(\xi(e) : e \in \mathbb{E}_{\Delta_r})$ for $r < R$, we construct Δ_R and the additional configuration $(\xi(e) : e \in \mathbb{E}_{\Delta_R} \setminus \mathbb{E}_{\Delta_{R-1}})$ by the following rule. By the remark above, there exists a box $\Lambda' \supseteq \Delta_{R-1}$ and a configuration ζ such that

$$\phi^{\xi_{n_{R-1}}}_{\Delta_{R-1},p,q}(A) = \phi^{\psi}_{\Delta_{R-1},p,q}(A)$$

for any ψ that agrees with ζ on $\mathbb{E}_{\Lambda'} \setminus \mathbb{E}_{\Delta_{R-1}}$. We find $m = n_R$ such that $m > n_{R-1}$ and $\Lambda_m \supseteq \Lambda'$, and we set $\Delta_R = \Lambda_m$ and $\xi(e) = \zeta(e)$ for $e \in \mathbb{E}_{\Delta_R} \setminus \mathbb{E}_{\Delta_{R-1}}$. By (4.16), $\phi^{\xi}_{\Delta_r,p,q}(A) \to \phi(A)$ as $r \to \infty$, whence $\phi^{\xi}_{\Delta_r,p,q} \Rightarrow \phi$.

The following claim is standard of its type. Part (b) is related to the so-called 'finite-energy property' to be discussed in the next section.

(4.17) Theorem. *Let $p \in [0, 1]$ and $q \in (0, \infty)$.*

(a) Existence. *The set $\mathcal{W}_{p,q}$ of limit-random-cluster measures is non-empty.*

(b) Finite-energy property. *Let $\phi \in \overline{\mathrm{co}\, \mathcal{W}_{p,q}}$ and $e \in \mathbb{E}^d$. We have that*

$$\min\left\{p, \frac{p}{p+q(1-p)}\right\} \le \phi(J_e \mid \mathcal{T}_e) \le \max\left\{p, \frac{p}{p+q(1-p)}\right\},$$

ϕ-almost-surely, where J_e is the event that e is open.

(c) Positive association. *If $q \in [1, \infty)$, any member of $\mathcal{W}_{p,q}$ is positively associated.*

Proof. (a) The metric space Ω is the product of discrete spaces, and is therefore compact. Any infinite family of probability measures on Ω is therefore tight, and hence relatively compact (by Prohorov's theorem, see [42]), which is to say that any infinite subsequence contains a weakly convergent subsubsequence. We apply this to the family $\{\phi^{\xi}_{\Lambda_n,p,q} : n = 1, 2, \dots\}$ for any given $\xi \in \Omega$ and any given sequence $\mathbf{\Lambda} = (\Lambda_n : n = 1, 2, \dots)$ with $\Lambda_n \uparrow \mathbb{Z}^d$ as $n \to \infty$.

(b) Let $\phi \in \mathcal{W}_{p,q}$, so that

$$\phi = \lim_{\Lambda \uparrow \mathbb{Z}^d} \phi^{\xi}_{\Lambda,p,q} \tag{4.18}$$

for some $\xi \in \Omega$ and some sequence of boxes Λ. For $\Sigma \subseteq \mathbb{Z}^d$ and $e \in \mathbb{E}^d$, let $\mathcal{F}_{\Sigma\setminus e}$ denote the σ-field generated by $\{\omega(f) : f \in \mathbb{E}_\Sigma,\ f \neq e\}$. By the martingale convergence theorem [164, eqn (12.3.10)] and weak convergence,

$$\begin{aligned}\phi(J_e \mid \mathcal{T}_e) &= \lim_{\Sigma\uparrow\mathbb{Z}^d} \phi(J_e \mid \mathcal{F}_{\Sigma\setminus e}) && \phi\text{-a.s.}\\ &= \lim_{\Sigma\uparrow\mathbb{Z}^d}\lim_{\Lambda\uparrow\mathbb{Z}^d} \phi^\xi_{\Lambda,p,q}(J_e \mid \mathcal{F}_{\Sigma\setminus e}) && \phi\text{-a.s.}\end{aligned}$$

The claim follows by Theorem 3.1(a). It is evident that any convex combination of measures in $\mathcal{W}_{p,q}$ satisfies the same inequalities. A similar argument yields the claim for weak limits of such combinations.

(c) Let $q \in [1, \infty)$, and let ϕ be expressed as in (4.18). By Lemma 4.14(a), each $\phi^\xi_{\Lambda,p,q}$ is positively associated, and the claim follows by Proposition 4.10(b). □

Let $\mathbf{\Lambda} = (\Lambda_n : n = 1, 2, \dots)$ be an increasing sequence of boxes such that $\Lambda_n \uparrow \mathbb{Z}^d$ as $n \to \infty$. When does the limit $\lim_{n\to\infty} \phi^\xi_{\Lambda_n,p,q}$ exist, and is it independent of the choice of the sequence $\mathbf{\Lambda}$? Only a limited amount is known when $q < 1$, and the reader is referred to Section 5.8 for this case. When $q \geq 1$, we may use positive association to prove the existence of the limit in the extremal cases with $\xi = 0, 1$. The next theorem comprises the basic existence result, together with some properties of the limit measures. It is preceded by some important definitions.

Let $G = (V, E)$ be a countable, locally finite[3] graph, and write $\Omega_E = \{0, 1\}^E$, and $\mathcal{F}_E$ for the σ-field generated by the cylinder subsets of Ω_E. An *automorphism* of G is a bijection $\tau : V \to V$ such that, for all $u, v \in V$, $\langle u, v\rangle \in E$ if and only if $\langle \tau(u), \tau(v)\rangle \in E$. We write $\mathrm{Aut}(G)$ for the group of all such automorphisms. The domain of an automorphism τ may be extended to the edge-set E by $\tau(\langle u, v\rangle) = \langle \tau(u), \tau(v)\rangle$. An automorphism τ generates an operator on Ω_E, denoted also by $\tau : \Omega_E \to \Omega_E$ and given by $\tau\omega(e) = \omega(\tau^{-1}e)$ for $e \in E$. A random variable $X : \Omega_E \to \mathbb{R}$ is called *τ-invariant* if $X(\omega) = X(\tau\omega)$ for all $\omega \in \Omega_E$. A probability measure μ on $(\Omega_E, \mathcal{F}_E)$ is called *τ-invariant* if $\mu(A) = \mu(\tau A)$ for all $A \in \mathcal{F}_E$.

Let Γ be a subgroup of $\mathrm{Aut}(G)$. A random variable $X : \Omega \to \mathbb{R}$ is called *Γ-invariant* if it is τ-invariant for all $\tau \in \Gamma$, and a similar definition holds for a probability measure μ on $(\Omega_E, \mathcal{F}_E)$. The measure μ is called *automorphism-invariant* if it is $\mathrm{Aut}(G)$-invariant. A probability measure μ on $(\Omega_E, \mathcal{F}_E)$ is called *Γ-ergodic* if every Γ-invariant random variable is μ-almost-surely constant, see [241, Chapter 6]. It is clear that, if $\Gamma' \subseteq \Gamma$, then μ is Γ-ergodic whenever it is Γ'-ergodic. In the case when Γ is the group generated by a single automorphism τ, we use the term *τ-ergodic* rather than Γ-ergodic.

We turn now to the graph $G = \mathbb{L}^d$, and to a class of automorphisms termed translations. Let $x \in \mathbb{Z}^d$, and define the function $\tau_x : \mathbb{Z}^d \to \mathbb{Z}^d$ by $\tau_x(y) = x + y$.

[3] A graph is called *locally finite* if every vertex has finite degree.

The automorphism τ_x is referred to as a *translation*. We denote the group of translations by $\mathbb{Z}^d$, noting that τ_0 is the identity map. A random variable $X : \Omega \to \mathbb{R}$ (respectively, a probability measure μ on $(\Omega, \mathcal{F})$) is called *translation-invariant* if it is $\mathbb{Z}^d$-invariant.

The probability measure μ on $(\Omega, \mathcal{F})$ is said to be *tail-trivial* if, for any tail event $A \in \mathcal{T}$, $\mu(A)$ equals either 0 or 1. The property of tail-triviality is important and useful for two reasons. First, tail-triviality implies mixing, see (4.22) and Corollary 4.23. Secondly, in statistical mechanics, for a given specification, tail-triviality is equivalent to extremality within the convex set of Gibbs states, see [134, Thm 7.7].

(4.19) Theorem (Thermodynamic limit) [8, 63, 122, 149, 150, 152].
Let $p \in [0,1]$ and $q \in [1,\infty)$.

(a) Existence. *Let $\mathbf{\Lambda} = (\Lambda_n : n = 1, 2, \dots)$ be an increasing sequence of boxes satisfying $\Lambda_n \uparrow \mathbb{Z}^d$ as $n \to \infty$. The weak limits*

$$\phi^b_{p,q} = \lim_{n\to\infty} \phi^b_{\Lambda_n,p,q}, \qquad b = 0, 1, \tag{4.20}$$

exist and are independent of the choice of $\mathbf{\Lambda}$.

(b) Automorphism-invariance. *The probability measure $\phi^b_{p,q}$ is automorphism-invariant, for $b = 0, 1$.*

(c) Extremality. *The $\phi^b_{p,q}$ are extremal in that*

$$\phi^0_{p,q} \le_{\rm st} \phi \le_{\rm st} \phi^1_{p,q}, \qquad \phi \in \mathcal{W}_{p,q}. \tag{4.21}$$

(d) Tail-triviality. *The $\phi^b_{p,q}$ are tail-trivial.*

A probability measure μ on $(\Omega, \mathcal{F})$ is said to be *mixing* if, for all $A, B \in \mathcal{F}$,

$$\lim_{|x|\to\infty} \mu(A \cap \tau_x B) = \mu(A)\mu(B), \tag{4.22}$$

which is to say that, for $\epsilon > 0$, there exists $N = N(\epsilon)$ such that

$$\big|\mu(A \cap \tau_x B) - \mu(A)\mu(B)\big| < \epsilon \qquad \text{if} \quad |x| \ge N.$$

(4.23) Corollary. *Let $p \in [0,1]$, $q \in [1,\infty)$, and $b \in \{0,1\}$. The probability measure $\phi^b_{p,q}$ is mixing, and is τ-ergodic for every translation τ of $\mathbb{L}^d$ other than the identity.*

Proof of Theorem 4.19. (a) Suppose first that $b = 0$. Let Λ and Δ be boxes satisfying $\Lambda \subseteq \Delta$, and let A be the event that all edges in $\mathbb{E}_\Delta \setminus \mathbb{E}_\Lambda$ have state 0. By Theorem 3.1(a), $\phi^0_{\Lambda,p,q}$ may be viewed as the marginal measure on $\mathbb{E}_\Lambda$ of $\phi^0_{\Delta,p,q}$ *conditioned* on the event A. Since A is a decreasing event, by positive association,

$$\phi^0_{\Lambda,p,q}(B) = \phi^0_{\Delta,p,q}(B \mid A) \le \phi^0_{\Delta,p,q}(B) \tag{4.24}$$

for any increasing $B \in \mathcal{F}_\Lambda$. Therefore, the increasing limit

$$\phi^0_{p,q}(B) = \lim_{\Lambda \uparrow \mathbb{Z}^d} \phi^0_{\Lambda,p,q}(B)$$

exists for all increasing cylinder events B, and the value of the limit does not depend on the way that $\Lambda \uparrow \mathbb{Z}^d$. The collection of all such events B is convergence-determining, [42, pp. 14–19], whence the limit probability measure $\phi^0_{p,q}$ exists. For the case $b = 1$, we let A be the event that all edges in $\mathbb{E}_\Delta \setminus \mathbb{E}_\Lambda$ are open, and we reverse the inequality in (4.24).

(b) The translation-invariance of $\phi^0_{p,q}$ is obtained as follows. Let F be a finite subset of $\mathbb{E}^d$, and let $B \in \mathcal{F}_F$ be increasing. Let τ be a translation of $\mathbb{L}^d$. For any box Λ containing all endvertices of all edges in F, we have by positive association as in (4.24) that

$$\phi^0_{p,q}(B) \ge \phi^0_{\Lambda,p,q}(B) = \phi^0_{\tau\Lambda,p,q}(\tau^{-1}B) \to \phi^0_{p,q}(\tau^{-1}B) \qquad \text{as } \Lambda \uparrow \mathbb{Z}^d.$$

Applying the same argument with B and τ replaced by $\tau^{-1}B$ and τ^{-1}, we obtain that $\phi^0_{p,q}(B) = \phi^0_{p,q}(\tau B)$. Similar arguments are valid for $\phi^1_{p,q}$.

Let $\mathcal{C}$ be the set of automorphisms that fix the origin. Each automorphism of $\mathbb{L}^d$ is a combination of a translation τ and an element $\sigma \in \mathcal{C}$. Every element of $\mathcal{C}$ preserves boxes of the form $\Lambda_n = [-n, n]^d$, and it follows by (4.20) that the $\phi^b_{p,q}$ are automorphism-invariant.

(c) By Lemma 4.14,

$$\phi^0_{\Lambda,p,q} \le_{\text{st}} \phi^\xi_{\Lambda,p,q} \le_{\text{st}} \phi^1_{\Lambda,p,q}, \qquad \xi \in \Omega,$$

and (4.21) follows by Proposition 4.10(a).

(d) We develop the proof of [31, 240] rather than the earlier approach of [152]. Let $b = 0$, an exactly analogous proof is valid for $b = 1$. Let Λ, Δ be boxes with $\Lambda \subseteq \Delta$, and let $A \in \mathcal{F}_\Lambda$ be increasing, and let $B \in \mathcal{F}_{\Delta\setminus\Lambda}$. By strong positive-association[4], Theorem 3.8(b),

$$\begin{aligned}\phi^0_{\Delta,p,q}(A \cap B) &= \phi^0_{\Delta,p,q}(A \mid B)\phi^0_{\Delta,p,q}(B)\\ &\ge \phi^0_{\Lambda,p,q}(A)\phi^0_{\Delta,p,q}(B).\end{aligned}$$

Let $\Delta \uparrow \mathbb{Z}^d$ to obtain that

$$\phi^0_{p,q}(A \cap B) \ge \phi^0_{\Lambda,p,q}(A)\phi^0_{p,q}(B).$$

Since this holds for $B \in \mathcal{F}_{\Delta\setminus\Lambda}$, it holds for $B \in \mathcal{T}_\Lambda$, and hence for $B \in \mathcal{T}$. Let $\Lambda \uparrow \mathbb{Z}^d$ to deduce that

$$\phi^0_{p,q}(A \cap B) \ge \phi^0_{p,q}(A)\phi^0_{p,q}(B), \qquad B \in \mathcal{T}. \tag{4.25}$$

[4] The case $\phi^0_{\Delta,p,q}(B) = 0$ should be handled separately.

Applying (4.25) to the complement $\overline{B}$, we have that

$$\phi^0_{p,q}(A \cap \overline{B}) \geq \phi^0_{p,q}(A)\phi^0_{p,q}(\overline{B}), \qquad B \in \mathcal{T}. \tag{4.26}$$

Since the sum of (4.25) and (4.26) holds with equality,

$$\phi^0_{p,q}(A \cap B) = \phi^0_{p,q}(A)\phi^0_{p,q}(B), \qquad B \in \mathcal{T}. \tag{4.27}$$

Since this holds for all increasing $A \in \mathcal{F}_\Lambda$, it holds (as in the proof of part (a)) for all $A \in \mathcal{F}$. Setting $A = B$ yields that $\phi^0_{p,q}(B)$ equals 0 or 1, which is to say that $\mathcal{T}$ is trivial. The same proof with several inequalities reversed is valid for $\phi^1_{p,q}$. □

Proof of Corollary 4.23. It is a general fact that tail-triviality implies mixing, see [134, Prop. 7.9] and the related discussion at [134, Remark 7.13, Prop. 14.9]. The τ-ergodicity of the $\phi^b_{p,q}$ is a standard application of mixing, as follows. Let $y \neq 0$ and $\tau = \tau_y$. Let B be a τ-invariant event, and apply (4.22) with $x = ny$ and $A = B$ to obtain, on letting $n \to \infty$, that $\phi^b_{p,q}(B) = \phi^b_{p,q}(B)^2$. Alternatively, note that the σ-field of τ-invariant events is contained in the completion of the tail σ-field $\mathcal{T}$, see the proof for $d = 1$ in [222, Prop. 4.5]. □

We close this section with the infinite-volume comparison inequalities and certain semicontinuity properties of the mean $\phi^b_{p,q}(X)$ of a random variable X.

(4.28) Proposition. *Let $p \in [0, 1]$ and $q \in [1, \infty)$.*

(a) Comparison inequalities. *For $b = 0, 1$, the measures $\phi^b_{p,q}$ satisfy the comparison inequalities*:

$$\begin{aligned} &\phi^b_{p_1,q_1} \leq_{\mathrm{st}} \phi^b_{p_2,q_2} \quad \text{if } q_1 \geq q_2 \geq 1, \text{ and } p_1 \leq p_2, \\ &\phi^b_{p_1,q_1} \geq_{\mathrm{st}} \phi^b_{p_2,q_2} \quad \text{if } q_1 \geq q_2 \geq 1, \text{ and } \frac{p_1}{q_1(1-p_1)} \geq \frac{p_2}{q_2(1-p_2)}. \end{aligned}$$

(b) Upper-semicontinuity. *Let X be an increasing upper-semicontinuous random variable. Then $\phi^1_{p,q}(X)$ is an upper-semicontinuous function of the vector (p, q), and is therefore a right-continuous function of p and a left-continuous function of q.*

(c) Lower-semicontinuity. *Let X be an increasing lower-semicontinuous random variable. Then $\phi^0_{p,q}(X)$ is a lower-semicontinuous function of the vector (p, q), and is therefore a left-continuous function of p and a right-continuous function of q.*

Conditions for the semicontinuity of an increasing random varable are given at (4.4). An important class of increasing upper-semicontinuous functions is provided by the indicator functions $X = 1_A$ of increasing closed events A. It is easily seen by (4.4) that such an indicator function is indeed upper-semicontinuous, and it follows by part (b) above that $\phi^1_{p,q}(A)$ is right-continuous in p and left-continuous in q. As an important example of such an event A, consider the event $\{0 \leftrightarrow \infty\}$, that there exists an infinite open path in $\mathbb{L}^d$ with endvertex 0.

Similarly, the indicator function of any increasing *open* event A is an increasing lower-semicontinuous random variable, and thus part (c) may be applied. We note that (b) and (c) apply to all increasing continuous random variables, and therefore to the indicator function $X = 1_B$ of any increasing cylinder B.

Proof of Proposition 4.28. (a) This is a consequence of Theorems 3.21 and 4.10(a).
(b) Let $\Lambda_n = [-n, n]^d$. Suppose X satisfies the given condition, and define X_n^b by $X_n^b(\omega) = X(\omega_{\Lambda_n}^b)$ for $b = 0, 1$, where ω_Λ^b is given in (4.2). Using stochastic orderings of measures and (4.5), we have for $m \le n$ that

$$\begin{aligned} \phi_{p,q}^1(X) \le \phi_{\Lambda_n,p,q}^1(X) \le \phi_{\Lambda_n,p,q}^1(X_m^1) &\quad \text{since } X \le X_m^1 \\ \to \phi_{p,q}^1(X_m^1) &\quad \text{as } n \to \infty \\ \to \phi_{p,q}^1(X) &\quad \text{as } m \to \infty, \end{aligned}$$

where we have used (4.4) and the monotone convergence theorem. Also,

$$\begin{aligned} \phi_{\Lambda_n,p,q}^1(X_n^1) \ge \phi_{\Lambda_{n+1},p,q}^1(X_n^1) &\quad \text{since } \Lambda_n \subseteq \Lambda_{n+1} \\ \ge \phi_{\Lambda_{n+1},p,q}^1(X_{n+1}^1) &\quad \text{since } X_n^1 \ge X_{n+1}^1. \end{aligned}$$

By the two inequalities above, the sequence $\phi_{\Lambda_n,p,q}^1(X_n^1)$, $n = 1, 2, \dots$, is non-increasing with limit $\phi_{p,q}^1(X)$. Each $\phi_{\Lambda_n,p,q}^1(X_n^1)$ is a continuous function of p and q, whence $\phi_{p,q}^1(X)$ is upper-semicontinuous.
(c) The argument of part (b) is valid with X_n^1 replaced by X_n^0, the boundary condition 1 replaced by 0, and with the inequalities reversed. □

4.4 Infinite-volume random-cluster measures

There is a second way to construct infinite-volume measures, this avoids weak limits and works directly on the infinite lattice. The following definition is based upon the well known Dobrushin–Lanford–Ruelle (DLR) definition of a Gibbs state, [102, 134, 226]. It was introduced in [111, 149, 150, 272] and discussed further in [63, 152].

(4.29) Definition. Let $p \in [0, 1]$ and $q \in (0, \infty)$. A probability measure ϕ on $(\Omega, \mathcal{F})$ is called a *DLR-random-cluster measure* with parameters p and q if:
(4.30)

$$\text{for all } A \in \mathcal{F} \text{ and boxes } \Lambda, \quad \phi(A \mid \mathcal{T}_\Lambda)(\xi) = \phi_{\Lambda,p,q}^\xi(A) \quad \text{for } \phi\text{-a.e. } \xi.$$

The set of such measures is denoted by $\mathcal{R}_{p,q}$.

The condition of this definition amounts to the following. Suppose we are given that the configuration off the finite box Λ is that of $\xi \in \Omega$. Then, for almost every ξ, the (conditional) measure on Λ is the finite-volume random-cluster measure $\phi^{\xi}_{\Lambda,p,q}$. It is not difficult to see, by a calculation of conditional probabilities, that no further generality may be gained by replacing the finite box Λ by a general finite subset of $\mathbb{Z}^d$. Indeed, we shall see in Proposition 4.37(b) that it suffices to have (4.30) for all pairs $\Lambda = \{x, y\}$ with $x \sim y$.

The structure of $\mathcal{R}_{p,q}$ relative to the set $\mathcal{W}_{p,q}$ remains somewhat obscure. It is not known, for example, whether or not $\mathcal{W}_{p,q} \subseteq \mathcal{R}_{p,q}$, and indeed one needs some work even to demonstrate that $\mathcal{R}_{p,q}$ is non-empty. The best result in this direction to date is restricted to probability measures having a certain additional property. For $\omega \in \Omega$, let $I(\omega)$ be the number of infinite open clusters of ω. We say that a probability measure ϕ on $(\Omega, \mathcal{F})$ has the *0/1-infinite-cluster property*[5] if $\phi(I \in \{0, 1\}) = 1$.

(4.31) Theorem [152, 153, 156, 272]. *Let $p \in [0, 1]$ and $q \in (0, \infty)$. If $\phi \in \overline{\text{co}\, \mathcal{W}_{p,q}}$ and ϕ has the 0/1-infinite-cluster property, then $\phi \in \mathcal{R}_{p,q}$.*

A sufficient condition for the 0/1-infinite-cluster property is provided by the uniqueness theorem of Burton–Keane, [72], namely translation-invariance[6] and so-called 'finite energy'. A probability measure ϕ on $(\Omega, \mathcal{F})$ is said to have the *finite-energy property* if

$$0 < \phi(J_e \mid \mathcal{T}_e) < 1 \quad \phi\text{-a.s., for all } e \in \mathbb{E}^d, \tag{4.32}$$

where, as before, J_e is the event that e is open.

(4.33) Theorem [152, 153, 156]. *Let $p \in [0, 1]$ and $q \in (0, \infty)$.*

(a) *The closed convex hull $\overline{\text{co}\, \mathcal{W}_{p,q}}$ contains some translation-invariant probability measure ϕ.*

(b) *Let $p \in (0, 1)$. Every $\phi \in \overline{\text{co}\, \mathcal{W}_{p,q}}$ has the finite-energy property.*

(c) *If $\phi \in \overline{\text{co}\, \mathcal{W}_{p,q}}$ is translation-invariant, then ϕ has the 0/1-infinite-cluster property.*

Theorems 4.31 and 4.33 imply jointly that $|\mathcal{R}_{p,q}| \neq \varnothing$ when $p \in (0, 1)$ and $q \in (0, \infty)$. [The cases $p = 0, 1$ are trivial.] We now present some of the basic properties of the set $\mathcal{R}_{p,q}$.

---

[5] The 0/1-infinite-cluster property is linked to the property of so-called 'almost-sure quasilocality', see Lemma 4.39 and [272].

[6] Rather less than full translation-invariance is in fact required. The proof in [72] uses ergodicity of the probability measure, rather than simply translation-invariance. Further comments about the extension to translation-invariant measures may be found in [73] and [136, p. 42]. See [158] for a general account of Burton–Keane uniqueness.

(4.34) Theorem [152]. *Let $p \in [0,1]$ and $q \in (0,\infty)$.*

(a) Existence. *The set $\mathcal{R}_{p,q}$ is non-empty and contains at least one translation-invariant member of $\overline{\text{co}\, \mathcal{W}_{p,q}}$. Furthermore, $\mathcal{R}_{p,q}$ is a convex set of measures.*

(b) Stochastic ordering. *If $q \in [1,\infty)$, then $\phi^b_{p,q} \in \mathcal{R}_{p,q}$ for $b = 0, 1$, and*

$$\phi^0_{p,q} \le_{\text{st}} \phi \le_{\text{st}} \phi^1_{p,q}, \qquad \phi \in \mathcal{R}_{p,q}. \tag{4.35}$$

(c) Extremality. *The $\phi^b_{p,q}$, $b = 0, 1$, are extremal elements of $\mathcal{R}_{p,q}$.*

It is an important open problem to identify all pairs (p,q) under which $\phi^0_{p,q} = \phi^1_{p,q}$. By (4.21) and (4.35), for $q \in [1,\infty)$,

$$(4.36) \qquad |\mathcal{W}_{p,q}| = |\mathcal{R}_{p,q}| = 1 \quad \text{if and only if} \quad \phi^0_{p,q} = \phi^1_{p,q},$$

so this amounts to asking for which pairs (p,q) there exist (simultaneously) a unique DLR-random-cluster measure and a unique limit-random-cluster measure. Various partial answers are known, see Theorems 4.63 and 5.33, and a conjecture appears at (5.34).

Let $q \in [1,\infty)$. Since the extremal measures $\phi^b_{p,q}$ are translation-invariant, they have the 0/1-infinite-cluster property, see Theorems 4.19(b) and 4.33(c). The ergodicity of the $\phi^b_{p,q}$ was proved in Corollary 4.23. We note two further properties of DLR-random-cluster measures, namely the finite-energy property, and positive association when $q \in [1,\infty)$. Let $e = \langle x, y\rangle$ be an edge, and let K_e be the event that x and y are joined by an open path of $\mathbb{E}^d \setminus \{e\}$.

(4.37) Proposition. *Let $p \in [0,1]$ and $q \in (0,\infty)$.*

(a) Finite-energy property. *Let $\phi \in \mathcal{R}_{p,q}$. For ϕ-almost-every ω,*

$$\phi(J_e \mid \mathcal{T}_e)(\omega) = \begin{cases} p & \text{if } \omega \in K_e, \\ \dfrac{p}{p+q(1-p)} & \text{if } \omega \notin K_e. \end{cases} \tag{4.38}$$

(b) *Conversely, if ϕ is a probability measure on $(\Omega, \mathcal{F})$ satisfying* (4.38) *for all $e \in \mathbb{E}^d$ and ϕ-almost-every ω, then $\phi \in \mathcal{R}_{p,q}$.*

(c) Positive association. *If $q \in [1,\infty)$ and $\phi \in \mathcal{R}_{p,q}$ is tail-trivial, then ϕ is positively associated.*

We shall use the following technical result in the proof of Theorem 4.31.

(4.39) Lemma [152]. *Let ϕ be a probability measure on $(\Omega, \mathcal{F})$ with the finite-energy property* (4.32) *and the* $0/1$*-infinite-cluster property. For any box Λ and any cylinder event $A \in \mathcal{F}_\Lambda$, the random variable $g(\omega) = \phi^{\omega}_{\Lambda,p,q}(A)$ is ϕ-almost-surely continuous.*

Proof. Let Λ be a finite box and $A \in \mathcal{F}_\Lambda$. The set D_g of discontinuities of the random variable $g(\omega) = \phi^{\omega}_{\Lambda,p,q}(A)$ is a subset of the set

$$D_g(\Lambda) = \bigcap_{\Delta:\, \Delta \supseteq \Lambda} \left\{ \omega : \sup_{\zeta:\, \zeta = \omega \text{ on } \Delta} |g(\zeta) - g(\omega)| > 0 \right\} \tag{4.40}$$

where the intersection is over all boxes Δ containing Λ, and we write '$\zeta = \omega$ on Δ' if $\zeta(e) = \omega(e)$ for $e \in \mathbb{E}_\Delta$. Let $D_{\Lambda,\Delta}$ be the set of all $\omega \in \Omega$ with the property: there exist two points $u, v \in \partial\Lambda$ such that both u and v are joined to $\partial\Delta$ by paths using ω-open edges of $\mathbb{E}_\Delta \setminus \mathbb{E}_\Lambda$, but u is not joined to v by such a path. If $D_{\Lambda,\Delta}$ does not occur, then $k(\zeta, \Lambda) = k(\omega, \Lambda)$ for all $\zeta \in \Omega$ such that $\zeta = \omega$ on Δ, implying that $g(\zeta) = g(\omega)$. It follows that

$$D_g(\Lambda) \subseteq \bigcap_{\Delta:\, \Delta \supseteq \Lambda} D_{\Lambda,\Delta}.$$

It easily seen that $\bigcap_\Delta D_{\Lambda,\Delta} = \{I_\Lambda \geq 2\}$, where I_Λ is the number of infinite open clusters of $\mathbb{E}^d \setminus \mathbb{E}_\Lambda$ intersecting $\partial\Lambda$. Therefore,

$$\phi(D_g) \leq \phi(D_g(\Lambda)) \leq \phi(I_\Lambda \geq 2). \tag{4.41}$$

By the finite-energy property (4.32),

$$\phi(I \geq 2) > 0 \quad \text{if} \quad \phi(I_\Lambda \geq 2) > 0. \tag{4.42}$$

By the $0/1$-infinite-cluster property, $\phi(I \geq 2) = 0$, and therefore $\phi(D_g) = 0$ as required. □

Proof of Theorem 4.33. (a) Since $\phi^0_{p,q} \in \mathcal{W}_{p,q}$ for $q \in [1, \infty)$, we shall consider the case when $q \in (0, 1)$ only. By Theorem 4.17(a), we may find $\phi \in \mathcal{W}_{p,q}$. Let

$$\psi_m = \frac{1}{|\Delta_m|} \sum_{x \in \Delta_m} \tau_x \circ \phi \tag{4.43}$$

where $\Delta_m = [-m, m]^d$, and $\tau_x \circ \phi$ is the probability measure on $(\Omega, \mathcal{F})$ given by $\tau_x \circ \phi(A) = \phi(\tau_x A)$ for the translation $\tau_x(y) = x + y$ of the lattice. Clearly, $\tau_x \circ \phi \in \mathcal{W}_{p,q}$ for all x, whence ψ_m belongs to the convex hull of $\mathcal{W}_{p,q}$. Let ψ be an accumulation point of the family $(\psi_m : m = 1, 2, \dots)$ of measures.

Let $\mathbf{e}$ be a unit vector of $\mathbb{Z}^d$. By (4.43), for any event A,

$$\left|\psi_m(A) - \tau_{\mathbf{e}} \circ \psi_m(A)\right| \leq \frac{|\partial\Delta_m|}{|\Delta_m|} \to 0 \qquad \text{as } m \to \infty, \tag{4.44}$$

whence ψ is $\tau_{\mathbf{e}}$-invariant. Certainly $\psi \in \overline{\mathrm{co}\,\mathcal{W}_{p,q}}$, and the proof of (a) is complete.
(b) This follows by Theorem 4.17(b).
(c) If $p = 0$ (respectively, $p = 1$), then ϕ is concentrated on the configuration 0 (respectively, 1), and the claim holds trivially. If $p \in (0, 1)$, it follows from (b) and the main theorem of [72]. See also the footnote on page 79. □

Proof of Theorem 4.31. The claim is trivial when $p = 0, 1$, and we assume that $p \in (0, 1)$. The proof is straightforward under the stronger hypothesis that $\phi \in \mathcal{W}_{p,q}$, and we begin with this special case. Suppose that $\mathbf{\Lambda} = (\Lambda_n : n = 1, 2, \dots)$, $\xi \in \Omega$, and $\phi \in \mathcal{W}_{p,q}$ are such that

$$\phi = \lim_{n\to\infty} \phi^{\xi}_{\Lambda_n,p,q},$$

and assume that ϕ has the 0/1-infinite-cluster property. Let Λ be a box and let $A \in \mathcal{F}_\Lambda$. By Lemma 4.13,

$$\text{(4.45)} \qquad \text{if } \Lambda \subseteq \Lambda_n, \quad \phi^{\omega}_{\Lambda,p,q}(A) = \phi^{\xi}_{\Lambda_n,p,q}(A \mid \mathcal{T}_\Lambda)(\omega) \quad \text{for } \phi^{\xi}_{\Lambda_n,p,q}\text{-a.e. } \omega.$$

Let B be a cylinder event in $\mathcal{T}_\Lambda$. By Theorem 4.33(b) and Lemma 4.39 applied to the measure ϕ, the function $1_B(\omega)\phi^{\omega}_{\Lambda,p,q}(A)$ is ϕ-almost-surely continuous, whence

$$\begin{aligned}
\phi\big(1_B(\cdot)\phi^{\cdot}_{\Lambda,p,q}(A)\big) &= \lim_{n\to\infty} \phi^{\xi}_{\Lambda_n,p,q}\big(1_B(\cdot)\phi^{\cdot}_{\Lambda,p,q}(A)\big) \\
&= \lim_{n\to\infty} \phi^{\xi}_{\Lambda_n,p,q}\big(1_B(\cdot)\phi^{\xi}_{\Lambda_n,p,q}(A \mid \mathcal{T}_\Lambda)\big) && \text{by (4.45)} \\
&= \lim_{n\to\infty} \phi^{\xi}_{\Lambda_n,p,q}(A \cap B) \\
&= \phi(A \cap B).
\end{aligned}$$

Since $\mathcal{T}_\Lambda$ is generated by its cylinder events, we deduce that

$$\text{(4.46)} \qquad \phi(A \mid \mathcal{T}_\Lambda) = \phi^{\cdot}_{\Lambda,p,q}(A) \qquad \phi\text{-a.s.},$$

whence $\phi \in \mathcal{R}_{p,q}$.

We require a further lemma for the general case. Let $X : \Omega \to \mathbb{R}$ be a bounded random variable, set

$$v(X) = \sup_{\omega,\omega' \in \Omega} |X(\omega) - X(\omega')|,$$

and let D_X be the discontinuity set of X, that is,

$$\text{(4.47)} \qquad D_X = \big\{\omega \in \Omega : X \text{ is discontinuous at } \omega\big\}.$$

(4.48) Lemma. *Let μ_n, μ be probability measures on $(\Omega, \mathcal{F})$ such that $\mu_n \Rightarrow \mu$ as $n \to \infty$. For any bounded random variable $X : \Omega \to \mathbb{R}$,*

$$\limsup_{n\to\infty} |\mu_n(X) - \mu(X)| \le v(X)\mu(D_X).$$

Proof. By [107, Thm 11.7.2], there exists a probability space $(\Sigma, \mathcal{G}, \mathbb{P})$ and random variables $\rho_n, \rho : \Sigma \to \Omega$ such that: ρ_n has law μ_n, ρ has law μ, and $\rho_n \to \rho$ almost surely. Therefore,

$$X(\rho_n)1_C(\rho) \to X(\rho)1_C(\rho) \qquad \mathbb{P}\text{-a.s.},$$

where $C = \Omega \setminus D_X$. By the bounded convergence theorem,

$$\begin{aligned}
|\mu_n(X) - \mu(X)| &= |\mathbb{P}(X(\rho_n) - X(\rho))| \\
&\le \mathbb{P}|X(\rho_n) - X(\rho)| \\
&= \mathbb{P}\big(|X(\rho_n) - X(\rho)|1_C(\rho)\big) + \mathbb{P}\big(|X(\rho_n) - X(\rho)|1_{\overline{C}}(\rho)\big) \\
&\le \mathbb{P}\big(|X(\rho_n) - X(\rho)|1_C(\rho)\big) + v(X)\mathbb{P}(1_{\overline{C}}(\rho)) \\
&\to 0 + v(X)\mu(\overline{C}) = v(X)\mu(D_X) \qquad \text{as } n \to \infty. \qquad \square
\end{aligned}$$

Let $\phi \in \overline{\text{co}\, \mathcal{W}_{p,q}}$ have the 0/1-infinite-cluster property, and write ϕ as $\phi = \lim_{n\to\infty} \phi_n$ where

$$\phi_n = \frac{1}{K_n}\sum_{i=1}^{K_n} \phi_{n,i}, \qquad \phi_{n,i} = \lim_{\Delta\uparrow\mathbb{Z}^d} \phi_{\Delta,p,q}^{\xi_{n,i}}. \tag{4.49}$$

The latter is actually a shorthand, since Δ will in general approach $\mathbb{Z}^d$ along some sequence of boxes which depends on the values of n and i, but this will not be important in what follows.

Let Λ be a box, and let $A \in \mathcal{F}_\Lambda$. Let B be a cylinder event in $\mathcal{T}_\Lambda$. Since $\mathcal{F}_\Lambda$ are $\mathcal{T}_\Lambda$ are generated by the classes of such cylinders, it is enough to prove that

$$\phi\big(1_B(\cdot)\phi^{\cdot}_{\Lambda,p,q}(A)\big) = \phi(A \cap B). \tag{4.50}$$

Let $D_{\Lambda,\Delta}$ be the event given after (4.40), noting as before that

$$D_{\Lambda,\Delta} \downarrow \{I_\Lambda \ge 2\} \qquad \text{as } \Delta \uparrow \mathbb{Z}^d, \tag{4.51}$$

where I_Λ is the number of infinite open clusters of $\mathbb{E}^d \setminus \mathbb{E}_\Lambda$ that intersect $\partial\Lambda$.

By (4.49) and Lemma 4.48,

$$\limsup_{\Delta\uparrow\mathbb{Z}^d} \big|\phi_{n,i}(1_B\phi^{\cdot}_{\Lambda,p,q}(A)) - \phi^{\xi_{n,i}}_{\Delta,p,q}(1_B\phi^{\cdot}_{\Lambda,p,q}(A))\big| \le \phi_{n,i}(I_\Lambda \ge 2), \tag{4.52}$$

as in (4.41). By Lemma 4.13,

$$\phi^{\xi_{n,i}}_{\Delta,p,q}(1_B\phi^{\cdot}_{\Lambda,p,q}(A)) = \phi^{\xi_{n,i}}_{\Delta,p,q}(A \cap B) \qquad \text{for all large } \Delta,$$

and therefore, on taking the limit as $\Delta \uparrow \mathbb{Z}^d$,

$$\big|\phi_{n,i}(1_B\phi^{\cdot}_{\Lambda,p,q}(A)) - \phi_{n,i}(A \cap B)\big| \le \phi_{n,i}(I_\Lambda \ge 2). \tag{4.53}$$

By (4.49) and Lemma 4.39,

$$\phi(1_B\phi^{\cdot}_{\Lambda,p,q}(A)) = \lim_{n\to\infty} \frac{1}{K_n}\sum_{i=1}^{K_n} \phi_{n,i}(1_B\phi^{\cdot}_{\Lambda,p,q}(A)),$$

$$\phi(A \cap B) = \lim_{n\to\infty} \frac{1}{K_n}\sum_{i=1}^{K_n} \phi_{n,i}(A \cap B),$$

whence

$$\begin{aligned}
&\big|\phi(1_B\phi^{\cdot}_{\Lambda,p,q}(A)) - \phi(A \cap B)\big| \\
&\qquad \le \limsup_{n\to\infty} \frac{1}{K_n}\sum_{i=1}^{K_n} \big|\phi_{n,i}(1_B\phi^{\cdot}_{\Lambda,p,q}(A)) - \phi_{n,i}(A \cap B)\big| \\
&\qquad \le \limsup_{n\to\infty} \frac{1}{K_n}\sum_{i=1}^{K_n} \phi_{n,i}(I_\Lambda \ge 2) \qquad \text{by (4.53)} \\
&\qquad \le \limsup_{n\to\infty} \frac{1}{K_n}\sum_{i=1}^{K_n} \phi_{n,i}(D_{\Lambda,\Delta}) \qquad \text{if } \Delta \supseteq \Lambda, \text{ by (4.51)} \\
&\qquad = \limsup_{n\to\infty} \phi_n(D_{\Lambda,\Delta}) \\
&\qquad = \phi(D_{\Lambda,\Delta}) \\
&\qquad \to \phi(I_\Lambda \ge 2) \qquad \text{as } \Delta \to \mathbb{Z}^d.
\end{aligned}$$

The final probability equals 0 as in (4.42), and therefore (4.50) holds. □

Proof of Theorem 4.34. (a) By Theorem 4.33, there exists $\phi \in \overline{\text{co}\,\mathcal{W}_{p,q}}$ with the 0/1-infinite-cluster property. By Theorem 4.31, $\phi \in \mathcal{R}_{p,q}$. Convexity follows immediately from Definition 4.29: for $\phi, \psi \in \mathcal{R}_{p,q}$ and $\alpha \in [0, 1]$, the measure $\alpha\phi + (1-\alpha)\psi$ satisfies the condition of the definition.
(b) Assume $q \in [1, \infty)$. By Theorem 4.19(b) the $\phi^b_{p,q}$ are translation-invariant, whence by Theorem 4.33(c) they have the 0/1-infinite-cluster property. By Theorem 4.31, each belongs to $\mathcal{R}_{p,q}$. Inequality (4.35) follows from Lemma 4.14(b) and Definition 4.29, on taking the limit as $\Lambda \to \mathbb{Z}^d$.

(c) The $\phi_{p,q}^b$ are tail-trivial by Theorem 4.19(d), and tail-triviality is equivalent to extremality, see [134, Thm 7.7]. There is a more direct proof using the stochastic ordering of part (b). If $\phi_{p,q}^0$ is not extremal, it may be written in the form $\phi_{p,q}^0 = \alpha\phi_1 + (1-\alpha)\phi_2$ for some $\alpha \in (0,1)$ and $\phi_1, \phi_2 \in \mathcal{R}_{p,q}$. For any increasing cylinder event A, $\phi_{p,q}^0(A) \le \min\{\phi_1(A), \phi_2(A)\}$ by (4.35), in contradiction of the above. A similar argument holds for $\phi_{p,q}^1$. □

Proof of Proposition 4.37. (a) This is a consequence of Definition 4.29 in conjunction with (3.3).

(b) Let ϕ satisfy (4.38) for all $e \in E$, and let Λ be a finite box. For ϕ-almost-every $\xi \in \Omega$, the conditional measure $\mu^\xi(\cdot) = \phi(\cdot \mid \mathcal{T}_\Lambda)(\xi)$ may be thought of as a probability measure on the finite set $\Omega_\Lambda = \{0,1\}^{\mathbb{E}_\Lambda}$ with an appropriate boundary condition. By (4.38) and Theorem 3.1(b), $\mu^\xi = \phi_{\Lambda,p,q}^\xi$ for ϕ-almost-every ξ, whence (4.30) holds and the claim follows.

(c) Let $q \in [1,\infty)$, and let $X, Y : \Omega \to \mathbb{R}$ be increasing, continuous random variables. For $\phi \in \mathcal{R}_{p,q}$,

$$\begin{aligned}
\phi(XY) &= \phi\big(\phi(XY \mid \mathcal{T}_\Lambda)\big) \\
&= \phi\big(\phi_{\Lambda,p,q}^{\cdot}(XY)\big) \\
&\ge \phi\big(\phi_{\Lambda,p,q}^{\cdot}(X)\phi_{\Lambda,p,q}^{\cdot}(Y)\big) && \text{by positive association} \\
&= \phi\big(\phi(X \mid \mathcal{T}_\Lambda)\phi(Y \mid \mathcal{T}_\Lambda)\big) \\
&\to \phi\big(\phi(X \mid \mathcal{T})\phi(Y \mid \mathcal{T})\big) && \text{as } \Lambda \uparrow \mathbb{Z}^d,
\end{aligned}$$

by the bounded convergence theorem and the backward martingale convergence theorem [107, Thm 10.6.1]. If ϕ is tail-trivial,

$$\phi(X \mid \mathcal{T}) = \phi(X), \quad \phi(Y \mid \mathcal{T}) = \phi(Y), \qquad \phi\text{-a.s.},$$

and the required positive-association inequality follows. □

4.5 Uniqueness via convexity of pressure

We address next the question of the uniqueness of limit- and DLR-random-cluster measures on $\mathbb{L}^d$ for given p and q. The main result of this section is the following. There exists a (possibly empty) countable subset $\mathcal{D}_q$ of the interval $[0,1]$ such that $\phi_{p,q}^0 = \phi_{p,q}^1$, and hence there exists a unique random-cluster measure in that $|\mathcal{W}_{p,q}| = |\mathcal{R}_{p,q}| = 1$, if and only if $p \notin \mathcal{D}_q$. Further results concerning the uniqueness of measures may be found at Theorems 5.33, 6.17, and 7.33.

The 'almost everywhere' uniqueness of random-cluster measures will be proved by showing that the asymptotic behaviour of the logarithm of the partition function does not depend on the choice of boundary condition, and then by relating the differentiability of the limit function to the uniqueness of measures. A certain convexity property of the limit function will play a role in studying its differentiability. Rather than working with the usual partition function $Z_\Lambda^\xi(p,q)$ of (4.12), we shall use the function $Y_\Lambda^\xi : \mathbb{R}^2 \to \mathbb{R}$ given by

$$Y_\Lambda^\xi(\pi,\kappa) = \sum_{\omega\in\Omega_\Lambda^\xi} \exp\{\pi|\mathbb{E}_\Lambda \cap \eta(\omega)| + \kappa k(\omega,\Lambda)\}, \tag{4.54}$$

and satisfying

$$Z_\Lambda^\xi(p,q) = (1-p)^{|\mathbb{E}_\Lambda|} Y_\Lambda^\xi(\pi,\kappa), \tag{4.55}$$

where $\pi = \pi(p)$ and $\kappa = \kappa(q)$ are given by

$$\pi(p) = \log\left(\frac{p}{1-p}\right), \qquad \kappa(q) = \log q. \tag{4.56}$$

Note that

$$Z_\Lambda^\xi(p,1) = 1, \qquad Y_\Lambda(\pi,0) = (1-p)^{-|\mathbb{E}_\Lambda|}. \tag{4.57}$$

We introduce next a function $G(\pi,\kappa)$ which describes the exponential asymptotics of $Y_\Lambda^\xi(\pi,\kappa)$ as $\Lambda \uparrow \mathbb{Z}^d$. In line with the terminology of statistical mechanics, we call this function the *pressure*. All logarithms will for convenience be natural logarithms.

(4.58) Theorem [145, 150, 152]. *Let Λ be a box of $\mathbb{L}^d$. The finite limits*

$$G(\pi,\kappa) = \lim_{\Lambda\uparrow\mathbb{Z}^d}\left\{\frac{1}{|\mathbb{E}_\Lambda|}\log Y_\Lambda^\xi(\pi,\kappa)\right\}, \qquad (\pi,\kappa)\in\mathbb{R}^2, \tag{4.59}$$

exist and are independent of $\xi \in \Omega$ and of the way in which $\Lambda \uparrow \mathbb{Z}^d$. The 'pressure' function G is a convex function on its domain $\mathbb{R}^2$.

In the proof, we shall see that G may be approximated from below and above to any required degree of accuracy by smooth functions of (π,κ), see (4.68)–(4.70).

We shall identify the set $\mathcal{D}_q$ mentioned at the start of this section as $\mathcal{D}_q = \mathcal{D}'_{\kappa(q)}$, a set given in the next theorem with $\kappa(q) = \log q$. This set corresponds to the points of non-differentiability of the convex function G. Recall that, by convexity, G is differentiable at (π,κ) if and only if G has both its partial derivatives at this point.

Let $\mathcal{D}'$ be the set of all (π,κ) at which G is not differentiable when viewed as a function from $\mathbb{R}^2$ to $\mathbb{R}$. Since G is convex, $\mathcal{D}'$ has Lebesgue measure 0, and indeed $\mathcal{D}'$ may be covered by a countable collection of rectifiable curves (see [115, Thm 8.18], [291, Thm 2.2.4]). For any line l of $\mathbb{R}^2$, the restriction of G to l is convex, whence G restricted to l is differentiable along l except at countably many points. Each such point of non-differentiability on l lies in $\mathcal{D}'$, but the converse may not generally be true.

The two partial derivatives of G have special physical significance for the random-cluster model, and one may show when $q > 1$ (that is, $\kappa > 0$) that G has one partial derivative at any given point (π,κ) if and only if it has both.

(4.60) Theorem.

(a) *For each $\kappa \in \mathbb{R}$, there exists a (possibly empty) countable subset $\mathcal{D}'_\kappa$ of reals such that $G(\pi, \kappa)$ is a differentiable function of π if and only if $\pi \notin \mathcal{D}'_\kappa$.*

(b) *For each $\pi \in \mathbb{R}$, there exists a (possibly empty) countable subset $\mathcal{D}''_\pi$ of reals such that $G(\pi, \kappa)$ is a differentiable function of κ if and only if $\kappa \notin \mathcal{D}''_\pi$.*

(c) *For $(\pi, \kappa) \in \mathbb{R} \times (0, \infty)$, exactly one of the following holds:*

  (i) *$(\pi, \kappa) \in \mathcal{D}'$, and G has neither partial derivative at (π, κ),*

  (ii) *$(\pi, \kappa) \notin \mathcal{D}'$, and G has both partial derivatives at (π, κ).*

Parts (a) and (b) follow from the remarks prior to the theorem. The proof of part (c) is deferred until later in this section. With $\mathcal{D}'_\kappa$ given in (a), we write $\mathcal{D}_q = \mathcal{D}'_{\kappa(q)}$.

For given $q \in (0, \infty)$, one thinks of $\mathcal{D}_q = \mathcal{D}'_{\kappa(q)}$ as the set of 'bad' values of p. The situation when $q \in (0, 1)$ is obscure. When $q \in (1, \infty)$, the set $\mathcal{D}_q$ is exactly the set of singularities of the random-cluster model in the sense of the next theorem. Here is some further notation. Let $q \in [1, \infty)$, and

$$h^b(p, q) = \phi^b_{p,q}(J_e), \qquad b = 0, 1, \tag{4.61}$$

where J_e is the event that e is open. Since the $\phi^b_{p,q}$ are automorphism-invariant[7], $h^b(p, q)$ does not depend on the choice of e, and therefore equals the edge-density under $\phi^b_{p,q}$. We write

$$F(p, q) = G(\pi, \kappa) \tag{4.62}$$

where (p, q) and (π, κ) are related by (4.56), and G is given in (4.59). We shall use the word 'pressure' for both F and G.

(4.63) Theorem. *Let $p \in (0, 1)$ and $q \in (1, \infty)$. The following five statements are equivalent.*

(a) *$p \notin \mathcal{D}_q$.*

(b) (i) *$h^0(x, q)$ is a continuous function of x at the point $x = p$.*

  (ii) *$h^1(x, q)$ is a continuous function of x at the point $x = p$.*

(c) *It is the case that $h^0(p, q) = h^1(p, q)$.*

(d) *There is a unique random-cluster measure with parameters p and q, that is, $|\mathcal{W}_{p,q}| = |\mathcal{R}_{p,q}| = 1$.*

What is the set $\mathcal{D}_q$? We shall return to this question in Section 5.3, but in the meantime we summarize the anticipated situation. Let $d \geq 2$ be given, and assume $q \in [1, \infty)$. It is thought to be the case that $\mathcal{D}_q$ is empty when $q - 1$ is

[7] There is an error in [152, Thm 4.5] in the case $q \in (0, 1)$. The correct condition there is that the measure ϕ be automorphism-invariant rather than translation-invariant.

small, and is a singleton point (that is, the critical value $p_c(q)$, see Section 5.1) when q is large. It is conjectured that there exists $Q = Q(d) > 1$ such that

$$\mathcal{D}_q = \begin{cases} \varnothing & \text{if } q \le Q, \\ \{p_c(q)\} & \text{if } q > Q. \end{cases} \tag{4.64}$$

This would imply in particular that $|\mathcal{R}_{p,q}| = 1$ unless $q > Q$ and $p = p_c(q)$. A further issue concerns the structure of $\mathcal{R}_{p,q}$ in situations where $|\mathcal{R}_{p,q}| > 1$. For further information about the *non-uniqueness* of random-cluster measures, the reader is directed to Sections 6.4 and 7.5.

Proof of Theorem 4.58. Let $p \in (0,1)$ and $q \in (0,\infty)$, and let (π, κ) be given by (4.56). We shall use a standard argument of statistical mechanics, namely the near-multiplicativity of $Y_\Lambda^\xi(\pi,\kappa)$ viewed as a function of Λ. The irrelevance to the limit of the boundary condition ξ hinges on the fact that $|\partial\Lambda|/|\Lambda| \to \infty$ as $\Lambda \uparrow \mathbb{Z}^d$.

We show first that the limit (4.59) exists with $\xi = 0$, and shall for the moment suppress explicit reference to the boundary condition. Let $\mathbf{n} = (n_1, n_2, \dots, n_d) \in \mathbb{N}^d$, write $|\mathbf{n}| = n_1 n_2 \cdots n_d$, and let $\Lambda_\mathbf{n}$ be the box $\prod_{i=1}^d [1, n_i]$. By the translation-invariance of $Z_\Lambda(p,q)$, we may restrict ourselves to boxes of this type.

We fix $\mathbf{k} \in \mathbb{N}^d$, and write

$$(\mathbf{n},\mathbf{k}) = \left(k_i \left\lfloor \frac{n_i}{k_i} \right\rfloor : i = 1,2,\dots,d\right), \qquad \left\lfloor \frac{\mathbf{n}}{\mathbf{k}} \right\rfloor = \prod_{i=1}^d \left\lfloor \frac{n_i}{k_i} \right\rfloor.$$

By Theorem 3.63, for $\mathbf{n} \ge \mathbf{k}$,

$$\begin{aligned} \left\lfloor \frac{\mathbf{n}}{\mathbf{k}} \right\rfloor & \left[\log Z_{\Lambda_\mathbf{k}} - d\left(\sum_{i=1}^d \frac{|\mathbf{k}|}{k_i}\right) \log(1 \vee q) \right] + \log Z_{\Lambda_\mathbf{n} \setminus \Lambda_{(\mathbf{n},\mathbf{k})}} \\ & \le \log Z_{\Lambda_\mathbf{n}} \\ & \le \left\lfloor \frac{\mathbf{n}}{\mathbf{k}} \right\rfloor \left[\log Z_{\Lambda_\mathbf{k}} - d\left(\sum_{i=1}^d \frac{|\mathbf{k}|}{k_i}\right) \log(1 \wedge q) \right] + \log Z_{\Lambda_\mathbf{n} \setminus \Lambda_{(\mathbf{n},\mathbf{k})}}, \end{aligned} \tag{4.65}$$

and furthermore,

$$\begin{aligned} \big(|\mathbf{n}| - |\mathbf{k}| \cdot \lfloor \mathbf{n}/\mathbf{k} \rfloor\big) \log\left(\frac{q}{(1 \vee q)^d}\right) & \le \log Z_{\Lambda_\mathbf{n} \setminus \Lambda_{(\mathbf{n},\mathbf{k})}} \\ & \le \big(|\mathbf{n}| - |\mathbf{k}| \cdot \lfloor \mathbf{n}/\mathbf{k} \rfloor\big) \log\left(\frac{q}{(1 \wedge q)^d}\right). \end{aligned}$$

Divide by $|\mathbf{n}|$ and let the n_i tend to ∞ to find that, for all $\mathbf{k}$,

$$
\begin{aligned}
(4.66)\quad & \frac{1}{|\mathbf{k}|}\log Z_{\Lambda_\mathbf{k}} - d\left(\sum_{i=1}^{d}\frac{1}{k_i}\right)\log(1\vee q) \\
&\qquad \le \liminf_{\mathbf{n}\to\infty}\left\{\frac{1}{|\mathbf{n}|}\log Z_{\Lambda_\mathbf{n}}\right\} \le \limsup_{\mathbf{n}\to\infty}\left\{\frac{1}{|\mathbf{n}|}\log Z_{\Lambda_\mathbf{n}}\right\} \\
&\qquad \le \frac{1}{|\mathbf{k}|}\log Z_{\Lambda_\mathbf{k}} - d\left(\sum_{i=1}^{d}\frac{1}{k_i}\right)\log(1\wedge q).
\end{aligned}
$$

Assume that $q \ge 1$, a similar argument is valid when $q < 1$. Therefore, the limit

$$
H(p,q) = \lim_{\mathbf{n}\to\infty}\left\{\frac{1}{|\mathbf{n}|}\log Z_{\Lambda_\mathbf{n}}\right\} \tag{4.67}
$$

exists, and furthermore

$$
\begin{aligned}
(4.68)\quad H(p,q) &= \sup_{\mathbf{k}}\left\{\frac{1}{|\mathbf{k}|}\log Z_{\Lambda_\mathbf{k}} - d\left(\sum_{i=1}^{d}\frac{1}{k_i}\right)\log(1\vee q)\right\} \\
&= \inf_{\mathbf{k}}\left\{\frac{1}{|\mathbf{k}|}\log Z_{\Lambda_\mathbf{k}} - d\left(\sum_{i=1}^{d}\frac{1}{k_i}\right)\log(1\wedge q)\right\}.
\end{aligned}
$$

Since $Z_{\Lambda_\mathbf{k}}$ is a continuous function of p and q, these equations imply that $H(p,q)$ may be approximated from below and above to any degree of accuracy by continuous functions, and is therefore continuous. We will obtain greater regularity from the claim of convexity to be proved soon. Evidently, as $\Lambda \uparrow \mathbb{Z}^d$,

$$
\frac{1}{|\mathbb{E}_\Lambda|}\log Z_\Lambda \to \frac{1}{d}H(p,q), \tag{4.69}
$$

and, by (4.55) and (4.62),

$$
\begin{aligned}
(4.70)\quad \frac{1}{|\mathbb{E}_\Lambda|}\log Y_\Lambda(\pi,\kappa) &\to -\log(1-p) + \frac{1}{d}H(p,q) \\
&= F(p,q) = G(\pi,\kappa).
\end{aligned}
$$

We show next that the same limit is valid with a general boundary condition. Let Λ be a finite box, and let

$$
G_\Lambda^\xi(\pi,\kappa) = \frac{1}{|\mathbb{E}_\Lambda|}\log Y_\Lambda^\xi(\pi,\kappa). \tag{4.71}
$$

For $\omega, \xi \in \Omega$, let $\omega_\Lambda^\xi \in \Omega_\Lambda^\xi$ be as in (4.2). Clearly,

$$
k(\omega_\Lambda^0,\Lambda) - |\partial\Lambda| \le k(\omega_\Lambda^1,\Lambda) \le k(\omega_\Lambda^\xi,\Lambda) \le k(\omega_\Lambda^0,\Lambda),
$$

whence

$$Y_\Lambda^0(\pi,\kappa)e^{-\kappa|\partial\Lambda|} \le Y_\Lambda^\xi(\pi,\kappa) \le Y_\Lambda^0(\pi,\kappa), \qquad \kappa \in [0,\infty),$$

and the same holds with the inequalities reversed when $\kappa \in (-\infty, 0)$. Therefore,

$$G_\Lambda^0(\pi,\kappa) - \kappa\frac{|\partial\Lambda|}{|\mathbb{E}_\Lambda|} \le G_\Lambda^\xi(\pi,\kappa) \le G_\Lambda^0(\pi,\kappa), \qquad \kappa \in [0,\infty),$$

and with the inequalities reversed if $\kappa \in (-\infty, 0)$. Since $|\partial\Lambda|/|\mathbb{E}_\Lambda| \to 0$ as $\Lambda \uparrow \mathbb{Z}^d$, the limit of G_Λ^ξ exists by (4.70), and is independent of the choice of ξ.

It is clear from its form that $G_\Lambda^\xi(\pi,\kappa)$ is a convex function on its domain $\mathbb{R}^2$. Indeed, Theorem 3.73(b) includes a representation of its second derivative in an arbitrary given direction as the variance of a random variable. We note from Theorem 3.73(a) for later use that

$$\nabla G_\Lambda^\xi(\pi,\kappa) = \frac{1}{|\mathbb{E}_\Lambda|}\big(\phi_{\Lambda,p,q}^\xi(|\eta(\omega)\cap\mathbb{E}_\Lambda|), \phi_{\Lambda,p,q}^\xi(k(\omega,\Lambda))\big). \tag{4.72}$$

Since, for any $\xi \in \Omega$, the $G_\Lambda^\xi(\pi,\kappa)$ are convex functions of (π,κ) which converge to the finite limit function $G(\pi,\kappa)$ as $\Lambda \uparrow \mathbb{Z}^d$, G is convex on $\mathbb{R}^2$. □

Proof of Theorem 4.63.
(c) $\iff$ (d). By (4.36), $|\mathcal{W}_{p,q}| = |\mathcal{R}_{p,q}| = 1$ if and only if $\phi_{p,q}^0 = \phi_{p,q}^1$. By Proposition 4.6 and the fact that $\phi_{p,q}^0 \le_{\mathrm{st}} \phi_{p,q}^1$, $\phi_{p,q}^0 = \phi_{p,q}^1$ if and only if $h^0(p,q) = h^1(p,q)$. Therefore, (c) and (d) are equivalent.
(a) $\iff$ (b) $\iff$ (c). This is inspired by a related computation for the Ising model, [233]. Let $p \in (0,1)$, $q \in (1,\infty)$, and let (π,κ) satisfy (4.56). Recall the functions G_Λ^ξ given in (4.71), and note from (4.72) that

$$\frac{dG_\Lambda^\xi}{d\pi} = \frac{1}{|\mathbb{E}_\Lambda|}\phi_{\Lambda,p,q}^\xi(|\eta(\omega)\cap\mathbb{E}_\Lambda|). \tag{4.73}$$

Since G is convex, $\mathcal{D}_q$ is countable. By the convexity of the G_Λ^ξ,

$$\frac{dG_\Lambda^\xi}{d\pi} \to \frac{dG}{d\pi} \quad \text{as } \Lambda \uparrow \mathbb{Z}^d, \qquad \xi \in \Omega,\ p \notin \mathcal{D}_q. \tag{4.74}$$

For any box Λ and any edge $e \in \mathbb{E}_\Lambda$,

$$\begin{aligned}\frac{1}{|\mathbb{E}_\Lambda|}\phi_{\Lambda,p,q}^0(|\eta(\omega)\cap\mathbb{E}_\Lambda|) &\le \phi_{p,q}^0(J_e)\\ &\le \phi_{p,q}^1(J_e)\\ &\le \frac{1}{|\mathbb{E}_\Lambda|}\phi_{\Lambda,p,q}^1(|\eta(\omega)\cap\mathbb{E}_\Lambda|),\end{aligned} \tag{4.75}$$

where we have used the automorphism-invariance of $\phi^0_{p,q}$ and $\phi^1_{p,q}$, together with the stochastic ordering of measures. We deduce on passing to the limit as $\Lambda \uparrow \mathbb{Z}^d$ that

$$\frac{dG}{d\pi} = \phi^0_{p,q}(J_e) = \phi^1_{p,q}(J_e), \qquad e \in \mathbb{E}^d,\ p \notin \mathcal{D}_q. \tag{4.76}$$

In particular, (a) implies (c).

Since $G(\pi, \kappa)$ is a convex function of π, it has right and left derivatives with respect to π, denoted respectively by $dG/d\pi^{\pm}$. Furthermore, $dG/d\pi^+$ (respectively, $dG/d\pi^-$) is right-continuous (respectively, left-continuous) and non-decreasing. We shall prove that

$$\frac{dG}{d\pi^+} - \frac{dG}{d\pi^-} = \phi^1_{p,q}(J_e) - \phi^0_{p,q}(J_e), \tag{4.77}$$

and that

$$\phi^1_{p,q}(J_e) = \lim_{p' \downarrow p} \phi^0_{p',q}(J_e), \quad \phi^0_{p,q}(J_e) = \lim_{p' \uparrow p} \phi^1_{p',q}(J_e). \tag{4.78}$$

In advance of proving (4.77) and (4.78), we note the following. By (4.77), (a) and (c) are equivalent. By (4.77)–(4.78), the following three statements are equivalent for any given π:

1. $p \notin \mathcal{D}_q$,
2. $h^0(x, q)$ is right-continuous at $x = p$,
3. $h^1(x, q)$ is left-continuous at $x = p$.

By Proposition 4.28, $h^0(\cdot, q)$ (respectively, $h^1(\cdot, q)$) is left-continuous (respectively, right-continuous), and therefore (a) is equivalent to each of (b)(i) and (b)(ii).

It remains to prove (4.77) and (4.78). We concentrate first on the first equation of (4.78). By Proposition 4.28(b), $h^1(\cdot, q)$ is right-continuous, whence

$$h^1(p, q) = \lim_{p' \downarrow p} h^1(p', q).$$

Now $\mathcal{D}_q$ is countable, whence $\phi^0_{p',q} = \phi^1_{p',q}$, and in particular $h^0(p', q) = h^1(p', q)$, for almost every p'. By the monotonicity of $h^0(\cdot, q)$,

$$h^1(p, q) = \lim_{p' \downarrow p} h^0(p', q).$$

as required. The second equation of (4.78) holds by a similar argument.

By the semicontinuity of the $dG/d\pi^{\pm}$, (4.76), and Proposition 4.28,

$$\frac{dG}{d\pi^+} = \lim_{\substack{x \downarrow \pi \\ x \notin \mathcal{D}'_\kappa}} \frac{d}{dx} G(x, \kappa) = \phi^1_{p,q}(J_e),$$

$$\frac{dG}{d\pi^-} = \lim_{\substack{x \uparrow \pi \\ x \notin \mathcal{D}'_\kappa}} \frac{d}{dx} G(x, \kappa) = \phi^0_{p,q}(J_e),$$

and (4.77) follows. □

Proof of Theorem 4.60. Parts (a) and (b) follow by the remarks prior to the statement of the theorem, and we turn to part (c). Recall first that G is differentiable at (π, κ), that is $(\pi, \kappa) \notin \mathcal{D}'$, if and only if G possesses both partial derivatives at (π, κ). It remains to show therefore that, for $\kappa \in (0, \infty)$, $\pi \in \mathcal{D}'_\kappa$ if and only if $\kappa \in \mathcal{D}''_\pi$. Let $\kappa \in (0, \infty)$. Since, by Theorem 4.63, $\mathcal{D}'_\kappa$ is exactly the set of $\pi = \pi(p)$ such that $\phi^0_{p,q} \neq \phi^1_{p,q}$, it suffices to show the following.

(4.79) Lemma. *Let* $p \in (0, 1)$, $q \in (1, \infty)$, *and let* (π, κ) *satisfy* (4.56). *Then* $\kappa \in \mathcal{D}''_\pi$ *if and only if* $\phi^0_{p,q} \neq \phi^1_{p,q}$.

Proof. The function G^ξ_Λ of (4.71) is convex in κ, whence

$$\frac{dG^\xi_\Lambda}{d\kappa} \to \frac{dG}{d\kappa} \quad \text{as } \Lambda \uparrow \mathbb{Z}^d, \qquad \xi \in \Omega,\ \kappa \notin \mathcal{D}''_\pi, \tag{4.80}$$

as in (4.74). Inequalities (4.75) become

$$\begin{aligned}\frac{1}{|\mathbb{E}_\Lambda|}\phi^0_{\Lambda,p,q}(k(\omega,\Lambda)) &\geq \frac{1}{|\mathbb{E}_\Lambda|}\phi^0_{p,q}(k(\omega,\Lambda))\\ &\geq \frac{1}{|\mathbb{E}_\Lambda|}\phi^1_{p,q}(k(\omega,\Lambda))\\ &\geq \frac{1}{|\mathbb{E}_\Lambda|}\phi^1_{\Lambda,p,q}(k(\omega,\Lambda)),\end{aligned} \tag{4.81}$$

since $k(\omega, \Lambda)$ is decreasing in ω. Therefore, by Theorem 3.73(a),

$$\begin{aligned}\frac{dG^0_\Lambda}{d\kappa} &\geq \phi^0_{p,q}\left(\frac{k(\omega,\Lambda)}{|\mathbb{E}_\Lambda|}\right)\\ &\geq \phi^1_{p,q}\left(\frac{k(\omega,\Lambda)}{|\mathbb{E}_\Lambda|}\right) \geq \frac{dG^1_\Lambda}{d\kappa}, \qquad \kappa \notin \mathcal{D}''_\pi.\end{aligned} \tag{4.82}$$

For $\omega \in \Omega$ and $x \in \mathbb{Z}^d$, let $C_x = C_x(\omega)$ be the ω-open cluster at x, and $|C_x|$ the number of its vertices. As in [154, Section 4.1],

$$k(\omega, \Lambda) = \sum_{x\in\Lambda} \frac{1}{|C_x \cap \Lambda|} \geq \sum_{x\in\Lambda} \frac{1}{|C_x|},$$

and

$$\begin{aligned}k(\omega,\Lambda) - \sum_{x\in\Lambda}\frac{1}{|C_x|} &= \sum_{x\in\Lambda}\left(\frac{1}{|C_x\cap\Lambda|} - \frac{1}{|C_x|}\right)\\ &\leq \sum_{\substack{x\in\Lambda:\\ x\leftrightarrow\partial\Lambda}} \frac{1}{|C_x\cap\Lambda|} \leq |\partial\Lambda|.\end{aligned}$$

The $\phi_{p,q}^b$ are τ-ergodic for all translations τ other than the identity. By the ergodic theorem applied to the family $\{|C_x|^{-1} : x \in \mathbb{Z}^d\}$ of bounded random variables,

$$\phi_{p,q}^b\left(\frac{k(\omega,\Lambda)}{|\Lambda|}\right) \to \phi_{p,q}^b(|C_0|^{-1}) \qquad \phi_{p,q}^b\text{-a.s. and in } L^1, \text{ as } \Lambda \uparrow \mathbb{Z}^d. \tag{4.83}$$

By (4.80), (4.82), and (4.83),

$$\frac{dG}{d\kappa} = \frac{1}{d}\phi_{p,q}^0(|C_0|^{-1}) = \frac{1}{d}\phi_{p,q}^1(|C_0|^{-1}), \qquad \kappa \notin \mathcal{D}_\pi''. \tag{4.84}$$

This implies by the next proposition that $\phi_{p,q}^0 = \phi_{p,q}^1$ for $\kappa \notin \mathcal{D}_\pi''$.

(4.85) Proposition. *Let $p \in (0,1)$ and $q \in [1,\infty)$. If*

$$\phi_{p,q}^0(|C_0|^{-1}) = \phi_{p,q}^1(|C_0|^{-1}) \tag{4.86}$$

then $\phi_{p,q}^0 = \phi_{p,q}^1$.

Proof. Suppose that (4.86) holds. There are two steps, in the first of which we show that the law of the vertex-set of C_0 is the same under $\phi_{p,q}^0$ and $\phi_{p,q}^1$. As in the proof of Proposition 4.6, there exists a probability measure μ on $(\Omega,\mathcal{F})\times(\Omega,\mathcal{F})$, with marginals $\phi_{p,q}^0$ and $\phi_{p,q}^1$, and such that

$$\mu\big(\{(\omega_0,\omega_1) \in \Omega^2 : \omega_0 \le \omega_1\}\big) = 1. \tag{4.87}$$

By (4.87), $C_x(\omega_0) \subseteq C_x(\omega_1)$ for all $x \in \mathbb{Z}^d$, μ-almost-surely. Let

$$E = \bigcap_{x\in\mathbb{Z}^d} \big\{(\omega_0,\omega_1) \in \Omega^2 : |C_x(\omega_0)|^{-1} = |C_x(\omega_1)|^{-1}\big\}.$$

By (4.86),

$$\mu(\overline{E}) \le \sum_{x\in\mathbb{Z}^d} \mu\big(|C_x(\omega_0)|^{-1} > |C_x(\omega_1)|^{-1}\big) = 0,$$

whence $\mu(E) = 1$.

If the vertex-set of $C_0(\omega_0)$ is a strict subset of that of $C_0(\omega_1)$, one of the two statements following must hold:

(i) $C_0(\omega_0)$ is finite and $|C_0(\omega_0)|^{-1} > |C_0(\omega_1)|^{-1}$,

(ii) $C_0(\omega_0)$ is infinite, $|C_0(\omega_0)|^{-1} = |C_0(\omega_1)|^{-1} = 0$, and there exists $x \in C_0(\omega_1) \setminus C_0(\omega_0)$.

By (4.86), the μ-probability of (i) is zero. By considering the two sub-cases of (ii) depending on whether $C_x(\omega_0)$ is finite or infinite, we find that the μ-probabiltiy of (ii) is no larger than

$$\mu(E) + \mu(I(\omega_0) \ge 2),$$

where $I(\omega)$ is the number of infinite open clusters of ω. By Theorem 4.33(c), $\mu(I(\omega_0) \geq 2) = \phi^0_{p,q}(I \geq 2) = 0$. We conclude as required that the vertex-sets of $C_0(\omega_0)$ and $C_0(\omega_1)$ are equal, μ-almost-surely. Therefore, by the translation-invariance of the $\phi^b_{p,q}$,

$$\phi^0_{p,q}(x \nleftrightarrow y) = \phi^1_{p,q}(x \nleftrightarrow y), \qquad x, y \in \mathbb{Z}^d. \tag{4.88}$$

We turn now to the second step. Let J_e be the event that edge $e = \langle x, y \rangle$ is open, and let K_e be the event that x and y are joined by an open path of $\mathbb{E}^d \setminus \{e\}$. By Proposition 4.37(a),

$$\begin{aligned} \phi^b_{p,q}(J_e) &= p\phi^b_{p,q}(K_e) + \frac{p}{p+q(1-p)}\phi^b_{p,q}(\overline{K_e}) \\ &= p + \left(\frac{p}{p+q(1-p)} - p\right)\phi^b_{p,q}(\overline{K_e}), \end{aligned}$$

and

$$\phi^b_{p,q}(\overline{K_e}) = \frac{\phi^b_{p,q}(\overline{J_e} \cap \overline{K_e})}{\phi^b_{p,q}(\overline{J_e} \mid \overline{K_e})} = \frac{\phi^b_{p,q}(x \nleftrightarrow y)}{q(1-p)/[p+q(1-p)]}.$$

Hence, by (4.88),

$$\phi^0_{p,q}(J_e) = \phi^1_{p,q}(J_e), \qquad e \in \mathbb{E}^d,$$

whence, by Proposition 4.6, $\phi^0_{p,q} = \phi^1_{p,q}$. □

We return now to the proof of Lemma 4.79. Suppose conversely that $\phi^0_{p,q} = \phi^1_{p,q}$, and let $q' < q < q''$. By Proposition 4.28(a) applied to the decreasing function $|C|^{-1}$,

$$\phi^1_{p,q'}(|C|^{-1}) \leq \phi^1_{p,q}(|C|^{-1}) = \phi^0_{p,q}(|C|^{-1}) \leq \phi^0_{p,q''}(|C|^{-1}).$$

Take the limits as $q' \uparrow q$ and $q'' \downarrow q$ along sequences satisfying $\kappa(q'), \kappa(q'') \notin \mathcal{D}''_\pi$, and use the monotonicity of these functions to find from (4.84) and Proposition 4.28 that

$$\frac{dG}{d\kappa^+} - \frac{dG}{d\kappa^-} = \frac{1}{d}\left[\phi^0_{p,q}(|C|^{-1}) - \phi^1_{p,q}(|C|^{-1})\right] = 0.$$

Therefore, G has the appropriate partial derivative at the point (π, κ), which is to say that $\kappa \notin \mathcal{D}''_\pi$ as required. □□

4.6 Potts and random-cluster models on infinite graphs

The random-cluster model provides a way to study the Potts model on finite graphs, as explained in Section 1.4. The method is valid for infinite graphs also, as summarized in this section in the context of the lattice $\mathbb{L}^d = (\mathbb{Z}^d, \mathbb{E}^d)$.

Let $p \in [0, 1)$, $q \in \{2, 3, \dots\}$, and $p = 1 - e^{-\beta}$ as usual, and consider the free and wired random-cluster measures, $\phi^0_{p,q}$ and $\phi^1_{p,q}$, respectively. The corresponding Potts measures on $\mathbb{L}^d$ are the free and '1' measures,

$$\pi_{\beta,q} = \lim_{\Lambda \uparrow \mathbb{Z}^d} \pi_{\Lambda,\beta,q}, \tag{4.89}$$

$$\pi^1_{\beta,q} = \lim_{\Lambda \uparrow \mathbb{Z}^d} \pi^1_{\Lambda,\beta,q}. \tag{4.90}$$

The measure $\pi_{\Lambda,\beta,q}$ is the Potts measure on Λ given in (1.5). The measure $\pi^1_{\Lambda,\beta,q}$ is the corresponding measure with '1' boundary conditions, given as in (1.5) but subject to the constraint that $\sigma_x = 1$ for all $x \in \partial\Lambda$. It is standard that the limits in (4.89)–(4.90) exist. Probably the easiest proof of this is to couple the Potts model with a random-cluster model on the same graph, and to use the stochastic monotonicity of the latter to prove the existence of the infinite-volume limit.

We explain this in the wired case, and a similar argument holds in the free case. Part (a) of the next theorem may be taken as the definition of the infinite-volume Potts measure $\pi^1_{\beta,q}$.

(4.91) Theorem [8].

(a) *Let ω be sampled from Ω with law $\phi^1_{p,q}$. Conditional on ω, each vertex $x \in \mathbb{Z}^d$ is assigned a random spin $\sigma_x \in \{1, 2, \dots, q\}$ in such a way that:*

(i) $\sigma_x = 1$ *if* $x \leftrightarrow \infty$,

(ii) σ_x *is uniformly distributed on* $\{1, 2, \dots, q\}$ *if* $x \not\leftrightarrow \infty$,

(iii) $\sigma_x = \sigma_y$ *if* $x \leftrightarrow y$,

(iv) $\sigma_{x_1}, \sigma_{x_2}, \dots, \sigma_{x_n}$ *are independent whenever* $x_1, x_2, \dots, x_n$ *are in different finite open clusters of* ω.

The law of the spin vector $\sigma = (\sigma_x : x \in \mathbb{Z}^d)$ *is denoted by* $\pi^1_{\beta,q}$ *and satisfies* (4.89).

(b) *Let σ be sampled from $\Sigma = \{1, 2, \dots, q\}^{\mathbb{Z}^d}$ with law $\pi^1_{\beta,q}$. Conditional on σ, each edge $e = \langle x, y\rangle \in \mathbb{E}^d$ is assigned a random state $\omega(e) \in \{0, 1\}$ in such a way that:*

(i) *the states of different edges are independent,*

(ii) $\omega(e) = 0$ *if* $\sigma_x \neq \sigma_y$,

(iii) *if* $\sigma_x = \sigma_y$, *then* $\omega(e) = 1$ *with probability* p,

The edge-configuration $\omega = (\omega(e) : e \in \mathbb{E}^d)$ *has law* $\phi^1_{p,q}$.

A similar theorem is valid for the pair $\phi^0_{p,q}, \pi_{\beta,q}$, with the difference that infinite open clusters are treated in the same way as finite clusters in part (a).

The Potts model has a useful property called 'reflection-positivity'. It is natural to ask whether a similar property is satisfied by general random-cluster measures. It was shown in [43] that the answer is negative for non-integer values of the parameter q.

Proof of Theorem 4.91. (a) Of the possible proofs we select one using coupling, another approach may be found in [142]. Let $\Lambda_n = [-n, n]^d$ and write $\Omega_n = \Omega^1_{\Lambda_n}$ and $\phi^1_n = \phi^1_{\Lambda_n,p,q}$. Let $\mathbf{\Omega}$ be the set of all vectors $\boldsymbol{\omega} = (\omega_1, \omega_2, \dots)$ such that: $\omega_n \in \Omega_n$ and $\omega_n \ge \omega_{n+1}$ for $n \ge 1$. Recall from the proof of Theorem 4.19(a) that $\phi_n \ge_{\text{st}} \phi_{n+1}$ for $n \ge 1$, and that $\phi_n \Rightarrow \phi^1_{p,q}$ as $n \to \infty$. By [237, Thm 6.1], there exists a measure μ on $\mathbf{\Omega}$ such that, for each $n \ge 1$, the law of the nth component ω_n is ϕ_n. For $\boldsymbol{\omega} \in \mathbf{\Omega}$, the limit $\omega_\infty = \lim_{n\to\infty} \omega_n$ exists by monotonicity and, by the weak convergence of the sequence $(\phi^1_n : n = 1, 2, \dots)$, ω_∞ has law $\phi^1_{p,q}$. Note that

$$\text{for } e \in \mathbb{E}^d, \quad \omega_n(e) = \omega_\infty(e) \quad \text{for all large } n. \tag{4.92}$$

Let $S = (S_x : x \in \mathbb{Z}^d)$ be independent random variables with the uniform distribution on the spin set $\{1, 2, \dots, q\}$. The S_x are chosen independently of the $\boldsymbol{\omega}$, and we abuse notation by writing μ for the required product measure on the product space $\mathbf{\Omega} \times \Sigma$.

Let $\omega \in \Omega$, and let the vector $\tau(\omega) = (\tau_w(\omega) : w \in \mathbb{Z}^d)$ be given by

$$\tau_w(\omega) = \begin{cases} 1 & \text{if } w \leftrightarrow \infty \text{ in the configuration } \omega, \\ S_{z_w} & \text{otherwise,} \end{cases}$$

where $z_w = z_w(\omega)$ is the earliest vertex in the lexicographic ordering of $\mathbb{Z}^d$ that belongs to the (finite) ω-open cluster at w.

Let us check that:

$$\text{for } w \in \mathbb{Z}^d, \quad \tau_w(\omega_n) = \tau_w(\omega_\infty) \quad \text{for all large } n. \tag{4.93}$$

If $w \leftrightarrow \infty$ in ω_∞, then $w \leftrightarrow \infty$ in ω_n for all large n, whence $\tau_w(\omega_n) = 1 = \tau_w(\omega_\infty)$ for all n. If, on the other hand, $w \not\leftrightarrow \infty$ in ω_∞ then, by (4.92), $C_w(\omega_n) = C_w(\omega_\infty)$ for all large n. Therefore, $\tau_w(\omega_n) = \tau_w(\omega_\infty)$ for all large n, and (4.93) is proved.

Let W be a finite subset of $\mathbb{Z}^d$ and, for $\omega \in \Omega$, define the vector $\tau_W(\omega) = (\tau_w(\omega) : w \in W)$. By Theorem 1.13(a), for n sufficiently large that $W \subseteq \Lambda_n$,

$$\mu(\tau_W(\omega_n) = \alpha) = \pi^1_{\Lambda_n,\beta,q}(\sigma_w = \alpha_w \text{ for } w \in W), \qquad \alpha \in \{1, 2, \dots, q\}^W. \tag{4.94}$$

By (4.93), the vector $\tau_W(\omega_n)$ is constant for all large (random) n. Therefore,

$$\tau_W(\omega_n) \to \tau_W(\omega_\infty) \qquad \text{as } n \to \infty,$$

and so, for $\alpha \in \{1, 2, \dots, q\}^W$,

$$\mu(\tau_W(\omega_n) = \alpha) \to \mu(\tau_W(\omega_\infty) = \alpha) \qquad \text{as } n \to \infty,$$

by the bounded convergence theorem. By (4.94), the vector $\tau_W(\omega_\infty)$ has as law the infinite-volume limit of the finite-volume measure $\pi^1_{\Lambda,\beta,q}$, and the claim is proved.

(b) We continue to employ the notation of the proof of part (a), where it was proved that the vector $\tau(\omega_\infty) = (\tau_x(\omega_\infty) : x \in \mathbb{Z}^d)$ has law $\pi^1_{\beta,q}$. Since ω_∞ has law $\phi^1_{p,q}$, it suffices to show that the conditional law of ω_∞ given $\tau(\omega_\infty)$ is that of the given recipe.

By the definition of $\tau(\omega_\infty)$, the edge $e = \langle x, y\rangle$ satisfies $\omega_\infty(e) = 0$ whenever $\tau_x(\omega_\infty) \neq \tau_y(\omega_\infty)$. Let $e_i = \langle x_i, y_i\rangle$, $i = 1, 2, \dots, n$, be a finite collection of distinct edges, and let D be the subset of $\Omega \times \Sigma$ given by

$$D = \big\{(\omega, S) : \tau_{x_i}(\omega) = \tau_{y_i}(\omega) \text{ for } i = 1, 2, \dots, n\big\}.$$

For any event A defined in terms of the states of the edges e_i, we have by (4.92)–(4.93) that

$$\mu\big(\omega_\infty \in A \,\big|\, (\omega_\infty, S) \in D\big) = \lim_{n\to\infty} \mu\big(\omega_n \in A \,\big|\, (\omega_n, S) \in D\big).$$

The law of ω_n is ϕ_n and, by Theorem 1.13(a), the vector $(\tau_x(\omega_n) : x \in \Lambda_n)$ has law $\pi^1_{\Lambda_n,\beta,q}$. By Theorem 1.13(b), the last probability equals $\psi_p(A)$ where ψ_p is product measure on $\{0, 1\}^n$ with density p. The claim follows. $\square$

Chapter 5

Phase Transition

Summary. When $q \in [1, \infty)$, there exists a critical value $p_c(q)$ of the edge-parameter p, separating the phase with no infinite cluster from the phase with one or more infinite clusters. Partial results are known for both phases, but important open problems remain. In the subcritical phase, exponential decay is proved for sufficiently small p, and is conjectured to hold for all $p < p_c(q)$. Much is known for the supercritical phase subject to the assumption that p exceeds a certain 'slab critical point' $\widehat{p}_c(q)$, conjectured to equal $p_c(q)$. The Wulff construction is a high point of the theory of the random-cluster model.

5.1 The critical point

The random-cluster model possesses an infinite open cluster if and only if p is sufficiently large. There is a critical value of p separating the regime in which all open clusters are finite from that in which infinite clusters exist. We explore this phase transition in this chapter. With the exception of the final Section 5.8, we shall assume for the entirety of the chapter that $q \in [1, \infty)$, and we shall concentrate on the extremal random-cluster measures $\phi^0_{p,q}$ and $\phi^1_{p,q}$. The quantities of principal interest are the $\phi^b_{p,q}$-*percolation-probabilities*,

(5.1) $$\theta^b(p, q) = \phi^b_{p,q}(0 \leftrightarrow \infty), \qquad b = 0, 1.$$

We define the critical points

(5.2) $$p^b_c(q) = \sup\{p : \theta^b(p, q) = 0\}, \qquad b = 0, 1.$$

By Proposition 4.28(a), the $\theta^b(\cdot, q)$ are non-decreasing functions, and therefore

(5.3) $$\theta^b(p, q) \begin{cases} = 0 & \text{if } p < p^b_c(q), \\ > 0 & \text{if } p > p^b_c(q). \end{cases}$$

By Theorem 4.63, $\phi^0_{p,q} = \phi^1_{p,q}$ for almost every $p \in [0,1]$. Therefore, $\theta^0(p,q) = \theta^1(p,q)$ for almost every p, and hence $p^0_c(q) = p^1_c(q)$. Henceforth, we use the abbreviated notation

$$p_c(q) = p^0_c(q) = p^1_c(q), \tag{5.4}$$

and we refer to $p_c(q)$ as the *critical point* of the random-cluster model.

It is almost trivial to prove that $p_c(q) = 1$ in the very special case when the number d of dimensions satisfies $d = 1$. In contrast, it is fundamental that $0 < p_c(q) < 1$ when $d \geq 2$. Not a great deal is known in general[1] about the way in which $p_c(q)$ behaves when viewed as a function of q. The following basic inequalities are consequences of the comparison inequalities of Proposition 4.28.

(5.5) Theorem [8]. *We have that*

$$\frac{1}{p_c(q)} \leq \frac{1}{p_c(q')} \leq \frac{q/q'}{p_c(q)} - \frac{q}{q'} + 1, \qquad 1 \leq q' \leq q. \tag{5.6}$$

From (5.6) we obtain that

$$0 \leq p_c(q) - p_c(q') \leq \frac{(q-q')p_c(q')(1-p_c(q'))}{q' + (q-q')p_c(q')}, \qquad 1 \leq q' \leq q, \tag{5.7}$$

whence, on setting $q' = 1$,

$$0 \leq p_c(q) - p_c(1) \leq \frac{(q-1)p_c(1)(1-p_c(1))}{1 + (q-1)p_c(1)}, \qquad q \geq 1. \tag{5.8}$$

Since $0 < p_c(1) < 1$ for $d \geq 2$, [154, Thm 1.10], we deduce the important fact that

$$0 < p_c(q) < 1, \qquad q \geq 1. \tag{5.9}$$

By (5.7), $p_c(q)$ is a continuous non-decreasing function of q. *Strict* monotonicity[2] requires the further comparison inequality of Theorem 3.24.

(5.10) Theorem [151]. *Let $d \geq 2$. When viewed as a function of q, the critical value $p_c(q)$ is Lipschitz-continuous and strictly increasing on the interval* $[1,\infty)$.

In advance of proving Theorems 5.5 and 5.10, we state and prove two facts of independent interest.

---

[1] Except for its behaviour for large q, see Theorem 7.34.

[2] The strict monotonicity of $p_c(q)$ as a function of the underlying lattice was proved in [39], see also [148].

(5.11) Proposition [8]. *For $p \in [0, 1]$ and $q \in [1, \infty)$,*

$$\phi^1_{\Lambda,p,q}(0 \leftrightarrow \partial\Lambda) \to \theta^1(p, q) \qquad \text{as } \Lambda \uparrow \mathbb{Z}^d.$$

There is no 'elementary' proof of the corresponding fact for the 0-boundary-condition measure $\phi^0_{\Lambda,p,q}$, and indeed this is unproven for general pairs (p, q).

(5.12) Proposition. *Let ϕ_Λ, $\Lambda \subseteq \mathbb{Z}^d$, be probability measures on $(\Omega, \mathcal{F})$ indexed by boxes Λ and satisfying $\phi_\Lambda \Rightarrow \phi$ as $\Lambda \uparrow \mathbb{Z}^d$. If ϕ has the $0/1$-infinite-cluster property, then*

$$\phi_\Lambda(x \leftrightarrow y) \to \phi(x \leftrightarrow y), \qquad x, y \in \mathbb{Z}^d.$$

Proof of Proposition 5.11. It is clear that

$$\phi^1_{p,q}(0 \leftrightarrow \partial\Lambda) \le \phi^1_{\Lambda,p,q}(0 \leftrightarrow \partial\Lambda) \le \phi^1_{\Lambda,p,q}(0 \leftrightarrow \partial\Delta) \qquad \text{for } \Delta \subseteq \Lambda,$$

by positive association and the fact that $\{0 \leftrightarrow \partial\Lambda\} \subseteq \{0 \leftrightarrow \partial\Delta\}$ when $\Delta \subseteq \Lambda$. We take the limits as $\Lambda \uparrow \mathbb{Z}^d$ and $\Delta \uparrow \mathbb{Z}^d$ in that order to obtain the claim. □

Proof of Proposition 5.12. Let x and y be vertices in a box Δ. Then,

$$\begin{aligned}
\phi_\Lambda(x \leftrightarrow y) &\ge \phi_\Lambda(x \leftrightarrow y \text{ in } \Delta) \\
&\to \phi(x \leftrightarrow y \text{ in } \Delta) \qquad \text{as } \Lambda \uparrow \mathbb{Z}^d \\
&\to \phi(x \leftrightarrow y) \qquad \text{as } \Delta \uparrow \mathbb{Z}^d.
\end{aligned}$$

Furthermore,

$$\begin{aligned}
\phi_\Lambda(x \leftrightarrow y,\ x \not\leftrightarrow y \text{ in } \Delta) &\le \phi_\Lambda(x, y \leftrightarrow \partial\Delta,\ x \not\leftrightarrow y \text{ in } \Delta) \\
&\to \phi(x, y \leftrightarrow \partial\Delta,\ x \not\leftrightarrow y \text{ in } \Delta) \qquad \text{as } \Lambda \uparrow \mathbb{Z}^d \\
&\to \phi(x, y \leftrightarrow \infty,\ x \not\leftrightarrow y) \qquad \text{as } \Delta \uparrow \mathbb{Z}^d.
\end{aligned}$$

The last probability equals 0 since ϕ has the $0/1$-infinite-cluster property. □

Proof of Theorem 5.5. Let $1 \le q' \le q$ and

$$\frac{p'}{q'(1-p')} = \frac{p}{q(1-p)}. \tag{5.13}$$

We apply Theorem 3.21 to the probability $\phi^1_{\Lambda,p,q}(0 \leftrightarrow \partial\Lambda)$. By Proposition 5.11, on letting $\Lambda \uparrow \mathbb{Z}^d$,

$$\theta^1(p', q) \le \theta^1(p', q') \le \theta^1(p, q).$$

If $p' < p_c(q')$, then $\theta^1(p', q') = 0$, so that $\theta^1(p', q) = 0$ and therefore $p' \le p_c(q)$. This implies that $p_c(q') \le p_c(q)$, the first inequality of (5.6). Similarly, if $p < p_c(q)$ then $p' \le p_c(q')$, whence

$$\frac{p_c(q)}{q(1-p_c(q))} \le \frac{p_c(q')}{q'(1-p_c(q'))},$$

and hence the second inequality of (5.6). □

Proof of Theorem 5.10. By (5.7),

$$0 \le \frac{p_c(q) - p_c(q')}{q - q'} \le \frac{1}{4q'}, \qquad 1 \le q' < q,$$

whence $p_c(q)$ is Lipschitz-continuous on the interval $[1, \infty)$. Turning to strict monotonicity, let γ be given as in Theorem 3.24 with $\Delta = 2d$, and let $1 \le q_2 < q_1$. Recall that $\gamma(p, q)$ is continuous, and is strictly increasing in p and strictly decreasing in q. We apply Theorem 3.24 to the graph obtained from Λ by identifying all vertices of $\partial\Lambda$, with spanning set $W = \Lambda \setminus \partial\Lambda$ satisfying $\deg(W) = 2d$, to obtain that

$$\phi^1_{\Lambda,p_1,q_1}(0 \leftrightarrow \partial\Lambda) \le \phi^1_{\Lambda,p_2,q_2}(0 \leftrightarrow \partial\Lambda) \quad \text{if} \quad \gamma(p_1, q_1) \le \gamma(p_2, q_2).$$

Let $\Lambda \uparrow \mathbb{Z}^d$ and deduce by Proposition 5.11 that

$$\theta^1(p_1, q_1) \le \theta^1(p_2, q_2) \quad \text{if} \quad \gamma(p_1, q_1) \le \gamma(p_2, q_2). \tag{5.14}$$

We claim that

$$\gamma(p_c(q_1), q_1) \ge \gamma(p_c(q_2), q_2). \tag{5.15}$$

Suppose on the contrary that $\gamma(p_c(q_1), q_1) < \gamma(p_c(q_2), q_2)$. By the continuity of γ, there exist $p_1 > p_c(q_1)$ and $p_2 < p_c(q_2)$ such that $\gamma(p_1, q_1) < \gamma(p_2, q_2)$. By (5.14),

$$\theta^1(p_1, q_1) \le \theta^1(p_2, q_2).$$

However, $\theta^1(p_1, q_1) > 0$ and $\theta^1(p_2, q_2) = 0$, a contradiction, and thus (5.15) holds. If $p_c(q_1) = p_c(q_2)$, then the strict monotonicity of $\gamma(\cdot, \cdot)$ and the fact that $q_2 < q_1$ are in contradiction of (5.15). Therefore $p_c(q_2) < p_c(q_1)$ as claimed. □

5.2 Percolation probabilities

The continuity of the percolation probabilities $\theta^b(p, q)$ is related to the uniqueness of random-cluster measures, in the sense that the $\theta^b(\cdot, q)$ are continuous at p if and only if there is a unique random-cluster measure at this value.

(5.16) Theorem. *Let $d \geq 2$ and $q \in [1, \infty)$.*

(a) *The function $\theta^0(\cdot, q)$ is left-continuous on $(0, 1] \setminus \{p_c(q)\}$.*

(b) *The function $\theta^1(\cdot, q)$ is right-continuous on $[0, 1)$.*

(c) *$\theta^0(p, q) = \theta^1(p, q)$ if and only if $p \notin \mathcal{D}_q$, where $\mathcal{D}_q$ is that of Theorem 4.63.*

(d) *Let $p \neq p_c(q)$. The functions $\theta^0(\cdot, q)$ and $\theta^1(\cdot, q)$ are continuous at the point p if and only if $p \notin \mathcal{D}_q$.*

Clearly, $\theta^0(p, q) = \theta^1(p, q) = 0$ if $q \in [1, \infty)$ and $p < p_c(q)$, and hence $\mathcal{D}_q \cap [0, p_c(q)) = \varnothing$, by part (c). It is presumably the case that $\theta^0(\cdot, q)$ and $\theta^1(\cdot, q)$ are continuous except possibly at $p = p_c(q)$. In addition it may be conjectured that $\theta^0(\cdot, q)$ is left-continuous on the entire interval $(0, 1]$. A verification of this conjecture would include a proof that

$$\theta^0(p_c(q), q) = \lim_{p \uparrow p_c(q)} \theta^0(p, q) = 0.$$

This would in particular solve one of the famous open problems of percolation theory, namely to show that $\theta(p_c(1), 1) = 0$, see [154, 161].

The functions $\theta^0(p, q)$ and $\theta^1(p, q)$ play, respectively, the roles of the magnetizations for Potts measures with free and constant-spin boundary conditions. We state this more fully as a theorem. As in Section 1.3, we write σ_u for the spin at vertex u of a Potts model with q local states (where q is now assumed to be integral). We denote by $\pi_{\beta,q}$ (respectively, $\pi^1_{\beta,q}$) the 'free' (respectively, '1') q-state Potts measure on $\mathbb{L}^d$ with parameter β, see (4.89)–(4.90).

(5.17) Theorem. *Let $d \geq 2$, $p \in (0, 1)$, $q \in \{2, 3, \dots\}$, and let β satisfy $p = 1 - e^{-\beta}$. We have that:*

$$(1 - q^{-1})\theta^0(p, q)^2 = \lim_{|u| \to \infty} \{\pi_{\beta,q}(\sigma_0 = \sigma_u) - q^{-1}\}, \tag{5.18}$$

$$(1 - q^{-1})\theta^1(p, q) = \pi^1_{\beta,q}(\sigma_0 = 1) - q^{-1}. \tag{5.19}$$

Equation (5.19) is standard (see [8, 108, 150]). Equation (5.18) is valid also with $\theta^0(p, q)$ and $\pi_{\beta,q}$ replaced, respectively, by $\theta^1(p, q)$ and $\pi^1_{\beta,q}$, and the proof is similar.

Proof of Theorem 5.16. We shall prove part (a) at the end of Section 8.8. Part (b) is a consequence of Proposition 4.28(b) applied to the indicator function of the increasing closed event $\{0 \leftrightarrow \infty\}$. Part (d) follows from (a)–(c) and Theorem 4.63, on noting that the $\theta^b(\cdot, q)$ are non-decreasing. It remains to prove (c). Certainly $\phi_{p,q}^0 = \phi_{p,q}^1$ if $p \notin \mathcal{D}_q$ (by Theorem 4.63), whence $\theta^0(p,q) = \theta^1(p,q)$ for $p \notin \mathcal{D}_q$. Suppose conversely that $q > 1$ and

$$\theta^0(p,q) = \theta^1(p,q). \tag{5.20}$$

We shall now give the main steps in a proof that

$$h^0(p,q) = h^1(p,q). \tag{5.21}$$

This will imply by Theorem 4.63 that $p \notin \mathcal{D}_q$.

Let $e = \langle u, v \rangle$ be an edge, and J_e the event that e is open. For $w \in \mathbb{Z}^d$, let $I_w = \{w \leftrightarrow \infty\}$, and let H_w be the event that w is in an infinite open path of $\mathbb{E}^d \setminus \{e\}$. As in the proof of Proposition 4.6, there exists a probability measure ψ on $(\Omega, \mathcal{F})^2$ with marginals $\phi_{p,q}^0$ and $\phi_{p,q}^1$, and assigning probability 1 to the set of pairs $(\omega_0, \omega_1) \in \Omega^2$ satisfying $\omega_0 \le \omega_1$. Let $F(\omega)$ be the set of vertices that are joined to infinity by open paths of the configuration $\omega \in \Omega$. We have that

$$0 \le \psi\big(F(\omega_0) \ne F(\omega_1)\big) \le \sum_{w \in \mathbb{Z}^d} \{\phi_{p,q}^1(I_w) - \phi_{p,q}^0(I_w)\} = 0, \tag{5.22}$$

by (5.20). The event $J_e \cap I_u \cap I_v$ is increasing, whence

$$\phi_{p,q}^0(J_e \cap I_u \cap I_v) \le \phi_{p,q}^1(J_e \cap I_u \cap I_v). \tag{5.23}$$

Also,

$$\begin{aligned}\phi_{p,q}^0(\overline{J_e} \cap I_u \cap I_v) &= \phi_{p,q}^0(\overline{J_e} \cap H_u \cap H_v)\\ &= \phi_{p,q}^0(\overline{J_e} \mid H_u \cap H_v)\phi_{p,q}^0(H_u \cap H_v).\end{aligned} \tag{5.24}$$

However,

$$\phi_{p,q}^0(\overline{J_e} \mid H_u \cap H_v) = \phi_{p,q}^1(\overline{J_e} \mid H_u \cap H_v)$$

by Proposition 4.37(a) and the fact (Theorem 4.34) that $\phi_{p,q}^0, \phi_{p,q}^1 \in \mathcal{R}_{p,q}$. In addition, $\phi_{p,q}^0(H_u \cap H_v) \le \phi_{p,q}^1(H_u \cap H_v)$ since $H_u \cap H_v$ is an increasing event. Therefore (5.24) implies that

$$\begin{aligned}\phi_{p,q}^0(\overline{J_e} \cap I_u \cap I_v) &\le \phi_{p,q}^1(\overline{J_e} \mid H_u \cap H_v)\phi_{p,q}^1(H_u \cap H_v)\\ &= \phi_{p,q}^1(\overline{J_e} \cap H_u \cap H_v) = \phi_{p,q}^1(\overline{J_e} \cap I_u \cap I_v).\end{aligned} \tag{5.25}$$

Adding (5.23) and (5.25), we obtain that

$$\phi_{p,q}^0(I_u \cap I_v) \le \phi_{p,q}^1(I_u \cap I_v).$$

Equality holds here by (5.22), and therefore equality holds in (5.23), which is to say that

$$\phi^0_{p,q}(J_e \cap I_u \cap I_v) = \phi^1_{p,q}(J_e \cap I_u \cap I_v). \tag{5.26}$$

It is obvious that

$$\phi^0_{p,q}(J_e \cap \overline{I_u} \cap I_v) = \phi^1_{p,q}(J_e \cap \overline{I_u} \cap I_v) \tag{5.27}$$

since both sides equal 0; the same equation holds with $\overline{I_u} \cap I_v$ replaced by $I_u \cap \overline{I_v}$.

Finally, we prove that

$$\phi^0_{p,q}(J_e \cap \overline{I_u} \cap \overline{I_v}) = \phi^1_{p,q}(J_e \cap \overline{I_u} \cap \overline{I_v}) \tag{5.28}$$

which, in conjunction with (5.26), (5.27), and the subsequent remark, implies the required (5.21) by addition. Let $\epsilon > 0$. Let Λ be a box containing u and v, and let $A_\Lambda = \{u \nleftrightarrow \partial\Lambda,\ v \nleftrightarrow \partial\Lambda\}$. We have that

$$\begin{aligned}
0 &\le \phi^0_{p,q}(A_\Lambda) - \phi^1_{p,q}(A_\Lambda) \\
&\to \phi^0_{p,q}(\overline{I_u} \cap \overline{I_v}) - \phi^1_{p,q}(\overline{I_u} \cap \overline{I_v}) \qquad \text{as } \Lambda \uparrow \mathbb{Z}^d \\
&\le \psi\big(F(\omega_0) \ne F(\omega_1)\big) = 0,
\end{aligned}$$

by (5.22). Therefore,

$$0 \le \phi^0_{p,q}(A_\Lambda) - \phi^1_{p,q}(A_\Lambda) < \epsilon \qquad \text{for all large } \Lambda,$$

and we pick Λ accordingly. The events $\{u \nleftrightarrow \partial\Lambda\}$ and $\{v \nleftrightarrow \partial\Lambda\}$ are cylinder events, whence

$$0 \le \phi^0_{\Delta,p,q}(A_\Lambda) - \phi^1_{\Delta,p,q}(A_\Lambda) < 2\epsilon \qquad \text{for all large } \Delta, \tag{5.29}$$

and we pick $\Delta \supseteq \Lambda$ accordingly. We now employ a certain coupling of $\phi^0_{\Delta,p,q}$ and $\phi^1_{\Delta,p,q}$. Similar couplings will be encountered later.

(5.30) Proposition. *Let $p \in (0,1)$ and $q \in [1,\infty)$, and let Λ, Δ be finite boxes of $\mathbb{Z}^d$ satisfying $\Lambda \subseteq \Delta$. For $\omega \in \Omega$, let $G = G(\omega) = \{x \in \Lambda : x \nleftrightarrow \partial\Lambda\}$. There exists a probability measure ψ_Δ on $\Omega^0_\Delta \times \Omega^1_\Delta$, with marginals $\phi^0_{\Delta,p,q}$ and $\phi^1_{\Delta,p,q}$, that assigns probability 1 to pairs (ω_0, ω_1) satisfying $\omega_0 \le \omega_1$, and with the additional property that, conditional on the set $G = G(\omega_1)$, both marginals of ψ_Δ on $\mathbb{E}_G$ equal the free random-cluster measure $\phi_{G,p,q}$.*

Writing $\mathcal{G}$ for the class of all subsets of Λ that contain both u and v, we have by the proposition that

$$\begin{aligned}\phi^1_{\Delta,p,q}(J_e \cap A_\Lambda) &= \sum_{g\in\mathcal{G}} \phi^1_{\Delta,p,q}(J_e,\ G=g) = \sum_{g\in\mathcal{G}} \psi_\Delta\big(\omega_1 \in J_e,\ G(\omega_1)=g\big)\\ &= \sum_{g\in\mathcal{G}} \psi_\Delta\big(\omega_1 \in J_e \,\big|\, G(\omega_1)=g\big)\psi_\Delta(G(\omega_1)=g)\\ &= \sum_{g\in\mathcal{G}} \psi_\Delta\big(\omega_0 \in J_e \,\big|\, G(\omega_1)=g\big)\psi_\Delta(G(\omega_1)=g)\\ &= \psi_\Delta(\omega_0 \in J_e,\ \omega_1 \in A_\Lambda)\\ &\le \psi_\Delta(\omega_0 \in J_e,\ \omega_0 \in A_\Lambda) = \phi^0_{\Delta,p,q}(J_e \cap A_\Lambda).\end{aligned}$$

Therefore,

$$\begin{aligned}0 &\le \phi^0_{\Delta,p,q}(J_e \cap A_\Lambda) - \phi^1_{\Delta,p,q}(J_e \cap A_\Lambda)\\ &= \psi_\Delta(\omega_0 \in J_e,\ \omega_0 \in A_\Lambda,\ \omega_1 \notin A_\Lambda)\\ &\le \psi_\Delta(\omega_0 \in A_\Lambda,\ \omega_1 \notin A_\Lambda) \le 2\epsilon,\end{aligned}$$

by (5.29). Let $\Delta \uparrow \mathbb{Z}^d$, $\Lambda \uparrow \mathbb{Z}^d$, and $\epsilon \downarrow 0$ in that order, to obtain (5.28). □

Proof of Proposition 5.30. Let $\phi^b = \phi^b_{\Delta,p,q}$. Since $\phi^0 \le_{\text{st}} \phi^1$, there exists a coupled probability measure on $\Omega^0_\Delta \times \Omega^1_\Delta$ with marginals ϕ^0, ϕ^1, and that allocates probability 1 to the set of pairs (ω_0, ω_1) with $\omega_0 \le \omega_1$. This fact is immediate from the stochastic ordering, but we require in addition the special property stated in the proposition, and to this end we shall develop a special coupling not dissimilar to those used in [38] and [259, p. 254]. We do this by building a random configuration $(\omega_0, \omega_1) \in \Omega^0_\Delta \times \Omega^1_\Delta$ in a sequential manner, and we shall speak of ω_0 (respectively, ω_1) as the *lower* (respectively, *upper*) configuration. We shall proceed edge by edge, and shall check the (conditional) stochastic ordering at each stage.

After stage n we will have found the (paired) states of edges belonging to some subset S_n of $\mathbb{E}_\Delta$. We begin with $S_0 = \varnothing$, and we build inwards starting at the boundary of Δ. Let $(e_l : l = 1, 2, \dots, L)$ be a deterministic ordering of the edges in $\mathbb{E}_\Delta$. Let e_{j_1} be the earliest edge in this ordering that is incident to some vertex in $\partial\Delta$, and let

$$I^b_0 = \{\text{every edge outside } \mathbb{E}_\Delta \text{ has state } b\}, \qquad b = 0, 1.$$

By the usual stochastic ordering,

$$\phi^0(e_{j_1} \text{ is open} \mid I^0_0) \le \phi^1(e_{j_1} \text{ is open} \mid I^1_0). \tag{5.31}$$

Therefore, we may find $\{0,1\}$-valued random variables $\omega_0(e_{j_1})$, $\omega_1(e_{j_1})$ with mean values as in (5.31) and satisfying $\omega_0(e_{j_1}) \le \omega_1(e_{j_1})$. We set $S_1 = \{e_{j_1}\}$ and

$$I^b_1 = I^b_0 \cap \{e_{j_1} \text{ has state } \omega_b(e_{j_1})\}, \qquad b = 0, 1.$$

We iterate this process. After stage r, we will have gathered the information I_r^0 (respectively, I_r^1) relevant to the lower (respectively, upper) process, and we will proceed to consider the state of some further edge $e_{j_{r+1}}$. The analogue of (5.31), namely

$$\phi^0(e_{j_{r+1}} \text{ is open} \mid I_r^0) \leq \phi^1(e_{j_{r+1}} \text{ is open} \mid I_r^1),$$

is valid since, by construction, $\omega_0(e_{j_s}) \leq \omega_1(e_{j_s})$ for $s = 1, 2, \dots, r$. Thus we may pick a pair of random states $\omega_0(e_{j_{r+1}})$, $\omega_1(e_{j_{r+1}})$ for the new edge, these having the correct marginals and satisfying $\omega_0(e_{j_{r+1}}) \leq \omega_1(e_{j_{r+1}})$.

Next is described how we choose the edges $e_{j_2}, e_{j_3}, \dots$. Suppose the first r stages of the above process are complete, and write $S_r = \{e_{j_s} : s = 1, 2, \dots, r\}$. Let K_r be the set of vertices $x \in \Delta$ such that there exists a path π joining x to some $z \in \partial\Delta$, with the property that $\omega_1(e) = 1$ for all $e \in \pi$. (This requires that every edge e in π has been considered in the first r stages, and that the ω_1-value of each such e was found to be 1.) We let $e_{j_{r+1}}$ be the earliest edge in the given ordering of $\mathbb{E}_\Delta$ that does not belong to S_r but possesses an endvertex in K_r.

Let us call a temporary halt at the stage when no new edge can be found. At this stage, R say, we have revealed the states of edges in a certain (random) set S_R. Let F_R be the set of edges in $\mathbb{E}_\Delta$ that are closed in the upper configuration. By construction, F_R contains exactly those edges of $\mathbb{E}_\Delta$ that have at least one endvertex in K_R and that have been determined to be closed in the upper configuration. By the ordering, the edges in F_R are closed in the lower configuration also. By construction, the lower (respectively, upper) configuration so far revealed is governed by the measure ϕ^0 (respectively, ϕ^1).

Suppose for the moment that $\Lambda = \Delta$, in which case $G(\omega_1) = \Delta \setminus K_R$. When extending the upper and lower configurations to edges in $\mathbb{E}_G$, the only relevant information gathered to date is that all edges in the edge-boundary $\Delta_e G$ are closed in both configurations. We may therefore complete ω_0, ω_1 at one stroke by taking them to be equal, with (common) law $\phi^0_{G,p,q}$. This proves the proposition in the special case when $\Lambda = \Delta$. Consider now the general case $\Lambda \subseteq \Delta$.

We explain next how to re-start the process at stage R. We began above with the 'seed' $\partial\Delta$ and we built a set of edges connected to $\partial\Delta$ by paths of open edges in the upper configuration, together with its closed edge-boundary. Having reached stage R, we choose a vertex $x \in \Delta$ satisfying $x \notin K_R \cup (\Lambda \setminus \partial\Lambda)$ that is incident to some edge of F_R. We then re-start the process with x as seed, and we continue until we have revealed the open cluster $C_x(\omega_1)$ at x in the upper configuration. We add the vertex-set of $C_x(\omega_1)$ to K_R to obtain a larger set K'. To F_R, we add all edges incident to vertices in this cluster that are closed in the upper configuration, obtaining thus a larger set F'. Next, we find another seed $y \notin K' \cup (\Lambda \setminus \partial\Lambda)$ incident to some edge in F'. This process is iterated until no new seed may be found.

At the end of all this, we have revealed the paired states of all edges in some set S. Let T be the union of the vertex-sets of the open clusters of all seeds. Since no further seed may be found, it is the case that $G(\omega_1) = \Delta \setminus T$. As before, the lower and upper configurations may be completed at one stroke by sampling the states of edges in $\mathbb{E}_G$ according to the free measure $\phi_{G,p,q}$. □

Proof of Theorem 5.17. Let μ be the (coupled) probability measure on $\Omega \times \Sigma$ given by the recipe of Theorem 4.91(a). We have that

$$\begin{aligned}\pi^1_{\beta,q}(\sigma_0 = 1) &= \mu(\sigma_0 = 1 \mid 0 \leftrightarrow \infty)\theta^1(p,q) \\ &\qquad + \mu(\sigma_0 = 1 \mid 0 \not\leftrightarrow \infty)[1 - \theta^1(p,q)] \\ &= \theta^1(p,q) + \frac{1}{q}[1 - \theta^1(p,q)],\end{aligned}$$

and (5.19) follows.

Turning to (5.18), we have similarly to the above that

$$(1 - q^{-1})\phi^0_{p,q}(0 \leftrightarrow u) = \pi_{\beta,q}(\sigma_0 = \sigma_u) - \frac{1}{q}, \qquad u \in \mathbb{Z}^d.$$

The claim is proven once we have shown that

$$\phi^0_{p,q}(0 \leftrightarrow u) \to \theta^0(p,q)^2 \qquad \text{as } |u| \to \infty. \tag{5.32}$$

By the 0/1-infinite-cluster property of $\phi^0_{p,q}$, see the remark after (4.36),

$$\phi^0_{p,q}(0 \leftrightarrow u) = \phi^0_{p,q}(0 \leftrightarrow \infty,\ u \leftrightarrow \infty) + \phi^0_{p,q}(u \in C,\ |C| < \infty).$$

The last probability tends to zero as $|u| \to \infty$. Also,

$$\phi^0_{p,q}(0 \leftrightarrow \infty,\ u \leftrightarrow \infty) \to \phi^0_{p,q}(0 \leftrightarrow \infty)^2 \qquad \text{as } |u| \to \infty,$$

since $\phi^0_{p,q}$ is mixing, see Corollary 4.23. □

5.3 Uniqueness of random-cluster measures

We record in this section some information about the set of values of p at which there exists a unique random-cluster measure.

(5.33) Theorem [8, 152]. *Let $q \in [1,\infty)$ and $d \geq 2$. There exists a unique random-cluster measure, in that $|\mathcal{W}_{p,q}| = |\mathcal{R}_{p,q}| = 1$, if either of the following holds:*

(a) *$\theta^0(p,q) = \theta^1(p,q)$, which is to say that $p \notin \mathcal{D}_q$,*

(b) *$p > p'$, where $p' = p'(q,d) \in [p_c(q), 1)$ is a certain real number.*

By part (a), there is a unique random-cluster measure for any p such that $\theta^1(p,q) = 0$, [8, Thm A.2]. In particular, there exists a unique random-cluster measure throughout the subcritical phase, that is, when $0 \leq p < p_c(q)$. It is an important open problem to establish the same conclusion when $p > p_c(q)$.

(5.34) Conjecture. *Let $q \in [1, \infty)$ and $d \geq 2$. We have that $\phi^0_{p,q} = \phi^1_{p,q}$, and therefore $|\mathcal{W}_{p,q}| = |\mathcal{R}_{p,q}| = 1$, if and only if either of the following holds:*

(i) *either $p < p_c(q)$ or $p > p_c(q)$,*

(ii) *$p = p_c(q)$ and $\theta^1(p_c(q), q) = 0$.*

Slightly more is known in the case of two dimensions. It is proved in Theorem 6.17 that there is a unique random-cluster measure when $d = 2$ and $p \neq p_{sd}(q)$, where $p_{sd}(q) = \sqrt{q}/(1 + \sqrt{q})$ is the 'self-dual' value of p. It is conjectured that $p_c(q) = p_{sd}(q)$ for $q \in [1, \infty)$.

Proof of Theorem 5.33. The sufficiency of (a) was proved in Theorem 5.16(c).

We sketch a proof that $\phi^0_{p,q} = \phi^1_{p,q}$ if p is sufficiently close to 1. There are certain topological complications in this[3], and we refrain from giving all the relevant details, most of which may be found in a closely related passage of [211, Section 2]. We begin by defining a lattice $\mathcal{L}$ with the same vertex set as $\mathbb{L}^d$ but with edge-relation

$$x \sim y \quad \text{if} \quad |x_i - y_i| \leq 1 \text{ for } 1 \leq i \leq d.$$

For $\omega \in \Omega$, we call a vertex x *white* if $\omega(e) = 1$ for all e incident with x in $\mathbb{L}^d$, and *black* otherwise. For any set V of vertices of $\mathcal{L}$, we define its *black cluster* $B(V)$ as the union of V together with the set of all vertices x_0 of $\mathcal{L}$ for which the following holds: there exists a path $x_0, e_0, x_1, e_1, \dots, e_{n-1}, x_n$ of alternating vertices and edges of $\mathcal{L}$ such that $x_0, x_1, \dots, x_{n-1} \notin V$, $x_n \in V$, and $x_0, x_1, \dots, x_{n-1}$ are black. Note that the colours of vertices in V have no effect on $B(V)$, and that $V \subseteq B(V)$. Let

$$\|B(V)\| = \sup \left\{ \sum_{i=1}^{d} |x_i - y_i| : x \in V,\ y \in B(V) \right\}.$$

When V is a singleton, $V = \{x\}$ say, we abbreviate $B(V)$ to $B(x)$.

For an integer n and a vertex x, the event $\{\|B(x)\| \geq n\}$ is a decreasing event, whence

$$\begin{aligned} (5.35) \qquad \phi^0_{p,q}(\|B(x)\| \geq n) &\leq \phi^0_{\Lambda,p,q}(\|B(x)\| \geq n) \qquad \text{for any box } \Lambda \\ &\leq \phi_{\Lambda,\pi}(\|B(x)\| \geq n), \end{aligned}$$

where $\phi_{\Lambda,\pi}$ is product measure on $\mathbb{E}_\Lambda$ with density $\pi = p/[p + q(1-p)]$, and we have used the comparison inequality of Proposition 4.28(a). By a Peierls argument (see [211, pp. 151–152]) there exists $\alpha(p)$ such that: the percolation (product) measure $\phi_\pi = \lim_{\Lambda \uparrow \mathbb{Z}^d} \phi_{\Lambda,\pi}$ satisfies

$$(5.36) \qquad \phi_\pi(\|B(x)\| \geq n) \leq e^{-n\alpha(p)}, \qquad n \geq 1,$$

[3] An alternative approach may be based on the methods of Section 7.2.

and furthermore $\alpha(p) > 0$ if p is sufficiently large, say $p > p'$ for some $p' \in [p_c(q), 1)$.

Let A be an increasing cylinder event, and find a finite box Λ such that $A \in \mathcal{F}_\Lambda$. Let Δ be a box satisfying $\Lambda \subseteq \Delta$. For any subset S of $\overline{\Lambda} = \mathbb{Z}^d \setminus \Lambda$ containing $\partial\Delta$, we define the 'internal boundary' $D(S)$ of S to be the set of all vertices x of $\mathcal{L}$ satisfying:

(a) $x \notin S$,

(b) x is adjacent in $\mathcal{L}$ to some vertex of S,

(c) there exists a path of $\mathbb{L}^d$ from x to some vertex in Λ, this path using no vertex of S.

Let $\widetilde{S} = S \cup D(S)$, and denote by $I(S)$ the set of vertices x_0 for which there exists a path $x_0, e_0, x_1, e_1, \dots, e_{n-1}, x_n$ of $\mathbb{L}^d$ with $x_n \in \Lambda$, $x_i \notin \widetilde{S}$ for all i. Note that every vertex of $\partial I(S)$ is adjacent to some vertex in $D(S)$. We shall concentrate on the case $S = B(\partial\Delta)$.

Let $\epsilon > 0$ and $p > p'$, where p' is given after (5.36). By (5.35)–(5.36), there exists a box Δ' sufficiently large that

$$\phi^0_{p,q}(K_{\Lambda,\Delta}) \geq 1 - \epsilon, \qquad \Delta \supseteq \Delta', \tag{5.37}$$

where $K_{\Lambda,\Delta} = \{\widetilde{B(\partial\Delta)} \cap \Lambda = \varnothing\}$. We pick Δ' accordingly, and let $\Delta \supseteq \Delta'$.

Assume that $K_{\Lambda,\Delta}$ occurs, so that $I = I(B(\partial\Delta))$ satisfies $I \supseteq \Lambda$. Let $\mathcal{H}_\Lambda$ be the set of all subsets h of $\overline{\Lambda}$ such that $\widetilde{h} \subseteq \overline{\Lambda}$. We note three facts about $B(\partial\Delta)$ and $D(B(\partial\Delta))$:

(a) $D(B(\partial\Delta))$ is $\mathbb{L}^d$-connected in that, for all pairs $x, y \in D(B(\partial\Delta))$, there exists a path of $\mathbb{L}^d$ joining x to y using vertices of $D(B(\partial\Delta))$ only,

(b) every vertex in $D(B(\partial\Delta))$ is white,

(c) $D(B(\partial\Delta))$ is measurable with respect to the colours of vertices in $\overline{I} = \mathbb{Z}^d \setminus I$, in the following sense: for any $h \in \mathcal{H}_\Lambda$, the event $\{B(\partial\Delta) = h, D(B(\partial\Delta)) = D(h)\}$ lies in the σ-field generated by the colours of vertices in $\overline{I(h)}$.

Claim (a) may be proved by adapting the argument used to prove [211, Lemma 2.23]; claim (b) is a consequence of the definition of $D(B(\partial\Delta))$; claim (c) holds since $D(B(\partial\Delta))$ is part of the boundary of the black cluster of $\mathcal{L}$ generated by $\partial\Delta$. Full proofs of (a) and (c) are not given here. They would be rather long, and would have much in common with [211, Section 2].

Let $h \in \mathcal{H}_\Lambda$. The $\phi^0_{p,q}$-probability of A, conditional on $\{B(\partial\Delta) = h\}$, is given by the wired measure $\phi^1_{I(h),p,q}$. This holds since: (a) every vertex in $\partial I(h)$ is adjacent to some vertex of $D(h)$, and (b) $D(h)$ is $\mathbb{L}^d$-connected and all vertices in $D(h)$ are white. Therefore, by conditional probability and positive association,

$$\begin{aligned} \phi^0_{p,q}(A) &\geq \phi^0_{p,q}\big(\phi^1_{I,p,q}(A) 1_{K_{\Lambda,\Delta}}\big) \\ &\geq \phi^0_{p,q}\big(\phi^1_{\Delta,p,q}(A) 1_{K_{\Lambda,\Delta}}\big) \qquad \text{since } I \subseteq \Delta \\ &\geq \phi^1_{\Delta,p,q}(A) - \epsilon \qquad \text{by (5.37).} \end{aligned} \tag{5.38}$$

Let $\Delta \uparrow \mathbb{Z}^d$ and $\epsilon \downarrow 0$ in that order, and deduce that $\phi^0_{p,q} \geq_{\text{st}} \phi^1_{p,q}$. Since $\phi^0_{p,q} \leq_{\text{st}} \phi^1_{p,q}$, we conclude that $\phi^0_{p,q} = \phi^1_{p,q}$. □

5.4 The subcritical phase

The random-cluster model is said to be in the subcritical phase when $p < p_c(q)$, and this phase is the subject of the next three sections. Let $q \in [1, \infty)$, $d \geq 2$, and $p < p_c(q)$. By Theorem 5.33(a), $\phi^0_{p,q} = \phi^1_{p,q}$, and hence $|\mathcal{W}_{p,q}| = |\mathcal{R}_{p,q}| = 1$. We shall denote the unique random-cluster measure by $\phi^0_{p,q}$.

The subcritical phase is characterized by the (almost-sure) absence of an infinite open cluster. Thus all open clusters are almost-surely finite, and one seeks estimates on the tail of the size of such a cluster. As described in [154, Chapter 6], one may study both the 'radius' and the 'volume' of a cluster C. We concentrate here on the cluster $C = C_0$ at the origin, and we define its *radius*[4] by

$$\text{rad}(C) = \max\{\|y\| : y \in C\} = \max\{\|y\| : 0 \leftrightarrow y\}. \tag{5.39}$$

It is immediate that $\text{rad}(C) \geq n$ if and only if $0 \leftrightarrow \partial\Lambda_n$, where $\Lambda_n = [-n, n]^d$. We note for later use that there exists a positive constant $\beta = \beta(d)$ such that

$$\beta|C|^{1/d} \leq \text{rad}(C) + 1 \leq |C|. \tag{5.40}$$

It is believed that the tails of both $\text{rad}(C)$ and $|C|$ decay exponentially when $p < p_c(q)$, but this is currently unproven. It is easy to prove that the appropriate limits exist, but the non-triviality of the limiting values remains open. That is, one may use subadditivity to show the existence of the constants

$$\psi(p, q) = \lim_{n\to\infty} \left\{ -\frac{1}{n} \log \phi^0_{p,q}(0 \leftrightarrow \partial\Lambda_n) \right\}, \tag{5.41}$$

$$\zeta(p, q) = \lim_{n\to\infty} \left\{ -\frac{1}{n} \log \phi^0_{p,q}(|C| \geq n) \right\}. \tag{5.42}$$

It is quite another matter to show as expected that

$$\psi(p, q) > 0, \ \zeta(p, q) > 0 \qquad \text{for } p < p_c(q). \tag{5.43}$$

We confine ourselves in this section to 'soft' arguments concerning the existence of ψ and ζ; the 'harder' arguments relevant to strict positivity are deferred to the next two sections. We begin by considering the radius of the cluster at the origin. The existence of the limit in (5.41) relies essentially on positive association. We write $e_n = (n, 0, 0, \dots, 0)$.

---

[4]Note the use of the distance function $\|\cdot\|$ rather than the function $|\cdot|$ of [154].

(5.44) Theorem. *Let μ be an automorphism-invariant probability measure on $(\Omega, \mathcal{F})$ which is positively associated. The limits*

$$\nu(\mu) = \lim_{n\to\infty}\left\{-\frac{1}{n}\log \mu(0 \leftrightarrow e_n)\right\},$$

$$\psi(\mu) = \lim_{n\to\infty}\left\{-\frac{1}{n}\log \mu(0 \leftrightarrow \partial\Lambda_n)\right\},$$

exist and satisfy $0 \le \nu(\mu) = \psi(\mu) \le \infty$, *and*

$$\mu(0 \leftrightarrow e_n) \le e^{-n\nu(\mu)}, \quad \mu(0 \leftrightarrow \partial\Lambda_n) \le \sigma n^{d-1} e^{-n\psi(\mu)}, \qquad n \ge 1,$$

where $\sigma = 2d^2 6^d$.

(5.45) Corollary. *Let $p \in (0,1]$ and $q \in [1,\infty)$. The limit*

$$\psi(p,q) = \lim_{n\to\infty}\left\{-\frac{1}{n}\log \phi^0_{p,q}(0 \leftrightarrow e_n)\right\} = \lim_{n\to\infty}\left\{-\frac{1}{n}\log \phi^0_{p,q}(0 \leftrightarrow \partial\Lambda_n)\right\}$$

exists and satisfies $0 \le \psi = \psi(p,q) < \infty$. *There exists a constant $\sigma = \sigma(d)$ such that*

$$\phi^0_{p,q}(0 \leftrightarrow e_n) \le e^{-n\psi}, \quad \phi^0_{p,q}(0 \leftrightarrow \partial\Lambda_n) \le \sigma n^{d-1} e^{-n\psi}, \qquad n \ge 1. \tag{5.46}$$

Proofs of Theorem 5.44 and Corollary 5.45. The proof of Theorem 5.44 follows exactly that of the corresponding parts of [154, Thms 6.10, 6.44], and the details are omitted. For the second proof, it suffices to check that $\phi^0_{p,q}$ satisfies the conditions of Theorem 5.44. □

We turn next to the volume $|C|$ of the open cluster at the origin.

(5.47) Theorem. *Let $\alpha \in (0,1)$, $\beta \in (0,1]$, and let μ be a translation-invariant probability measure on $(\Omega, \mathcal{F})$ satisfying the 'uniform insertion-tolerance condition'*

$$\alpha \le \mu(J_e \mid \mathcal{T}_e) \le \beta, \qquad \mu\text{-almost-surely, for } e \in \mathbb{E}^d,$$

where J_e is the event that e is open. The limit

$$\zeta(\mu) = \lim_{n\to\infty}\left\{-\frac{1}{n}\log \mu(|C| = n)\right\} \tag{5.48}$$

exists and satisfies

$$\mu(|C| = n) \le \frac{(1-\alpha)^2}{\alpha} n e^{-n\zeta(\mu)}, \qquad n \ge 1. \tag{5.49}$$

Furthermore, $0 \le \zeta(\mu) \le -\log[\alpha(1-\beta)^{2(d-1)}]$.

It is an easy consequence of (5.48)–(5.49) that

$$-\frac{1}{n}\log \mu(n \le |C| < \infty) \to \zeta(\mu) \qquad \text{as } n \to \infty. \tag{5.50}$$

(5.51) Corollary. *Let $p \in (0,1)$ and $q \in [1,\infty)$. The limit*

$$\zeta(p,q) = \lim_{n\to\infty} \left\{ -\frac{1}{n} \log \phi^0_{p,q}(|C| = n) \right\}$$

exists and satisfies

$$\phi^1_{p,q}(|C| = n) \le \frac{1}{p}(1-p)^2[p + q(1-p)]ne^{-n\zeta}, \qquad n \ge 1.$$

Proofs of Theorem 5.47 and Corollary 5.51. These are obtained by following the proof of [154, Thm 6.78], and the details are omitted. □

Since $\phi^0_{p,q}(0 \leftrightarrow \infty) > 0$ when $p > p_c(q)$, it is elementary that

$$\psi(p,q) = 0 \qquad \text{for } p > p_c(q). \tag{5.52}$$

It is rather less obvious that

$$\zeta(p,q) = 0 \qquad \text{for } p > p_c(q), \tag{5.53}$$

and this is implied (for sufficiently large p) by Theorem 5.108. It is an important open problem to prove that $\psi(p,q) > 0$ and $\zeta(p,q) > 0$ when $p < p_c(q)$.

(5.54) Conjecture (Exponential decay). *Let $q \in [1,\infty)$. Then $\psi(p,q) > 0$ and $\zeta(p,q) > 0$ whenever $p < p_c(q)$.*

A partial result in this direction is the following rather weak statement; related results may be obtained via Theorem 3.79 as in Theorem 6.30.

(5.55) Theorem. *Let $q \in [1,\infty)$ and $0 < p < p_c(1)$. Then $\psi(p,q) > 0$ and $\zeta(p,q) > 0$.*

Proof. Let $q \in [1,\infty)$, while noting in passing that the method of proof is valid even when $q \in (0,1)$, using the comparison inequalities of Theorem 3.21 as in (5.118). By Proposition 4.28(a), $\phi^0_{p,q} \le_{\text{st}} \phi_p$. Therefore,

$$\phi^0_{p,q}(0 \leftrightarrow \partial\Lambda_n) \le \phi_p(0 \leftrightarrow \partial\Lambda_n),$$

whence $\psi(p,q) \ge \psi(p,1)$, and the strict positivity of ψ follows by the corresponding statement for percolation, [154, Thm 6.14].

Similarly,

$$\phi^0_{p,q}(|C| = n) \le \phi^0_{p,q}(|C| \ge n) \le \phi_p(|C| \ge n).$$

By [154, eqn (6.82)],

$$-\frac{1}{n}\log \phi_p(|C| \ge n) \to \zeta(p,1) \qquad \text{as } n \to \infty.$$

Furthermore, $\zeta(p,1) > 0$ when $p < p_c(1)$, by [154, Thm 6.78]. □

5.5 Exponential decay of radius

We address next the exponential decay of the radius of an open cluster. The existence of the limit

$$(5.56) \qquad \psi(p,q) = \lim_{n\to\infty} \left\{ -\frac{1}{n} \log \phi_{p,q}^0(0 \leftrightarrow \partial\Lambda_n) \right\}$$

follows from Corollary 5.45, and the problem is to determine for which p, q it is the case that $\psi(p,q) > 0$. See Conjecture 5.54.

In the case of percolation, a useful intermediate step was the proof by Hammersley [177] that $\psi(p,1) > 0$ whenever the two-point connectivity function is summable, that is,

$$\phi_p(|C|) = \sum_{x\in\mathbb{Z}^d} \phi_p(0 \leftrightarrow x) < \infty.$$

Similarly, Simon [300] and Lieb [234] proved the exponential decay of the two-point function of Ising and other models under a summability assumption on this function, see Section 9.3. Such conclusions provoke the following question in the current context: if $\phi_{p,q}^0(0 \leftrightarrow \partial\Lambda_n)$ decays at some polynomial rate as $n \to \infty$, then must it necessarily decay at an exponential rate? An affirmative answer is provided in the discussion that follows.

We concentrate here on the quantity

$$(5.57) \qquad L(p,q) = \limsup_{n\to\infty} \left\{ n^{d-1} \phi_{p,q}^0(0 \leftrightarrow \partial\Lambda_n) \right\}.$$

By the comparison inequality, Proposition 4.28, $L(p,q)$ is non-decreasing in p, and therefore,

$$L(p,q) \begin{cases} < \infty & \text{if } p < \widetilde{p}_c(q), \\ = \infty & \text{if } p > \widetilde{p}_c(q), \end{cases}$$

where

$$(5.58) \qquad \widetilde{p}_c(q) = \sup\left\{ p : L(p,q) < \infty \right\}.$$

Clearly $\widetilde{p}_c(q) \le p_c(q)$, and equality is believed to hold.

(5.59) Conjecture [163]. *If $q \in [1,\infty)$, then $\widetilde{p}_c(q) = p_c(q)$.*

Certainly $\widetilde{p}_c(q) = p_c(q)$ when $q = 1$, see [154], and we shall see at Theorem 7.33 that this holds also when q is sufficiently large. It is in addition valid for $q = 2$, see Theorem 9.53 and the remarks thereafter.

The condition $L(p,q) < \infty$ amounts to the statement that the radius $R = \text{rad}(C)$ has a tail decaying at least as fast as $n^{-(d-1)}$. This is slightly weaker than the moment condition $\phi_{p,q}^0(R^{d-1}) < \infty$. In fact, $L(p,q) = 0$ if $\phi_{p,q}^0(R^{d-1}) < \infty$, since

$$n^{d-1}\phi_{p,q}^0(0 \leftrightarrow \partial\Lambda_n) = n^{d-1}\phi_{p,q}^0(R \ge n) \le \sum_{k=n}^{\infty} k^{d-1}\phi_{p,q}^0(R = k).$$

There is a converse statement. If $p < \widetilde{p}_c(q)$ then $L(p,q) < \infty$, implying that

$$n^c \phi^0_{p,q}(0 \leftrightarrow \partial\Lambda_n) \to 0 \quad \text{for all } c \text{ satisfying } c < d-1.$$

This in turn implies, as in [164, Exercise 5.6.4], that $\phi^0_{p,q}(R^c) < \infty$ for all $c < d - 1$.

We state next the main conclusion of this section. A related result is to be found at Theorem 5.86.

(5.60) Theorem. *Let $q \in [1,\infty)$. The function ψ in (5.56) satisfies $\psi(p,q) > 0$ whenever $0 < p < \widetilde{p}_c(q)$.*

The proof, which is delayed until later in the section, uses the method of exponential steepness described in Section 3.5. Let A be an event, and recall from (2.51) the definition of the random variable H_A,

$$H_A(\omega) = \inf\left\{\sum_e |\omega'(e) - \omega(e)| : \omega' \in A\right\}, \qquad \omega \in \Omega.$$

We shall consider the event $A_n = \{0 \leftrightarrow \partial\Lambda_n\}$, and we write H_n for H_{A_n}. The question of ascertaining the asymptotics of H_n may be viewed as a first-passage problem. Imagine you are travelling from 0 to $\partial\Lambda_n$; travel along open edges is instantaneous, but along each closed edge requires time 1. The fastest route requires time H_n, and one is interested in the *time-constant* η, defined as $\eta = \lim_{n\to\infty}\{n^{-1}H_n\}$.

(5.61) Theorem (Existence of time-constant). *Let μ be a probability measure on $(\Omega, \mathcal{F})$ that is automorphism-invariant and $\mathbb{Z}^d$-ergodic. The deterministic limit*

$$\eta(\mu) = \lim_{n\to\infty}\left\{\frac{1}{n}H_n\right\}$$

exists μ-almost-surely and in $L^1(\mu)$.

The constant $\eta(\mu)$ is called the *time-constant* associated with μ.

Proof. See the comments in [119, 211], and the later paper [58]. The existence of the limit η is a consequence of a theorem attributed in [119, p. 748, Erratum] and [211, p. 259] to Derrienic. □

We apply this to the measure $\mu = \phi^0_{p,q}$ to deduce the existence, $\phi^0_{p,q}$-almost-surely, of the associated (deterministic) time-constant

$$\eta(p,q) = \lim_{n\to\infty}\left\{\frac{1}{n}H_n\right\}. \tag{5.62}$$

By Proposition 4.28, $\eta(p,q)$ is non-increasing in p, and we define

$$p_{\text{tc}}(q) = \sup\{p : \eta(p,q) > 0\}. \tag{5.63}$$

We seek a condition under which $\eta(p,q) > 0$. As usual, C denotes the vertex set of the open cluster at the origin.

(5.64) Theorem (Positivity of time-constant) [163]. *Let $p \in (0, 1)$ and $q \in [1, \infty)$. If $\phi^0_{p,q}(|C|^{2d+\epsilon}) < \infty$ for some $\epsilon > 0$, then $\eta(p, q) > 0$.*

We define the further critical point

$$p_g(q) = \sup\{p : \psi(p, q) > 0\}, \tag{5.65}$$

with $\psi(p, q)$ as in (5.56). The *correlation length* $\xi(p, q)$ is defined by

$$\xi(p, q) = \psi(p, q)^{-1},$$

subject to the convention that $0^{-1} = \infty$. Note that $\xi(p, q)$ is non-decreasing in p. Thus $\phi^0_{p,q}(0 \leftrightarrow \partial\Lambda_n)$ decays exponentially as $n \to \infty$ if and only if $\xi(p, q) < \infty$.

(5.66) Theorem. *Let $q \in [1, \infty)$. It is the case that $p_{tc}(q) = p_g(q)$.*

By this theorem and the prior observations,

$$p_{tc}(q) = p_g(q) = \widetilde{p}_c(q) \le p_c(q), \tag{5.67}$$

with equality conjectured. From the next section onwards, we shall use the expression $\widetilde{p}_c(q)$ for the common value of $p_{tc}(q)$, $p_g(q)$, $\widetilde{p}_c(q)$.

In the percolation case (when $q = 1$), the above first-passage problem and the associated time-constant $\eta(p, q)$ have been studied in detail; see [208, 211]. Several authors have given serious attention to a closely related question when $q = 2$ and $d = 2$, namely, the corresponding question for the two-dimensional Ising model with the 'passage time' H_n replaced by the minimum number of changes of spin along paths from the origin to $\partial\Lambda_n$, see [1, 90, 119]. The time-constant in the Ising case cannot exceed the corresponding random-cluster time-constant $\eta(p, 2)$, since each edge of the Ising model having endvertices with unlike spins gives rise to a closed edge in the (coupled) random-cluster model.

We turn now to the proofs of Theorems 5.60 and 5.66, and shall use the 'exponential-steepness' Theorems 3.42 and 3.45. Let A be an increasing cylinder event. We apply (3.44) and (3.47) to the random-cluster measure $\phi^0_{\Lambda_m,p,q}$, noting that

$$\frac{q(1-r)}{s-r} < C < \frac{q}{s-r}.$$

Let $m \to \infty$ to obtain that, for $0 < r < s < 1$,

$$\phi^0_{r,q}(A) \le \phi^0_{s,q}(A)\exp\{-4(s-r)\phi^0_{s,q}(H_A)\}, \tag{5.68}$$

$$\phi^0_{r,q}(H_A) \ge \frac{-\log\phi^0_{s,q}(A)}{\log[q/(s-r)]} - \frac{C}{C-1}. \tag{5.69}$$

Note in passing that inequalities (5.68) and (5.69), with $A = A_n = \{0 \leftrightarrow \partial\Lambda_n\}$, imply that the correlation length $\xi(p, q)$ is *strictly* increasing in p whenever it is finite, cf. [154, Thm 6.14].

Proof of Theorem 5.66. Let $r < s < p_{\mathrm{tc}}(q)$. Since $s < p_{\mathrm{tc}}(q)$, there exists $\gamma = \gamma(s, q) > 0$ such that

$$\phi_{s,q}(H_n) \geq n\gamma, \qquad n \geq 1. \tag{5.70}$$

Let $A = A_n = \{0 \leftrightarrow \partial\Lambda_n\}$. In conjunction with (5.70), (5.68) implies the exponential decay of $\phi_{r,q}(A_n)$, whence $r \leq p_{\mathrm{g}}(q)$. Therefore $p_{\mathrm{tc}}(q) \leq p_{\mathrm{g}}(q)$.

Conversely, suppose that $r < s < p_{\mathrm{g}}(q)$. There exists $\alpha = \alpha(s, q) > 0$ such that $\phi^0_{s,q}(A_n) \leq e^{-\alpha n}$. By (5.69) with $A = A_n$ and some $\beta = \beta(r, s, q) > 0$,

$$\phi^0_{r,q}(H_n) \geq \frac{-\log(e^{-\alpha n})}{\log[q/(s-r)]} - \beta = \frac{\alpha n}{\log[q/(s-r)]} - \beta,$$

whence $r \leq p_{\mathrm{tc}}(q)$. Therefore $p_{\mathrm{g}}(q) \leq p_{\mathrm{tc}}(q)$. □

There are two stages in the proof of Theorem 5.60. In the first, we apply (5.68)–(5.69) with $A = A_n$, and we utilize an iterative scheme to prove that $\phi^0_{p,q}(A_n)$ decays 'near-exponentially' when $p < \widetilde{p}_{\mathrm{c}}(q)$. In the second stage, we use Theorems 5.64 and 5.66 to deduce full exponential decay. The conclusion of the first stage may be summarized as follows.

(5.71) Lemma. *Let $q \in [1, \infty)$, and let $0 < p < \widetilde{p}_{\mathrm{c}}(q)$. There exist constants $c = c(p, q) \in (0, \infty)$, $\Delta = \Delta(p, q) \in (0, 1)$ such that*

$$\phi^0_{p,q}(A_n) \leq \exp(-cn^{\Delta}), \qquad n \geq 1. \tag{5.72}$$

Lemma 5.71 will be proved by an iterative scheme which may be continued further. If this is done, one obtains that $\phi^0_{p,q}(A_n)$ decays at least as fast as $\exp(-\alpha_k n/\log^k n)$ for any $k \geq 1$, where $\alpha_k = \alpha_k(p, q) > 0$ and $\log^k n$ is the kth iterate of logarithm, that is,

$$\log^1 x = \log x, \qquad \log^k x = \log(1 \vee \log^{k-1} x), \quad k \geq 2.$$

Proof of Lemma 5.71. We shall use (5.68) and (5.69) in an iterative scheme. In the following, we shall sometimes use *real* quantities when *integers* are required. All terms of the form o(1) or O(1) are to be interpreted in the limit as $n \to \infty$.

Fix $q \in [1, \infty)$. For $p < \widetilde{p}_{\mathrm{c}}(q)$, there exists $c_1(p) > 0$ such that

$$\phi^0_{p,q}(A_n) \leq \frac{c_1(p)}{n^{d-1}}, \qquad n \geq 1. \tag{5.73}$$

Let $0 < r < s < t < \widetilde{p}_{\mathrm{c}}(q)$. By (5.69),

$$\phi^0_{s,q}(H_n) \geq \frac{-\log \phi^0_{t,q}(A_n)}{\log D} + \mathrm{O}(1) \geq \frac{(d-1)\log n}{\log D} + \mathrm{O}(1)$$

where $1 < D = q/(t-s) < \infty$. We substitute this into (5.68) to obtain that

$$\phi^0_{r,q}(A_n) \le \frac{c_2(r)}{n^{d-1+\Delta_2(r)}}, \qquad n \ge 1, \tag{5.74}$$

for some strictly positive and finite $c_2(r)$ and $\Delta_2(r)$. This holds for all $r < \widetilde{p}_c(q)$, and is an improvement in order of magnitude over (5.73).

We obtain next an improvement of (5.74). Let m be a positive integer, and let $R_i = im$ for $i = 0, 1, \dots, K$, where $K = \lfloor n/m \rfloor$. Let $L_i = \{\partial\Lambda_{R_i} \leftrightarrow \partial\Lambda_{R_{i+1}}\}$, and let $F_i = H_{L_i}$, the minimal number of extra edges needed for L_i to occur. Clearly,

$$H_n \ge \sum_{i=0}^{K-1} F_i, \tag{5.75}$$

since every path from 0 to $\partial\Lambda_n$ traverses each annulus $\Lambda_{R_{i+1}} \setminus \Lambda_{R_i}$. There exists a constant $\rho \in [1, \infty)$ such that $|\partial\Lambda_R| \le \rho R^{d-1}$ for all R. By the translation-invariance of $\phi^0_{p,q}$,

$$\phi^0_{p,q}(L_i) \le |\partial\Lambda_{R_i}|\phi^0_{p,q}(A_m) \le \rho n^{d-1}\phi^0_{p,q}(A_m). \tag{5.76}$$

Let $0 < r < s < \widetilde{p}_c(q)$, and let $c_2 = c_2(s)$, $\Delta_2 = \Delta_2(s)$ where the functions $c_2(p)$ and $\Delta_2(p)$ are chosen as in (5.74). By (5.74) and (5.76),

$$\phi^0_{s,q}(L_i) \le \rho n^{d-1}\frac{c_2}{m^{d-1+\Delta_2}} \le \frac{1}{2} \tag{5.77}$$

if

$$m = [(3\rho c_2)n^{d-1}]^{1/(d-1+\Delta_2)}, \tag{5.78}$$

and we choose m accordingly (here and later, we take n to be large). Now $F_i \ge 1$ if L_i does not occur, whence

$$\phi^0_{s,q}(H_n) \ge \sum_{i=0}^{K-1}[1 - \phi^0_{s,q}(L_i)] \ge \tfrac{1}{2}K \tag{5.79}$$

by (5.75) and (5.77). Also,

$$K = \lfloor n/m \rfloor \ge an^{\Delta_3} \tag{5.80}$$

by (5.78), for appropriate constants $a \in (0, \infty)$ and $\Delta_3 \in (0, 1)$. By (5.68) and (5.79),

$$\phi^0_{r,q}(A_n) \le \exp(-c_3 n^{\Delta_3}), \qquad n \ge 1, \tag{5.81}$$

where $c_3 = c_3(r) \in (0, \infty)$ and $\Delta_3 = \Delta_3(r) \in (0, 1)$. □

Proof of Theorem 5.64. Assume the given hypothesis. We shall use an extension of an argument taken from [119]. Let Π_n be the set of all paths of $\mathbb{L}^d$ joining the origin to $\partial\Lambda_n$. With $T(\pi)$ denoting the number of closed edges in a path π, we have that

$$T(\pi) + 1 \geq \sum_{x \in \pi} \frac{1}{|C_x \cap \pi|}$$

where the summation is over all vertices x of π, C_x is the open cluster at x, and $|C_x \cap \pi|$ is the number of vertices common to C_x and π. By Jensen's inequality,

$$\frac{T(\pi)+1}{|\pi|} \geq \frac{1}{|\pi|} \sum_{x \in \pi} \frac{1}{|C_x|} \geq \left\{ \frac{1}{|\pi|} \sum_{x \in \pi} |C_x| \right\}^{-1}.$$

Therefore,

$$\frac{H_n + 1}{n} \geq \inf_{\pi \in \Pi_n} \left\{ \frac{T(\pi)+1}{|\pi|} \right\} \geq \frac{1}{K_n},$$

where

$$K_n = \sup_{\pi \in \Pi_n} \left\{ \frac{1}{|\pi|} \sum_{x \in \pi} |C_x| \right\}.$$

By (5.62), $\phi_{p,q}^0(\eta \geq K^{-1}) = 1$, where

$$K = \limsup_{m \to \infty} \left[\sup \left\{ \frac{1}{|\pi|} \sum_{x \in \pi} |C_x| : |\pi| = m \right\} \right]. \tag{5.82}$$

The (inner) supremum is over all paths from the origin containing m vertices. We propose to show that $\phi_{p,q}^0(K < \infty) = 1$, whence $\eta > 0$ as required.

Let $\{\widetilde{C}_x : x \in \mathbb{Z}^d\}$ be a collection of independent subsets of $\mathbb{Z}^d$ with the property that $\widetilde{C}_x$ has the same law as C_x. We claim, as in [119], that the family $\{|C_x| : x \in \mathbb{Z}^d\}$ is dominated stochastically by $\{M_x : x \in \mathbb{Z}^d\}$, where

$$M_x = \sup\big\{|\widetilde{C}_y| : y \in \mathbb{Z}^d,\ x \in \widetilde{C}_y\big\},$$

and we shall prove this inductively. Let $v_1, v_2, \dots$ be a deterministic ordering of $\mathbb{Z}^d$. Given the random variables $\{\widetilde{C}_x : x \in \mathbb{Z}^d\}$, we shall construct a family $\{D_x : x \in \mathbb{Z}^d\}$ having the same joint law as $\{C_x : x \in \mathbb{Z}^d\}$ and satisfying: for each x, there exists y such that $D_x \subseteq \widetilde{C}_y$. First, we set $D_{v_1} = \widetilde{C}_{v_1}$. Given $D_{v_1}, D_{v_2}, \dots, D_{v_n}$, we define $E = \bigcup_{i=1}^n D_{v_i}$. If $v_{n+1} \in E$, we set $D_{v_{n+1}} = D_{v_j}$ for some j such that $v_{n+1} \in D_{v_j}$. If $v_{n+1} \notin E$, we proceed as follows. Let $\Delta_e E$ be the set of edges of $\mathbb{Z}^d$ having exactly one endvertex in E. We may find a (random) subset F of $\widetilde{C}_{v_{n+1}}$ such that F has the conditional law of $C_{v_{n+1}}$ *given that* all edges in $\Delta_e E$ are closed; we now set $D_{v_{n+1}} = F$. We are using two properties of $\phi^0_{p,q}$ here. Firstly, the law of $C_{v_{n+1}}$ given $C_{v_1}, C_{v_2}, \dots, C_{v_n}$ depends only on $\Delta_e E$, and secondly, $\phi^0_{p,q}$ is positively associated. We obtain the required stochastic domination accordingly.

By (5.82) (and subject to K and the $\widetilde{C}_x$ being defined on the same probability space),

$$K \le \limsup_{m\to\infty} \left[\sup\left\{ \frac{1}{|\pi|} \sum_{x\in\pi} M_x : |\pi| = m \right\} \right] \qquad \text{a.s.}$$

By [119, Lemma 2],

$$K \le 2 \limsup_{m\to\infty} \left[\sup\left\{ \frac{1}{|\Gamma|} \sum_{x\in\Gamma} |\widetilde{C}_x|^2 : |\Gamma| = m \right\} \right] \qquad \text{a.s.}$$

where the (inner) supremum is over all animals Γ of $\mathbb{L}^d$ having m vertices and containing the origin. By the main result of [97], the right side is almost-surely finite so long as each $|\widetilde{C}_x|^2$ has finite $(d+\epsilon)$th moment for some $\epsilon > 0$. The required conclusion follows. □

Proof of Theorem 5.60. Let $q \in [1,\infty)$ and $p < \widetilde{p}_c(q)$. Find r such that $p < r < \widetilde{p}_c(q)$. By Lemma 5.71, there exist $c, \Delta > 0$ such that

$$\phi^0_{r,q}(A_n) \le \exp(-cn^\Delta), \qquad n \ge 1.$$

This implies that $\phi^0_{r,q}(|C|^{2d+1}) < \infty$. By Theorem 5.64, $\eta(r,q) > 0$, and so $r \le p_{tc}(q)$. By Theorem 5.66, $r \le p_g(q)$, and the claim follows. □

5.6 Exponential decay of volume

For percolation, there is a beautiful proof of the exponential decay of volume using only that of radius. This proof hinges on the independence of the states of different edges, and may therefore not be extended at present to general random-cluster models, see [154, Thm 6.78]. We shall instead make use here of the block arguments of [209], obtaining thereby the exponential decay of volume subject to a condition on p believed but not known to hold throughout the subcritical phase. This condition differs slightly from that of the last section in that it involves the decay rate of certain *finite-volume* probabilities.

Let $a \ge 1$, and let

$$L^a(p,q) = \limsup_{n\to\infty} \left\{ n^{d-1} \phi^1_{\Lambda_{an},p,q}(0 \leftrightarrow \partial\Lambda_n) \right\}. \tag{5.83}$$

As at (5.57), $L^a(p,q)$ is non-decreasing in p, and therefore,

$$L^a(p,q) \begin{cases} = 0 & \text{if } p < \widetilde{p}^a_c(q), \\ \in (0,\infty] & \text{if } p > \widetilde{p}^a_c(q), \end{cases}$$

where

$$\widetilde{p}_c^a(q) = \sup\{p : L^a(p,q) = 0\}. \tag{5.84}$$

Clearly $\widetilde{p}_c^a(q)$ is non-decreasing in a, and furthermore $\widetilde{p}_c^a(q) \le \widetilde{p}_c(q)$ for all $a \ge 1$. We set

$$\widetilde{p}_c^\infty(q) = \lim_{a\to\infty} \widetilde{p}_c^a(q).$$

It is elementary that $\widetilde{p}_c^\infty(q) \le \widetilde{p}_c(q)$, and we conjecture that equality holds.

(5.85) Conjecture. *If $q \in [1,\infty)$, then $\widetilde{p}_c^\infty(q) = \widetilde{p}_c(q)$.*

Here is the main result of this section.

(5.86) Theorem. *Let $q \in [1,\infty)$. There exists $\rho(p,q)$, satisfying $\rho(p,q) > 0$ when $p < \widetilde{p}_c^\infty(q)$, such that*

$$\phi_{p,q}^0(|C| \ge n) \le e^{-n\rho}, \qquad n \ge 1.$$

The hypothesis $p < \widetilde{p}_c^\infty(q)$ is slightly stronger than that of Theorem 5.60, and so is the conclusion, since $\phi_{p,q}^0(\mathrm{rad}(C) \ge n) \le \phi_{p,q}^0(|C| \ge n)$.

Proof. We adapt the arguments of [209, Section 2], from which we extract the main steps. For $N \ge 1$ and $i = 1, 2, \dots, d$, we define the box

$$T_N(i) = [0, 3N]^{i-1} \times [0, N] \times [0, 3N]^{d-i}.$$

For $\omega \in \Omega$, an *i-crossing* of $T_N(i)$ is an open path $x_0, e_0, x_1, e_1, \dots, e_m$ of alternating vertices and edges of $T_N(i)$ such that the ith coordinate of x_0 (respectively, x_m) is 0 (respectively, N). Such crossings are in the short direction. For $b > 3$, we define

$$\tau_N^b = \phi_{\Lambda_{bN},p,q}^1\big(T_N(i) \text{ has an } i\text{-crossing}\big), \tag{5.87}$$

noting by rotation-invariance that τ_N^b does not depend on the value of i.

Let N be a fixed positive integer. From $\mathbb{L}^d$ we construct a new lattice $\mathcal{L}$ as follows. First, $\mathcal{L}$ has vertex set $\mathbb{Z}^d$. Two vertices $\mathbf{x}, \mathbf{y}$ of $\mathcal{L}$ are deemed adjacent in $\mathcal{L}$ if and only if $|x_i - y_i| \le 3$ for all $i = 1, 2, \dots, d$. The better to distinguish vertices of $\mathbb{L}^d$ and $\mathcal{L}$, we shall use bold letters to indicate the latter. Let $\omega \in \Omega$. Vertex $\mathbf{x}$ of $\mathcal{L}$ is coloured *white* if there exists $i \in \{1, 2, \dots, d\}$ such that $N\mathbf{x} + T_N(i)$ has an i-crossing, and is coloured *black* otherwise. The event $\{\mathbf{x} \text{ is white}\}$ is increasing, and is defined in terms of the states of edges in the box $\Delta(\mathbf{x}) = N\mathbf{x} + [0, 3N]^d$.

The following lemma relates the size of the open cluster C at the origin of $\mathbb{L}^d$ to the sizes of white clusters of $\mathcal{L}$. For $\mathbf{x} \in \mathbb{Z}^d$, we write $W_{\mathbf{x}}$ for the connected cluster of white vertices of $\mathcal{L}$ containing $\mathbf{x}$.

(5.88) Lemma [209]. *Let $\omega \in \Omega$. Assume that C contains some vertex v with*

$$v \in \prod_{j=1}^{d} \big[x_j N, (x_j + 1)N - 1\big],$$

for some $\mathbf{x} = (x_1, x_2, \ldots, x_d) \in \mathbb{Z}^d$ satisfying

$$|x_j| \geq 2 \quad \text{for some} \quad j \in \{1, 2, \ldots, d\}. \tag{5.89}$$

There exists a neighbour $\mathbf{y}$ of the origin $\mathbf{0}$ of $\mathcal{L}$ such that

$$|W_{\mathbf{y}}| \geq 7^{-2d} \left(\frac{|C| - (4N)^d}{N^d} \right). \tag{5.90}$$

Proof. This may be derived from that of [209, Lemma 2]. □

Since $\mathbf{0}$ has fewer than 7^d neighbours on $\mathcal{L}$,

$$\phi^0_{p,q}(|C| \geq n) \leq 7^d \phi^0_{p,q}(|W_{\mathbf{0}}| \geq An - 1), \tag{5.91}$$

where $A = 7^{-2d} N^{-d}$. Therefore,

$$\phi^0_{p,q}(|C| \geq n) \leq 7^d \sum_{m \geq An-1} a_m M_{p,q}(m), \tag{5.92}$$

where a_m is the number of connected sets w of m vertices of $\mathcal{L}$ containing $\mathbf{0}$, and

$$M_{p,q}(m) = \max\big\{\phi^0_{p,q}(\text{all vertices in } w \text{ are white}) : |w| = m\big\}.$$

By the final display of the proof of [209, Lemma 3],

$$a_m \leq 7^{2d}(7^d e)^m, \tag{5.93}$$

and it remains to bound $\phi^0_{p,q}$(all vertices in w are white).

Fix $b > 3$, to be chosen later. Let w be as above with $|w| = m$. There exists a constant $c = c(b) > 0$ such that: w contains at least t vertices $\mathbf{y}(1), \mathbf{y}(2), \ldots, \mathbf{y}(t)$ such that $t \geq cm$ and the boxes $N\mathbf{y}(r) + \Lambda_{bN}$, $r = 1, 2, \ldots, t$, of $\mathbb{L}^d$ are disjoint. We may choose such a set $\{\mathbf{y}(r) : r = 1, 2, \ldots, t\}$ in a way which depends only on the set w. Then

$$\phi^0_{p,q}(\text{all vertices in } w \text{ are white}) \leq \phi^0_{p,q}\big(\mathbf{y}(r) \text{ is white}, r = 1, 2, \ldots, t\big). \tag{5.94}$$

The events $\{\mathbf{y}(r) \text{ is white}\}$, $r = 1, 2, \ldots, t$, are dependent under $\phi^0_{p,q}$. However, by positive association,

(5.95)

$$\phi^0_{p,q}\big(\mathbf{y}(r) \text{ is white}, r = 1, 2, \ldots, t\big) \leq \phi^0_{p,q}\big(\mathbf{y}(r) \text{ is white}, r = 1, 2, \ldots, t \,\big|\, E\big),$$

where E is the event that every edge e having both endvertices in $N\mathbf{y}(r)+\partial\Lambda_{bN}$, for any given $r \in \{1, 2, \dots, t\}$, is open. Under the conditional measure $\phi^0_{p,q}(\cdot \mid E)$, the events $\{\mathbf{y}(r)$ is white$\}$, $r = 1, 2, \dots, t$, are independent, whence by symmetry

$$\text{(5.96)} \quad \phi^0_{p,q}\big(\mathbf{y}(r) \text{ is white}, r = 1, 2, \dots, t\big) \le \big\{\phi^1_{\Lambda_{bN},p,q}(\mathbf{0} \text{ is white})\big\}^t \le (d\tau^b_N)^t.$$

By (5.92)–(5.96),

$$\text{(5.97)} \qquad \phi^0_{p,q}(|C| \ge n) \le 7^d \sum_{m \ge An-1} 7^{2d}(7^d e)^m (d\tau^b_N)^{\lfloor cm \rfloor}.$$

Let $a > 1$ and choose $b > 3 + a$, noting that $x + \Lambda_{aN} \subseteq \Lambda_{bN}$ for all $x \in \mathbb{Z}^d$ lying in the region $R = [0, 3N]^{d-1} \times \{0\}$. If $T_N(d)$ has a d-crossing, there exists $x \in R$ such that $x \leftrightarrow x + \partial\Lambda_N$. Since $\phi^1_{\Lambda_{bN},p,q} \le_{\text{st}} \phi^1_{\Lambda_{aN},p,q}$,

$$\begin{aligned} \text{(5.98)} \qquad \tau^b_N &\le \sum_{x \in R} \phi^1_{\Lambda_{bN},p,q}(x \leftrightarrow x + \partial\Lambda_N) \\ &\le |R|\phi^1_{\Lambda_{aN},p,q}(0 \leftrightarrow \partial\Lambda_N) \\ &= (3N+1)^{d-1}\phi^1_{\Lambda_{aN},p,q}(0 \leftrightarrow \partial\Lambda_N). \end{aligned}$$

Let $p < \widetilde{p}^{\infty}_{c}(q)$, and choose $a > 1$ such that $p < \widetilde{p}^{a}_{c}(q)$. With $b > 3 + a$, the right side of (5.98) may be made as small as required by a suitably large choice of N, and we choose N in such a way that $7^d e(d\tau^b_N)^c < \frac{1}{2}$. Inequality (5.97) provides the required exponential bound. □

5.7 The supercritical phase and the Wulff crystal

Percolation theory is a source of intuition for the more general random-cluster model, but it has not always been possible to make such intuition rigorous. This is certainly so in the supercritical phase, where several of the basic questions remain unanswered to date. We shall work in this section with the free and wired measures, $\phi^0_{p,q}$ and $\phi^1_{p,q}$, and we assume throughout that $q \in [1, \infty)$.

The first property of note is the almost-sure uniqueness of the infinite open cluster. A probability measure ϕ on $(\Omega, \mathcal{F})$ is said to have the 0/1-*infinite-cluster property* if the number I of infinite open clusters satisfies $\phi(I \in \{0, 1\}) = 1$. We recall from Theorem 4.33(c) that every translation-invariant member of the closed convex hull of $\mathcal{W}_{p,q}$ has the 0/1-infinite-cluster property. By ergodicity, see Corollary 4.23, we arrive at the following.

(5.99) Theorem (Uniqueness of infinite open cluster). *Let $p \in [0, 1]$ and $q \in [1, \infty)$. We have for $b = 0, 1$ that*

$$\phi^b_{p,q}(I = 1) = 1 \quad \textit{whenever} \quad \theta^b(p, q) > 0. \tag{5.100}$$

Let $q \in [1, \infty)$ and $p > p_c(q)$. There exists ($\phi^b_{p,q}$-almost-surely) a unique infinite open cluster. What may be said about the shapes and sizes of *finite* open clusters? One expects finite clusters to have properties broadly similar to those of supercritical percolation. Much progress has been made in recent years towards proofs of such statements, but a vital step remains unresolved. As was true formerly for percolation, the results in question are proved only for p exceeding a certain 'slab critical point' $\widehat{p_c}(q)$, and it is an important open problem to prove that $\widehat{p_c}(q) = p_c(q)$ for all $q \in [1, \infty)$.

Here is an illustration. It is fundamental for supercritical percolation that the tails of the radius and volume of a finite open cluster decay exponentially in n and $n^{(d-1)/d}$ respectively, see [154, Thms 8.18, 8.65]. This provokes an important problem for the random-cluster model whose full resolution remains open. Partial results are known when $p > \widehat{p_c}(q)$, see Theorems 5.104 and 5.108.

(5.101) Conjecture. *Let $p \in [0, 1]$ and $q \in [1, \infty)$. There exist $\sigma = \sigma(p, q)$, $\gamma = \gamma(p, q)$, satisfying $\sigma(p, q), \gamma(p, q) > 0$ when $p > p_c(q)$, such that*

$$\phi^1_{p,q}(n \le \mathrm{rad}(C) < \infty) \le e^{-n\sigma},$$
$$\phi^1_{p,q}(n \le |C| < \infty) \le e^{-\gamma n^{(d-1)/d}}, \qquad n \ge 1.$$

We turn next to a discussion of the so-called 'Wulff construction'. Much attention has been paid to the sizes and shapes of clusters formed in models of statistical mechanics. When a cluster C is infinite with a strictly positive probability, but is constrained to have some large *finite* size N, then C is said to form a large 'droplet'. The asymptotic shape of such a droplet, in the limit of large N, is prescribed in general terms by the theory of the so-called Wulff crystal[5]. In the case of the random-cluster model, we ask for properties of the open cluster C at the origin, conditional on the event $\{N \le |C| < \infty\}$ for large N. The rigorous picture is not yet complete, but techniques have emerged through the work of Cerf and Pisztora, [83, 84, 276], which may be expected to reveal in due course a complete account of the Wulff theory of large finite clusters in the random-cluster model. A full account of this work would be too lengthy for inclusion here, and we content ourselves with a brief summary.

The study of the Wulff crystal is bound up with the law of the volume of a finite cluster, see Conjecture 5.101. It is straightforward to adapt the corresponding percolation proof (see [154, Thm 8.61]) to obtain that

$$\phi^1_{p,q}(|C| = n) \ge e^{-\gamma n^{(d-1)/d}},$$

[5]Such shapes are named after the author of [325]. The first mathematical results on Wulff shapes were proved for the two-dimensional Ising model in [104], see the review [55].

for some γ satisfying $\gamma < \infty$ when $p_c(q) < p < 1$. It is believed as noted above that this is the correct order for the rate of decay of $\phi^1_{p,q}(|C| = n)$ when $p > p_c(q)$.

Before continuing, we make a comment concerning the number of dimensions. The case $d = 2$ is special (see Chapter 6). By the duality theory for planar graphs, the dual of a supercritical random-cluster measure is a subcritical random-cluster measure, and this permits the use of special arguments. We shall therefore suppose for the majority of the rest of this section that $d \geq 3$; some remarks about the two-dimensional case are made after Theorem 5.108.

A partial account of the Wulff construction and the decay of volume of a finite cluster is provided in [83], where the asymptotic shape of droplets is studied in the special case of the Ising model. The proofs to date rely on two assumptions on the value of p, namely that p is such that $\phi^0_{p,q} = \phi^1_{p,q}$, cf. Conjecture 5.34, and secondly that p exceeds a certain 'slab critical point' $\widehat{p}_c(q)$ which we introduce next.

Fix $q \in [1, \infty)$ and let $d \geq 3$. Let $S(L, n)$ be the *slab* given as

$$S(L, n) = [0, L - 1] \times [-n, n]^{d-1},$$

and let $\psi^{L,n}_{p,q} = \phi^0_{S(L,n),p,q}$ be the random-cluster measure on $S(L, n)$ with parameters p, q, and with free boundary conditions. We denote by $\Pi(p, L)$ the property that:

$$\text{there exists } \alpha > 0 \text{ such that, for all } n \text{ and all } x \in S(L, n),\ \psi^{L,n}_{p,q}(0 \leftrightarrow x) > \alpha.$$

It is not hard to see that $\Pi(p, L) \Rightarrow \Pi(p', L')$ if $p \leq p'$ and $L \leq L'$, and it is thus natural to define the quantities

$$(5.102) \qquad \widehat{p}_c(q, L) = \inf\{p : \Pi(p, L) \text{ occurs}\}, \quad \widehat{p}_c(q) = \lim_{L\to\infty} \widehat{p}_c(q, L).$$

Clearly, $p_c(q) \leq \widehat{p}_c(q) < 1$. It is believed that equality holds in that $\widehat{p}_c(q) = p_c(q)$, and it is a major open problem to prove this[6].

(5.103) Conjecture [276]. *Let $q \in [1, \infty)$ and $d \geq 3$. Then $\widehat{p}_c(q) = p_c(q)$.*

The case $q = 1$ of Conjecture 5.103 is special, since percolation enjoys a spatial independence not shared with general random-cluster models. This additional property has been used in the formulation of a type of 'dynamic renormalization', which has in turn yielded a proof that $\widehat{p}_c(1) = p_c(1)$ for percolation in three or more dimensions, see [24], [154, Chapter 7], [161]. Such arguments have been adapted by Bodineau to the Ising model, resulting in proofs that $\widehat{p}_c(2) = p_c(2)$ and that the pure phases are the unique extremal Gibbs states when $p \neq p_c(2)$, see

[6]One may expect the methods of Section 7.5 to yield a proof that $\widehat{p}_c(q) = p_c(q)$ for sufficiently large q.

[53, 54]. Such arguments do not to date have a full random-cluster counterpart. Instead, in the random-cluster setting, one exploits what might be termed 'static renormalization' methods, or 'block arguments', see [83, 276]. One divides space into blocks, constructs events of an appropriate nature on such blocks, having large probabilities, and then allows these events to combine across space. There have been substantial successes using this technique, of which the most striking is the resolution (subject to side conditions) of the Wulff construction for the Ising model.

We state next an exponential-decay theorem for the radius of a finite cluster; the proof is given at the end of this section. It is an immediate corollary that the 'truncated two-point connectivity function' $\phi^1_{p,q}(x \leftrightarrow y,\ x \nleftrightarrow \infty)$ decays exponentially in the distance $\|x - y\|$, whenever $p > \widehat{p}_c(q)$.

(5.104) Theorem. *Let* $q \in [1, \infty)$, $d \geq 3$, *and* $p > \widehat{p}_c(q)$. *There exists* $\sigma = \sigma(p, q) > 0$ *such that*

$$\phi^1_{p,q}(n \leq \mathrm{rad}(C) < \infty) \leq e^{-n\sigma}, \qquad n \geq 1.$$

We turn now to the Wulff construction. Subject to a verification of Conjecture 5.103, and of a positive answer to the question of the uniqueness of random-cluster measures when $p > p_c(q)$, the block arguments of Cerf and Pisztora yield a largely complete picture of the Wulff theory of random-cluster models with $q \in [1, \infty)$, see [83, 276] and also [84]. Paper [81] is a fine review of Wulff constructions for percolation, Ising, and random-cluster models.

The reader is referred to [81] for an introductory discussion to the physical background of the Wulff construction. It may be summarized as follows for random-cluster models. Let $\Lambda_n = [-n, n]^d$, and consider the wired random-cluster measure $\phi^1_{\Lambda_n,p,q}$ with $p > p_c(q)$. The larger an open cluster, the more likely it is to be joined to the boundary $\partial\Lambda_n$. Suppose that we condition on the event that there exists in Λ_n an open cluster C that does not intersect $\partial\Lambda_n$ and that has volume of the order of the volume n^d of the box. What can be said about the shape of C? Since $p > p_c(q)$, there is little cost in having large *volume*, and the price of such a cluster accumulates around its *external boundary*. It turns out that the price may be expressed as a surface integral of an appropriate function termed 'surface tension'. This 'surface tension' may be specified as the exponential rate of decay of a certain probability. The Wulff prediction for the shape of C is that, when re-scaled in the limit of large n, it converges to the solution of a certain variational problem, that is, the limit shape is obtained by minimizing a certain surface integral subject to a constraint on its volume.

For $A \subseteq \mathbb{Z}^d$, let $\rho(A)$ be the number of vertices $x \in A$ such that $x \leftrightarrow \partial A$. When $p > p_c(q)$, $\rho(\Lambda_n)$ has order $|\Lambda_n|$. Let C be the open cluster at the origin, and suppose we condition on the event $\{|C| \geq \alpha n^d,\ C \cap \partial\Lambda_n = \varnothing\}$ where $\alpha > 0$. This conditioning implies a change in value of $\rho(\Lambda_n)/|\Lambda_n|$ amounting to a large deviation. The link between Wulff theory and large deviations is made more concrete in the next theorem. The set $\mathcal{D}_q$ is given in Theorem 4.63 as the (at most

Figure 5.1. Images of the Wulff crystal for the two-dimensional Ising model at two distinct temperatures, produced by simulation in time, and reproduced by courtesy of Raphaël Cerf. The simulations were for finite time, and the images are therefore only approximations to the true crystals. The pictures are 1024 pixels square, and the inverse-temperatures are $\beta = \frac{4}{3}, \frac{10}{11}$. The corresponding random-cluster models have $q = 2$ and $p = 1 - e^{-4/3}, 1 - e^{-10/11}$.

countable) set of values of p at which there is non-uniqueness of random-cluster measures with cluster-weighting factor q.

(5.105) Theorem [81, 83]. *Let $q \in [1, \infty)$ and $d \geq 3$. Let $p \in (\widehat{p}_c(q), 1)$ be such that $p \notin \mathcal{D}_q$. There exists a bounded, closed, convex set $\mathcal{W}$ of $\mathbb{R}^d$ containing the origin in its interior such that the following holds. Let $\theta = \theta^1(p, q)$, and let $\alpha \in (0, \theta)$ be sufficiently close to θ that the re-scaled crystal*

$$\mathcal{W}(\alpha) = \left(1 - \frac{\alpha}{\theta}\right)^{1/d} \frac{\mathcal{W}}{|\mathcal{W}|^{1/d}}$$

is a subset of the unit cube $[-\frac{1}{2}, \frac{1}{2}]^d$. Then, as $n \to \infty$,

$$\frac{1}{n^{d-1}} \log \phi^1_{\Lambda_n, p, q}\big(\rho(\Lambda_n) \leq \alpha |\Lambda_n|\big) \to -d \left(1 - \frac{\alpha}{\theta}\right)^{(d-1)/d} |\mathcal{W}|^{1/d}.$$

The set $\mathcal{W}$ is termed the 'Wulff crystal', and $|\mathcal{W}|$ denotes its d-dimensional Lebesgue measure. For proofs of this and later theorems, the reader is referred to the original papers. The Wulff crystal for the closely related Ising model is illustrated in Figure 5.1.

Theorem 5.105 may be stated without explicit reference to the set $\mathcal{W}$. The geometry of $\mathcal{W}$ becomes important in the complementary theorem, following. It is framed in terms of the convergence of random measures, and the point mass on the point $x \in \mathbb{R}$ is denoted by δ_x.

(5.106) Theorem [81]. *Let $q \in [1, \infty)$ and $d \geq 3$. Let $p \in (\widehat{p}_c(q), 1)$ be such that $p \notin \mathcal{D}_q$. There exists a bounded, closed, convex set $\mathcal{W}$ of $\mathbb{R}^d$ containing the origin in its interior such that the following holds. Under the conditional measure obtained from $\phi^1_{p,q}$ by conditioning on the event $\{n^d \leq |C| < \infty\}$, the random measure*

$$\frac{1}{n^d} \sum_{x \in C} \delta_{x/n}$$

converges in probability, with respect to the bounded, uniformly continuous functions, towards the set $\{\theta 1_{\mathcal{W}}(a+x)\, dx : a \in \mathbb{R}\}$ of measures, where $\theta = \theta^1(p, q)$. The probabilities of deviations are of order $\exp(-cn^{d-1})$.

The meaning of the conclusion is as follows. For $k \geq 1$, for any bounded, uniformly continuous function $f : \mathbb{R}^d \to \mathbb{R}^k$, and for any $\epsilon > 0$, there exists $c = c(d, k, p, q, f, \epsilon) > 0$ such that

$$(5.107) \quad \overline{\phi}^1_{p,q}\left(\exists a \in \mathbb{R}^d \text{ s.t. } \left| \frac{1}{n^d} \sum_{x \in C} f(x/n) - \theta \int_{x \in \mathcal{W}} f(a+x)\, dx \right| \leq \epsilon \right) \geq 1 - e^{-cn^{d-1}}, \qquad n \geq 1,$$

where $\overline{\phi}^1_{p,q}$ is the measure obtained from $\phi^1_{p,q}$ by conditioning on the event $\{n^d \leq |C| < \infty\}$, and $|\cdot|$ is the Euclidean norm on $\mathbb{R}^k$. This is a way of saying that the external boundary of a large finite open cluster with cardinality approximately n^d resembles the boundary of a translate of $n\mathcal{W}$. Within this boundary, the open cluster has density approximately θ, whilst the density outside is zero. It is presumably the case that the a in (5.107) may be chosen independently of f and ϵ, but this has not yet been proved.

One important consequence of the analysis of [83] is an exact asymptotic for the probability that $|C|$ is large.

(5.108) Theorem [81]. *Let $q \in [1, \infty)$ and $d \geq 3$. Let $p \in (\widehat{p}_c(q), 1)$ be such that $p \notin \mathcal{D}_q$. There exists $\gamma = \gamma(p, q) \in (0, \infty)$ such that*

$$\frac{1}{n^{d-1}} \log \phi^1_{p,q}(n^d \leq |C| < \infty) \to -\gamma \qquad \text{as } n \to \infty.$$

The above results are valid in two dimensions also although, as noted earlier, this case is special. When $d = 2$, the slab critical point $\widehat{p}_c(q)$ is replaced by the infimum of values p at which the dual process has exponential decay of connections (see (6.5) for the relation between the dual and primal parameter-values). That is, when $d = 2$,

$$\widehat{p}_c(q) = \frac{q(1 - p_g(q))}{p_g(q) + q(1 - p_g(q))}$$

where $p_g(q)$ is given at (5.65). Fluctuations in droplet shape for random-cluster models in two dimensions have been studied in [17, 18].

Proof of Theorem 5.104. We adapt the proof of [87] as reported in [154, Thm 8.21]. We shall build the cluster C at the origin (viewed as a set of open edges) step by step, in a manner akin to the proof of Proposition 5.30. First, we order the edges of $\mathbb{L}^d$ in some arbitrary but deterministic way, and we write e_i for the ith edge in this ordering. Let $\omega \in \Omega$. We shall construct a sequence $(C_0, D_0), (C_1, D_1), \dots$ of pairs of (random) edge-sets such that $C_i \subseteq C_{i+1}$ and $D_i \subseteq D_{i+1}$ for each i. Every edge in each C_i (respectively, D_i) will be open (respectively, closed). Let $C_0 = D_0 = \varnothing$. Having found (C_m, D_m) for $m = 0, 1, \dots, n$, we find the earliest edge $e \notin C_n \cup D_n$ in the above ordering such that e has an endvertex in common with some member of C_n; if $C_n = \varnothing$ we take $e \notin D_n$ to be the earliest edge incident to the origin if such an edge exists. We now define

$$(C_{n+1}, D_{n+1}) = \begin{cases} (C_n \cup \{e\}, D_n) & \text{if } e \text{ is open,} \\ (C_n, D_n \cup \{e\}) & \text{if } e \text{ is closed.} \end{cases}$$

This process is continued until no candidate edge e may be found, which is to say that we have exhausted the open cluster C. If $C_n = C$ for some n then we define $C_l = C$ for $l \geq n$, so that

$$C = \lim_{n\to\infty} C_n. \tag{5.109}$$

Let $H_n = \{x \in \mathbb{Z}^d : x_1 = n\}$, and let G_n be the event that the origin belongs to a finite cluster that intersects H_n. The box Λ_n has $2d$ faces, whence, by the rotation-invariance of $\phi^0_{p,q}$,

$$\phi^1_{p,q}\big(0 \leftrightarrow \partial\Lambda_n,\ |C| < \infty\big) \leq 2d\phi^1_{p,q}(G_n). \tag{5.110}$$

We shall prove that, for $p > \widehat{p}_c(q)$, there exists $\gamma > 0$ such that

$$\phi^1_{p,q}(G_n) \leq e^{-n\gamma}, \qquad n \geq 1, \tag{5.111}$$

and the claim of the theorem is an immediate consequence.

The idea of the proof of (5.111) is as follows. Since $p > \widehat{p}_c(q)$ by assumption, we may find an integer L such that $p > \widehat{p}_c(q, L)$. Write $S(L) = [0, L) \times \mathbb{Z}^{d-1}$, and

$$S_i(L) = S(L) + (i-1)Le_1 = \big[(i-1)L, iL\big) \times \mathbb{Z}^{d-1}, \tag{5.112}$$

where $e_1 = (1, 0, 0, \dots, 0)$. Suppose that G_{mL} occurs for some $m \geq 1$. Then each of the regions $S_i(L)$, $i = 1, 2, \dots, m$, is traversed by an open path π from the origin. Since $p > \widehat{p}_c(q, L)$, there is $\phi^0_{S_i(L),p,q}$-probability 1 that $S_i(L)$ contains an infinite open cluster, and π must avoid all such clusters for $i = 1, 2, \dots, m$.

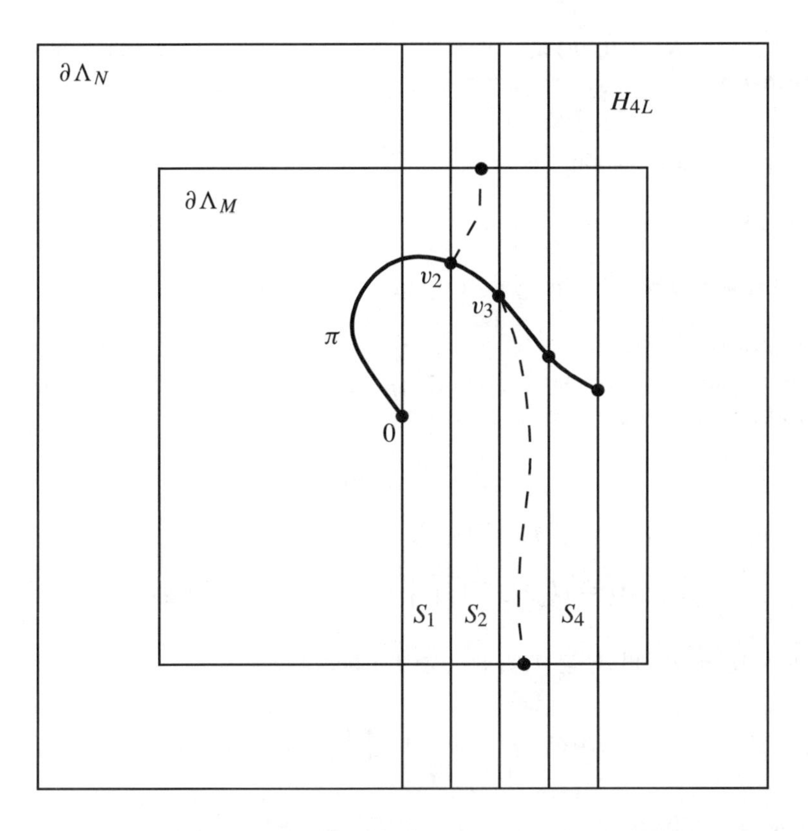

Figure 5.2. Any path π from the origin to H_n (with $n = 4L$ in this picture) traverses the regions $S_i = S_i(L)$, $i = 1, 2, 3, 4$. The vertex v_j where π first hits the slab S_j may be joined (with strictly positive conditional probability) within the slab to $\partial\Lambda_M$.

By a suitable coupling argument, the chance of this is smaller than α^m for some $\alpha = \alpha(p, q, L) < 1$.

We have to do a certain amount of work to make this argument rigorous. First, we construct the open cluster C at the origin as the limit of the sequence $(C_m : m = 1, 2, \dots)$ in the manner described above. Next, we construct a sequence $v_1, v_2, \dots$ of vertices in the following manner. We set $v_1 = 0$, the origin. For $i \geq 2$, we let

$$m_i = \min\{m \geq 1 : C_m \text{ contains some vertex of } S_i(L)\},$$

and we denote the (unique) vertex in question by v_i. Such a v_i exists if and only if some vertex in $S_i(L)$ lies in the open cluster at the origin. We obtain thus a sequence $v_1, v_2, \dots, v_T$ of vertices where

$$T = \sup\{i : C \cap S_i(L) \neq \varnothing\}.$$

Let $S(L,N) = [0,L)\times[-N,N]^{d-1}$. Since $p > \widehat{p}_c(q,L)$, we may find $\alpha > 0$ such that

$$\phi^0_{S(L,N),p,q}(0 \leftrightarrow v) > \alpha, \qquad v \in S(L,N).$$

By positive association,

$$\phi^0_{S(L,N),p,q}(v \leftrightarrow \partial\Lambda_M) > \alpha^2, \qquad v \in S(L,N),\ 0 < M \le N. \tag{5.113}$$

Let n be a positive integer satisfying $n \ge L$, and write $n = rL + s$ where $0 \le s < L$. Let $n < M < N$, and consider the probability $\psi_N(G_{n,M})$ where $\psi_N = \phi^1_{\Lambda_N,p,q}$ and $G_{n,M} = \{0 \leftrightarrow H_n,\ 0 \not\leftrightarrow \partial\Lambda_M\}$. Later, we shall take the limit as $M, N \to \infty$. On the event $G_{n,M}$, we have that $T \ge r$, and $v_i \not\leftrightarrow \partial\Lambda_M$ in $S_i(L)$, for $i = 1, 2, \dots, r$. Therefore,

$$\psi_N(G_{n,M}) \le \psi_N(A_r), \tag{5.114}$$

where

$$A_j = \{T \ge j\} \cap \left\{ \bigcap_{i=1}^{j} \{v_i \not\leftrightarrow \partial\Lambda_M \text{ in } S_i(L)\} \right\}.$$

Now $A_0 = \Omega$, and $A_j \supseteq A_{j+1}$ for $j \ge 1$, whence

$$\psi_N(G_{n,M}) \le \psi_N(A_1) \prod_{j=2}^{r} \psi_N(A_j \mid A_{j-1}). \tag{5.115}$$

Let $j \in \{2, 3, \dots, r\}$, and consider the conditional probability $\psi_N(A_j \mid A_{j-1})$; the case $j = 1$ is similar. We have that

$$\begin{aligned} &\psi_N(A_j \mid A_{j-1}) \\ &\quad \le \sum_{v \in H_{(j-1)L}} \psi_N\big(v \not\leftrightarrow \partial\Lambda_M \text{ in } S_j(L) \,\big|\, v_j = v,\ T \ge j,\ A_{j-1}\big) \\ &\qquad\qquad \times \psi_N(v_j = v \mid T \ge j,\ A_{j-1})\psi_N(T \ge j \mid A_{j-1}). \end{aligned}$$

We claim that

$$\psi_N\big(v \not\leftrightarrow \partial\Lambda_M \text{ in } S_j(L) \,\big|\, v_j = v,\ T \ge j,\ A_{j-1}\big) \le 1 - \alpha^2. \tag{5.116}$$

This will imply that

$$\begin{aligned} \psi_N(A_j \mid A_{j-1}) &\le (1-\alpha^2) \sum_{v \in H_{(j-1)L}} \psi_N(v_j = v \mid T \ge j,\ A_{j-1}) \\ &= 1 - \alpha^2, \end{aligned}$$

yielding in turn by (5.114)–(5.115) that

$$\psi_N(G_{n,M}) \le (1-\alpha^2)^r.$$

Let $N \to \infty$ and $M \to \infty$ to obtain that

$$\phi^1_{p,q}(0 \leftrightarrow H_n,\ 0 \nleftrightarrow \infty) \le (1-\alpha^2)^{\lfloor n/L \rfloor}, \qquad n \ge 1,$$

and (5.111) follows as required.

It remains to prove (5.116), which we do by a coupling argument. Suppose that we have 'built' the cluster at the origin until the first epoch $m = m_j$ at which C_m touches $S_j(L)$ and, in so doing, we have discovered that $v_j = v$, $T \ge j$, and A_{j-1} occurs. The event $E_v = \{v \nleftrightarrow \partial\Lambda_M \text{ in } S_j(L)\}$ is measurable on the σ-field generated by the edge-states in $S_j(L)$, and the configuration on $S_j(L)$ is governed by a certain conditional probability measure, namely that featuring in (5.116). This conditional measure on $S_j(L)$ dominates (stochastically) the free random-cluster measure on $S_j(L) \cap \Lambda_N = S(L, N) + (j-1)Le_1$. Since the last region is a translate of $S(L, N)$,

$$\psi_N(E_v \mid v_j = v,\ T \ge j,\ A_{j-1}) \le 1 - \alpha^2,$$

by (5.113), and (5.116) is proved. □

5.8 Uniqueness when $q < 1$

Only a limited amount is known about the (non-)uniqueness of random-cluster measures on $\mathbb{L}^d$ when $q < 1$, owing to the absence of stochastic ordering and the failure of positive association. By Theorems 4.31 and 4.33, there exists at least one translation-invariant member of $\overline{\text{co}\, \mathcal{W}_{p,q}}$, and this measure is a DLR-random-cluster measure. One may glean a little concerning uniqueness from the comparison inequalities, Theorem 3.21, from which we extract the facts that, for the random-cluster measure $\phi_{G,p,q}$ on a finite graph $G = (V, E)$,

$$\phi_{G,p_1,1} \le_{\text{st}} \phi_{G,p_2,q} \quad \text{if} \quad q \le 1,\ p_1 \le p_2, \tag{5.117}$$

$$\phi_{G,p_1,1} \ge_{\text{st}} \phi_{G,p_2,q} \quad \text{if} \quad q \le 1,\ \frac{p_1}{1-p_1} \ge \frac{p_2}{q(1-p_2)}. \tag{5.118}$$

One may deduce the following by making comparisons with the percolation model.

(5.119) Theorem. *For $d \ge 2$, there exists $p' = p'(d) < 1$ such that the following holds. Let $p \in (0, 1)$, $q \in (0, 1]$, and write $\pi = p/[p + q(1-p)]$. We have that $|\mathcal{W}_{p,q}| = |\mathcal{R}_{p,q}| = 1$ whenever either $\theta(\pi, 1) = 0$ or $p > p'$.*

Exponential decay holds similarly when $q \in (0, 1)$ and $\pi < p_c(1)$. That is, there exists $\psi = \psi(p, q) > 0$ such that $\phi^0_{p,q}(|C| = n) \le e^{-n\psi}$, see the comment in the proof of Theorem 5.55.

Proof[7]. The proof is similar to that of Proposition 5.30 and is therefore only sketched. Let p, q be such that $q \in (0, 1)$ and $\theta(\pi, 1) = 0$, and let Λ and Δ be

[7] See also [8, 156, 281].

boxes satisfying $\Lambda \subseteq \Delta$. A *cutset* is defined to be a subset S of $\mathbb{E}_\Delta \setminus \mathbb{E}_\Lambda$ such that: every path joining Λ to $\partial\Delta$ uses at least one edge of S, and S is minimal with this property. For a cutset S, we write $\text{int}\, S$ for the set of edges of $\mathbb{E}_\Delta$ possessing no endvertex x such that $x \leftrightarrow \partial\Delta$ off S, and we write $\widetilde{S} = S \cup \text{int}\, S$. There is a partial order on cutsets given by $S_1 \le S_2$ if $\widetilde{S_1} \subseteq \widetilde{S_2}$.

Let $A \in \mathcal{F}_\Lambda$. Let Δ, Σ be boxes such that $\Lambda \subseteq \Delta \subseteq \Sigma$, and let $\xi, \tau \in \Omega$. By (5.118), there exists a probability measure ψ_Σ on $\{0,1\}^{\mathbb{E}_\Sigma} \times \{0,1\}^{\mathbb{E}_\Sigma} \times \{0,1\}^{\mathbb{E}_\Sigma}$ such that the following hold.

(i) The set of triples $(\omega_1, \omega_2, \omega_3)$ satisfying $\omega_1 \le \omega_3$ and $\omega_2 \le \omega_3$ has ψ_Σ-probability 1.

(ii) The first marginal of ψ_Σ is $\phi^\xi_{\Sigma,p,q}$, the second marginal restricted to $\mathbb{E}_\Delta$ is $\phi^\tau_{\Delta,p,q}$, and the third marginal is the product measure $\phi_{\Sigma,\pi}$.

(iii) Let M denote the maximal cutset of Δ every edge of which is closed in ω_3, and note that M exists if and only if $\omega_3 \in \{\partial\Lambda \nleftrightarrow \partial\Delta\}$. Conditional on M, the marginal law of both $\{\omega_1(e) : e \in \text{int}\, M\}$ and $\{\omega_2(e) : e \in \text{int}\, M\}$ is the free measure $\phi^0_{\text{int}\, M,p,q}$.

By conditioning on M,

$$\left|\phi^\xi_{\Sigma,p,q}(A) - \phi^\tau_{\Delta,p,q}(A)\right| \le \phi_{\Sigma,\pi}(\partial\Lambda \leftrightarrow \partial\Delta). \tag{5.120}$$

By Theorem 4.17(a), there exists a probability measure $\rho \in \mathcal{W}_{p,q}$, and we choose $\tau \in \Omega$ and an increasing sequence $\boldsymbol{\Delta} = (\Delta_n : n = 1, 2, \dots)$ such that $\phi^\tau_{\Delta_n,p,q} \Rightarrow \rho$ as $n \to \infty$. Suppose that $\rho' \in \mathcal{W}_{p,q}$ and $\rho' \ne \rho$. There exists $\xi \in \Omega$ and an increasing sequence $\boldsymbol{\Sigma} = (\Sigma_n : n = 1, 2, \dots)$ such that $\phi^\xi_{\Sigma_n,p,q} \Rightarrow \rho'$.

For m sufficiently large that $\Lambda \subseteq \Delta_m$, let $n = n_m$ satisfy $\Delta_m \subseteq \Sigma_n$. By (5.120) with $\Delta = \Delta_m$, $\Sigma = \Sigma_n$,

$$\left|\phi^\xi_{\Sigma_n,p,q}(A) - \phi^\tau_{\Delta_m,p,q}(A)\right| \le \phi_{\Sigma_n,\pi}(\partial\Lambda \leftrightarrow \partial\Delta_m). \tag{5.121}$$

Let $n \to \infty$ and $m \to \infty$ in that order. Since $\theta(\pi, 1) = 0$, the right side tends to zero, and therefore $\rho'(A) = \rho(A)$. This holds for all cylinders A, and therefore $\rho' = \rho$, a contradiction. It follows that $|\mathcal{W}_{p,q}| = 1$. An alternative argument uses the method of [117].

Suppose next that $\phi \in \mathcal{R}_{p,q}$ so that, for any box Δ,

$$\phi(A \mid \mathcal{T}_\Delta)(\xi) = \phi^\xi_{\Delta,p,q}(A) \qquad \phi\text{-a.s.}$$

By (5.120) with $\Sigma = \Delta = \Delta_m$ and ρ as above,

$$\begin{aligned} |\phi(A) - \rho(A)| &= \lim_{m\to\infty} \left|\phi(\phi(A \mid \mathcal{T}_{\Delta_m})) - \phi^\tau_{\Delta_m,p,q}(A)\right| \\ &\le \lim_{m\to\infty} \phi_{\Delta_m,\pi}(\partial\Lambda \leftrightarrow \partial\Delta_m) = 0, \end{aligned}$$

whence $\mathcal{R}_{p,q} = \{\rho\}$.

A similar proof of uniqueness is valid for large p, using (5.117) and the approach taken for Theorem 5.33(b). □

Chapter 6

In Two Dimensions

Summary. The dual of the random-cluster model on a planar graph is a random-cluster model also. The self-duality of the square lattice gives rise to the conjecture that $p_c(q) = p_{sd}(q)$ for $q \in [1, \infty)$, where $p_{sd}(q)$ denotes the self-dual point $\sqrt{q}/(1+\sqrt{q})$. Using duality, one obtains the uniqueness of random-cluster measures for $p \neq p_{sd}(q)$ and $q \in [1, \infty)$. The phase transition is discontinuous if q is sufficiently large. Results similar to those for the square lattice may be obtained for the triangular and hexagonal lattices, using the star–triangle transformation. It is expected when $q \in [1, 4)$ that the critical process may be described by a stochastic Löwner evolution.

6.1 Planar duality

The duality theory of planar graphs provides a technique for studying random-cluster models in two dimensions. We shall see that, for a dual pair (G, G_d) of finite planar graphs, the measures $\phi_{G,p,q}$ and $\phi_{G_d,p_d,q}$ are dual measures in a certain sense to be explained soon, where p and p_d are related by $p_d/(1-p_d) = q(1-p)/p$. Such a duality survives the passage to a thermodynamic limit, and may therefore be applied also to infinite planar graphs including the square lattice $\mathbb{L}^2$. The square lattice has the further property of being isomorphic to its (infinite) dual, and this observation leads to many results of significance for the associated model. We begin with an account of planar duality in the random-cluster context.

A graph is called *planar* if it may be embedded in $\mathbb{R}^2$ in such a way that two edges intersect only at a common endvertex. Let $G = (V, E)$ be a planar (finite or infinite) graph embedded in $\mathbb{R}^2$. We obtain its dual graph $G_d = (V_d, E_d)$ as follows[1]. We place a dual vertex within each face of G, including any infinite face of G if such exist. For each $e \in E$ we place a dual edge $e_d = \langle x_d, y_d \rangle$ joining the two dual vertices lying in the two faces of G abutting e; if these two faces are the same, then $x_d = y_d$ and e_d is a loop. Thus, V_d is in one–one correspondence with

[1] The roman letter 'd' denotes 'dual' rather than 'dimension'.

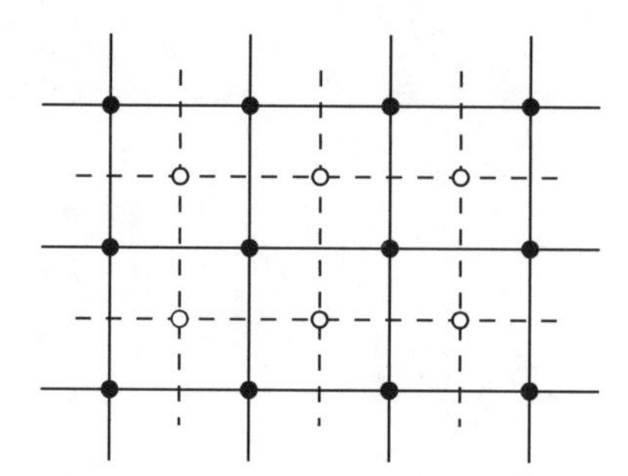

Figure 6.1. The planar dual of the square lattice $\mathbb{L}^2$ is isomorphic to $\mathbb{L}^2$.

the set of faces of G, and E_d is in one–one correspondence with E. It is easy to see as in Figure 6.1 that the dual $\mathbb{L}^2_\mathrm{d}$ of the square lattice $\mathbb{L}^2$ is isomorphic to $\mathbb{L}^2$.

What is the relevance of graphical duality to random-cluster measures on G. Suppose that G is finite. A configuration $\omega \in \Omega = \{0,1\}^E$ gives rise to a dual configuration $\omega_\mathrm{d} \in \Omega_\mathrm{d} = \{0,1\}^{E_\mathrm{d}}$ given by $\omega_\mathrm{d}(e_\mathrm{d}) = 1 - \omega(e)$. That is, e_d is declared open if and only if e is closed[2]. As before, to each configuration ω_d there corresponds the set $\eta(\omega_\mathrm{d}) = \{e_\mathrm{d} \in E_\mathrm{d} : \omega_\mathrm{d}(e_\mathrm{d}) = 1\}$ of its 'open edges', so that $\eta(\omega_\mathrm{d})$ is in one–one correspondence with $E \setminus \eta(\omega)$. Let $f(\omega_\mathrm{d})$ be the number of faces of the graph $(V_\mathrm{d}, \eta(\omega_\mathrm{d}))$, including the unique infinite face. By drawing a picture, one may easily be convinced (see Figure 6.2) that the faces of $(V_\mathrm{d}, \eta(\omega_\mathrm{d}))$ are in one–one correspondence with the components of $(V, \eta(\omega))$, and therefore

$$f(\omega_\mathrm{d}) = k(\omega). \tag{6.1}$$

We shall make use of Euler's formula (see [320]), namely

$$k(\omega) = |V| - |\eta(\omega)| + f(\omega) - 1, \qquad \omega \in \Omega, \tag{6.2}$$

and we note also for later use that

$$|\eta(\omega)| + |\eta(\omega_\mathrm{d})| = |E|. \tag{6.3}$$

Let $q \in (0,\infty)$ and $p \in (0,1)$. The random-cluster measure on G is given by

$$\phi_{G,p,q}(\omega) \propto \left(\frac{p}{1-p}\right)^{|\eta(\omega)|} q^{k(\omega)}, \qquad \omega \in \Omega,$$

where the constant of proportionality depends on G, p, and q. Therefore,

$$\begin{aligned} \phi_{G,p,q}(\omega) &\propto \left(\frac{p}{1-p}\right)^{-|\eta(\omega_\mathrm{d})|} q^{f(\omega_\mathrm{d})} && \text{by (6.1) and (6.3)} \\ &\propto \left(\frac{q(1-p)}{p}\right)^{|\eta(\omega_\mathrm{d})|} q^{k(\omega_\mathrm{d})} && \text{by (6.2) applied to } \omega_\mathrm{d} \\ &\propto \phi_{G_\mathrm{d},p_\mathrm{d},q}(\omega_\mathrm{d}), \end{aligned} \tag{6.4}$$

[2]This differs from the convention used in [154].

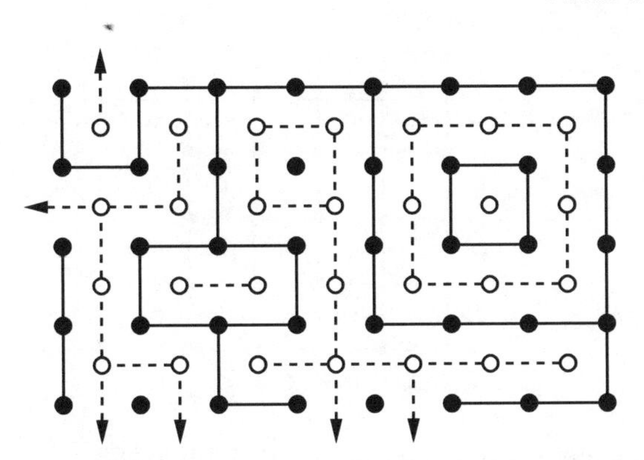

Figure 6.2. A primal configuration ω (with solid lines and vertices) and its dual configuration ω_{d} (with dashed lines and hollow vertices). The arrows join the given vertices of the dual to a dual vertex in the infinite face. Note that each face of the dual graph corresponds to a unique component of the primal graph lying 'just within'.

where the dual parameter p_{d} is given by

$$\frac{p_{\mathrm{d}}}{1-p_{\mathrm{d}}} = \frac{q(1-p)}{p}. \tag{6.5}$$

Note that the dual value of p_{d} satisfies $(p_{\mathrm{d}})_{\mathrm{d}} = p$. Since (6.4) involves probability measures, we deduce that

$$\phi_{G,p,q}(\omega) = \phi_{G_{\mathrm{d}},p_{\mathrm{d}},q}(\omega_{\mathrm{d}}), \qquad \omega \in \Omega. \tag{6.6}$$

It will later be convenient to work with the edge-parameter[3]

$$x = \frac{q^{-\frac{1}{2}}p}{1-p}, \tag{6.7}$$

for which the primal/dual transformation (6.5) becomes

$$xx_{\mathrm{d}} = 1. \tag{6.8}$$

The unique fixed point of the mapping $p \mapsto p_{\mathrm{d}}$ is easily seen from (6.5) to be the *self-dual point* $p_{\mathrm{sd}}(q)$ given by

$$p_{\mathrm{sd}}(q) = \frac{\sqrt{q}}{1+\sqrt{q}}. \tag{6.9}$$

We note that

$$\phi_{G,p_{\mathrm{sd}}(q),q}(\omega) \propto q^{\frac{1}{2}|\eta(\omega)|+k(\omega)} \propto q^{\frac{1}{2}(k(\omega_{\mathrm{d}})+k(\omega))},$$

[3]We shall work at a later stage with the parameter $y = p/(1-p)$, for which the primal/dual relation is $yy_{\mathrm{d}} = q$.

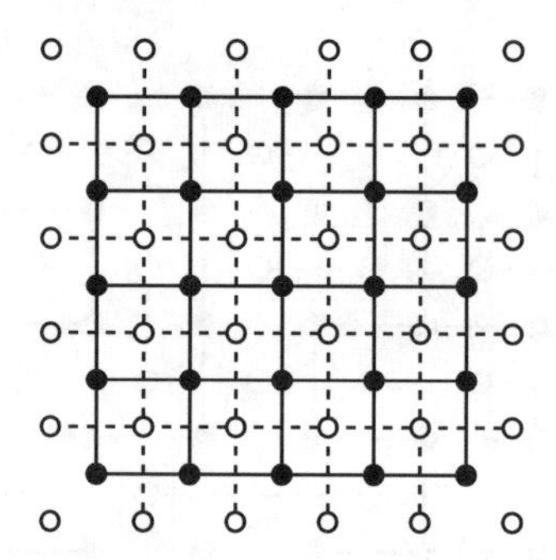

Figure 6.3. The dual of the box $\Lambda(n) = [-n, n]^2$ is obtained from the box $[-n-1, n]^2 + (\frac{1}{2}, \frac{1}{2})$ by identifying all vertices in its boundary.

by (6.1)–(6.2). This representation at the self-dual point $p_{\rm sd}(q)$ highlights the duality of measures.

When we keep track of the constants of proportionality in (6.4), we find that the partition function

$$Z_G(p, q) = \sum_{\omega \in \Omega} p^{|\eta(\omega)|}(1-p)^{|E \setminus \eta(\omega)|} q^{k(\omega)}$$

satisfies the duality relation

$$Z_G(p, q) = q^{|V|-1} \left(\frac{1-p}{p_{\rm d}} \right)^{|E|} Z_{G_{\rm d}}(p_{\rm d}, q). \tag{6.10}$$

Therefore,

$$Z_G(p_{\rm sd}(q), q) = q^{|V|-1-\frac{1}{2}|E|} Z_{G_{\rm d}}(p_{\rm sd}(q), q). \tag{6.11}$$

We consider now the square lattice $\mathbb{L}^2 = (\mathbb{Z}^2, \mathbb{E}^2)$. Let $\Lambda(n) = [-n, n]^2$, viewed as a subgraph of $\mathbb{L}^2$, and note from Figure 6.3 that its dual graph $\Lambda(n)_{\rm d}$ may be obtained from the box $[-n-1, n]^2 + (\frac{1}{2}, \frac{1}{2})$ by identifying all boundary vertices. By (6.6), and with a small adjustment on the boundary of $\Lambda(n)_{\rm d}$,

$$\phi^0_{\Lambda(n),p,q}(\omega) = \phi^1_{\Lambda(n)_{\rm d},p_{\rm d},q}(\omega_{\rm d}) \tag{6.12}$$

for configurations ω on $\Lambda(n)$. Let A be a cylinder event of $\Omega = \{0, 1\}^{\mathbb{E}^2}$, and write $A_{\rm d}$ for the dual event of $\Omega_{\rm d} = \{0, 1\}^{\mathbb{E}^2_{\rm d}}$, that is, $A_{\rm d} = \{\omega_{\rm d} \in \Omega_{\rm d} : \omega \in A\}$. On letting $n \to \infty$ in (6.12), we obtain by Theorem 4.19(a) that

$$\phi^0_{p,q}(A) = \overline{\phi}^1_{p_{\rm d},q}(A_{\rm d}),$$

where the notation $\overline{\phi}$ is used to indicate the random-cluster measure on the dual configuration space $\Omega_{\rm d}$. By a similar argument,

$$\phi^1_{p,q}(A) = \overline{\phi}^0_{p_{\rm d},q}(A_{\rm d}).$$

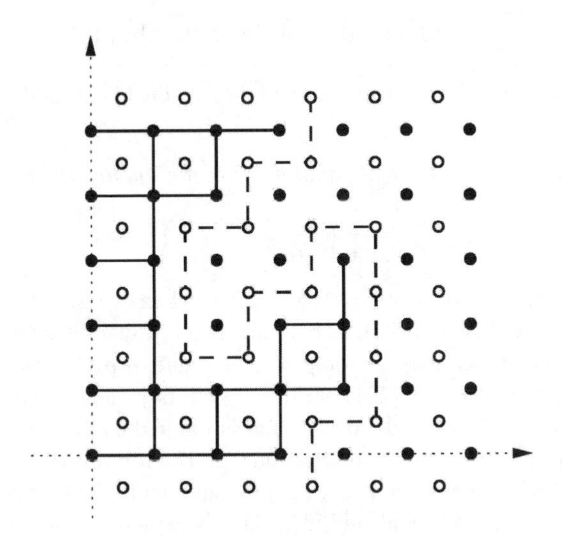

Figure 6.4. The box $S(5)$ and its dual $S(5)_{\rm d}$. There exists no open crossing of $S(5)$ from left to right if and only if there exists an open dual crossing of $S(5)_{\rm d}$ from top to bottom.

We summarize the above in a theorem.

(6.13) Theorem. *Consider the square lattice $\mathbb{L}^2$, and let $q \in [1, \infty)$. For any cylinder event A,*

$$\phi^b_{p,q}(A) = \overline{\phi}^{1-b}_{p_{\rm d},q}(A_{\rm d}), \qquad b = 0, 1,$$

where $A_{\rm d} = \{\omega_{\rm d} \in \Omega_{\rm d} : \omega \in A\}$.

There is a key application of duality to the existence of open crossings of a box. Let $S(n) = [0, n+1] \times [0, n]$ and let $S(n)_{\rm d}$ be its dual box $[0, n] \times [-1, n] + (\frac{1}{2}, \frac{1}{2})$. Let $\mathrm{LR}(n)$ be the event that there exists an open path of $S(n)$ joining some vertex on its left side to some vertex on its right side. It is standard that $\mathrm{LR}(n)_{\rm d}$ is the event that there exists no open dual crossing from the top to the bottom of $S(n)_{\rm d}$. This is explained further in [154, Section 11.3] and illustrated in Figure 6.4.

(6.14) Theorem. *Let $q \in [1, \infty)$. We have that*

$$\phi^0_{p_{\rm sd}(q),q}(\mathrm{LR}(n)) + \phi^1_{p_{\rm sd}(q),q}(\mathrm{LR}(n)) = 1, \qquad n \ge 1.$$

Proof. Apply Theorem 6.13 with $b = 0$ to the event $A = \mathrm{LR}(n)$, and use the fact that $\overline{\phi}^1_{p,q}(\mathrm{LR}(n)_{\rm d}) = \phi^1_{p,q}(\overline{\mathrm{LR}(n)}) = 1 - \phi^1_{p,q}(\mathrm{LR}(n))$. □

6.2 The value of the critical point

It is conjectured that the critical point and the self-dual point of the square lattice are equal.

(6.15) Conjecture. *The critical value $p_c(q)$ of the square lattice $\mathbb{L}^2$ is given by*

$$p_c(q) = \frac{\sqrt{q}}{1+\sqrt{q}}, \qquad q \in [1, \infty). \tag{6.16}$$

This has been proved when $q = 1, q = 2$, and when $q \geq 25.72$. The $q = 1$ case was answered by Kesten, [207], in his famous proof that the critical probability of bond percolation on $\mathbb{L}^2$ is $\frac{1}{2}$. For $q = 2$, the value of $p_c(2)$ given above agrees with the Kramers–Wannier [221] and Onsager [264] calculations of the critical temperature of the Ising model on $\mathbb{Z}^2$, and is implied by probabilistic results in the modern vernacular, see [5] and Section 9.3. The formula (6.16) for $p_c(q)$ has been established rigorously in [224, 225] for sufficiently large (real) values of q, specifically $q \geq 25.72$ (see also [153]). This is explored further in Section 6.4, see Theorem 6.35.

Several other remarkable conjectures about the phase transition on $\mathbb{L}^2$ may be found in the physics literature as consequences of 'exact' but non-rigorous arguments involving ice-type models, see [26]. These include exact formulae for the asymptotic behaviour of the partition function $\lim_{\Lambda\uparrow\mathbb{Z}^2}\{Z_\Lambda(p,q)\}^{1/|\Lambda|}$, and also for the edge-densities at the self-dual point $p_{sd}(q)$, that is, the quantities $h^b(q) = \phi^b_{p_{sd}(q),q}(e \text{ is open})$ for $b = 0, 1$. These formulae are summarized in Section 6.6.

Conjecture 6.15 asserts that $p_c(q) = p_{sd}(q)$ for $q \in [1, \infty)$. One part of this equality is known. Recall that $\theta^0(p, q) = \phi^0_{p,q}(0 \leftrightarrow \infty)$.

(6.17) Theorem [152, 314]. *Consider the square lattice $\mathbb{L}^2$, and let $q \in [1, \infty)$.*

(a) *We have that $\theta^0(p_{sd}(q), q) = 0$, whence $p_c(q) \geq p_{sd}(q)$.*

(b) *There exists a unique random-cluster measure if $p \neq p_{sd}(q)$, that is,*

$$|\mathcal{R}_{p,q}| = |\mathcal{W}_{p,q}| = 1 \quad \text{if} \quad p \neq \frac{\sqrt{q}}{1+\sqrt{q}}.$$

The complementary inequality $p_c(q) \leq p_{sd}(q)$ has eluded mathematicians despite progress by physicists, [183]. Here is an intuitive argument to justify the latter inequality. Suppose on the contrary that $p_c(q) > p_{sd}(q)$, so that $p_c(q)_d < p_{sd}(q)$. For $p \in (p_c(q)_d, p_c(q))$ we have also that $p_d \in (p_c(q)_d, p_c(q))$. Therefore, for $p \in (p_c(q)_d, p_c(q))$, both primal and dual processes comprise (almost surely) the union of *finite* open clusters. This contradicts the intuitive picture, supported for $p \neq p_c(q)$ by our knowledge of percolation, of finite open clusters of one process floating in an infinite open ocean of the other process.

Conjecture 6.15 would be proven if one could show the sufficiently fast decay of $\phi^0_{p,q}(0 \leftrightarrow \partial\Lambda(n))$ as $n \to \infty$. An example of such a statement may be found at Lemma 6.28, and another follows. Recall from Section 5.5 the quantity $\widetilde{p}_c(q)$.

(6.18) Theorem [163]. *Let $q \in [1, \infty)$ and suppose that, for all $p < p_c(q)$, there exists $A = A(p, q) < \infty$ with*

$$\phi^0_{p,q}(0 \leftrightarrow \partial\Lambda(n)) \le \frac{A}{n}, \qquad n \ge 1. \tag{6.19}$$

Then $\widetilde{p}_c(q) = p_c(q) = p_{sd}(q)$.

Rigorous numerical upper bounds of impressive accuracy have been achieved for the square lattice and certain other two-dimensional lattices.

(6.20) Theorem [15]. *The critical point $p_c(q)$ of the square lattice $\mathbb{L}^2$ satisfies*

$$p_c(q) \le \frac{\sqrt{q}}{\sqrt{1 - q^{-1}} + \sqrt{q}}, \qquad q \in [2, \infty). \tag{6.21}$$

For example, when $q = 10$, we have that $0.760 \le p_c(10) \le 0.769$, to be compared with the conjecture that $p_c(10) = \sqrt{10}/(1 + \sqrt{10}) \simeq 0.760$. The upper bound in (6.21) is the dual value of $p_{sd}(q - 1)$. See also Theorem 6.30.

Exact values for the critical points of the triangular and hexagonal lattices may be conjectured similarly, using graphical duality together with the star–triangle transformation; see Section 6.6.

Proof of Theorem 6.17. (a) There are at least two ways of proving this. One way is to use the circuit-construction argument pioneered by Harris, [181], and developed further in [47, 130], see Theorem 6.47. We shall instead adapt an argument of Zhang using the 0/1-infinite-cluster property, see [154, p. 289]. Let $p = p_{sd}(q)$, so that $\phi^0_{p,q}$ and $\phi^1_{p,q}$ are dual measures in the sense of Theorem 6.13.

For $n \ge 1$, let $A^l(n)$ (respectively $A^r(n)$, $A^t(n)$, $A^b(n)$) be the event that some vertex on the left (respectively right, top, bottom) side of the square $T(n) = [0, n]^2$ lies in an infinite open path of $\mathbb{L}^2$ using no other vertex of $T(n)$. Clearly $A^l(n)$, $A^r(n)$, $A^t(n)$, and $A^b(n)$ are increasing events whose union equals the event $\{T(n) \leftrightarrow \infty\}$. Furthermore, by rotation-invariance,

$$\text{for } b = 0, 1 \text{ and } n \ge 1, \quad \phi^b_{p,q}(A^u(n)) \text{ is constant for } u = \text{l, r, t, b}. \tag{6.22}$$

Suppose that $\theta^0(p, q) > 0$, whence by stochastic ordering $\theta^1(p, q) > 0$. Since the $\phi^b_{p,q}$ have the 0/1-infinite-cluster property,

$$\phi^b_{p,q}\big(A^l(n) \cup A^r(n) \cup A^t(n) \cup A^b(n)\big) \to 1 \qquad \text{as } n \to \infty.$$

By positive association,

$$\phi^b_{p,q}(T(n) \not\leftrightarrow \infty) \ge \phi^b_{p,q}\big(\overline{A^l(n)}\big)\phi^b_{p,q}\big(\overline{A^r(n)}\big)\phi^b_{p,q}\big(\overline{A^t(n)}\big)\phi^b_{p,q}\big(\overline{A^b(n)}\big),$$

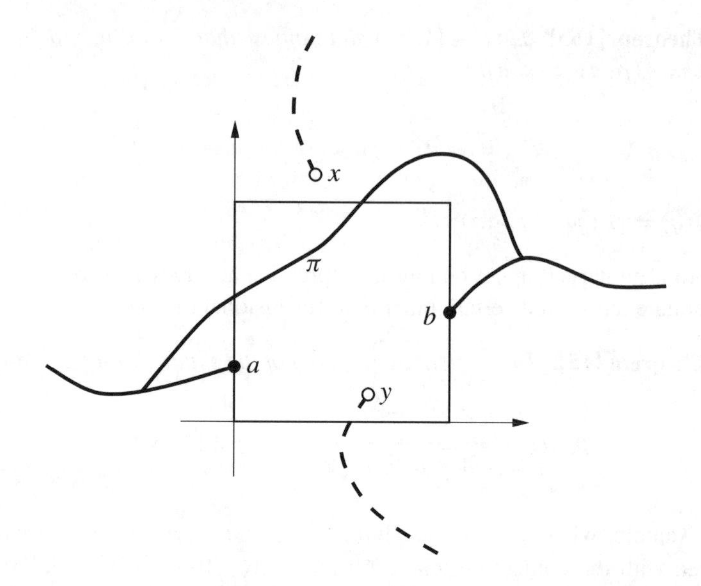

Figure 6.5. Vertices a and b lie in infinite open clusters of $\mathbb{L}^2 \setminus T(N)$, and vertices x and y lie in infinite open clusters of the dual lattice $\mathbb{L}^2_{\mathrm{d}} \setminus T(N)_{\mathrm{d}}$. If there exists a unique infinite open cluster of $\mathbb{L}^2$, then there exists an open path π joining a to b, and thus the infinite dual clusters at x and y are disjoint.

implying by (6.22) that

$$\phi^b_{p,q}(A^u(n)) \to 1 \qquad \text{as } n \to \infty, \text{ for } u = \mathrm{l, r, t, b}. \tag{6.23}$$

We choose N such that

$$\phi^b_{p,q}(A^u(N)) > \tfrac{7}{8} \qquad \text{for } u = \mathrm{l, r, t, b}, \text{ and } b = 0, 1. \tag{6.24}$$

Moving to the dual lattice, we define the dual box

$$T(n)_{\mathrm{d}} = [0, n]^2 + (\tfrac{1}{2}, \tfrac{1}{2}). \tag{6.25}$$

Let $B^{\mathrm{l}}(n)$ (respectively, $B^{\mathrm{r}}(n)$, $B^{\mathrm{t}}(n)$, $B^{\mathrm{b}}(n)$) be the event (in Ω_{d}) that some vertex on the left (respectively right, top, bottom) side of $T(n)_{\mathrm{d}}$ lies in an infinite open path of the dual lattice $\mathbb{L}^2_{\mathrm{d}}$ using no other vertex of $T(n)_{\mathrm{d}}$. Clearly,

$$\overline{\phi}^1_{p,q}(B^u(N)) = \phi^1_{p,q}(A^u(N)) > \tfrac{7}{8} \qquad \text{for } u = \mathrm{l, r, t, b}. \tag{6.26}$$

Consider now the event $A = A^{\mathrm{l}}(N) \cap A^{\mathrm{r}}(N) \cap B^{\mathrm{t}}(N) \cap B^{\mathrm{b}}(N)$, viewed as a subset of Ω and illustrated in Figure 6.5. The probability that A does not occur satisfies

$$\begin{aligned}
\phi^0_{p,q}(\overline{A}) &\le \phi^0_{p,q}\big(\overline{A^{\mathrm{l}}(N)}\big) + \phi^0_{p,q}\big(\overline{A^{\mathrm{r}}(N)}\big) + \overline{\phi}^1_{p,q}\big(\overline{B^{\mathrm{t}}(N)}\big) + \overline{\phi}^1_{p,q}\big(\overline{B^{\mathrm{b}}(N)}\big) \\
&\le \tfrac{1}{2} \qquad \text{by (6.24) and (6.26),}
\end{aligned}$$

giving that $\phi^b_{p,q}(A) \geq \frac{1}{2}$ for $b = 0, 1$.

We now use the fact that every random-cluster measure $\phi^b_{p,q}$ has the 0/1-infinite-cluster property, see Theorem 4.33(c). If A occurs, then $\mathbb{L}^2 \setminus T(N)$ contains two disjoint infinite open clusters, since the clusters in questions are separated by infinite open paths of the dual; any open path of $\mathbb{L}^2 \setminus T(N)$ joining these two clusters would contain an edge which crosses an open edge of the dual, and no such edge can exist. Similarly, on A, the graph $\mathbb{L}^2_{\mathrm{d}} \setminus T(N)_{\mathrm{d}}$ contains two disjoint infinite open clusters, separated physically by infinite open paths of $\mathbb{L}^2 \setminus T(N)$. The whole lattice $\mathbb{L}^2$ contains (almost surely) a *unique* infinite open cluster, and it follows that there exists (almost surely on A) an open connection π of $\mathbb{L}^2$ between the fore-mentioned infinite open clusters. By the geometry of the situation (see Figure 6.5), this connection forms a barrier to possible open connections of the dual joining the two infinite open dual clusters. Therefore, almost surely on A, the dual lattice contains two or more infinite open clusters. Since the latter event has probability 0, it follows that $\phi^b_{p,q}(A) = 0$ in contradiction of the inequality $\phi^b_{p,q}(A) \geq \frac{1}{2}$. The initial hypothesis that $\theta^0(p, q) > 0$ is therefore incorrect, and the proof is complete.

(b) By part (a), $\theta^1(p, q) = 0$ for $p < p_{\mathrm{sd}}(q)$, whence, by Theorem 5.33(a), $|\mathcal{R}_{p,q}| = |\mathcal{W}_{p,q}| = 1$ for $p < p_{\mathrm{sd}}(q)$.

Suppose now that $p > p_{\mathrm{sd}}(q)$ so that, by (6.5), $p^{\mathrm{d}} < p_{\mathrm{sd}}(q)$. By part (a) and Theorem 4.63,

$$\overline{\phi}^0_{p^{\mathrm{d}},q}(e_{\mathrm{d}} \text{ is closed}) = \overline{\phi}^1_{p^{\mathrm{d}},q}(e_{\mathrm{d}} \text{ is closed}), \qquad e \in \mathbb{E}^2,$$

and by Theorem 6.13,

$$\phi^b_{p,q}(e \text{ is open}) = \overline{\phi}^{1-b}_{p^{\mathrm{d}},q}(e_{\mathrm{d}} \text{ is closed}), \qquad b = 0, 1.$$

Therefore, $\phi^0_{p,q}(e \text{ is open}) = \phi^1_{p,q}(e \text{ is open})$, and the claim follows by Theorem 4.63. □

Proof of Theorem 6.18. Under the given hypothesis, $\widetilde{p}_{\mathrm{c}}(q) = p_{\mathrm{c}}(q)$. Suppose that $p_{\mathrm{sd}}(q) < p_{\mathrm{c}}(q)$, and that (6.19) holds with $p = p_{\mathrm{sd}}(q)$ and $A = A(p_{\mathrm{sd}}(q), q)$. By Theorem 5.33, $\phi^0_{p_{\mathrm{sd}}(q),q} = \phi^1_{p_{\mathrm{sd}}(q),q}$, implying (6.19) with $\phi^0_{p_{\mathrm{sd}}(q),q}$ replaced by $\phi^1_{p_{\mathrm{sd}}(q),q}$.

If, for illustration, $A < \frac{1}{2}$, then, by a consideration of the left endvertex of a crossing of $S(n)$,

$$\phi^0_{p_{\mathrm{sd}}(q),q}(\mathrm{LR}(n)) = \phi^1_{p_{\mathrm{sd}}(q),q}(\mathrm{LR}(n)) < (n+1)\frac{A}{n+1} < \frac{1}{2}, \tag{6.27}$$

in contradiction of Theorem 6.14. Therefore $p_{\mathrm{sd}}(q) = p_{\mathrm{c}}(q)$.

More generally, by Theorem 5.60, $\phi^0_{p_{\mathrm{sd}}(q),q}(0 \leftrightarrow \partial\Lambda(n))$ decays exponentially as $n \to \infty$. Exponential decay holds for $\phi^1_{p_{\mathrm{sd}}(q),q}$ also, as above, and (6.27) follows for large n. Therefore, $p_{\mathrm{sd}}(q) = p_{\mathrm{c}}(q)$ as claimed. □

We precede the proof of Theorem 6.20 with a lemma.

(6.28) Lemma. *Let $q \in [1, \infty)$, and let p and p_{d} satisfy* (6.5). *With C the open cluster at the origin and $b \in \{0, 1\}$,*

$$\text{if} \quad \phi^b_{p_{\mathrm{d}},q}(\mathrm{rad}(C)) < \infty \quad \text{then} \quad \theta^{1-b}(p, q) > 0.$$

In particular, $p_{\mathrm{c}}(q) = p_{\mathrm{sd}}(q)$ under the condition:

$$\phi^0_{p,q}(\mathrm{rad}(C)) < \infty, \qquad p < p_{\mathrm{sd}}(q).$$

Proof[4]. Let $\Lambda(n) = [-n, n]^2$, and let $1 \le r < t < \infty$. By Theorem 6.13,

$$\phi^{1-b}_{p,q}\big(\partial\Lambda(r) \nleftrightarrow \partial\Lambda(t)\big) = \overline{\phi}^b_{p_{\mathrm{d}},q}\left(\bigcup_{s=r}^{t-1} A_s(r, t)\right),$$

where $A_s(r, t)$ is the event that $(s + \frac{1}{2}, \frac{1}{2})$ belongs to an open circuit of the dual, lying in $\Lambda(t)_{\mathrm{d}} \setminus \Lambda(r)_{\mathrm{d}}$ and having $\Lambda(r)$ in its interior. By the translation-invariance of random-cluster measures, Theorem 4.19(b),

$$\phi^{1-b}_{p,q}\big(\partial\Lambda(r) \nleftrightarrow \partial\Lambda(t)\big) \le \sum_{s=r}^{t-1} \overline{\phi}^b_{p_{\mathrm{d}},q}\big(\mathrm{rad}(C_{\mathrm{d}}) \ge r + s\big),$$

where C_{d} is the open cluster at the origin of the dual lattice. Letting $t \to \infty$,

$$\phi^{1-b}_{p,q}\big(\partial\Lambda(r) \nleftrightarrow \infty\big) \le \sum_{s=r}^{\infty} \phi^b_{p_{\mathrm{d}},q}\big(\mathrm{rad}(C) \ge r + s\big). \tag{6.29}$$

Suppose that $\phi^b_{p_{\mathrm{d}},q}(\mathrm{rad}(C)) < \infty$, and pick R such that

$$\sum_{s=2R}^{\infty} \phi^b_{p_{\mathrm{d}},q}\big(\mathrm{rad}(C) \ge s\big) < 1.$$

By (6.29), $\phi^{1-b}_{p,q}(\partial\Lambda(R) \nleftrightarrow \infty) < 1$, whence $\theta^{1-b}(p, q) > 0$ as required. □

Proof of Theorem 6.20. Let $q \in [2, \infty)$ and $b \in \{0, 1\}$. By the forthcoming Theorem 6.30, $\phi^1_{p_{\mathrm{d}},q}(\mathrm{rad}(C)) < \infty$ when $p_{\mathrm{d}} < p_{\mathrm{sd}}(q - 1)$. By Lemma 6.28, $\theta^0(p, q) > 0$ whenever

$$p > \big(p_{\mathrm{sd}}(q - 1)\big)_{\mathrm{d}} = \frac{\sqrt{q}}{\sqrt{1 - q^{-1}} + \sqrt{q}}.$$ □

[4] An alternative proof appears in [141]. See also [314].

6.3 Exponential decay

A valuable consequence of the comparison methods developed in [15] is the exponential decay of connectivity functions when $q \in [2, \infty)$ and

$$p < p_{\mathrm{sd}}(q-1) = \frac{\sqrt{q-1}}{1+\sqrt{q-1}}.$$

(6.30) Theorem (Exponential decay) [15]. *Let $q \in [2, \infty)$, and consider the random-cluster model on the box $\Lambda(n) = [-n, n]^2$. There exists $\alpha = \alpha(p, q)$ satisfying $\alpha(p, q) > 0$ when $p < p_{\mathrm{sd}}(q-1)$ such that*

$$\phi^1_{\Lambda(n),p,q}(0 \leftrightarrow \partial\Lambda(n)) \le e^{-\alpha n}, \qquad n \ge 1.$$

By stochastic ordering,

$$\phi^1_{\Lambda(n),p,q}(0 \leftrightarrow \partial\Lambda(m)) \le \phi^1_{\Lambda(m),p,q}(0 \leftrightarrow \partial\Lambda(m)), \qquad m \le n,$$

and therefore, on taking the limit as $n \to \infty$,

$$\phi^1_{p,q}(0 \leftrightarrow \partial\Lambda(m)) \le e^{-\alpha m}, \qquad p < p_{\mathrm{sd}}(q-1),\ q \ge 2,\ m \ge 1,$$

by the above theorem. In summary,

$$p_{\mathrm{sd}}(q-1) \le \widetilde{p}_{\mathrm{c}}(q) \le p_{\mathrm{sd}}(q), \qquad q \ge 2,$$

where $\widetilde{p}_{\mathrm{c}}(q)$ is the threshold for exponential decay, see (5.65) and (5.67). We recall the conjecture that $\widetilde{p}_{\mathrm{c}}(q) = p_{\mathrm{c}}(q)$.

Proof. We use the comparison between the random-cluster model and the Ising model with external field, as described in Section 3.7. Consider the wired random-cluster measure on a box Λ with $q \in [2, \infty)$. By Theorem 3.79 and the note following (3.83), the set of vertices that are joined to $\partial\Lambda$ by open paths is stochastically smaller than the set of $+$ spins in the Ising model on Λ with $+$ boundary conditions and parameters β', h' satisfying (3.80). The maximum vertex degree of $\mathbb{L}^2$ is $\Delta = 4$ and, by (3.82),

$$e^{\beta_4} = \frac{q-2}{\sqrt{q-1}-1} = 1 + \sqrt{q-1},$$

so that

$$1 - e^{-\beta_4} = \frac{\sqrt{q-1}}{1+\sqrt{q-1}} = p_{\mathrm{sd}}(q-1).$$

Let $p = 1 - e^{-\beta} < p_{\mathrm{sd}}(q-1)$. By (3.83), $h' < 0$. By stochastic domination,

$$\phi^1_{\Lambda,p,q}(0 \leftrightarrow \partial\Lambda) \le \pi^+_{\Lambda,\beta',h'}(0 \leftrightarrow^+ \partial\Lambda), \tag{6.31}$$

where $\{0 \leftrightarrow^+ \partial\Lambda\}$ is the event that there exists a path of Λ joining 0 to some vertex of $\partial\Lambda$ all of whose vertices have spin $+1$. By results of [88, 182] (see the discussion in [15, p. 438]), the right side of (6.31) decays exponentially in the shortest side-length of Λ. □

6.4 First-order phase transition

The $q = 1$ case of the random-cluster measure is the percolation model, with associated product measure $\phi_p = \phi_{p,1}$. One of the outstanding problems for percolation is to prove the continuity for all d of the percolation probability $\theta(p) = \phi_p(0 \leftrightarrow \infty)$ at the critical point $p_c = p_c(1)$, see [154, Section 8.3]. By a standard argument of semi-continuity, this amounts to proving that $\theta(p_c) = 0$, which is to say that there exists (almost surely) no infinite open cluster *at* the critical point. The situation for general q is quite different. It turns out that $\theta^1(p_c(q), q) > 0$ for all large q.

(6.32) Conjecture. *Consider the d-dimensional lattice $\mathbb{L}^d$ where $d \geq 2$.*

(a) $\theta^0(p_c(q), q) = 0$ *for* $q \in [1, \infty)$.

(b) *There exists* $Q = Q(d) \in (1, \infty)$ *such that*

$$\theta^1(p_c(q), q) \begin{cases} = 0 & \text{if } q < Q, \\ > 0 & \text{if } q > Q. \end{cases}$$

In the vernacular of statistical physics, we speak of the phase transition as being of *second order* if $\theta^1(p_c(q), q) = 0$, and of *first order* otherwise. Thus the random-cluster transition is expected to be of first order if and only if q is sufficiently large. There are two issues: to prove the existence of a 'sharp transition in q', and to calculate the 'critical value' $Q(d)$ of q. The first problem is strangely difficult. It is natural to seek some monotonicity, perhaps of the function $f(q) = \theta^1(p_c(q), q)$, but this has proved elusive even in two dimensions. As for the value of $Q(d)$, it is believed[5] that $Q(d)$ is non-increasing in d and satisfies

$$Q(d) = \begin{cases} 4 & \text{if } d = 2, \\ 2 & \text{if } d \geq 6. \end{cases} \tag{6.33}$$

A first-order transition is characterized by a discontinuity in the order-parameter $\theta^1(p, q)$. Two further indicators of first-order transition are: discontinuity of the edge-densities $h^b(p, q) = \phi^b_{p,q}(e \text{ is open})$, $b = 0, 1$, and the existence of a so-called 'non-vanishing mass gap'. The edge-densities are sometimes termed the 'energy' functions, since they arise thus in the Potts model.

The term 'mass gap' arises in the study of the exponential decay of correlations in the subcritical phase, in the limit as $p \uparrow p_c(q)$. Of the various ways of expressing this, we choose to work with the probability $\phi^0_{p,q}(0 \leftrightarrow \partial\Lambda(n))$, where $\Lambda(n) = [-n, n]^d$. Recall from Theorem 5.45 that there exists a function $\psi = \psi(p, q)$ such that

$$\phi^0_{p,q}(0 \leftrightarrow \partial\Lambda(n)) \approx e^{-n\psi} \qquad \text{as } n \to \infty,$$

[5] See [26, 324] and the footnote on page 183.

where '$\approx$' denotes logarithmic asymptotics. Clearly, $\psi(p, q)$ is a non-increasing function of p, and $\psi(p, q) = 0$ if $\theta^0(p, q) > 0$. It is believed that $\psi(p, q) > 0$ if $p < p_c(q)$. We speak of the limit

$$\mu(q) = \lim_{p \uparrow p_c(q)} \psi(p, q)$$

as the *mass gap*. It is believed that the transition is of first order if and only if there is a non-vanishing mass gap, that is, if $\mu(q) > 0$.

(6.34) Conjecture. *Consider the d-dimensional lattice $\mathbb{L}^d$ where $d \geq 2$. Then*

$$\mu(q) \begin{cases} = 0 & \text{if } q < Q(d), \\ > 0 & \text{if } q > Q(d), \end{cases}$$

where $Q(d)$ is given in Conjecture 6.32.

The first proof of first-order phase transition for the Potts model with large q was discovered by Kotecký and Shlosman, [220]. Amongst the later proofs is that of [225], and this is best formulated in the language of the random-cluster model, [224]. It takes a very simple form in the special case $d = 2$, as shown in this section. The general case of $d \geq 2$ is treated in Chapter 7.

There follows a reminder concerning the number a_n of self-avoiding walks on $\mathbb{L}^2$ beginning at the origin. It is standard, [244], that $a_n^{1/n} \to \kappa$ as $n \to \infty$, for some constant κ termed the *connective constant* of the lattice. Let

$$Q = \left\{ \tfrac{1}{2} \left(\kappa + \sqrt{\kappa^2 - 4} \right) \right\}^4.$$

We have that $2.620 < \kappa < 2.696$, see [302], whence $21.61 < Q < 25.72$. Let

$$\psi(q) = \frac{1}{24} \log \left\{ \frac{(1 + \sqrt{q})^4}{q\kappa^4} \right\},$$

noting that $\psi(q) > 0$ if and only if $q > Q$.

(6.35) Theorem (Discontinuous phase transition when $d = 2$) [153, 225]. *Consider the square lattice $\mathbb{L}^2$, and let $q > Q$.*

(a) Critical point. *The critical point is given by $p_c(q) = \sqrt{q}/(1 + \sqrt{q})$.*

(b) Discontinuous transition. *We have that $\theta^1(p_c(q), q) > 0$.*

(c) Non-vanishing mass gap. *For any $\psi < \psi(q)$ and all large n,*

$$\phi^0_{p_c(q),q}(0 \leftrightarrow \partial\Lambda(n)) \leq e^{-n\psi}.$$

(d) Discontinuous edge-densities. *The functions $h^b(p, q) = \phi^b_{p,q}(e \text{ is open})$, $b = 0, 1$, are discontinuous functions of p at $p = p_c(q)$.*

Similar conclusions may be obtained for general $d \geq 2$ when q is sufficiently large ($q > Q(d)$ for suitable $Q(d)$). Whereas, in the case $d = 2$, planar duality provides an especially simple proof, the proof for general d utilizes nested sequences of surfaces of $\mathbb{R}^d$ and requires a control of the effective boundary conditions within the surfaces. See Section 7.5.

By Theorem 6.17(b), whenever q is such that the phase transition is of first order, then necessarily $p_c(q) = p_{sd}(q)$.

The idea of the proof of the theorem is as follows. There is a partial order on circuits Γ of $\mathbb{L}^2$ given by: $\Gamma \le \Gamma'$ if the bounded component of $\mathbb{R}^2 \setminus \Gamma$ is a subset of that of $\mathbb{R}^2 \setminus \Gamma'$. We work at the self-dual point $p = p_{sd}(q)$, and with the box $\Lambda(n)$ with wired boundary conditions. Roughly speaking, an 'outer contour' is defined to be a circuit Γ of the dual graph $\Lambda(n)_d$ all of whose edges are open in the dual (that is, they traverse closed edges in the primal graph $\Lambda(n)$), and that is maximal with this property. Using self-duality, one may show that

$$\phi^1_{\Lambda(n),p_{sd}(q),q}(\Gamma \text{ is an outer circuit}) \le \frac{1}{q}\left(\frac{q}{(1+\sqrt{q})^4}\right)^{|\Gamma|/4},$$

for any given circuit Γ of $\Lambda(n)_d$. Combined with a circuit-counting argument of Peierls-type involving the connective constant, this estimate implies after a little work the claims of Theorem 6.35. The idea of the proof appeared in [225] in the context of Potts models, and the random-cluster formulation may be found in [153]; see also Section 7.5 of the current work.

Proof of Theorem 6.35. This proof carries a health warning. The use of two-dimensional duality raises certain issues which are tedious to resolve with complete rigour, and we choose not to do so here. Such issues may be resolved either by the methods of [210, p. 386] when $d = 2$, or by those expounded in Section 7.2 for general $d \ge 2$. Let $n \ge 1$, let $\Lambda = \Lambda(n) = [-n, n]^2$, and let $\Lambda_d = [-n, n-1]^2 + (\frac{1}{2}, \frac{1}{2})$ be those vertices of the dual of Λ that lie inside Λ (that is, we omit the dual vertex in the infinite face of Λ). We shall work with 'wired' boundary conditions on Λ, and we let $\omega \in \Omega_\Lambda = \{0, 1\}^{\mathbb{E}_\Lambda}$. The *exterior* (respectively, *interior*) of a given circuit Γ of either $\mathbb{L}^2$ or its dual $\mathbb{L}^2_d$ is defined to be the unbounded (respectively, bounded) component of $\mathbb{R}^2 \setminus \Gamma$. A circuit Γ of Λ_d is called an *outer circuit* of a configuration $\omega \in \Omega_\Lambda$ if the following hold:

(a) all edges of Γ are open in the dual configuration ω_d, which is to say that they traverse closed edges of Λ,

(b) the origin of $\mathbb{L}^2$ is in the interior of Γ,

(c) every vertex of Λ lying in the exterior of Γ, but within distance of $1/\sqrt{2}$ of some vertex of Γ, belongs to the same component of ω.

See Figure 6.6 for an illustration of the meaning of 'outer circuit'.

Each circuit Γ of Λ_d partitions the set $\mathbb{E}_\Lambda$ of edges of Λ into three sets, namely

$$\begin{aligned} E &= \{e \in \mathbb{E}_\Lambda : e \text{ lies in the exterior of } \Gamma\}, \\ I &= \{e \in \mathbb{E}_\Lambda : e \text{ lies in the interior of } \Gamma\}, \\ \Gamma' &= \{e \in \mathbb{E}_\Lambda : e_d \in \Gamma\}. \end{aligned}$$

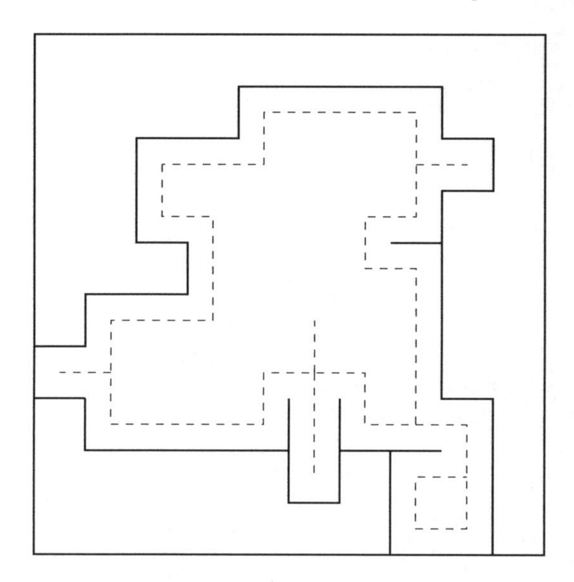

Figure 6.6. The solid lines represent open edges of Λ. The dashed lines include an outer circuit Γ of the dual Λ_{d}.

The set I forms a connected subgraph of Λ. We write $G_\Gamma \subseteq \{0,1\}^E$ for the set of configurations of edges in E satisfying property (c) above.

Let $F \subseteq \mathbb{E}_\Lambda$, and write V_F for the set of vertices incident to members of F. For $\omega \in \{0,1\}^F$, let

$$\pi_F(\omega) = p^{|\eta(\omega)|}(1-p)^{|F\setminus\eta(\omega)|}q^{k(\omega)},$$

where $k(\omega)$ is the number of components of the graph $(V_F, \eta(\omega))$. We shall sometimes impose a boundary condition on F as follows. Let $\partial_{\mathrm{ext}}F$ be the set of vertices in V_F that belong to infinite paths of $\mathbb{L}^2$ using no other vertex of V_F. We write

$$\pi_F^1(\omega) = p^{|\eta(\omega)|}(1-p)^{|F\setminus\eta(\omega)|}q^{k^1(\omega)}, \qquad \omega \in \{0,1\}^F,$$

where $k^1(\omega)$ is the number of components of $(V_F, \eta(\omega))$ counted according to the convention that components that intersect $\partial_{\mathrm{ext}}F$ are counted only as one in total.

Our target is to obtain an upper bound for the probability that a given circuit Γ is an outer circuit. Let Γ be a circuit of Λ_{d} with 0 in its interior. Since no open component of ω contains points lying in both the exterior and interior of an outer circuit, the event $\mathrm{OC}(\Gamma) = \{\Gamma \text{ is an outer circuit}\}$ satisfies

$$\begin{aligned}\phi^1_{\Lambda,p,q}(\mathrm{OC}(\Gamma)) &= \frac{1}{Z^1_\Lambda}\sum_\omega 1_{\mathrm{OC}(\Gamma)}(\omega)\pi^1_\Lambda(\omega) \\ &= \frac{1}{Z^1_\Lambda}(1-p)^{|\Gamma|}Z^1_E Z_I \end{aligned} \tag{6.36}$$

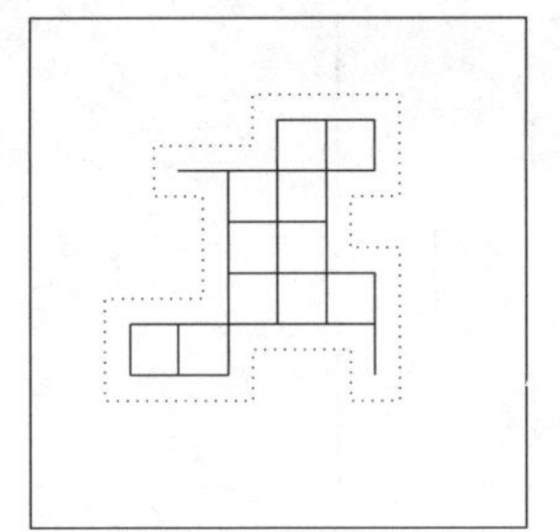
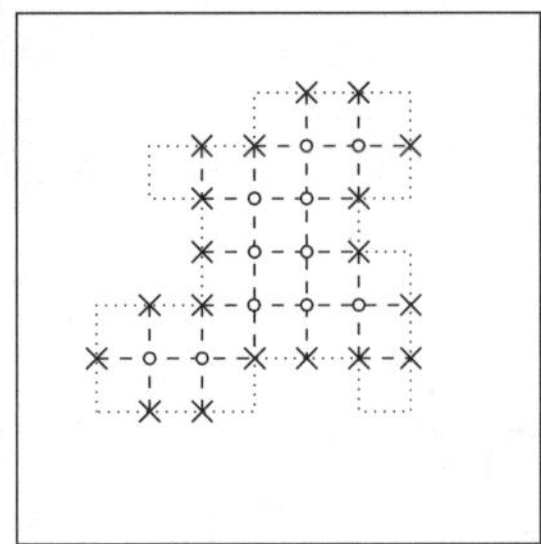
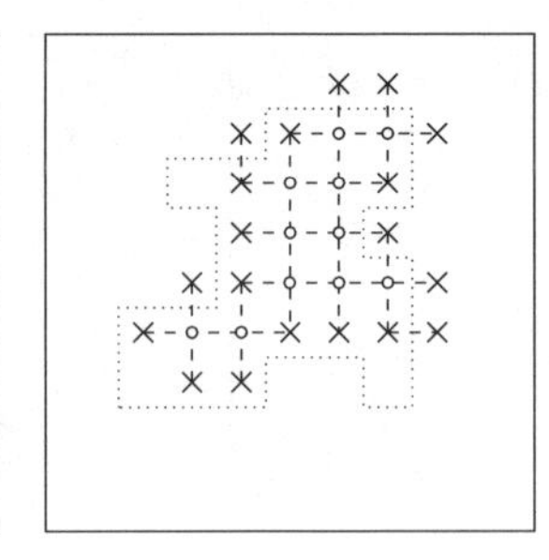

Figure 6.7. The interior edges I of Γ are marked in the leftmost picture, and the dual $I_{\rm d}$ in the centre picture (the vertices marked with a cross are identified as a single vertex). The shifted set $I^* = I_{\rm d} + (\frac{1}{2}, \frac{1}{2})$ is drawn in the rightmost picture. Note that $I^* \subseteq I \cup \Gamma'$.

where $\pi^1_\Lambda = \pi^1_{\Lambda,p,q}$, Z^1_Λ is the wired partition function of Λ, and

$$Z^1_E = \sum_{\omega' \in G_\Gamma} \pi^1_E(\omega'), \qquad Z_I = \sum_{\omega'' \in \{0,1\}^I} \pi_I(\omega'').$$

We use duality next. Let $I_{\rm d}$ be the set of dual edges that cross the primal edges in I, and let m be the number of vertices of Λ inside Γ. By (6.10),

$$Z_I = q^{m-1} \left(\frac{1-p}{p^{\rm d}} \right)^{|I|} Z^1_{I_{\rm d}}(p_{\rm d}, q), \tag{6.37}$$

where $p_{\rm d}$ satisfies (6.5), $Z^1_{I_{\rm d}}(p_{\rm d}, q)$ is the partition function for dual configurations on $(V_{\rm d}, I_{\rm d})$ with wired boundary conditions, and $V_{\rm d}$ is the set of vertices incident to $I_{\rm d}$ (with the convention that all vertices of $V_{\rm d}$ on its boundary are identified, as indicated in Figure 6.7).

The partition functions have the following property of supermultiplicativity when $q \in [1, \infty)$. For any dual circuit Γ with the origin in its interior,

$$\begin{aligned} Z^1_\Lambda &= \sum_{\omega \in G_\Gamma \times \{0,1\}^{I \cup \Gamma'}} \pi^1_\Lambda(\omega) \\ &\ge \sum_{\omega' \in G_\Gamma} \pi^1_E(\omega') \sum_{\omega'' \in \{0,1\}^{I \cup \Gamma'}} \pi^1_{I \cup \Gamma'}(\omega'') \\ &= Z^1_E Z^1_{I \cup \Gamma'}. \end{aligned} \tag{6.38}$$

Let $I^* = I_{\rm d} + (\frac{1}{2}, \frac{1}{2})$, where $I_{\rm d}$ is viewed as a subset of $\mathbb{R}^2$. Note from Figure 6.7 that $I^* \subseteq I \cup \Gamma'$, and therefore

$$Z^1_{I \cup \Gamma'} \ge Z^1_{I^*} \tag{6.39}$$

by Lemma 3.69 and inequality (3.70). By (6.38)–(6.39),

$$Z^1_\Lambda \ge Z^1_E Z^1_{I^*} = Z^1_E Z^1_{I_{\rm d}}. \tag{6.40}$$

Set $p = p_{\mathrm{sd}}(q) = \sqrt{q}/(1+\sqrt{q})$. By (6.36)–(6.40) and (6.11),

$$\begin{aligned}(6.41)\qquad \phi^1_{\Lambda,p,q}(\mathrm{OC}(\Gamma)) &= (1-p)^{|\Gamma|}\frac{Z^1_E Z_I}{Z^1_\Lambda}\\ &= (1-p)^{|\Gamma|}q^{m-1-\frac{1}{2}|I|}\frac{Z^1_E Z^1_{I_{\mathrm{d}}}}{Z^1_\Lambda}\\ &\le (1-p)^{|\Gamma|}q^{m-1-\frac{1}{2}|I|}.\end{aligned}$$

Since each vertex of Λ (inside Γ) has degree 4,

$$4m = 2|I| + |\Gamma|,$$

whence

$$(6.42)\qquad \phi^1_{\Lambda,p,q}(\mathrm{OC}(\Gamma)) \le (1-p)^{|\Gamma|}q^{\frac{1}{4}|\Gamma|-1} = \frac{1}{q}\left(\frac{q}{(1+\sqrt{q})^4}\right)^{|\Gamma|/4}.$$

The number of dual circuits of Λ having length l and containing the origin in their interior is no greater than la_l, where a_l is the number of self-avoiding walks of $\mathbb{L}^2$ beginning at the origin with length l. Therefore,

$$\sum_\Gamma \phi^1_{\Lambda,p,q}(\mathrm{OC}(\Gamma)) \le \sum_{l=4}^\infty \frac{la_l}{q}\left(\frac{q}{(1+\sqrt{q})^4}\right)^{l/4}.$$

Now $l^{-1}\log a_l \to \kappa$ as $l \to \infty$, where κ is the connective constant of $\mathbb{L}^2$. Suppose that $q > Q$, so that $q\kappa^4 < (1+\sqrt{q})^4$. There exists $A(q) < \infty$ such that

$$\sum_\Gamma \phi^1_{\Lambda,p,q}(\mathrm{OC}(\Gamma)) < A(q), \qquad n \ge 1.$$

If $A(q) < 1$ (which holds for sufficiently large q) then, by the assumption of wired boundary conditions,

$$\begin{aligned}\phi^1_{\Lambda,p,q}(0 \leftrightarrow \partial\Lambda) &= \phi^1_{\Lambda,p,q}(\mathrm{OC}(\Gamma) \text{ occurs for no } \Gamma)\\ &\ge 1 - A(q) > 0.\end{aligned}$$

On letting $n \to \infty$, we obtain by Proposition 5.11 that $\theta^1(p,q) > 0$ when $p = \sqrt{q}/(1+\sqrt{q})$. By Theorem 6.17(a), this implies parts (a) and (b) of the theorem when q is sufficiently large.

For general $q > Q$, we have only that $A(q) < \infty$. In this case, we find $N < n$ such that

$$\sum_{\Gamma:\,\Gamma \text{ outside } \Lambda(N)} \phi^1_{\Lambda,p,q}(\mathrm{OC}(\Gamma)) < \tfrac{1}{2},$$

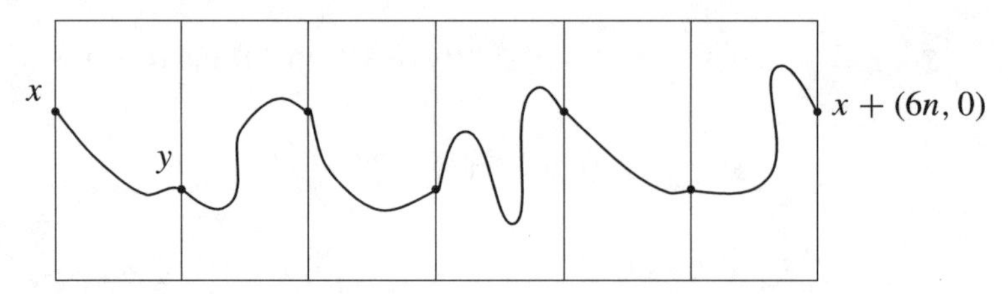

Figure 6.8. Six copies of a rectangle having width n and height $2n$ may be put together to form a rectangle with size $6n$ by $2n$. If each is crossed by an open path joining the images of x and y, then the larger rectangle is crossed between its shorter sides.

where Γ is said to be outside $\Lambda(N)$ if it contains $\Lambda(N)$ in its interior. Then

$$\phi^1_{\Lambda,p,q}(\Lambda(N) \leftrightarrow \partial\Lambda) \geq \tfrac{1}{2}.$$

Let $n \to \infty$ to find that $\phi^1_{p,q}(\Lambda(N) \leftrightarrow \infty) \geq \frac{1}{2}$, implying that $\theta^1(p,q) > 0$ as required.

Turning to part (c), let $p = p_{\mathrm{d}} = p_{\mathrm{sd}}(q) = \sqrt{q}/(1+\sqrt{q})$. Let A_n be the event that the annulus $\Delta_n = \Lambda(3n) \setminus \Lambda(n-1)$ contains an open circuit with 0 in its interior. By Theorem 6.13 and (6.42),

$$\phi^0_{\Lambda(r),p,q}(A_n) \leq \sum_{m=8n}^{\infty} \frac{m a_m}{q} \left(\frac{q}{(1+\sqrt{q})^4} \right)^{m/4}, \qquad r > 3m.$$

We have used the fact that, if A_n occurs, there exists a maximal open circuit Γ of $\Lambda(r)$ containing 0 and with length at least $8n$. In the dual of $\Lambda(r)$, Γ constitutes an outer circuit. Let $r \to \infty$ to obtain that

$$\phi^0_{p,q}(A_n) \leq \sum_{m=8n}^{\infty} \frac{m a_m}{q} \left(\frac{q}{(1+\sqrt{q})^4} \right)^{m/4}, \qquad r > 3m. \tag{6.43}$$

Let LR_n denote the event that there exists an open crossing of the rectangle $R_n = [0,n] \times [0,2n]$ from its left to its right side, and set $\lambda_n = \phi^0_{p,q}(\mathrm{LR}_n)$. There exists a point x on the left side of R_n and a point y on its right side such that

$$\phi^0_{p,q}(x \leftrightarrow y \text{ in } R_n) \geq \frac{\lambda_n}{(2n+1)^2}.$$

By placing six of these rectangles side by side (as in Figure 6.8), we have by positive association that

$$\phi^0_{p,q}\big(x \leftrightarrow x + (6n,0) \text{ in } [0,6n] \times [0,2n]\big) \geq \left(\frac{\lambda_n}{(2n+1)^2} \right)^6. \tag{6.44}$$

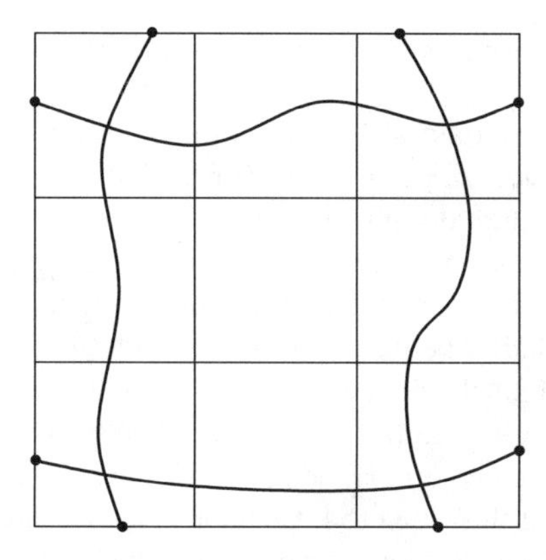

Figure 6.9. If each of four rectangles with dimensions $6n$ by $2n$ is crossed by an open path between its shorter sides, then the annulus Δ_n contains an open circuit having the origin in its interior.

We now use four copies of the rectangle $[0, 6n] \times [0, 2n]$ to construct the annulus Δ_n (see Figure 6.9). If each of these copies contains an open crossing, then the annulus contains a circuit around 0. By positive association again,

$$\phi^0_{p,q}(A_n) \geq \left(\frac{\lambda_n}{(2n+1)^2}\right)^{24}. \tag{6.45}$$

Finally, if $0 \leftrightarrow \partial\Lambda(n)$, then one of the four rectangles $[0, n] \times [-n, n]$, $[-n, n] \times [0, n]$, $[-n, 0] \times [-n, n]$, $[-n, n] \times [-n, 0]$ is traversed by an open path between its two longer sides. Therefore,

$$\phi^0_{p,q}(0 \leftrightarrow \partial\Lambda(n)) \leq 4\lambda_n. \tag{6.46}$$

Combining (6.43)–(6.46), we obtain that

$$\begin{aligned}\phi^0_{p,q}(0 \leftrightarrow \partial\Lambda(n)) &\leq 4(2n+1)^2[\phi^0_{p,q}(A_n)]^{1/24} \\ &\leq 4(2n+1)^2 \left\{ \sum_{m=8n}^{\infty} \frac{ma_m}{q} \left(\frac{q}{(1+\sqrt{q})^4} \right)^{m/4} \right\}^{1/24}.\end{aligned}$$

Now, $m^{-1} \log a_m \to \kappa$ as $m \to \infty$, and part (c) follows.

By parts (b) and (c), $\phi^0_{p_c(q),q} \neq \phi^1_{p_c(q),q}$. Part (d) follows by Theorem 4.63. □

6.5 General lattices in two dimensions

Planar duality is an important technique in the study of interacting systems on a two-dimensional lattice $\mathcal{L}$, but it is no panacea. It may be summarized in the two statements: the external boundary of a bounded connected subgraph of $\mathcal{L}$ is topologically one-dimensional, and the statistical mechanics of the boundary may be studied via an appropriate stochastic model on a certain dual lattice $\mathcal{L}_{\mathrm{d}}$. Duality provides a relation between a primal model on $\mathcal{L}$ and a dual model on $\mathcal{L}_{\mathrm{d}}$. In situations in which the dual model is related to the primal, or to some other known system, one may sometimes obtain exact results. The exact calculations of critical probabilities of percolation models on the square, triangular, and hexagonal lattices are examples of this, see [154, Chapter 11]. Individuals less burdened by the pulse for mathematical rigour have exploited duality to obtain exact but non-rigorous predictions for other two-dimensional processes (see, for example, [26]), of which a major example is the conjecture that $p_{\mathrm{c}}(q) = \sqrt{q}/(1+\sqrt{q})$ for the random-cluster model on $\mathbb{L}^2$. Such predictions are often beautiful and usually provocative to mathematicians.

We shall not explore duality in general here, noting only in passing the existence of many open problems of significance in extending known results for, say, the square lattice to general primal/dual pairs. We discuss instead two specific issues relating, in turn, to the critical points of a general primal/dual pair, and in the next section to exact calculations for the triangular and hexagonal lattices.

Here is our definition of a lattice, [154, Section 12.1]. A *lattice in d dimensions* is a connected loopless graph $\mathcal{L}$, with bounded vertex degrees, that is embedded in $\mathbb{R}^d$ in such a way that:

(a) the translations $x \mapsto x + e$ are automorphisms of $\mathcal{L}$ for each unit vector e parallel to a coordinate axis,

(b) all edges are of non-zero length, and

(c) every compact subset of $\mathbb{R}^d$ intersects only finitely many edges.

Let $\mathcal{L} = (\mathbb{V}, \mathbb{E})$ be a planar two-dimensional lattice, and let $\mathcal{L}_{\mathrm{d}}$ be its dual lattice, defined as in Section 6.1. We shall require some further symmetries of $\mathcal{L}$, namely that:

(d) the reflection mappings $\rho_{\mathrm{h}}, \rho_{\mathrm{v}} : \mathbb{R}^2 \to \mathbb{R}^2$ given by

$$\rho_{\mathrm{h}}(x, y) = (-x, y), \quad \rho_{\mathrm{v}}(x, y) = (x, -y), \qquad (x, y) \in \mathbb{R}^2,$$

are automorphisms of $\mathcal{L}$.

Let $p \in [0, 1]$ and $q \in [1, \infty)$. Under the above conditions, the random-cluster measures $\phi^b_{\mathcal{L},p,q}$ exist for $b = 0, 1$, and are invariant under horizontal and vertical translations, and under horizontal and vertical axis-reflection. They are in addition ergodic with respect to horizontal and vertical translation (separately), and they are positively associated. Such facts may be proved in exactly the same manner as were the corresponding statements for the hypercubic lattice $\mathbb{L}^d$ in Chapter 4.

Let $p_{\mathrm{c}}(q, \mathcal{L})$ denote the critical value of the random-cluster model on $\mathcal{L}$.

(6.47) Theorem. *The critical points $p_c(q, \mathcal{L})$, $p_c(q, \mathcal{L}_d)$ satisfy the inequality*

$$p_c(q, \mathcal{L}) \geq \big(p_c(q, \mathcal{L}_d)\big)_d. \tag{6.48}$$

Proof. Let $p > p_c(q, \mathcal{L})$, so that $\phi^1_{\mathcal{L},p,q}(0 \leftrightarrow \infty) > 0$. The arguments leading to the main result of [130] may be adapted to the current setting[6] to show that all open clusters in the dual lattice $\mathcal{L}_d$ are almost-surely finite. Therefore, $p_d \leq p_c(q, \mathcal{L}_d)$, whence $p \geq \big(p_c(q, \mathcal{L}_d)\big)_d$ as required. □

Equality may be conjectured in (6.48). Suppose that $\mathcal{L}$ and $\mathcal{L}_d$ are isomorphic or, weaker, that $p_c(q, \mathcal{L}) = p_c(q, \mathcal{L}_d)$. Inequality (6.48) implies then that $p_c(q, \mathcal{L}) \geq p_{sd}(q)$ (see Theorem 6.17(a) for the case of the square lattice). If (6.48) were to hold with equality, we would obtain that $p_c(q, \mathcal{L}) = p_{sd}(q)$.

Theorem 6.47 may be used to prove the uniqueness of random-cluster measures for $p \neq p_c(q, \mathcal{L})$. Some further notation must first be introduced to deal with case when $\mathcal{L}$ is not edge-transitive[7]. Let $S = [0, 1)^2 \subseteq \mathbb{R}^2$. Let I_S be the set of edges of $\mathcal{L}$ with both endvertices in S, and E_S the set of edges with exactly one endvertex in S. Let

$$N_S(\omega) = \sum_{e \in I_S} \omega(e) + \sum_{e \in E_S} \tfrac{1}{2}\omega(e), \qquad \omega \in \Omega = \{0, 1\}^{\mathbb{E}}, \tag{6.49}$$

and define the *edge-density* by

$$h^b_{\mathcal{L}}(p, q) = \frac{1}{N_S(1)} \phi^b_{\mathcal{L},p,q}(N_S), \qquad b = 0, 1. \tag{6.50}$$

If $\mathcal{L}$ is edge-transitive, it is easily seen that $h^b_{\mathcal{L}}(p, q)$ is simply the probability under $\phi^1_{\mathcal{L},p,q}$ that a given edge is open.

(6.51) Theorem. *Let $\mathcal{L}$, $\mathcal{L}_d$ be a primal/dual pair of planar lattices in two dimensions and suppose $\mathcal{L}$ satisfies* (a)–(d) *above. Let $p \in [0, 1]$ and $q \in [1, \infty)$, and assume that $p \neq p_c(q, \mathcal{L})$.*

(i) *The edge-density $h^b_{\mathcal{L}}(x, q)$ is a continuous function of x at the point $x = p$, for $b = 0, 1$.*

(ii) *It is the case that $h^0_{\mathcal{L}}(p, q) = h^1_{\mathcal{L}}(p, q)$.*

(iii) *There is a unique random-cluster measure on $\mathcal{L}$ with parameters p and q, that is, $|\mathcal{W}_{p,q}(\mathcal{L})| = |\mathcal{R}_{p,q}(\mathcal{L})| = 1$, in the natural notation.*

In the notation of Theorem 4.63, we have that $\mathcal{D}_q \subseteq \{p_c(q, \mathcal{L})\}$. In particular[8], if there exists a first-order phase transition at some value p, then necessarily

---

[6]Paper [130] treats vertex-models on $\mathbb{Z}^2$ governed by measures with certain properties of translation/rotation-invariance, ergodicity, and positive association. The arguments are however more general and apply also to edge-models on planar graphs with corresponding properties.

[7]A graph $G = (V, E)$ is called *edge-transitive* if: for every pair $e, f \in E$, there exists an automorphism of G mapping e to f. See Sections 3.3 and 10.12 for a related notion of transitivity.

[8]Related matters for Potts models are discussed in [47].

$p = p_c(q, \mathcal{L})$. As in Theorem 5.16, the percolation probabilities $\theta^b_{\mathcal{L}}(\cdot, q) = \phi^b_{\mathcal{L},p,q}(0 \leftrightarrow \infty)$, $b = 0, 1$, are continuous[9] except possibly at the value $p = p_c(q, \mathcal{L})$.

Proof. (i) For $p < p_c(q, \mathcal{L})$, this follows as in Theorems 4.63 and 5.33(a). When $p > p_c(q, \mathcal{L})$, we have from (6.48) that $p_d < p_c(q, \mathcal{L}_d)$. As in Theorem 6.13,

$$h^b_{\mathcal{L}}(p, q) + h^{1-b}_{\mathcal{L}_d}(p_d, q) = 1.$$

By part (i) applied to the dual lattice $\mathcal{L}_d$, each $h^b_{\mathcal{L}}(x, q)$ is continuous at the point $x = p_d$. Parts (ii) and (iii) follow as in Theorem 4.63, see also the proof of Theorem 6.17(b). □

6.6 Square, triangular, and hexagonal lattices

There is a host of exact but non-rigorous 'results' for two-dimensional models which, while widely accepted by physicists, continue to be subjected to mathematical investigations. Some of these claims have been made rigorous and, in so doing, mathematicians have discovered new structures of beauty and complexity. The outstanding contemporary example of new structure provoked by physics is the theory of stochastic Löwner evolutions (SLE). This has had considerable impact on percolation, Brownian motion, and on other systems with a property of conformal invariance; see Section 6.7 for a short account of SLE in the random-cluster context.

Amongst 'exact' but non-rigorous results for the random-cluster model is the claim that, for the square lattice, $p_c(q) = \sqrt{q}/(1 + \sqrt{q})$. Baxter's 1982 book [26] remains a good source for this and related statements, usually in the context of Potts models but extendable to random-cluster models with $q \in [1, \infty)$. Such statements are achieved typically by following a sequence of transformations between models, arriving thus at a 'soluble ice-type model' on a new graph termed the 'medial graph'. It has proved difficult to ascertain whether such methods are entirely rigorous, since they involve chains of argument which may seem individually innocuous but which omit significant analytical details. We attempt no more here than brief accounts of some of the conclusions together with a partial mathematical commentary.

Consider the square lattice $\mathbb{L}^2$. Instead of working with a single edge-parameter p, we allow greater generality by associating with each horizontal (respectively, vertical) edge the parameter p_h (respectively, p_v), and we write $\mathbf{p} = (p_h, p_v)$. It will be convenient as in (6.7)–(6.8) to work instead with the parameters $\mathbf{x} = (x_h, x_v)$ given by

$$x_h = \frac{q^{-\frac{1}{2}} p_h}{1 - p_h}, \qquad x_v = \frac{q^{-\frac{1}{2}} p_v}{1 - p_v},$$

[9]This may also be proved directly for a primal/dual pair, using the arguments of Theorems 5.33 and 6.47.

and their dual values $x_{\mathrm{h,d}}$, $x_{\mathrm{v,d}}$ satisfying

$$x_{\mathrm{h}}x_{\mathrm{h,d}} = 1, \qquad x_{\mathrm{v}}x_{\mathrm{v,d}} = 1.$$

Write $\phi^b_{G,\mathbf{x},q}$ for a corresponding random-cluster measure on a graph G, and moreover

$$\phi^b_{\mathbf{x},q} = \lim_{\Lambda\uparrow\mathbb{L}^2} \phi^b_{\Lambda,\mathbf{x},q}, \qquad \theta^b(\mathbf{x},q) = \phi^b_{\mathbf{x},q}(0 \leftrightarrow \infty).$$

The duality map of Section 6.1 maps a random-cluster model on $\mathbb{L}^2$ with parameter $\mathbf{x} = (x_{\mathrm{h}}, x_{\mathrm{v}})$ to a random-cluster model on $\mathbb{L}^2_{\mathrm{d}}$ with parameter $\mathbf{x}_{\mathrm{d}} = (x_{\mathrm{v,d}}, x_{\mathrm{h,d}})$. The primal and dual models have the same parameters whenever $x_{\mathrm{h}} = x_{\mathrm{v,d}}$ and $x_{\mathrm{v}} = x_{\mathrm{h,d}}$, which is to say that

$$x_{\mathrm{h}}x_{\mathrm{v}} = 1, \tag{6.52}$$

and we refer to the model as 'self-dual' if (6.52) holds. The following conjecture generalizes Conjecture 6.15.

(6.53) Conjecture. *Let* $x_{\mathrm{h}}, x_{\mathrm{v}} \in (0,\infty)$ *and* $q \in [1,\infty)$. *For* $b = 0, 1$,

$$\theta^b(\mathbf{x},q) \begin{cases} = 0 & \textit{if } x_{\mathrm{h}}x_{\mathrm{v}} < 1, \\ > 0 & \textit{if } x_{\mathrm{h}}x_{\mathrm{v}} > 1. \end{cases}$$

The proof in the case of percolation (when $q = 1$) may be found at [154, Thm 11.115]. Partial progress in the direction of the general conjecture is provided by the next theorem.

(6.54) Theorem. *Let* $x_{\mathrm{h}}, x_{\mathrm{v}} \in (0,\infty)$ *and* $q \in [1,\infty)$. *Then*

$$\theta^0(\mathbf{x},q) = 0 \quad \textit{if} \quad x_{\mathrm{h}}x_{\mathrm{v}} \le 1.$$

Proof. Let $n \ge 1$, and let

$$D(n) = \left\{ y \in \mathbb{Z}^2 : |y_1| + |y_2 - \tfrac{1}{2}| \le n + \tfrac{1}{2} \right\}$$

be the 'offset diamond' illustrated in Figure 6.10. The proof follows that of Theorem 6.17(a), but working with $D(n)$ in place of $T(n)$. We omit the details, noting only that the proof uses the 0/1-infinite-cluster property of the measures $\phi^b_{\mathbf{x},q}$, and the symmetry of the model under reflection in both the vertical axis of $\mathbb{R}^2$ and the line $\{(y_1, \frac{1}{2}) : y_1 \in \mathbb{R}\}$. □

Here are two exact but non-rigorous claims for this model. We recall from Theorem 4.58 the 'pressure' function G given in the current context as

$$G(\mathbf{x},q) = \lim_{\Lambda\uparrow\mathbb{Z}^2} \left\{ \frac{1}{|\mathbb{E}_\Lambda|} \log Y_\Lambda(\mathbf{p},q) \right\}, \qquad (\mathbf{x},q) \in (0,\infty)^2 \times [1,\infty),$$

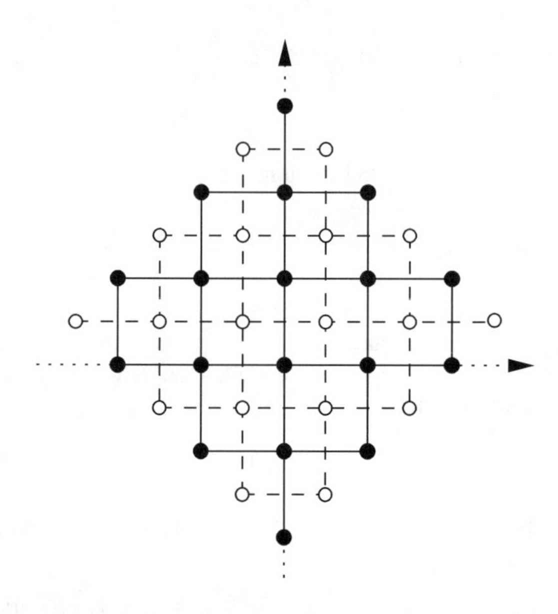

Figure 6.10. The diamond $D(n)$ of the square lattice $\mathbb{L}^2$ when $n = 2$, and the associated 'dual' diamond $D(n)_{\text{d}}$ of the dual lattice $\mathbb{L}^2_{\text{d}}$.

where

$$\begin{aligned} Y_\Lambda(\mathbf{p}, q) &= (1-p_{\text{h}})^{-\frac{1}{2}|\mathbb{E}_\Lambda|}(1-p_{\text{v}})^{-\frac{1}{2}|\mathbb{E}_\Lambda|} Z_\Lambda(\mathbf{p}, q) \\ &= \sum_{\omega\in\Omega_\Lambda} (x_{\text{h}}\sqrt{q})^{|\eta_{\text{h}}(\omega)|}(x_{\text{v}}\sqrt{q})^{|\eta_{\text{v}}(\omega)|} q^{k(\omega)}, \end{aligned}$$

and $\eta_{\text{h}}(\omega)$ (respectively, $\eta_{\text{v}}(\omega)$) is the set of open horizontal (respectively, vertical) edges of the configuration $\omega \in \Omega_\Lambda = \{0, 1\}^{\mathbb{E}_\Lambda}$. By duality as in Section 6.1,

$$G(\mathbf{x}, q) = G(\mathbf{x}_{\text{d}}, q) + \tfrac{1}{2}\log(x_{\text{h}}x_{\text{v}}), \qquad \mathbf{x} \in (0, \infty)^2. \tag{6.55}$$

By mapping the random-cluster model onto an ice-type model as in [26, Section 12.5], one obtains the following exact computation,

$$G(\mathbf{x}, q) = \tfrac{1}{2}\psi(x_{\text{h}}) + \tfrac{1}{2}\psi(x_{\text{v}}) + \tfrac{1}{4}\log q, \qquad x_{\text{h}}x_{\text{v}} = 1,$$

where the function $\psi : (0, \infty) \to \mathbb{R}$ is given as follows.

(i) When $0 < q < 4$, choose $\mu \in (0, \frac{1}{2}\pi)$ and $\gamma \in (0, \mu)$ by

$$2\cos\mu = \sqrt{q}, \quad x = \frac{\sin\gamma}{\sin(\mu - \gamma)},$$

and then

$$\psi(x) = \frac{1}{2}\int_{-\infty}^{\infty} \frac{\sinh((\pi - \mu)t)\sinh(2\gamma t)}{t\sinh(\pi t)\cosh(\mu t)}\,dt.$$

(ii) When $q = 4$, let $\tau = x/(1+x)$, and then

$$\psi(x) = \int_0^\infty \frac{e^{-y}}{y} \operatorname{sech} y \sinh(2\tau y)\, dy.$$

(iii) When $q > 4$, choose $\lambda > 0$ and $\beta \in (0, \lambda)$ by

$$2\cosh\lambda = \sqrt{q}, \qquad x = \frac{\sinh\beta}{\sinh(\lambda - \beta)},$$

and then

$$\psi(x) = \beta + \sum_{n=1}^{\infty} \frac{e^{-n\lambda}}{n} \operatorname{sech}(n\lambda)\sinh(2n\beta).$$

Our second exact asymptotic relation concerns the mean density of open edges,

$$h^b(\mathbf{p}, q) = \lim_{\Lambda \uparrow \mathbb{Z}^2} \left\{ \frac{1}{|\mathbb{E}_\Lambda|} \phi^b_{\Lambda,\mathbf{p},q}(|\eta|) \right\}, \qquad b = 0, 1.$$

By the translation-invariance of the infinite-volume measures, the mean numbers of open horizontal and vertical edges satisfy

(6.56)
$$\begin{aligned} \frac{2}{|\mathbb{E}_\Lambda|}\phi^b_{\Lambda,\mathbf{p},q}(|\eta_{\mathrm{h}}|) &\to h^b_{\mathrm{h}}(\mathbf{p}, q) = \phi^b_{\mathbf{p},q}(e_{\mathrm{h}} \text{ is open}),\\ \frac{2}{|\mathbb{E}_\Lambda|}\phi^b_{\Lambda,\mathbf{p},q}(|\eta_{\mathrm{v}}|) &\to h^b_{\mathrm{v}}(\mathbf{p}, q) = \phi^b_{\mathbf{p},q}(e_{\mathrm{v}} \text{ is open}),\end{aligned}$$

as $\Lambda \uparrow \mathbb{Z}^2$, where e_{h} (respectively, e_{v}) is a representative horizontal (respectively, vertical) edge of $\mathbb{L}^2$. Therefore,

$$h^b(\mathbf{p}, q) = \tfrac{1}{2}\big[h^b_{\mathrm{h}}(\mathbf{p}, q) + h^b_{\mathrm{v}}(\mathbf{p}, q)\big].$$

As before, except possibly on the self-dual curve $x_{\mathrm{h}}x_{\mathrm{v}} = 1$, the functions $h^b_{\mathrm{h}}(\cdot, q)$, $h^b_{\mathrm{v}}(\cdot, q)$ are continuous and $h^0_{\mathrm{h/v}}(\mathbf{p}, q) = h^1_{\mathrm{h/v}}(\mathbf{p}, q)$. [We write h/v to indicate that either possibility, chosen consistently within a given equation, is valid.] It is believed that the $h^0_{\mathrm{h/v}}(\cdot, q)$ are left-continuous, and the $h^1_{\mathrm{h/v}}(\cdot, q)$ right-continuous, on the self-dual curve, in that

$$h^0_{\mathrm{h/v}}(\mathbf{p}, q) = \lim_{\mathbf{p}' \uparrow \mathbf{p}} h^0_{\mathrm{h/v}}(\mathbf{p}', q), \quad h^1_{\mathrm{h/v}}(\mathbf{p}, q) = \lim_{\mathbf{p}' \downarrow \mathbf{p}} h^1_{\mathrm{h/v}}(\mathbf{p}', q), \qquad \mathbf{p} \in (0, 1)^2.$$

By duality as in Theorem 6.13,

(6.57)
$$h^0_{\mathrm{h/v}}(\mathbf{p}, q) + h^1_{\mathrm{v/h}}(\mathbf{p}_{\mathrm{d}}, q) = 1, \qquad \mathbf{p} \in (0, 1)^2.$$

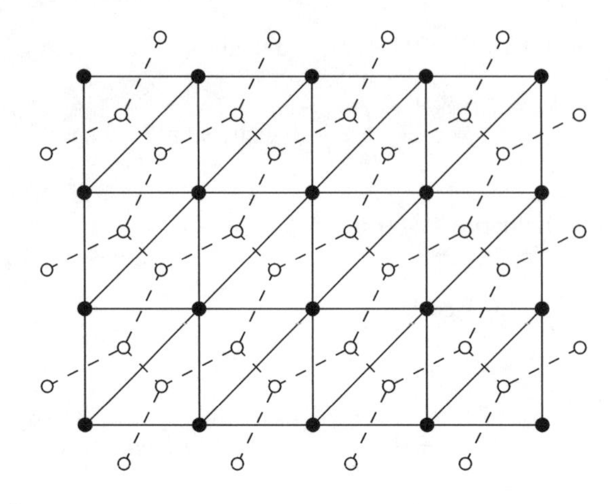

Figure 6.11. The triangular lattice $\mathbb{T}$ and its dual (hexagonal) lattice $\mathbb{H}$.

It is believed when $q \in [1, 4)$ that the transition is of second order, and thus in particular that the $h^b_{\mathrm{h/v}}(\mathbf{p}, q)$ are continuous *on* the self-dual curve. This implies that

$$h^0_{\mathrm{h/v}}(\mathbf{p}, q) = h^1_{\mathrm{h/v}}(\mathbf{p}, q), \qquad \mathbf{p} \in (0, 1)^2,\ q \in [1, 4).$$

It follows from this and (6.57) that one should have

$$h^b(\mathbf{p}, q) = \tfrac{1}{2}\big[h^b_{\mathrm{h}}(\mathbf{p}, q) + h^b_{\mathrm{v}}(\mathbf{p}, q)\big] = \tfrac{1}{2}, \qquad b = 0, 1, \tag{6.58}$$

when $x_{\mathrm{h}} x_{\mathrm{v}} = 1$ and $q \in [1, 4)$.

The transition is expected to be of first order when $q \in (4, \infty)$, and the exact computations reported in [26, Section 12.5] yield when $x_{\mathrm{h}} x_{\mathrm{v}} = 1$ that

$$\begin{aligned} h^0_{\mathrm{h/v}}(\mathbf{p}, q) &= x_{\mathrm{h/v}}\big[\psi'(x_{\mathrm{h/v}}) - \zeta(x_{\mathrm{h/v}}) P_0\big], \\ h^1_{\mathrm{h/v}}(\mathbf{p}, q) &= x_{\mathrm{h/v}}\big[\psi'(x_{\mathrm{h/v}}) + \zeta(x_{\mathrm{h/v}}) P_0\big], \end{aligned} \tag{6.59}$$

where

$$\zeta(x) = \frac{\sinh \lambda}{1 + x^2 + 2x \cosh \lambda}, \qquad P_0 = \prod_{m=1}^{\infty} [\tanh(m\lambda)]^2,$$

with λ given as in case (iii) above. Since $2\cosh\lambda = \sqrt{q}$,

$$\zeta(x) = \frac{\frac{1}{2}\sqrt{q-4}}{1 + x^2 + x\sqrt{q}},$$

a formula which underscores the relevance of the condition $q > 4$. In the symmetric case $x_{\mathrm{h}} = x_{\mathrm{v}} = 1$, we deduce when $q > 4$ that the discontinuity of the edge-density at the critical point equals

$$h^1(\mathbf{p}, q) - h^0(\mathbf{p}, q) = \frac{\sqrt{q-4}}{2 + \sqrt{q}} \prod_{m=1}^{\infty} [\tanh(m\lambda)]^2.$$

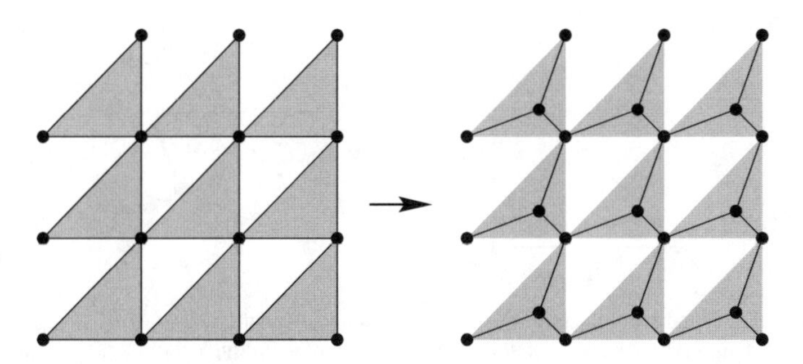

Figure 6.12. In the star–triangle transformation, alternate triangles of $\mathbb{T}$ are replaced as shown by stars. The triangular lattice $\mathbb{T}$ is thereby transformed into a copy of the hexagonal lattice $\mathbb{H}$. The shaded triangles are referred to henceforth as *grey* triangles.

We turn now from speculation concerning the (self-dual) square lattice towards rigorous mathematics concerning the triangular/hexagonal pair of lattices, denoted by $\mathbb{T}$ and $\mathbb{H}$ respectively. It is elementary that $\mathbb{T}$ is the planar dual of $\mathbb{H}$, and *vice versa* (see Figure 6.11). This fact permits a relation as in Theorem 6.13 between the random-cluster model on $\mathbb{T}$ with parameters p, q and that on $\mathbb{H}$ with parameters p_{d}, q. There is a second transformation between $\mathbb{T}$ and $\mathbb{H}$ called the 'star–triangle transformation' and illustrated in Figure 6.12. Alternate triangles of $\mathbb{T}$ are replaced by stars, and the resulting graph is isomorphic to $\mathbb{H}$. We shall henceforth refer to the shaded triangles in Figure 6.12 as *grey triangles*.

We explain next the use of the star–triangle transformation[10], and this we do with the extra generality allowed by assigning to each edge e the individual edge-parameter p_e. Rather than working with the p_e, we shall work with the variables

$$y_e = \frac{p_e}{1 - p_e},$$

see the footnote on page 135. For any finite subgraph $G = (V, E)$ of $\mathbb{T}$,

$$\phi_{G,p,q}(\omega) = \frac{1}{Y_G(p,q)} \left\{ \prod_{e \in E} y_e^{\omega(e)} \right\} q^{k(\omega)}, \qquad \omega \in \{0,1\}^E.$$

Suppose that G contains some grey triangle $T = ABC$ with edge-set $E_T = \{e_1, e_2, e_3\}$, drawn on the left side of Figure 6.13. We propose to replace T by the star S on the right side, adding thereby a supplementary vertex, and with the edge-parameter values y_1', y_2', y_3' as shown. We shall see that that, under certain conditions on the y_i and y_i', the probabilities of a large family of 'connection events' are not altered by this transformation. The conditions in question are as follows:

(6.60) $$\psi_{\mathbb{T}}(y_1, y_2, y_3) = 0,$$

(6.61) $$y_i y_i' = q \quad \text{for} \quad i = 1, 2, 3,$$

[10]References to the star–triangle transformation in the context of the Potts model may be found in [26, 324]. See also [92].

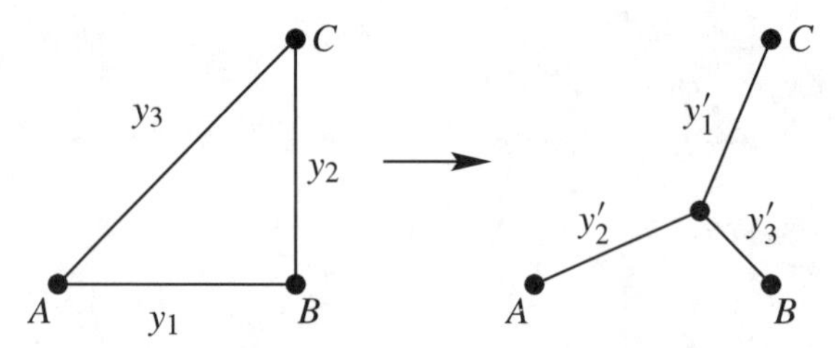

Figure 6.13. Bond parameters in the star–triangle transformation. The 'grey' triangle T on the left is replaced by the star S on the right, and the corresponding parameters are as marked.

where

$$\psi_{\mathbb{T}}(y_1, y_2, y_3) = y_1y_2y_3 + y_1y_2 + y_2y_3 + y_3y_1 - q. \tag{6.62}$$

We note for later use that $\psi_{\mathbb{H}}(y'_1, y'_2, y'_3) = 0$ under (6.60)–(6.61), where

$$\psi_{\mathbb{H}}(y_1, y_2, y_3) = y_1y_2y_3 - q(y_1 + y_2 + y_3) - q^2. \tag{6.63}$$

Let $\omega \in \{0, 1\}^E$, and define the equivalence (connection) relation $\leftrightarrow_\omega$ on V in the usual way, that is, $u \leftrightarrow v$ if and only if there exists a open path of ω from u to v. We think of $\leftrightarrow_\omega$ as a random equivalence relation. Write $G^S = (V^S, E^S)$ for the graph obtained from G after the replacement of T by S, noting that V^S is obtained from V by the addition of a vertex in the interior of T. Each $\omega^S \in \{0, 1\}^{E^S}$ gives rise similarly to an equivalence relation on V which we denote as $\leftrightarrow_{\omega^S}$.

(6.64) Lemma. *Let $q \in (0, \infty)$. Let $G = (V, E)$ be a finite subgraph of $\mathbb{T}$ and let $T = ABC$ be a grey triangle of G as above. Let $\mathbf{p} \in (0, 1)^E$, and let $\mathbf{p}^S \in (0, 1)^{E^S}$ be such that: $p^S_e = p_e$ for $e \in E \setminus E_T$, and on T and S the corresponding parameters y_i, y'_i satisfy* (6.60)–(6.61). *The law of $\leftrightarrow_\omega$ under $\phi_{G,\mathbf{p},q}$ is the same as the law of $\leftrightarrow_{\omega^S}$ under $\phi_{G^S,\mathbf{p}^S,q}$.*

Proof. Let e_1, e_2, e_3 (respectively, e'_i) be the edges of T (respectively, S), and write y_i (respectively, y'_i) for the corresponding parameters as in Figure 6.13. Let $(\omega(e) : e \in E\setminus\{e_1, e_2, e_3\})$ be given, and consider the conditional random-cluster measures P^T_ω (respectively, P^S_ω) on the e_i (respectively, on the e'_i). There are three disjoint classes of configuration ω which must be considered, depending on which of the following holds:

(a) A, B, C are in distinct open clusters of ω restricted to $E \setminus \{e_1, e_2, e_3\}$,

(b) two members of $\{A, B, C\}$ are in the same such cluster, and the third is not,

(c) A, B, and C are in the same such cluster.

In each case, we propose to show that, under (6.60)–(6.61), the connections between A, B, C have the same conditional probabilities under both P^T_ω and P^S_ω. The required calculations are simple in principle, and we shall omit many details. In particular, we shall verify the claim under case (b) only, the other two cases being similar. Assume then that (b) holds, and suppose for definiteness that the configuration ω is such that: A and B are joined off T, but C is joined off T to neither A nor B. By Theorem 3.1, the probabilities of connections internal to T are given as follows:

$$
\begin{aligned}
P_\omega^T(A \leftrightarrow B \text{ and } B \not\leftrightarrow C \text{ in } T) &= \frac{1}{Y} y_1 q^2,\\
P_\omega^T(A \not\leftrightarrow B \text{ and } B \leftrightarrow C \text{ in } T) &= \frac{1}{Y} y_2 q,\\
P_\omega^T(A \not\leftrightarrow B \text{ and } A \leftrightarrow C \text{ in } T) &= \frac{1}{Y} y_3 q,\\
P_\omega^T(A \leftrightarrow B \leftrightarrow C \text{ in } T) &= \frac{1}{Y}(y_1y_2y_3 + y_1y_2 + y_2y_3 + y_3y_1)q,
\end{aligned}
\tag{6.65}
$$

where

$$
Y = (y_1y_2y_3 + y_1y_2 + y_2y_3 + y_3y_1 + y_2 + y_3)q + (1 + y_1)q^2. \tag{6.66}
$$

Note that the events in question concern the existence (or not) of open paths within T only. The remaining term $P_\omega^T(A \not\leftrightarrow B \not\leftrightarrow C \text{ in } T)$ is given by the fact that the sum of the probabilities of all such configurations on T equals 1.

The corresponding probabilities for connections internal to S are:

$$
\begin{aligned}
P_\omega^S(A \leftrightarrow B \text{ and } B \not\leftrightarrow C \text{ in } S) &= \frac{1}{Y'} y_2'y_3' q^2,\\
P_\omega^S(A \not\leftrightarrow B \text{ and } B \leftrightarrow C \text{ in } S) &= \frac{1}{Y'} y_1'y_3' q,\\
P_\omega^S(A \not\leftrightarrow B \text{ and } A \leftrightarrow C \text{ in } S) &= \frac{1}{Y'} y_1'y_2' q,\\
P_\omega^S(A \leftrightarrow B \leftrightarrow C \text{ in } S) &= \frac{1}{Y'} y_1'y_2'y_3' q,
\end{aligned}
\tag{6.67}
$$

where

$$
Y' = (y_1'y_2'y_3' + y_1'y_2' + y_1'y_3')q + (y_2'y_3' + y_1' + y_2' + y_3')q^2 + q^3. \tag{6.68}
$$

It is left to the reader to check that, under (6.60)–(6.61), the probabilities in (6.65) and (6.67) are equal. Similar computations are valid in cases (a) and (c) also, and it follows that, in loose terms, the replacement of T by S is 'invisible' to connections elsewhere in the graph G. □

Lemma 6.64 allows us to replace one grey triangle of G by a star. This process may be iterated until every grey triangle of G has been thus replaced. If G is itself a union of grey triangles, then the resulting graph is a subgraph of the hexagonal lattice $\mathbb{H}$. By working on a square region Λ of $\mathbb{T}$ and passing to the limit as $\Lambda \uparrow \mathbb{T}$, we find in particular that connections on $\mathbb{T}$ have the same probabilities as connections on $\mathbb{H}$ so long as the edge-parameters on $\mathbb{T}$ satisfy (6.60) and the corresponding parameters on $\mathbb{H}$ satisfy (6.61). In particular the percolation probabilities are the same. We now make the last statement more specific.

Write $\mathbb{E}_{\mathbb{T}}$ (respectively, $\mathbb{E}_{\mathbb{H}}$) for the edge-set of $\mathbb{T}$ (respectively, $\mathbb{H}$). Let $\mathbf{p} = (p_e : e \in \mathbb{E}_{\mathbb{T}}) \in (0,1)^{\mathbb{E}_{\mathbb{T}}}$, and let $y_e = p_e/(1-p_e)$. We speak of $\mathbf{p}$ as being of *type* γ if, for every grey triangle T, the three parameters y_1, y_2, y_3 of the edges of T satisfy $\psi_{\mathbb{T}}(y_1, y_2, y_3) = \gamma$. Suppose that $\mathbf{p}$ is of type 0, as in (6.60). Applying the star–triangle transformation to every grey triangle of $\mathbb{T}$, we obtain a copy $\mathbb{H}$ of the hexagonal lattice, and we choose the parameters $\mathbf{p}' = (p'_e : e \in \mathbb{E}_{\mathbb{H}})$ of edges of this lattice in such a way that (6.61) holds. By the above discussion, the percolation probabilities $\theta^b_{\mathbb{T}}$ and $\theta^b_{\mathbb{H}}$ satisfy

$$\theta^b_{\mathbb{T}}(\mathbf{p}, q) = \theta^b_{\mathbb{H}}(\mathbf{p}', q), \qquad b = 0, 1, \tag{6.69}$$

whenever $q \in [1, \infty)$.

A *labelled lattice* is a lattice $\mathcal{L}$ together with a real vector $\mathbf{p}$ indexed by the edge-set of $\mathcal{L}$. An *automorphism* of a labelled lattice $(\mathcal{L}, \mathbf{p})$ is a graph automorphism τ of $\mathcal{L}$ such that $p_{\tau(e)} = p_e$ for every edge e.

Equation (6.69) leads to a proposal for the so-called 'critical surfaces' of the triangular and hexagonal lattices. The crude argument is as follows. Suppose that $\mathbf{p}, \mathbf{p}'$ are as above. If $\theta^0_{\mathbb{T}}(\mathbf{p}, q) > 0$ then, by (6.69), $\theta^0_{\mathbb{H}}(\mathbf{p}', q) > 0$ also. If we accept a picture of an infinite primal ocean of $\mathbb{H}$ encompassing bounded islands of its dual, then it follows that $\theta^1_{\mathbb{H}_d}((\mathbf{p}')_d, q) = 0$. If the initial labelled lattice $(\mathbb{T}, \mathbf{p})$ has a sufficiently large automorphism group then it may, by (6.61), be the case that $(\mathbb{H}_d, (\mathbf{p}')_d)$ is isomorphic to $(\mathbb{T}, \mathbf{p})$, in which case

$$0 = \theta^1_{\mathbb{H}_d}((\mathbf{p}')_d, q) = \theta^1_{\mathbb{T}}(\mathbf{p}, q).$$

This is a contradiction, and we deduce that $\theta^0_{\mathbb{T}}(\mathbf{p}, q) = 0$ whenever $\mathbf{p}$ is of type 0.

On the other hand, some readers may be able to convince themselves that there should exist no non-empty interval $(\alpha, \beta) \subseteq \mathbb{R}$ such that: neither $\mathbb{T}$ nor its dual lattice possesses an infinite cluster whenever the type of $\mathbf{p}$ lies in (α, β). One arrives via these non-rigorous arguments at the (unproven) statement that

$$\theta^0_{\mathbb{T}}(\mathbf{p}, q) \begin{cases} = 0 & \text{if } \mathbf{p} \text{ is of non-positive type,} \\ > 0 & \text{if } \mathbf{p} \text{ is of strictly positive type,} \end{cases} \tag{6.70}$$

with a similar conjecture for the hexagonal lattice.

Let $p_1, p_2, p_3 \in (0,1)$ and let $y_i = p_i/(1-p_i)$. We restrict the discussion now to the situation in which every grey triangle of $\mathbb{T}$ has three edges with parameters p_1, p_2, p_3, in some order. The corresponding process on $\mathbb{H}$ has parameters p'_i where the $y'_i = p'_i/(1-p'_i)$ satisfy (6.61). The assertions above motivate the proposals that:

$$\begin{aligned} &\mathbb{T} \text{ has critical surface} \quad y_1y_2y_3 + y_1y_2 + y_2y_3 + y_3y_1 - q = 0, \\ &\mathbb{H} \text{ has critical surface} \quad y'_1y'_2y'_3 - q(y'_1 + y'_2 + y'_3) - q^2 = 0, \end{aligned} \tag{6.71}$$

in the sense that

$$\theta^0_{\mathbb{T}}(\mathbf{p}, q) \begin{cases} = 0 & \text{if } \psi_{\mathbb{T}}(\mathbf{y}) \le 0, \\ > 0 & \text{if } \psi_{\mathbb{T}}(\mathbf{y}) > 0, \end{cases}$$

with a similar statement for $\mathbb{H}$. It is not known how to make (6.71) rigorous, neither is it even accepted that the above statements are true in generality, since no explicit assumption has been made about the automorphism groups of the labelled lattices in question.

We move now to the special case of the homogeneous random-cluster model on $\mathbb{T}$, with constant edge-parameter $p_e = p$ for every edge e. One part of the above discussion may be made rigorous, as follows.

(6.72) Theorem. *Let $q \in [1, \infty)$.*

(a) *Consider the random-cluster model on the triangular lattice $\mathbb{T}$, and let p be such that $y = p/(1-p)$ satisfies $y^3 + 3y^2 - q = 0$. Then $\theta^0_{\mathbb{T}}(p, q) = 0$, and therefore $p_c(q, \mathbb{T}) \ge p$.*

(b) *Consider the random-cluster model on the hexagonal lattice $\mathbb{H}$, and let p' be such that $y = p'/(1-p')$ satisfies $y^3 - 3qy - q^2 = 0$. Then $\theta^0_{\mathbb{H}}(p', q) = 0$, and therefore $p_c(q, \mathbb{H}) \ge p'$.*

Proof. This may be proved either by adapting the argument used to prove Theorems 6.17(a) and 6.54, or by following the proof of Theorem 6.47. The former approach utilizes the 0/1-infinite-cluster property, and the latter approach makes use of the circuit-generation procedure pioneered in [181] and extended in [130]. Under either method, it is important that the labelled lattices be invariant under translations and possess axes of mirror-symmetry. □

It is generally believed that the critical values of $\mathbb{T}$ and $\mathbb{H}$ are the values given in Theorem 6.72. To prove this, it would suffice to have a reasonable upper bound for $\phi^0_{\mathbb{T},p,q}(0 \leftrightarrow \partial\Lambda(n))$, where $\Lambda(n) = [-n, n]^2$. See the related Theorem 6.18 and Lemma 6.28.

We close this section with an open problem. Arguably the simplest system on the triangular lattice which possesses insufficient symmetry for the above proof is that in which every horizontal (respectively, vertical, diagonal) edge of $\mathbb{T}$ has edge-parameter p_h (respectively, p_v, p_d). The ensuing labelled lattice has properties of translation-invariance but has no axis of mirror-symmetry. Instead, it is symmetric under reflections in the origin. We conjecture that the equivalent of Theorem 6.72 holds for this process, namely that

$$\theta^0_{\mathbb{T}}(\mathbf{p}, q) = 0 \quad \text{if} \quad \psi_{\mathbb{T}}(p_h, p_v, p_d) = 0. \tag{6.73}$$

Indeed, one expects that the critical surface is given by $\psi_{\mathbb{T}}(p_h, p_v, p_d) = 0$. The proof of the corresponding statement for the percolation model may be found at [154, Thm 11.116].

6.7 Stochastic Löwner evolutions

Many exact calculations are 'known' for critical processes in two dimensions, but the required physical arguments have sometimes appeared in varying degrees magical or revelationary to mathematicians. The recently developed technology of stochastic Löwner evolutions (SLE), discovered by Schramm [294], promises a rigorous underpinning of many such arguments in a manner consonant with modern probability theory. Roughly speaking, the theory of SLE informs us of the correct weak limit of a critical process in the limit of large spatial scales, and in addition provides a mechanism for performing calculations for the limit process.

Let $\mathbb{U} = (-\infty, \infty) \times (0, \infty)$ denote the upper half-plane of $\mathbb{R}^2$, with closure $\overline{\mathbb{U}}$. We view $\mathbb{U}$ and $\overline{\mathbb{U}}$ as subsets of the complex plane. Consider the ordinary differential equation

$$\frac{d}{dt} g_t(z) = \frac{2}{g_t(z) - B_{\kappa t}}, \qquad z \in \overline{\mathbb{U}} \setminus \{0\},$$

subject to the boundary condition $g_0(z) = z$, where $t \in [0, \infty)$, κ is a positive constant, and $(B_t : t \geq 0)$ is a standard Brownian motion. The solution exists when $g_t(z)$ is bounded away from $B_{\kappa t}$. More specifically, for $z \in \overline{\mathbb{U}}$, let τ_z be the infimum of all times τ such that 0 is a limit point of $g_s(z) - B_{\kappa s}$ in the limit as $s \uparrow \tau$. We let

$$H_t = \{z \in \mathbb{U} : \tau_z > t\}, \qquad K_t = \{z \in \overline{\mathbb{U}} : \tau_z \leq t\},$$

so that H_t is open, and K_t is compact. It may now be seen that g_t is a conformal homeomorphism from H_t to $\mathbb{U}$.

We call $(g_t : t \geq 0)$ a *stochastic Löwner evolution* (SLE) with parameter κ, written SLE_κ, and we call the K_t the *hulls* of the process. There is good reason to believe that the family $K = (K_t : t \geq 0)$ provides the correct scaling limit of a variety of random spatial processes, the value of κ being chosen according to the process in question. General properties of SLE_κ, viewed as a function of κ, have been studied in [284, 316], and a beautiful theory has emerged. For example, the hulls K form (almost surely) a simple path if and only if $\kappa \leq 4$. If $\kappa > 8$, then SLE_κ generates (almost surely) a space-filling curve.

Schramm [294, 295] has identified the relevant value of κ for several different processes, and has indicated that percolation has scaling limit SLE_6. Full rigorous proofs are not yet known even for general percolation models. For the special case of site percolation on the triangular lattice $\mathbb{T}$, Smirnov [304, 305] has proved the very remarkable result that the crossing probabilities of re-scaled regions of $\mathbb{R}^2$ satisfy Cardy's formula, and he has outlined a connection to a 'full scaling limit' and to the process SLE_6. (This last statement is illustrated and partly explained in Figure 6.14.) The full scaling limit for critical percolation on $\mathbb{T}$ as an SLE_6-based loop process was announced by Camia and Newman in [75] and the proofs may be found in [76].

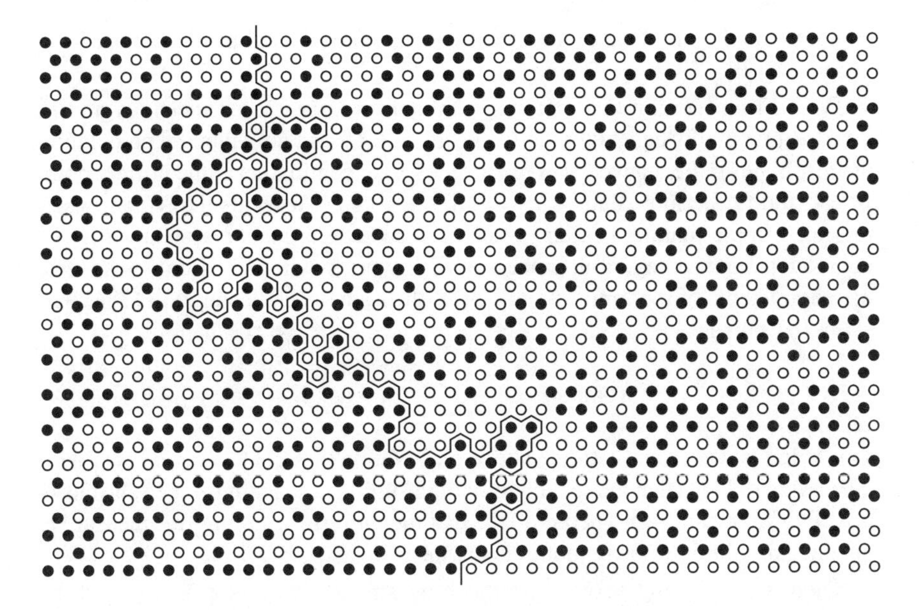

Figure 6.14. Site percolation on the triangular lattice with p equal to the critical point $\frac{1}{2}$, and with a mixed boundary condition along the lower side. The interface traces the boundary between the white and the black clusters touching the boundary, and is termed the 'exploration process'. In the limit of small lattice-spacing, the interface converges in a certain manner to the graph of a function that satisfies the Löwner differential equation driven by a Brownian motion with variance parameter $\kappa = 6$.

It is possible to perform calculations for stochastic Löwner evolutions, and in particular to confirm, [230, 307], the values of many critical exponents associated with site percolation on the triangular lattice. The outcomes are in agreement with predictions of mathematical physicists considered previously to be near-miraculous, see [154, Chapter 9]. In addition, SLE_6 satisfies the appropriate version of Cardy's formula, [80, 227].

The technology of SLE is a major piece of contemporary mathematics which promises to explain phase transitions in an important class of two-dimensional disordered systems, and to help bridge the gap between probability theory and conformal field theory. It has in addition provided complete explanations of conjectures made by mathematicians and physicists concerning the intersection exponents and fractionality of frontier of two-dimensional Brownian motion, see [228, 229].

Further work is needed to prove the validity of the limiting operation for other percolation models and random processes. Lawler, Schramm, and Werner have verified in [231] the existence of the scaling limit for loop-erased random walk and for the uniform spanning-tree Peano curve, and have shown them to be SLE_2 and SLE_8 respectively. It is believed that self-avoiding walk on $\mathbb{L}^2$, [244], has scaling limit $SLE_{8/3}$. Schramm and Sheffield have proved that the so-called harmonic explorer and the interface of the discrete Gaussian free field have common limit SLE_4, see [296, 297].

We turn now to the random-cluster model on $\mathbb{L}^2$ with parameters p and q. For $q \in [1, 4)$, it is believed as in Conjectures 6.15 and 6.32 that the percolation probability $\theta(p, q)$, viewed as a function of p, is continuous at the critical point $p_c(q)$, and furthermore that $p_c(q) = \sqrt{q}/(1 + \sqrt{q})$. It seems likely that, when re-scaled in the manner similar to that of percolation, the cluster-boundaries of the model converge to a limit process of SLE type. It will remain only to specify the parameter κ of the limit in terms of q. It has been conjectured in [284] that $\kappa = \kappa(q)$ satisfies

$$\cos(4\pi/\kappa) = -\tfrac{1}{2}\sqrt{q}, \qquad \kappa \in (4, 8).$$

This value is consistent with the above observation that $\kappa(1) = 6$, and also with the finding of [231] that the scaling limit of the uniform spanning-tree Peano curve is SLE_8. We recall from Theorem 1.23 that the uniform spanning-tree measure is obtained as a limit of the random-cluster measure as $p, q \downarrow 0$.

There are uncertainties over how this programme will develop. For a start, the theory of random-cluster models is not so complete as that of percolation and of the uniform spanning tree. Secondly, the existence of spatial limits is currently known only in certain special cases. The programme is however ambitious and promising, and may ultimately yield a full picture of the critical behaviour, including the numerical values of critical exponents, of random-cluster models with $q \in [1, 4)$, and hence of Ising/Potts models also. There is good reason to expect early progress for the case $q = 2$, for which the random-cluster interface should converge to $SLE_{16/3}$, and the Ising (spin) interface to SLE_3, [306]. The reader is referred to [295] for a survey of open problems and conjectures concerning SLE.

Chapter 7

Duality in Higher Dimensions

Summary. The boundaries of clusters in d dimensions are (topologically) $(d-1)$-dimensional and, in their study, one encounters new geometrical difficulties when $d \geq 3$. By representing the random-cluster model as a sequence of nested contours with alternately wired and free boundary conditions, one arrives at the proof that the phase transition is discontinuous for sufficiently large q. There is a random-cluster analysis of non-translation-invariant states of Dobrushin-type when $d \geq 3$, $q \in [1, \infty)$, and p is sufficiently large.

7.1 Surfaces and plaquettes

Duality is a fundamental technique in the study of a number of stochastic models on a planar graph $G = (V, E)$. Domains of G which are 'switched-on' in the model are surrounded by contours of the dual graph G_d which are 'switched-off'. We make this more concrete as follows. We take as sample space the set $\Omega = \{0, 1\}^E$ where, as usual, an edge e is called *open* in $\omega \in \Omega$ if $\omega(e) = 1$. There exists no open path between two vertices x, y of G if and only if there exists a contour in the dual graph that separates x and y and that traverses closed edges only. Such facts have been especially fruitful in the case of percolation, because the dual process of closed edges is itself a percolation process. We saw similarly in Section 6.1 that the dual of a random-cluster model on a planar graph G is a random-cluster model on the dual graph G_d, and this observation led to a largely complete theory of the random-cluster model on the square lattice. When $d = 2$, one may summarize this with the facile remark that $2 = 1 + 1$, viewed as an expression of the fact that the co-dimension of a line in $\mathbb{R}^2$ is 1. The situation in three and more dimensions is much more complicated since the co-dimension of a line in $\mathbb{R}^d$ is $d - 1$, and one is led therefore to a consideration of surfaces and their geometry.

We begin with a general description of duality in three dimensions (see, for example, [6, 139]) and we consider for the moment the three-dimensional cubic lattice $\mathbb{L}^3$. The dual lattice $\mathbb{L}^3_d$ is obtained by translating $\mathbb{L}^3$ by the vector

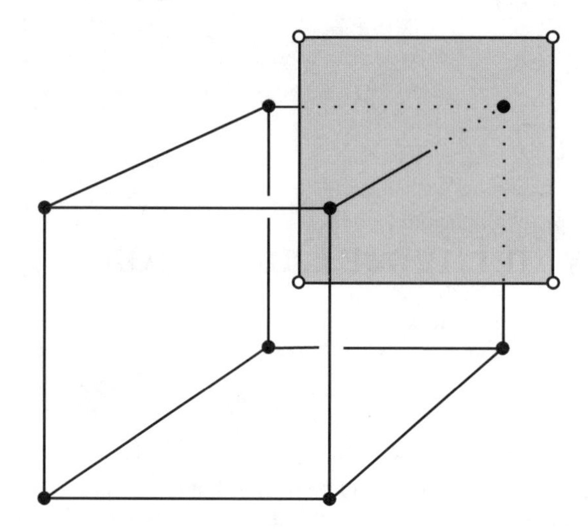

Figure 7.1. A unit cube of the primal lattice $\mathbb{L}^3$, and a plaquette of the dual lattice. The open circles are vertices of the dual lattice $\mathbb{L}^3_{\mathrm{d}}$ placed at the centres of the primal unit cubes. Each edge e of $\mathbb{L}^3$ passes through the centre of some face common to two unit cubes of the dual lattice, illustrated here by the shaded region, and this face is the 'plaquette' associated with e.

$\frac{1}{2} = (\frac{1}{2}, \frac{1}{2}, \frac{1}{2})$; each vertex of $\mathbb{L}^3_{\mathrm{d}}$ lies at the centre of a unit cube of the primal lattice $\mathbb{L}^3$. We define a *plaquette* to be a (topologically) closed unit square in $\mathbb{R}^3$ with corners lying in the dual vertex set $\mathbb{Z}^3 + \frac{1}{2}$. That is, plaquettes are the bounding faces of the unit cubes of the dual lattice $\mathbb{L}^3_{\mathrm{d}}$. Each edge e of $\mathbb{L}^3$ passes through the centre of a dual plaquette, namely the plaquette that is perpendicular to e and passes through its centre, see Figure 7.1.

Two distinct plaquettes h_1 and h_2 are called 1-*connected*, written $h_1 \stackrel{1}{\sim} h_2$, if: either $h_1 = h_2$, or $h_1 \cap h_2$ is homeomorphic to the unit interval [0, 1]. A set of plaquettes is called 1-connected if they are connected when viewed as the vertex-set of a graph with adjacency relation $\stackrel{1}{\sim}$. Consider a finite, connected, open cluster C of $\mathbb{L}^3$. It has an external edge-boundary $\Delta_{\mathrm{e}}C$ comprising all closed edges with exactly one endvertex in C. Edges in $\Delta_{\mathrm{e}}C$ correspond to plaquettes of the dual $\mathbb{L}^3_{\mathrm{d}}$, and it turns out that this set of plaquettes contains a 1-connected surface that separates C from ∞. Thus, connectivity in the primal lattice is constrained by the existence of 1-connected 'surfaces' of dual plaquettes.

Here is a plan of this chapter. There appears in Section 7.2 a topological argument which is fundamental to the study of the random-cluster model with $d \geq 3$. This extends the two-dimensional duality results of [210, Appendix] to three and more dimensions. There are two principal components in the remainder of the chapter. In Sections 7.3–7.5, a representation of the wired and free random-cluster models as polymer models (in the sense of statistical mechanics) is established and developed. This leads to the famous result that the phase transition of the random-cluster model is discontinuous when $d \geq 2$ and q is sufficiently large, [224]. The second component is the proof in Sections 7.6–7.11 of the existence of 'Dobrushin interfaces' for all random-cluster models with $d \geq 3$, $q \in [1, \infty)$, and sufficiently large p. This generalizes Dobrushin's work on non-translation-invariant Gibbs states for the Ising model, [103], and extends even to the percolation model. A considerable amount of geometry is required for this, and the account given here draws heavily on the original paper, [139].

7.2 Basic properties of surfaces

The principal target of this section is to study the geometry of the dual surface corresponding to the external boundary of a finite connected subgraph of $\mathbb{L}^d$. The results are presented for $d \geq 3$, but the reader is advised to concentrate on the case $d = 3$. We write $\mathbb{L}^d_{\mathrm{d}}$ for the dual lattice of $\mathbb{L}^d$, being the translate of $\mathbb{L}^d$ by the vector $\frac{1}{2} = (\frac{1}{2}, \frac{1}{2}, \dots, \frac{1}{2})$.

Let $d \geq 3$ and let $B_0 = [0, 1]^d$, viewed as a subset of $\mathbb{R}^d$. The elementary cubes of $\mathbb{L}^d_{\mathrm{d}}$ are translates by integer vectors of the cube $B_0 - \frac{1}{2} = [-\frac{1}{2}, \frac{1}{2}]^d$. The boundary of $B_0 - \frac{1}{2}$ is the union of the $2d$ sets $P_{i,u}$ given by

$$P_{i,u} = [-\tfrac{1}{2}, \tfrac{1}{2}]^{i-1} \times \{u - \tfrac{1}{2}\} \times [-\tfrac{1}{2}, \tfrac{1}{2}]^{d-i}, \tag{7.1}$$

for $i = 1, 2, \dots, d$ and $u = 0, 1$. A *plaquette* (in $\mathbb{R}^d$) is defined to be a translate by an integer vector of some $P_{i,u}$. We point out that plaquettes are (topologically) closed $(d-1)$-dimensional subsets of $\mathbb{R}^d$, and that plaquettes are lines when $d = 2$, and are unit squares when $d = 3$ (see Figure 7.1). Let $\mathbb{H}$ denote the set of all plaquettes in $\mathbb{R}^d$. The straight line-segment joining the vertices of an edge $e = \langle x, y \rangle$ passes through the middle of a plaquette denoted by $h(e)$, which we call the *dual* plaquette of e. More precisely, if $y = x + e_i$ where $e_i = (0, \dots, 0, 1, 0, \dots, 0)$ is the unit vector in the direction of increasing ith coordinate, then $h(e) = P_{i,1} + x$.

Let $s \in \{1, 2, \dots, d-2\}$. Two distinct plaquettes h_1 and h_2 are said to be *s-connected*, written $h_1 \overset{s}{\sim} h_2$, if $h_1 \cap h_2$ contains a homeomorphic image of the s-dimensional unit cube $[0, 1]^s$. We say that h_1 and h_2 are 0-connected, written $h_1 \overset{0}{\sim} h_2$, if $h_1 \cap h_2 \neq \varnothing$. Note that $h_1 \overset{d-2}{\sim} h_2$ if and only if $h_1 \cap h_2$ is homeomorphic to $[0, 1]^{d-2}$. A set of plaquettes is said to be s-connected if they are connected when viewed as the vertex-set of a graph with adjacency relation $\overset{s}{\sim}$. Of particular importance is the case $s = d-2$. The distance $\|h_1, h_2\|$ between two plaquettes h_1, h_2 is defined to be the L^∞ distance between their centres. For any set H of plaquettes, we write $E(H)$ for the set of edges of $\mathbb{L}^d$ to which they are dual.

We consider next some geometrical matters. The words 'connected' and 'component' should be interpreted for the moment in their topological sense. Let $T \subseteq \mathbb{R}^d$, and write $\overline{T}$ for the closure of T in $\mathbb{R}^d$. We define the *inside* $\mathrm{ins}(T)$ to be the union of all bounded connected components of $\mathbb{R}^d \setminus T$; the *outside* $\mathrm{out}(T)$ is the union of all unbounded connected components of $\mathbb{R}^d \setminus T$. The set T is said to *separate* $\mathbb{R}^d$ if $\mathbb{R}^d \setminus T$ has more than one connected component. For a set $H \subseteq \mathbb{H}$ of plaquettes, we define the set $[H] \subseteq \mathbb{R}^d$ by

$$(7.2) \qquad [H] = \{x \in \mathbb{R}^d : x \in h \text{ for some } h \in H\}.$$

We call a finite set H of plaquettes a *splitting set* if it is $(d-2)$-connected and $\mathbb{R}^d \setminus [H]$ contains at least one bounded connected component.

The two theorems that follow are in a sense dual to one another. The first is an analogue[1] in a general number of dimensions of Proposition 2.1 of [210, Appendix], where two-dimensional mosaics were considered.

(7.3) Theorem [139]. *Let $d \geq 3$, and let $G = (V, E)$ be a finite connected subgraph of $\mathbb{L}^d$. There exists a splitting set Q of plaquettes such that:*

(i) $V \subseteq \text{ins}([Q])$,

(ii) *every plaquette in Q is dual to some edge of $\mathbb{E}^d$ with exactly one endvertex in V,*

(iii) *if W is a connected set of vertices of $\mathbb{L}^d$ such that $V \cap W = \varnothing$, and there exists an infinite path on $\mathbb{L}^d$ starting in W that uses no vertex in V, then $W \subseteq \text{out}([Q])$.*

For any set δ of plaquettes, we define its *closure* $\overline{\delta}$ to be the set

$$(7.4) \qquad \overline{\delta} = \delta \cup \big\{h \in \mathbb{H} : h \text{ is } (d-2)\text{-connected to some member of } \delta\big\}.$$

Let $\delta = \{h(e) : e \in D\}$ be a $(d-2)$-connected set of plaquettes. Consider the subgraph $(\mathbb{Z}^d, \mathbb{E}^d \setminus D)$ of $\mathbb{L}^d$, and let C be a component of this graph. Let $\Delta_{\text{v},\delta}C$ denote the set of all vertices v in C for which there exists $w \in \mathbb{Z}^d$ with $h(\langle v, w\rangle) \in \overline{\delta}$, and let $\Delta_{\text{e},\delta}C$ denote the set of edges f of C for which $h(f) \in \overline{\delta} \setminus \delta$. Note that edges in $\Delta_{\text{e},\delta}C$ have both endvertices belonging to $\Delta_{\text{v},\delta}C$.

(7.5) Theorem [139]. *Let $d \geq 3$. Let $\delta = \{h(e) : e \in D\}$ be a $(d-2)$-connected set of plaquettes, and let $C = (V_C, E_C)$ be a finite connected component of the graph $(\mathbb{Z}^d, \mathbb{E}^d \setminus D)$. There exists a splitting set $Q = Q_C$ of plaquettes such that:*

(i) $V_C \subseteq \text{ins}([Q])$,

(ii) $Q \subseteq \delta$,

(iii) *every plaquette in Q is dual to some edge of $\mathbb{E}^d$ with exactly one endvertex in C.*

Furthermore, the graph $(\Delta_{\text{v},\delta}C, \Delta_{\text{e},\delta}C)$ is connected.

This theorem will be used later to show that, for a suitable (random) set δ of plaquettes, the random-cluster measure within a bounded connected component of $\mathbb{R}^d \setminus [\delta]$ is that with *wired* boundary condition. The argument is roughly as follows. Let $\omega \in \Omega$, and let $\delta = \{h(e) : e \in D\}$ be a maximal $(d-2)$-connected

[1] This answers a question which arose in 1980 during a conversation with H. Kesten.

set of plaquettes that are open (in the sense that they are dual to ω-closed edges of $\mathbb{L}^d$, see (7.9)). Let $h = h(f) \in \overline{\delta} \setminus \delta$. It must be the case that f is open, since if f were closed then $h(f)$ would be open, which would in turn imply that $h(f) \in \delta$, a contradiction. That is to say, for any finite connected component C of $(\mathbb{Z}^d, \mathbb{E}^d \setminus D)$, every edge in $\Delta_{e,\delta}C$ is open. By Theorem 7.5, the boundary $\Delta_{v,\delta}C$, when augmented by the set $\Delta_{e,\delta}C$ of edges, is a connected graph. The random-cluster measure on C, conditional on the set δ, is therefore a wired measure.

We shall require one further theorem of similar type.

(7.6) Theorem. *Let $d \geq 3$ and let $\delta = \{h(e) : e \in D\}$ be a finite $(d-2)$-connected set of plaquettes. Let $C = (V, E)$ be the subgraph of $(\mathbb{Z}^d, \mathbb{E}^d \setminus D)$ comprising all vertices and edges lying in* out$([\delta])$. *There exists a subset Q of δ such that:*

(i) *Q is $(d-2)$-connected,*

(ii) *every plaquette in Q is dual to some edge of $\mathbb{E}^d$ with at least one endvertex in C.*

Furthermore, the graph $(\Delta_{v,\delta}C, \Delta_{e,\delta}C)$ is connected.

Proof of Theorem 7.3. Related results may be found in [82, 101, 159]. The theorem may be proved by extending the proof of [159, Lemma 7.2], but instead we adapt the proof given for three dimensions in [139]. Consider the set of edges of $\mathbb{L}^d$ with exactly one endvertex in V, and let P be the corresponding set of plaquettes.

Let $x \in V$. We show first that $x \in \text{ins}([P])$. Let $\mathcal{U}$ be the set of all closed unit cubes of $\mathbb{R}^d$ having centres in V. Since all relevant sets in this proof are simplicial, the notions of path-connectedness and arc-connectedness coincide. Recall that an unbounded path of $\mathbb{R}^d$ from x is a continuous mapping $\gamma : [0, \infty) \to \mathbb{R}^d$ with $\gamma(0) = x$ and unbounded image. For any such path γ satisfying $|\gamma(t)| \to \infty$ as $t \to \infty$, γ has a final point $z(\gamma)$ belonging to the (closed) union of all cubes in $\mathcal{U}$. Now $z(\gamma) \in [P]$ for all such γ, and therefore $x \in \text{ins}([P])$.

Let λ_s denote s-dimensional Lebesgue measure, so that, in particular, $\lambda_0(S) = |S|$. A subset S of $\mathbb{R}^d$ is called:

$$\begin{cases} \textit{thin} & \text{if } \lambda_{d-3}(S) < \infty, \\ \textit{fat} & \text{if } \lambda_{d-2}(S) > 0. \end{cases}$$

Let $P_1, P_2, \dots, P_n$ be the $(d-2)$-connected components of P. Note that $[P_i] \cap [P_j]$ is thin, for $i \neq j$. We show next that there exists i such that $x \in \text{ins}([P_i])$. Suppose for the sake of contradiction that this is false, which is to say that $x \notin \text{ins}([P_i])$ for all i. Then $x \notin \overline{P}_i = [P_i] \cup \text{ins}([P_i])$ for $i = 1, 2, \dots, n$. Note that each $\overline{P}_i$ is a closed set which does not separate $\mathbb{R}^d$.

Let $i \neq j$. We claim that:

(7.7) either $\overline{P}_i \cap \overline{P}_j$ is thin, or one of the sets $\overline{P}_i, \overline{P}_j$ is a subset of the other.

To see this, suppose that $\overline{P}_i \cap \overline{P}_j$ is fat; we shall deduce as required that either $\overline{P}_i \subseteq \overline{P}_j$ or $\overline{P}_i \supseteq \overline{P}_j$.

Suppose further that $\overline{P}_i \cap [P_j]$ is fat. Since $[P_j]$ is a union of plaquettes and $\overline{P}_i$ is a union of plaquettes and cubes, all with corners in $\mathbb{Z}^d + \frac{1}{2}$, there exists a pair h_1, h_2 of plaquettes of $\mathbb{L}^d_{\mathrm{d}}$ such that $h_1 \stackrel{d-2}{\sim} h_2$, and $g = h_1 \cap h_2$ satisfies $g \subseteq \overline{P}_i \cap [P_j]$. We cannot have $g \subseteq [P_i]$ since $[P_i] \cap [P_j]$ is thin, whence $\mathrm{int}(g) \subseteq \mathrm{ins}([P_i])$, where $\mathrm{int}(g)$ denotes the interior of g viewed as a subset of $\mathbb{R}^{d-2}$. Now, $[P_j]$ is $(d-2)$-connected and $[P_i] \cap [P_j]$ is thin, so that $[P_j]$ is contained in the closure of $\mathrm{ins}([P_i])$, implying that $[P_j] \subseteq \overline{P}_i$ and therefore $\overline{P}_j \subseteq \overline{P}_i$.

Suppose next that $\overline{P}_i \cap [P_j]$ is thin but $\overline{P}_i \cap \mathrm{ins}([P_j])$ is fat. Sincc $[P_i]$ is $(d-2)$-connected, it has by definition no thin cutset. Since $[P_i] \cap [P_j]$ is thin, either $[P_i] \subseteq \overline{P}_j$ or $[P_i]$ is contained in the closure of the unbounded component of $\mathbb{R}^d \setminus [P_j]$. The latter cannot hold since $\overline{P}_i \cap \mathrm{ins}([P_j])$ is fat, whence $[P_i] \subseteq \overline{P}_j$ and therefore $\overline{P}_i \subseteq \overline{P}_j$. Statement (7.7) has been proved.

By (7.7), we may write $R = \bigcup_{i=1}^n \overline{P}_i$ as the union of distinct closed bounded sets $\widetilde{P}_i$, $i = 1, 2, \dots, k$, where $k \le n$, that do not separate $\mathbb{R}^d$ and such that $\widetilde{P}_i \cap \widetilde{P}_j$ is thin for $i \neq j$. By Theorem 11 of [223, §59, Section II][2], R does not separate $\mathbb{R}^d$. By assumption, $x \notin R$, whence x lies in the unique component of the complement $\mathbb{R}^d \setminus R$, in contradiction of the assumption that $x \in \mathrm{ins}([P])$. We deduce that there exists k such that $x \in \mathrm{ins}([P_k])$, and we define $Q = P_k$.

Consider now a vertex $y \in V$. Since $G = (V, E)$ is connected, there exists a path in $\mathbb{L}^d$ that connects x with y using only edges in E. Whenever u and v are two consecutive vertices on this path, $h(\langle u, v\rangle)$ does not belong to P. Therefore, y lies in the inside of $[Q]$. Claims (i) and (ii) are now proved with Q as given, and it remains to prove (iii).

Let W be as in (iii), and let $w \in W$ be such that: there exists an infinite path on $\mathbb{L}^d$ with endvertex w and using no vertex of V. Whenever u and v are two consecutive vertices on such a path, the plaquette $h(\langle u, v\rangle)$ does not lie in P. It follows that $w \in \mathrm{out}([P])$, and therefore $w \in \mathrm{out}([Q])$. □

Proof of Theorem 7.5. Let $H = (\Delta_{\mathrm{v},\delta}C, \Delta_{\mathrm{e},\delta}C)$. Let $x \in \Delta_{\mathrm{v},\delta}C$, and write H_x for the connected component of H containing x. We claim that there exists a plaquette $h_x = h(\langle y, z\rangle) \in \delta$ such that $y \in H_x$.

The claim holds with $y = x$ and $h_x = h(\langle x, z\rangle)$ if x has a neighbour z with $h(\langle x, z\rangle) \in \delta$. Assume therefore that x has no such neighbour z. Since $x \in \Delta_{\mathrm{v},\delta}C$, x has some neighbour u in $\mathbb{L}^d$ with $h(\langle x, u\rangle) \in \overline{\delta} \setminus \delta$. Following a consideration of the various possibilities, there exists $\widetilde{h} \in \delta$ such that $\widetilde{h} \stackrel{d-2}{\sim} h(\langle x, u\rangle)$, and

either (a) $\widetilde{h} = h(\langle u, z\rangle)$ for some z,

or (b) $\widetilde{h} = h(\langle v, z\rangle)$ for some v, z satisfying $v \sim x,\ z \sim u$.

---

[2]This theorem states, subject to a mild change of notation, that: "If none of the closed sets F_0 and F_1 cuts $\mathcal{S}_d$ between the points p and q and if $\dim(F_0 \cap F_1) \le d-3$, their union $F_0 \cup F_1$ does it neither". Here, $\mathcal{S}_d$ denotes the d-sphere.

If (a) holds, we take $y = u$ ($\in H_x$) and $h_x = \widetilde{h}$. If (a) does not hold but (b) holds for some v, z, we take $y = v$ ($\in H_x$) and $h_x = \widetilde{h}$.

We apply Theorem 7.3 with $G = H_x$ to obtain a splitting set Q_x, and we claim that

$$Q_x \cap \delta \neq \varnothing. \tag{7.8}$$

This we prove as follows. If $h_x \in Q_x$, the claim is immediate. Suppose that $h_x \notin Q_x$, so that $[h_x] \cap \mathrm{ins}([Q_x]) \neq \varnothing$, implying that δ intersects both $\mathrm{ins}([Q_x])$ and $\mathrm{out}([Q_x])$. Since both δ and Q_x are $(d-2)$-connected sets of plaquettes, it follows that $\delta \cup Q_x$ is $(d-2)$-connected. Therefore, there exist $h' \in \delta$, $h'' \in Q_x$ such that $h' \stackrel{d-2}{\sim} h''$. If $h'' \in \delta$, then (7.8) holds, so we may assume that $h'' \notin \delta$, and hence $h'' \in \overline{\delta} \setminus \delta$. Then $h'' = h(\langle v, w \rangle)$ for some $v \in H_x$, and therefore $w \in H_x$, a contradiction. We conclude that (7.8) holds.

Now, (7.8) implies that $Q_x \subseteq \delta$. Suppose on the contrary that $Q_x \not\subseteq \delta$, so that there exist $h' \in \delta$, $h'' \in Q_x \setminus \delta$ such that $h' \stackrel{d-2}{\sim} h''$. This leads to a contradiction by the argument just given, whence $Q_x \subseteq \delta$.

Suppose now that x and y are vertices of H such that H_x and H_y are distinct connected components. Either H_x lies in $\mathrm{out}([Q_y])$, or H_y lies in $\mathrm{out}([Q_x])$. Since $Q_x, Q_y \subseteq \delta$, either possibility contradicts the assumption that x and y are connected in C. Therefore, $H_x = H_y$ as claimed. Part (i) of the theorem holds with $Q = Q_x$. □

Proof of Theorem 7.6. This makes use the methods of the last two proofs, and is only sketched. Let $Q \subseteq \mathbb{H}$ be the set of plaquettes that are dual to edges of $\mathbb{E}^d \setminus E$ with at least one endvertex in V. By the definition of the graph $C = (V, E)$, $Q \subseteq \delta$. Let $Q_1, Q_2, \dots, Q_m$ be the $(d-2)$-connected components of Q. If $m \geq 2$, there exists a non-empty subset $H \subseteq \delta \setminus Q$ such that $Q \cup H$ is $(d-2)$-connected but no strict subset of $Q \cup H$ is $(d-2)$-connected. Each $h = h(e) \in H$ must be such that at least one vertex of e lies in $\mathrm{out}(Q)$, in contradiction of the definition of Q. It follows that Q is $(d-2)$-connected.

The connectivity of $(\Delta_{\mathrm{v},\delta} C, \Delta_{\mathrm{e},\delta} C)$ may be proved in very much the same way as in the proof of Theorem 7.5. □

7.3 A contour representation

The dual of a two-dimensional random-cluster model is itself a random-cluster model, as explained in Chapter 6. The corresponding statement is plainly false in three or more dimensions, since the geometry of plaquettes differs from that of edges. Consider an edge-configuration $\omega \in \Omega = \{0, 1\}^{\mathbb{E}^d}$, and the corresponding plaquette-configuration $\pi = (\pi(h) : h \in \mathbb{H})$ given by

$$\pi(h(e)) = 1 - \omega(e), \qquad e \in \mathbb{E}^d. \tag{7.9}$$

Thus, $h(e)$ is open if and only if e is closed. The open plaquettes form surfaces, or 'contours', and one seeks to understand the geometry of the original process through a study of the probable structure of such contours. Contours are objects of some geometrical complexity, and they demand a proper study in their own right, of which the results of Section 7.2 form part.

The study of contours for the random-cluster model has as principal triumph a fairly complete analysis of the model for large q. The central feature of this analysis is the proof that, at the critical point $p = p_c(q)$ for sufficiently large q, the contour measures of both free and wired models have convergent cluster expansions. This implies a discontinuous phase transition, the existence of a mass gap, and a number of other facts presented in Section 7.5.

Cluster (or 'polymer') expansions form a classical topic of statistical mechanics, and their theory is extensive and well understood by experts. Rather than developing the theory from scratch here, we shall in the next section abstract those ingredients that are relevant for the current application. Meanwhile, we concentrate on formulating the random-cluster model in a manner resonant with polymer expansions. The account given here is an expansion and elaboration of that found in [224]. A further treatment may be found in [65, 66].

Henceforth in this chapter we shall assume, unless otherwise stated, that $d = 3$. Similar results are valid whenever $d \geq 3$, and stronger results hold when $d = 2$. A *plaquette* is taken to be a closed unit square of the dual lattice $\mathbb{L}^3_{\mathrm{d}}$, and each plaquette $h = h(e)$ is pierced by a unique edge e of $\mathbb{L}^3$.

Since the random-cluster model involves probability measures on the set of *edge*-configurations, we shall consider functions on the power set of the edge-set $\mathbb{E}^3$ rather than of the vertex-set $\mathbb{Z}^3$. Let E be a finite subset of $\mathbb{E}^3$, and let $\mathbb{L}_E = (V_E, E)$ denote the induced subgraph of $\mathbb{L}^3$. We shall consider the partition functions of the wired and free random-cluster measures on this graph, and to this end we introduce various notions of 'boundary'. Let D be a (finite or infinite) subset of $\mathbb{E}^3$, and write $\overline{D} = \mathbb{E}^3 \setminus D$ for its complement.

(i) The *vertex-boundary* ∂D is the set of all $x \in V_D$ such that there exists an edge $e = \langle x, z \rangle$ with $e \notin D$. Note that $\partial D = \partial \overline{D}$.

We shall require three (related) types of 'edge-boundaries' of D.

(ii) The 1-*edge-boundary* $\partial_e D$ is defined[3] to be the set of all edges $e \in D$ such that there exists $f \notin D$ with the property that $h(e) \overset{1}{\sim} h(f)$.

(iii) The *external edge-boundary* $\Delta_{\mathrm{ext}} D$ is the set of all edges $e \notin D$ that are incident to some vertex in ∂D.

(iv) The *internal edge-boundary* $\Delta_{\mathrm{int}} D$ is the external edge-boundary of the complement $\overline{D}$, that is, $\Delta_{\mathrm{int}} D = \Delta_{\mathrm{ext}} \overline{D}$. In other words, $\Delta_{\mathrm{int}} D$ includes every edge $e \in D$ that is incident to some $x \in \partial D$.

[3]When working with $\mathbb{L}^d$ for general d, $\partial_e D$ would be taken to be the $(d-2)$-edge-boundary, given similarly but with 1 replaced by $d - 2$.

Let $p \in (0,1)$, $q \in (0,\infty)$, and $r = p/(1-p)$. As is usual in classical statistical mechanics, it is the partition functions which play leading roles. Henceforth, we take E to be a finite subset of $\mathbb{E}^3$. We consider first the wired measure on $\mathbb{L}_E$, which we define via its partition function[4]

$$Z^1(E) = \sum_{\substack{D:\, D\subseteq E \\ D\supseteq \partial_e E}} r^{|D|} q^{k^1(D,E)}, \tag{7.10}$$

where $k^1(D,E)$ denotes the number of connected components (including the infinite cluster and any isolated vertices) of $\mathbb{L}^3$ after the removal of edges in $E\setminus D$. This definition (7.10) differs slightly from that of (4.12) with $\xi = 1$, but it may be seen via Theorem 7.5 that the corresponding probability measure amounts to the wired measure on the edge-set $E \setminus \partial_e E$. It is presented in the above manner in order to facilitate certain relations to be derived soon.

We define similarly the free partition function on $\mathbb{L}_E$ by

$$Z^0(E) = \sum_{\substack{D:\, D\subseteq E \\ D\cap\Delta_{\mathrm{int}} E =}} r^{|D|} q^{k^0(V_E\setminus\partial E, D)}, \tag{7.11}$$

where $k^0(G)$ denotes the number of connected components of a graph G including isolated vertices[5]. Since $\Delta_{\mathrm{int}} E$ includes every edge $e \in E$ that is incident to some vertex $x \in \partial E$, every $x \in \partial E$ is isolated for all sets D contributing to the summation in (7.11), and these vertices are not included in the cluster-count $k(V_E \setminus \partial E, D)$. The measure defined by (7.11) differs slightly from that given at (4.11)–(4.12) with $\xi = 0$, but it may be seen that the corresponding probability measure amounts to the free measure on the graph $(V_E, E \setminus \Delta_{\mathrm{int}} E)$.

By an argument similar to that of Theorem 4.58, there exists a function F, termed the *pressure*, such that

$$F(p,q) = \lim_{E\uparrow\mathbb{E}^3}\left\{\frac{1}{|E|}\log Z^1(E)\right\} = \lim_{E\uparrow\mathbb{E}^3}\left\{\frac{1}{|E|}\log Z^0(E)\right\}, \tag{7.12}$$

where the limit is taken in a suitable 'van Hove' sense.

We introduce next the classes of 'wired' and 'free' contours of the lattice $\mathbb{L}^3$. For $s \in \{0,1\}$ and $e, f \in \mathbb{E}^3$, we write $e \stackrel{s}{\sim} f$ if $h(e) \stackrel{s}{\sim} h(f)$. A subset D of $\mathbb{E}^3$ is said to be *s-connected* if it is connected when viewed as a graph with adjacency relation $\stackrel{s}{\sim}$. Thus, D is s-connected if and only if the set $\{h(f) : f \in D\}$ of plaquettes is s-connected. Let $D \subseteq \mathbb{E}^3$, and consider its external edge-boundary $\gamma = \Delta_{\mathrm{ext}} D$. We call the set γ a *wired contour* (respectively, *free contour*) if it is

[4]It is convenient in the present setting to think of a configuration as a subset of edges rather than as a 0/1-vector. We adopt the convention that $Z^1(\varnothing) = 1$.

[5]We set $Z^0(E) = 1$ if $E \setminus \Delta_{\mathrm{int}} E = \varnothing$. In particular, $Z^0(\varnothing) = 1$.

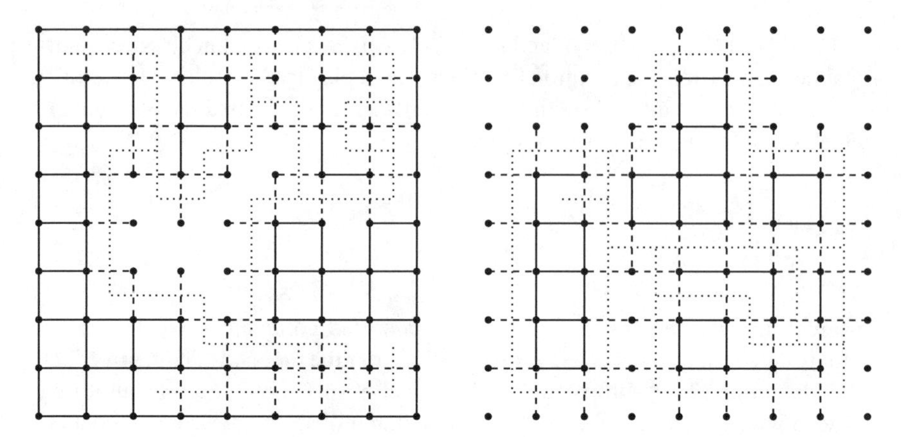

Figure 7.2. Examples of wired and free contours in two dimensions. The solid lines comprise D, the dashed lines are the contours, and the dotted lines the dual plaquettes. A wired contour resembles an archipelago joined by causeways, a free contour resembles a single island traversed by canals.

1-connected and $\mathbb{E}^3 \setminus D$ is finite (respectively, it is 1-connected and D is finite). Illustrations of wired and free contours are presented in Figure 7.2. For any (wired or free) contour γ, the unique infinite connected component of $\mathbb{E}^3 \setminus \gamma$ is denoted by $\text{ext}(\gamma)$ (or $\text{ext}\,\gamma$), and we define also $\overline{\gamma} = \mathbb{E}^3 \setminus \text{ext}(\gamma)$ and $\text{int}(\gamma) = \overline{\gamma} \setminus \gamma$. Note by Theorem 7.5 that every finite connected cluster C of $\mathbb{E}^3 \setminus \gamma$ lies in the inside of some splitting set $Q = Q_C$ of plaquettes drawn from $\{h(e) : e \in \gamma\}$.

The set of all wired (respectively, free) contours of $\mathbb{L}^3$ is denoted by $\mathfrak{C}_{\text{w}}$ (respectively, $\mathfrak{C}_{\text{f}}$), and we write γ_{w} (respectively, γ_{f}) for a typical wired (respectively, free) contour. The length $\|\gamma\|$ of a contour is defined as

$$\|\gamma\| = \begin{cases} \left|\left\{(x, y) : x \in \partial\overline{\gamma},\ \langle x, y\rangle \in \overline{\gamma}\right\}\right|, & \gamma \in \mathfrak{C}_{\text{w}}, \\ \left|\left\{(x, y) : x \in V_{\text{int}\,\gamma},\ \langle x, y\rangle \notin \text{int}\,\gamma\right\}\right|, & \gamma \in \mathfrak{C}_{\text{f}}. \end{cases} \tag{7.13}$$

For $\gamma \in \mathfrak{C}_{\text{w}} \cap \mathfrak{C}_{\text{f}}$, the appropriate choice of $\|\gamma\|$ will be clear from the context. In each case, we count the number of *ordered* pairs (x, y); for example, for $\gamma \in \mathfrak{C}_{\text{w}}$, if $x, y \in \partial\overline{\gamma}$ and $e = \langle x, y\rangle \in \overline{\gamma}$, then e contributes a total of 2 to $\|\gamma\|$. We note for later use that, by elementary counting arguments,

$$2d|V_{\overline{\gamma}} \setminus \partial\overline{\gamma}| = 2|\overline{\gamma}| - \|\gamma\|, \qquad \gamma \in \mathfrak{C}_{\text{w}}, \tag{7.14}$$

$$2d|V_{\text{int}\,\gamma}| = 2|\text{int}\,\gamma| + \|\gamma\|, \qquad \gamma \in \mathfrak{C}_{\text{f}}, \tag{7.15}$$

and furthermore[6],

$$|\partial(\text{int}\,\gamma)| \le |\gamma| \le \|\gamma\|, \qquad \gamma \in \mathfrak{C}_{\text{w}}, \tag{7.16}$$

$$|\partial(\text{int}\,\gamma)| \le \|\gamma\|, \qquad \gamma \in \mathfrak{C}_{\text{f}}. \tag{7.17}$$

---

[6]The second inequality of (7.16) follows from the fact that every edge in a wired contour γ is incident to some vertex in $\partial\overline{\gamma}$. See (7.18).

It may be seen by Theorem 7.5 that

$$\Delta_{\text{int}}\overline{\gamma} = \gamma, \qquad \gamma \in \mathfrak{C}_{\text{w}}. \tag{7.18}$$

Two contours γ_1, γ_2 of the same class are said to be *compatible* if $\gamma_1 \cup \gamma_2$ is not 1-connected. We call the pair γ_1, γ_2 *externally compatible* if they are compatible and in addition $\overline{\gamma_1} \subseteq \text{ext}(\gamma_2)$ and $\overline{\gamma_2} \subseteq \text{ext}(\gamma_1)$. If $\Gamma = \{\gamma_1, \gamma_2, \dots, \gamma_n\}$ is a family of pairwise externally-compatible contours of the same class, we write $\overline{\Gamma} = \bigcup_i \overline{\gamma_i}$, $\text{ext}(\Gamma) = \mathbb{E}^3 \setminus \overline{\Gamma}$, and $\text{int}(\Gamma) = \overline{\Gamma} \setminus \Gamma$. Here, we have used Γ to denote the set of edges in the union of the γ_i.

Let $\Gamma_{\text{w}} = \{\gamma_1, \gamma_2, \dots, \gamma_m\}$ be a family of pairwise externally-compatible wired contours. It may be seen that

$$\Delta_{\text{int}}\overline{\Gamma} = \bigcup_{i=1}^{m} \Delta_{\text{int}}\overline{\gamma_i},$$

and, by (7.11),

$$Z^0(\overline{\Gamma_{\text{w}}}) = \prod_{i=1}^{m} Z^0(\overline{\gamma_i}). \tag{7.19}$$

Similarly, if $\Gamma_{\text{f}} = \{\gamma_1, \gamma_2, \dots, \gamma_n\}$ is a family of pairwise externally-compatible free contours, then

$$\partial_{\text{e}}(\text{int}\,\Gamma_{\text{f}}) = \bigcup_{i=1}^{n} \partial_{\text{e}}(\text{int}\,\gamma_i),$$

and, by (7.10),

$$q^{n-1} Z^1(\text{int}\,\Gamma_{\text{f}}) = \prod_{i=1}^{n} Z^1(\text{int}\,\gamma_i). \tag{7.20}$$

A key step in the transformation of the random-cluster model to a polymer model is the derivation of recursive expressions for $Z^1(E)$ and $Z^0(E)$ in terms of partition functions of subsets of E. We describe this first for the wired partition function $Z^1(E)$. The subset $E \subseteq \mathbb{E}^3$ is called *co-connected* if $|E| < \infty$ and $\mathbb{E}^3 \setminus E$ is connected. Let E be co-connected. Let $D \subseteq E$ be such that $\partial_{\text{e}} E \subseteq D$. Let D_∞ be the set of edges in the unique infinite connected component of $D \cup E^{\text{c}}$, and let $\Gamma(D) = \Delta_{\text{ext}} D_\infty$. The set $\Gamma(D)$ may be expressed as a union of maximal 1-connected sets γ_i, $i = 1, 2, \dots, m$, which are pairwise externally-compatible wired contours, and we write $\Gamma_{\text{w}}(D) = \{\gamma_1, \gamma_2, \dots, \gamma_m\}$. Note that every edge in $\Gamma(D)$ belongs to $E \setminus \partial_{\text{e}} E$. Thus, to each set D there corresponds a collection $\Gamma_{\text{w}}(D)$, and the summation in (7.10) may be partitioned according to the value of $\Gamma_{\text{w}}(D)$. For a given family $\Gamma_{\text{w}} = \{\gamma_1, \gamma_2, \dots, \gamma_m\}$ of pairwise externally-compatible wired contours in $E \setminus \partial_{\text{e}} E$, the corresponding part of the summation in (7.10) is over sets D with $\Gamma_w(D) = \Gamma_w$, and the constraints on such D are as follows:

1. D contains no edge in any γ_i,
2. D contains every edge of E not belonging to $\overline{\Gamma_w}$.

This leads via (7.18) and (7.19) to the formula

$$Z^1(E) = \sum_{\Gamma_w \subset E \setminus \partial_e E} r^{|E \setminus \overline{\Gamma_w}|} q Z^0(\overline{\Gamma_w}) \tag{7.21}$$

where the summation is over all families Γ_w of pairwise externally-compatible wired contours contained in $E \setminus \partial_e E$. By Theorems 7.3 and 7.5, each such $\overline{\Gamma_w}$ is co-connected.

We turn now to the free partition function $Z^0(E)$. Let $D \subseteq E \setminus \Delta_{\mathrm{int}} E$. Let D^c_∞ be the set of edges in the unique infinite 1-connected component of $D^c = \mathbb{E}^3 \setminus D$, and let $\Gamma(D) = \Delta_{\mathrm{int}} D^c_\infty$. The set $\Gamma(D)$ may be expressed as a union of maximal 1-connected sets γ_i, $i = 1, 2, \dots, n$, which are pairwise externally-compatible free contours, and we write $\Gamma_f(D) = \{\gamma_1, \gamma_2, \dots, \gamma_n\}$. We note that every edge in $\Gamma(D)$ belongs to E. Thus, to each set D there corresponds a collection $\Gamma_f(D)$, and the summation in (7.11) may be partitioned according to the value of $\Gamma_f(D)$. For a given family $\Gamma_f = \{\gamma_1, \gamma_2, \dots, \gamma_n\}$ of pairwise externally-compatible free contours in E, one sums over sets D with $\Gamma_f(D) = \Gamma_f$, and the constraints on such D are as follows:

1. $D \subseteq \mathrm{int}\, \Gamma_f$,
2. for $i = 1, 2, \dots, n$, D contains every edge in $\mathrm{int}\, \gamma_i$ that is 1-connected to some edge in γ_i.

This leads by (7.11), (7.20), and Theorem 7.5 to the formula

$$Z^0(E) = \sum_{\Gamma_f \subset E} q^{|V_E \setminus \partial E| - |V_{\mathrm{int}\, \Gamma_f}|} q^{n-1} Z^1(\mathrm{int}\, \Gamma_f), \tag{7.22}$$

where the summation is over all families Γ_f of pairwise externally-compatible free contours γ contained in E. By Theorems 7.3 and 7.5, each such $\mathrm{int}\, \Gamma_f$ is co-connected.

The next step is to transform the random-cluster model into a so-called polymer model of statistical mechanics. To the latter model we shall apply certain standard results summarized in the next section, and we shall return to the random-cluster application in Section 7.5.

7.4 Polymer models

The partition function of a lattice model in a finite volume Λ of $\mathbb{R}^d$ may generally be written in the form

$$Z(\Lambda) = \sum_{\Sigma \subset \Lambda} \prod_{\gamma \in \Sigma} \Phi(\gamma), \tag{7.23}$$

where the summation[7] is over all compatible families Σ in Λ (including the empty family, which contributes 1) comprising certain types of geometrical objects γ called 'polymers'. The nature of these polymers, of the weight function Φ (which we shall assume to be non-negative), and of the meaning of 'compatibility', depend on the particular model in question. We summarize some basic properties of such polymer models in this section, and shall apply these results to random-cluster models in the next section. The current target is to communicate the theory in the broad. The details of this theory have the potential to complicate the message, and they will therefore be omitted in almost their entirety. In the interests of brevity, certain liberties will be taken with the level of rigour. The theory of polymer models is well developed in the literature of statistical mechanics, and the reader may consult the papers [85, 216, 219, 274, 275, 326], the book [301], and the references therein.

The discontinuity of the Potts phase transition was proved first in [220] via a so-called chessboard estimate. This striking result, combined with the work of [218], inspired the proof via polymer models of the discontinuity of the random-cluster phase transition, [224]. The last paper is the basis for the present account.

The study of polymer models is wider than is required for our specific applications, and a general approach may be found in [219]. For the sake of concreteness, we note the following. Our applications will involve co-connected subsets Λ of $\mathbb{E}^3$. Our polymers will be either wired or free contours in the sense of the last section, and 'compatible' shall be interpreted in the sense of that section. Our weight functions Φ will be assumed henceforth to be strictly positive and automorphism-invariant, in that $\Phi(\gamma) = \Phi(\tau\gamma)$ for any automorphism τ of $\mathbb{L}^3$.

One seeks conditions under which the limit

$$f(\Phi) = \lim_{\Lambda \uparrow \mathbb{E}^3} \left\{ \frac{1}{|\Lambda|} \log Z(\Lambda) \right\} \tag{7.24}$$

exists, together with bounds on the deviation

$$\sigma(\Lambda, \Phi) = |\Lambda| f(\Phi) - \log Z(\Lambda). \tag{7.25}$$

These are obtained by elementary arguments under the assumption that the $\Phi(\gamma)$ decay exponentially in the size of γ, with a sufficiently negative exponent. With

[7] We adopt the convention that $Z(\varnothing) = 1$.

each polymer γ we associate a natural measure of 'size' denoted by $\|\gamma\|$ and, for $\tau \in (0, \infty)$, we call Φ a τ-*functional* if

$$\Phi(\gamma) \le e^{-\tau\|\gamma\|} \qquad \text{for all } \gamma. \tag{7.26}$$

The principal conclusions that follow are not stated unambiguously as a theorem since their exact hypotheses will not be specified. Throughout this and the next section, the terms c and c_i are positive finite constants which depend only on the particular type of model and not on the function Φ. These constants may depend on the underlying lattice (which we shall take to be $\mathbb{L}^3$), and may therefore vary with the number d of dimensions.

(7.27) 'Theorem'. *There exist $c, c_1, c_2 \in (0, \infty)$ such that the following holds. Let Φ be a τ-functional with $\tau > c$.*

(a) *The limit $f(\Phi)$ exists in* (7.24)*, and satisfies $0 \le f(\Phi) \le e^{-c_1\tau}$.*

(b) *The deviation in* (7.25) *satisfies $|\sigma(\Lambda, \Phi)| \le |\partial\Lambda| e^{-c_2\tau}$ for all finite Λ.*

The polymer model is said to be *convergent* when the condition of the above 'Theorem' is satisfied.

Sketch proof. Here are some comments on the proof. The existence of the pressure $f(\Phi)$ in part (a) may be shown using subadditivity in a manner similar to the proof of Theorem 4.58. This part of the conclusion is valid irrespective of the assumption that Φ be a τ-functional, although it may in general be the case that $f(\Phi) = \infty$. One obtains a formula for the limit function $f(\Phi)$ in the following manner. Let

$$\psi(E) = \sum_{\Lambda \subseteq E} (-1)^{|E\setminus\Lambda|} \log Z(\Lambda), \qquad E \subseteq \mathbb{E}^3,\ |E| < \infty. \tag{7.28}$$

By the inclusion–exclusion principle[8],

$$\log Z(\Lambda) = \sum_{E \subseteq \Lambda} \psi(E). \tag{7.29}$$

By (7.23), $Z(\Lambda_1 \cup \Lambda_2) = Z(\Lambda_1)Z(\Lambda_2)$ if Λ_1 and Λ_2 have no common vertex. By (7.28), ψ is automorphism-invariant and satisfies

$$\psi(E) = 0 \qquad \text{if } E \text{ is not connected.} \tag{7.30}$$

Under the assumption of 'Theorem' 7.27, one may obtain after a calculation that

$$|\psi(E)| \le e^{-c_3\tau\|E\|} \tag{7.31}$$

for a suitable definition of the size $\|E\|$ and for some $c_3 \in (0, \infty)$.

[8] As in [144].

Formula (7.29) motivates the proposal that, for any given $e \in \mathbb{E}^3$,

$$f(\Phi) = \lim_{\Lambda \uparrow \mathbb{E}^3} \left\{ \frac{1}{|\Lambda|} \log \mathcal{Z}(\Lambda) \right\} = \sum_{E:\, e \in E} \frac{\psi(E)}{|E|},$$

and this may be proved rigorously by use of (7.31) with sufficiently large τ. The inequality of part (a) follows. By (7.29) again,

$$\sigma(\Lambda, \Phi) = \sum_{e \in \Lambda} \sum_{E:\, e \in E} 1_{\{E \cap \Lambda^c \neq \varnothing\}} \frac{\psi(E)}{|E|},$$

and, by (7.30),

$$|\sigma(\Lambda, \Phi)| \leq \sum_{x \in \partial\Lambda} \sum_{E:\, x \in V_E} |\psi(E)|.$$

Part (b) follows by (7.31) and a combinatorial estimate. □

Turning to probabilities, the partition function $\mathcal{Z}(\Lambda)$ gives rise to a probability measure κ on the set of compatible families in Λ, namely

$$\kappa(\Sigma) = \frac{1}{\mathcal{Z}(\Lambda)} \Phi(\Sigma), \qquad \Sigma \subset \Lambda,$$

where $\Phi(\Sigma) = \prod_{\gamma \in \Sigma} \Phi(\gamma)$. The following elementary result will be useful later.

(7.32) Theorem (Peierls estimate). *Let γ be a polymer of Λ. The κ-probability that γ belongs to a randomly chosen compatible family satisfies*

$$\kappa\big(\{\Sigma : \gamma \in \Sigma\}\big) \leq \Phi(\gamma).$$

Proof. We write $\Sigma \perp \gamma$ to mean that Σ is a compatible family satisfying: $\gamma \notin \Sigma$, and $\Sigma \cup \{\gamma\}$ is a compatible family. Then,

$$\begin{aligned} \kappa\big(\{\Sigma : \gamma \in \Sigma\}\big) &= \frac{1}{\mathcal{Z}(\Lambda)} \sum_{\Sigma:\, \Sigma \perp \gamma} \Phi(\Sigma)\Phi(\gamma) \\ &\leq \frac{\displaystyle\sum_{\Sigma:\, \Sigma \perp \gamma} \Phi(\Sigma)\Phi(\gamma)}{\displaystyle\sum_{\Sigma:\, \Sigma \perp \gamma} \Phi(\Sigma)[1 + \Phi(\gamma)]} \leq \Phi(\gamma). \end{aligned}$$

□

7.5 Discontinuous phase transition for large q

It is a principal theorem for Potts and random-cluster models that the phase transition is discontinuous when q is sufficiently large, see [68, 220, 251] for Potts models and [224] for random-cluster models. This is proved for random-cluster models by showing that the maximal contours of both wired and free models at $p = p_c(q)$ have the same laws as those of certain convergent polymer models. Such use of contour expansions is normally termed a 'Pirogov–Sinai' approach[9], after the authors of [274, 275].

Here are the main results, expressed for a general number d of dimensions.

(7.33) Theorem (Discontinuous phase transition) [224]. *Let $d \geq 2$. There exists $Q = Q(d)$ such that following hold when $q > Q$.*

(a) *The edge-densities*

$$h^b(p,q) = \phi^b_{p,q}(e \text{ is open}), \qquad b = 0, 1,$$

are discontinuous functions of p at the critical point $p_c(q)$.

(b) *The percolation probabilities satisfy*

$$\theta^0(p_c(q), q) = 0, \quad \theta^1(p_c(q), q) > 0.$$

(c) *There is a unique random-cluster measure when $p \neq p_c(q)$, and at least two random-cluster measures when $p = p_c(q)$, in that*

$$\phi^0_{p_c(q),q} \neq \phi^1_{p_c(q),q}.$$

(d) *If $p < p_c(q)$, there is exponential decay and a non-vanishing mass gap, in that the unique random-cluster measure satisfies*

$$\phi_{p,q}(0 \leftrightarrow x) \leq e^{-\alpha|x|}, \qquad x \in \mathbb{Z}^d,$$

for some $\alpha = \alpha(p,q)$ satisfying $\alpha \in (0, \infty)$ and

$$\lim_{p \uparrow p_c(q)} \alpha(p,q) > 0.$$

The large-q behaviour of $p_c(q)$ is given as follows. One may obtain an expansion of $p_c(q)$ in powers of $q^{-1/d}$ by pursuing the proof further.

(7.34) Theorem [224]. *For $d \geq 3$,*

$$p_c(q) = 1 - q^{-1/d} + O(q^{-2/d}) \qquad \text{as } q \to \infty.$$

This may be compared to the exact value $p_c(q) = \sqrt{q}/(1+\sqrt{q})$ when $d = 2$ and q is large, see Theorem 6.35. For $d \geq 3$ and large q, there exist non-translation-invariant random-cluster measures at the critical point $p_c(q)$.

[9] An overview of contour methods may be found in [217].

(7.35) Theorem (Non-translation-invariant measure at $p_c(q)$) [85, 254]. *Let $d \geq 3$. There exists $Q = Q(d)$ such that there exists a non-translation-invariant DLR-random-cluster measure when $p = p_c(q)$ and $q > Q$.*

It is not especially fruitful to seek numerical estimates on the $Q(d)$ above. Such estimates may be computed, but they turn out to be fairly distant from those anticipated, namely[10]

$$Q(2) = 4, \quad Q(d) = 2 \quad \text{for } d \geq 6. \tag{7.36}$$

No proof of Theorem 7.35 is included here, and the reader is referred for more details to the given references.

Numerous facts for Potts models with large q follow from the above. Let $d \geq 2$ and $p = 1 - e^{-\beta}$, and consider the q-state Potts model on $\mathbb{L}^d$ with inverse-temperature β. Let q be large. When $\beta < \beta_c(q)$ (respectively, $\beta > \beta_c(q)$), the number of distinct translation-invariant Gibbs states is 1 (respectively, q). When $\beta = \beta_c(q)$, there are $q + 1$ distinct extremal translation-invariant Gibbs states, corresponding to the free measure and the 'b-boundary-condition' measure for $b \in \{1, 2, \dots, q\}$, and every translation-invariant Gibbs state is a convex combination of these $q + 1$ states. When $d \geq 3$, there exist in addition an infinity of non-translation-invariant Gibbs states at the critical point $\beta_c(q)$. Further discussion may be found in [65, 66, 68, 136, 224, 251, 254].

In preparation for the proofs of Theorems 7.33 and 7.34, we introduce an extension of the polymer model of the last section, in the context of the wired and free contours of Section 7.3. For a finite subset E of $\mathbb{E}^3$, let

$$\mathcal{Z}(E; \Phi) = \sum_{\Sigma \subset E} \prod_{\gamma \in \Sigma} \Phi(\gamma) \tag{7.37}$$

be the partition function of a polymer model on E. The admissible families Σ of polymers will be either families of wired contours (lying in $E \setminus \partial_e E$) or families of free contours (lying in E); in either case they are required to be pairwise compatible. By a standard iterative argument, the sum in (7.37) may be restricted to families Γ of *pairwise externally-compatible* contours, and (7.37) becomes

$$\mathcal{Z}(E; \Phi) = \sum_{\Gamma \subset E} \prod_{\gamma \in \Gamma} \Psi(\gamma) \tag{7.38}$$

where

$$\Psi(\gamma) = \Phi(\gamma)\mathcal{Z}(\text{int}\,\gamma; \Phi). \tag{7.39}$$

[10] Some progress has been made towards bounds on the value of $Q(d)$. It is proved in [45] that the 3-state Potts model has a discontinuous transition for large d, and in [46] that discontinuity occurs when $d = 3$ for a long-range Potts model with exponentially decaying interactions. See [140] for related work when $d = 2$.

The letter Σ (respectively, Γ) will always denote a family of pairwise compatible contours (respectively, pairwise externally-compatible contours).

Let $\beta \in \mathbb{R}$. In either of the cases above, we define

$$(7.40)\qquad \begin{aligned} \mathcal{Z}(E;\Phi,\beta) &= \sum_{\Gamma\subset E}\prod_{\gamma\in\Gamma} e^{\beta|\overline{\gamma}|}\Psi(\gamma)\\ &= \sum_{\Gamma\subset E}\prod_{\gamma\in\Gamma} e^{\beta|\overline{\gamma}|}\Phi(\gamma)\mathcal{Z}(\operatorname{int}\gamma;\Phi), \end{aligned}$$

and we say that this new model has parameters (β, Φ).

We shall consider a pair of such models. The first has parameters $(\beta_{\mathrm{w}}, \Phi_{\mathrm{w}})$, and its polymer families comprise pairwise compatible wired contours; the second has parameters $(\beta_{\mathrm{f}}, \Phi_{\mathrm{f}})$ and it involves free contours. They are defined as follows. Let $p \in (0,1)$, $q \in [1,\infty)$, $r = p/(1-p)$, and $\beta_{\mathrm{w}}, \beta_{\mathrm{f}} \in [0,\infty)$. The weight functions $\Phi_{\mathrm{w}}(\gamma) = \Phi_{\mathrm{w}}^{\beta_{\mathrm{w}}}(\gamma)$, $\Phi_{\mathrm{f}}(\gamma) = \Phi_{\mathrm{f}}^{\beta_{\mathrm{f}}}(\gamma)$ are defined inductively on the size of $\overline{\gamma}$ by:

(7.41)

$$\begin{aligned} \Phi_{\mathrm{w}}^{\beta_{\mathrm{w}}}(\gamma)\mathcal{Z}(\operatorname{int}\gamma;\Phi_{\mathrm{w}}^{\beta_{\mathrm{w}}}) &= \Psi_{\mathrm{w}}^{\beta_{\mathrm{w}}}(\gamma) = (re^{\beta_{\mathrm{w}}})^{-|\overline{\gamma}|}Z^0(\overline{\gamma}), && \gamma\in\mathfrak{C}_{\mathrm{w}},\\ \Phi_{\mathrm{f}}^{\beta_{\mathrm{f}}}(\gamma)\mathcal{Z}(\operatorname{int}\gamma;\Phi_{\mathrm{f}}^{\beta_{\mathrm{f}}}) &= \Psi_{\mathrm{f}}^{\beta_{\mathrm{f}}}(\gamma) = e^{-\beta_{\mathrm{f}}|\overline{\gamma}|}q^{-|V_{\operatorname{int}\gamma}|}Z^1(\operatorname{int}\gamma), && \gamma\in\mathfrak{C}_{\mathrm{f}}. \end{aligned}$$

These functions give rise to polymer models which are related to the free and wired random-cluster models, as described in the first part of the next theorem. They have related pressure functions $f(\Phi_{\mathrm{w}}^{\beta_{\mathrm{w}}})$, $f(\Phi_{\mathrm{f}}^{\beta_{\mathrm{f}}})$ given as in (7.24). The theorem is stated for general $d \ge 2$, but the reader is advised to concentrate on the case $d = 3$.

(7.42) Theorem [224]. *Let $d \ge 2$, $p \in (0,1)$, $q \in [1,\infty)$, and $r = p/(1-p)$. For $\beta_{\mathrm{w}}, \beta_{\mathrm{f}} \in [0,\infty)$ and a co-connected set E,*

$$(7.43)\qquad \begin{aligned} Z^1(E) &= r^{|E|}q\mathcal{Z}(E;\Phi_{\mathrm{w}}^{\beta_{\mathrm{w}}},\beta_{\mathrm{w}}),\\ Z^0(E) &= q^{|V_E\setminus\partial E|}\mathcal{Z}(E;\Phi_{\mathrm{f}}^{\beta_{\mathrm{f}}},\beta_{\mathrm{f}}). \end{aligned}$$

Let

$$(7.44)\qquad \tau = \frac{1}{8d}\log q - 5.$$

There exists $Q = Q(d)$ such that the following hold when $q > Q$.

(a) *There exist reals $b_{\mathrm{w}}, b_{\mathrm{f}} \in [0,\infty)$ such that $\Phi_{\mathrm{w}}^{b_{\mathrm{w}}}$ and $\Phi_{\mathrm{f}}^{b_{\mathrm{f}}}$ are τ-functionals with $\tau > c$, with c as in the hypothesis of 'Theorem' 7.27, and that the pressure $F(p,q)$ of (7.12) satisfies*

$$F(p,q) = f(\Phi_{\mathrm{w}}^{b_{\mathrm{w}}}) + b_{\mathrm{w}} + \log r = f(\Phi_{\mathrm{f}}^{b_{\mathrm{f}}}) + b_{\mathrm{f}} + \frac{1}{d}\log q. \qquad (7.45)$$

(b) *There exists a unique value* $\widetilde{p} = \widetilde{p}(q)$ *such that the values* b_{w}, b_{f} *in part* (a) *satisfy*:

$$\begin{aligned} &\textit{if } p < \widetilde{p}, && \textit{then } b_{\mathrm{w}} > 0,\ b_{\mathrm{f}} = 0, \\ &\textit{if } p = \widetilde{p}, && \textit{then } b_{\mathrm{w}} = 0,\ b_{\mathrm{f}} = 0, \\ &\textit{if } p > \widetilde{p}, && \textit{then } b_{\mathrm{w}} = 0,\ b_{\mathrm{f}} > 0. \end{aligned} \tag{7.46}$$

Proof of Theorem 7.42. We follow the scheme of [224] which in turn makes use of [218, 326]. For any given $\beta_{\mathrm{w}}, \beta_{\mathrm{f}} \in [0, \infty)$, equations (7.41) may be combined with (7.19)–(7.22) to obtain (7.43).

For $\beta_{\mathrm{w}}, \beta_{\mathrm{f}} \in [0, \infty)$, let $\Phi_{\mathrm{w}} = \Phi_{\mathrm{w}}^{\beta_{\mathrm{w}}}$, $\Phi_{\mathrm{f}} = \Phi_{\mathrm{f}}^{\beta_{\mathrm{f}}}$ be given by (7.41). Let $\tau = \tau(q)$ be as in (7.44), and choose Q' such that $\tau(Q') > c$ where c is the constant in the hypothesis of 'Theorem' 7.27. We assume henceforth that

$$q > Q'. \tag{7.47}$$

We define the τ-functionals

$$\overline{\Phi}_{\mathrm{w}}^{\beta_{\mathrm{w}}}(\gamma) = \min\{\Phi_{\mathrm{w}}^{\beta_{\mathrm{w}}}(\gamma), e^{-\tau\|\gamma\|}\}, \qquad \gamma \in \mathfrak{C}_{\mathrm{w}}, \tag{7.48}$$

$$\overline{\Phi}_{\mathrm{f}}^{\beta_{\mathrm{f}}}(\gamma) = \min\{\Phi_{\mathrm{f}}^{\beta_{\mathrm{f}}}(\gamma), e^{-\tau\|\gamma\|}\}, \qquad \gamma \in \mathfrak{C}_{\mathrm{f}}, \tag{7.49}$$

and let
(7.50)

$$\begin{aligned} b_{\mathrm{w}} = \sup B_{\mathrm{w}} \quad &\text{where} \quad B_{\mathrm{w}} = \bigl\{\beta_{\mathrm{w}} \ge 0 : f(\overline{\Phi}_{\mathrm{w}}^{\beta_{\mathrm{w}}}) + \beta_{\mathrm{w}} + \log r \le F(p,q)\bigr\}, \\ b_{\mathrm{f}} = \sup B_{\mathrm{f}} \quad &\text{where} \quad B_{\mathrm{f}} = \bigl\{\beta_{\mathrm{f}} \ge 0 : f(\overline{\Phi}_{\mathrm{f}}^{\beta_{\mathrm{f}}}) + \beta_{\mathrm{f}} + d^{-1}\log q \le F(p,q)\bigr\}. \end{aligned}$$

We make three observations concerning the definition of b_{w}; similar reasoning applies to b_{f}. Firstly, since $\overline{\Phi}_{\mathrm{w}}^{0} \le \Phi_{\mathrm{w}}^{0}$,

$$Z^1(E) \ge r^{|E|} \mathcal{Z}(E; \overline{\Phi}_{\mathrm{w}}^{0}, 0) = r^{|E|}\mathcal{Z}(E; \overline{\Phi}_{\mathrm{w}}^{0}),$$

by (7.43). Applying 'Theorem' 7.27 to the τ-functional $\overline{\Phi}_{\mathrm{w}}^{0}$,

$$F(p,q) \ge \log r + f(\overline{\Phi}_{\mathrm{w}}^{0}),$$

whence $0 \in B_{\mathrm{w}}$. Secondly, by 'Theorem' 7.27 again, $f(\overline{\Phi}_{\mathrm{w}}^{0}) \le e^{-c_1\tau}$, whence $\beta \notin B_{\mathrm{w}}$ for large β. The third observation is contained in the next lemma which is based on the corresponding step of [218]. The lemma will be used later also, and its proof is deferred until that of Theorem 7.42 is otherwise complete.

(7.51) Lemma. *Let* $\alpha \in (0, \infty)$. *There exists* $Q'' = Q''(\alpha) \ge Q'$ *such that the following holds. If* $q > Q''$, *the functions* $h(\beta, r) = f(\overline{\Phi}_{\mathrm{w}}^{\beta})$, $f(\overline{\Phi}_{\mathrm{f}}^{\beta})$ *have the Lipschitz property*: *for* $\beta, \beta' \in [0, \infty)$ *and* $r, r' \in (0, \infty)$,

$$\bigl|h(\beta, r) - h(\beta', r')\bigr| \le \alpha \left\{ |\beta - \beta'| + \frac{|r - r'|}{r \wedge r'} \right\}.$$

Assume henceforth that

$$q > Q'' = Q''(\tfrac{1}{2}). \tag{7.52}$$

By Lemma 7.51, the pressure $f(\overline{\Phi}_{\mathrm{w}}^{\beta_{\mathrm{w}}})$ (respectively, $f(\overline{\Phi}_{\mathrm{f}}^{\beta_{\mathrm{f}}})$) is continuous in β_{w} (respectively, β_{f}), and it follows by the prior observations that the suprema in (7.50) are attained, and hence

$$F(p,q) = f(\overline{\Phi}_{\mathrm{w}}^{b_{\mathrm{w}}}) + b_{\mathrm{w}} + \log r = f(\overline{\Phi}_{\mathrm{f}}^{b_{\mathrm{f}}}) + b_{\mathrm{f}} + \frac{1}{d}\log q. \tag{7.53}$$

By Lemma 7.51 and the continuity in p of $F(p,q)$, Theorem 4.58,

(7.54) $\quad b_{\mathrm{w}} = b_{\mathrm{w}}(p)$ and $b_{\mathrm{f}} = b_{\mathrm{f}}(p)$ are continuous functions of $p \in (0,1)$.

Having chosen the values b_{w} and b_{f}, we shall henceforth suppress their reference in the notation for the weight functions $\Phi_{\mathrm{w}}, \Phi_{\mathrm{f}}, \overline{\Phi}_{\mathrm{w}}, \overline{\Phi}_{\mathrm{f}}$, and we prove next that

$$\begin{aligned} \Phi_{\mathrm{w}}(\gamma) &\le e^{-\tau\|\gamma\|}, \qquad \gamma \in \mathcal{C}_{\mathrm{w}},\\ \Phi_{\mathrm{f}}(\gamma) &\le e^{-\tau\|\gamma\|}, \qquad \gamma \in \mathcal{C}_{\mathrm{f}}. \end{aligned} \tag{7.55}$$

This implies in particular that $\Phi_{\mathrm{w}} = \overline{\Phi}_{\mathrm{w}}$ and $\Phi_{\mathrm{f}} = \overline{\Phi}_{\mathrm{f}}$, and then (7.45) follows from (7.53). We shall prove (7.55) by induction on $|\overline{\gamma}|$.

It is not difficult to see that (7.55) holds for $\gamma_{\mathrm{w}} \in \mathcal{C}_{\mathrm{w}}$ with $|\overline{\gamma_{\mathrm{w}}}| \le 1$, and for $\gamma_{\mathrm{f}} \in \mathcal{C}_{\mathrm{f}}$ with $|\overline{\gamma_{\mathrm{f}}}| \le 2$. This is trivial in the latter case since the free contour γ_{f} with smallest $\|\gamma_{\mathrm{f}}\|$ has $\|\gamma_{\mathrm{f}}\| = 2(2d-1)$, and it is proved in the former case as follows. Let $\gamma_{\mathrm{w}} \in \mathcal{C}_{\mathrm{w}}$ be such that $|\overline{\gamma_{\mathrm{w}}}| = 1$, which is to say that γ_{w} comprises a single edge. By (7.41), $\Phi_{\mathrm{w}}(\gamma_{\mathrm{w}}) = (re^{b_{\mathrm{w}}})^{-1}$. By (7.12), $F(p,q) \ge d^{-1}\log q$, and the claim follows by (7.53) and the fact that $f(\overline{\Phi}_{\mathrm{w}}) \le 1$, see 'Theorem' 7.27(a).

Let $k \ge 1$ and assume that (7.55) holds for all $\gamma_{\mathrm{w}} \in \mathcal{C}_{\mathrm{w}}$ satisfying $|\overline{\gamma_{\mathrm{w}}}| \le k$ and all $\gamma_{\mathrm{f}} \in \mathcal{C}_{\mathrm{f}}$ satisfying $|\overline{\gamma_{\mathrm{f}}}| \le k+1$. Let γ_{w} be a wired contour with $|\overline{\gamma_{\mathrm{w}}}| = k+1$.

Any contour $\gamma'_{\mathrm{w}} \in \mathcal{C}_{\mathrm{w}}$ contributing to $\mathcal{Z}(\mathrm{int}\,\gamma_{\mathrm{w}}; \Phi_{\mathrm{w}})$ satisfies $|\overline{\gamma'_{\mathrm{w}}}| \le k$. By the induction hypothesis,

$$\begin{aligned} \mathcal{Z}(\mathrm{int}\,\gamma_{\mathrm{w}}; \Phi_{\mathrm{w}}) &= \mathcal{Z}(\mathrm{int}\,\gamma_{\mathrm{w}}; \overline{\Phi}_{\mathrm{w}})\\ &= \exp\big\{|\mathrm{int}\,\gamma_{\mathrm{w}}| f(\overline{\Phi}_{\mathrm{w}}) - \sigma(\mathrm{int}\,\gamma_{\mathrm{w}}, \overline{\Phi}_{\mathrm{w}})\big\}, \end{aligned} \tag{7.56}$$

where

$$\sigma(E,\Phi) = |E| f(\Phi) - \log \mathcal{Z}(E;\Phi)$$

as in (7.25). Any contour $\gamma_{\mathrm{f}} \in \mathcal{C}_{\mathrm{f}}$ contributing to $\mathcal{Z}(\overline{\gamma_{\mathrm{w}}}; \Phi_{\mathrm{f}})$ is a subset of $\overline{\gamma_{\mathrm{w}}}$, and therefore satisfies $|\overline{\gamma_{\mathrm{f}}}| \le k+1$. By the induction hypothesis as above,

$$\begin{aligned} \mathcal{Z}(\overline{\gamma_{\mathrm{w}}}; \Phi_{\mathrm{f}}) &= \mathcal{Z}(\overline{\gamma_{\mathrm{w}}}; \overline{\Phi}_{\mathrm{f}})\\ &= \exp\big\{|\overline{\gamma_{\mathrm{w}}}| f(\overline{\Phi}_{\mathrm{f}}) - \sigma(\overline{\gamma_{\mathrm{w}}}, \overline{\Phi}_{\mathrm{f}})\big\}. \end{aligned} \tag{7.57}$$

By (7.41),

$$\begin{aligned}
\Phi_{\mathrm{w}}(\gamma_{\mathrm{w}}) &= (re^{b_{\mathrm{w}}})^{-|\overline{\gamma_{\mathrm{w}}}|} \frac{Z^0(\overline{\gamma_{\mathrm{w}}})}{\mathcal{Z}(\operatorname{int}\gamma_{\mathrm{w}};\, \Phi_{\mathrm{w}})} \\
&= (re^{b_{\mathrm{w}}})^{-|\overline{\gamma_{\mathrm{w}}}|} q^{|V(\overline{\gamma_{\mathrm{w}}})\setminus\partial\overline{\gamma_{\mathrm{w}}}|} \frac{\mathcal{Z}(\overline{\gamma_{\mathrm{w}}};\, \Phi_{\mathrm{f}}, b_{\mathrm{f}})}{\mathcal{Z}(\operatorname{int}\gamma_{\mathrm{w}};\, \Phi_{\mathrm{w}})} \qquad \text{by (7.43)} \\
&\le (re^{b_{\mathrm{w}}})^{-|\overline{\gamma_{\mathrm{w}}}|} q^{|V(\overline{\gamma_{\mathrm{w}}})\setminus\partial\overline{\gamma_{\mathrm{w}}}|} e^{b_{\mathrm{f}}|\overline{\gamma_{\mathrm{w}}}|} \frac{\mathcal{Z}(\overline{\gamma_{\mathrm{w}}};\, \Phi_{\mathrm{f}})}{\mathcal{Z}(\operatorname{int}\gamma_{\mathrm{w}};\, \Phi_{\mathrm{w}})} \\
&= \exp\Big\{-|\overline{\gamma_{\mathrm{w}}}|\big(\log r + b_{\mathrm{w}} - b_{\mathrm{f}} - f(\overline{\Phi}_{\mathrm{f}})\big) \\
&\qquad\qquad + |V(\overline{\gamma_{\mathrm{w}}}) \setminus \partial\overline{\gamma_{\mathrm{w}}}| \log q - |\operatorname{int}\gamma_{\mathrm{w}}| f(\overline{\Phi}_{\mathrm{w}})\Big\} \\
&\qquad \times \exp\big\{\sigma(\operatorname{int}\gamma_{\mathrm{w}}, \overline{\Phi}_{\mathrm{w}}) - \sigma(\overline{\gamma_{\mathrm{w}}}, \overline{\Phi}_{\mathrm{f}})\big\} \qquad \text{by (7.56)–(7.57).}
\end{aligned}$$

We use (7.13)–(7.14) and (7.53) to obtain that

$$\Phi_{\mathrm{w}}(\gamma_{\mathrm{w}}) \le q^{-\|\gamma_{\mathrm{w}}\|/(2d)} \exp\big\{|\gamma_{\mathrm{w}}| f(\overline{\Phi}_{\mathrm{w}}) + \sigma(\operatorname{int}\gamma_{\mathrm{w}}, \overline{\Phi}_{\mathrm{w}}) - \sigma(\overline{\gamma_{\mathrm{w}}}, \overline{\Phi}_{\mathrm{f}})\big\}. \tag{7.58}$$

By 'Theorem' 7.27, $f(\overline{\Phi}_{\mathrm{w}}) \le e^{-c_1\tau} \le 1$, and

$$|\sigma(E, \overline{\Phi}_{\mathrm{w}})| \le |\partial E| e^{-c_2\tau}, \quad |\sigma(E, \overline{\Phi}_{\mathrm{f}})| \le |\partial E| e^{-c_2\tau}$$

for co-connected sets E. By (7.58), (7.16), and (7.44),

$$\Phi_{\mathrm{w}}(\gamma_{\mathrm{w}}) \le q^{-\|\gamma_{\mathrm{w}}\|/(2d)} e^{5\|\gamma_{\mathrm{w}}\|} \le e^{-\tau\|\gamma_{\mathrm{w}}\|}, \tag{7.59}$$

as required in the induction step.

We consider now a free contour γ_{f} with $|\overline{\gamma_{\mathrm{f}}}| = k + 2$. By an elementary geometric argument,

$$\|\gamma_{\mathrm{f}}\| \ge 2(2d - 1). \tag{7.60}$$

Arguing as in the wired case above, we obtain subject to the induction hypothesis that

$$\Phi_{\mathrm{f}}(\gamma_{\mathrm{f}}) \le q \cdot q^{-\|\gamma_{\mathrm{f}}\|/(2d)} \exp\big\{\sigma(\operatorname{int}\gamma_{\mathrm{f}}, \overline{\Phi}_{\mathrm{f}}) - \sigma(\operatorname{int}\gamma_{\mathrm{f}}, \overline{\Phi}_{\mathrm{w}})\big\}, \tag{7.61}$$

by (7.15). By (7.17),

$$\Phi_{\mathrm{f}}(\gamma_{\mathrm{f}}) \le q \cdot q^{-\|\gamma_{\mathrm{f}}\|/(2d)} e^{5\|\gamma_{\mathrm{f}}\|}.$$

By (7.60) and the fact that $d \ge 2$,

$$\|\gamma_{\mathrm{f}}\| - 2d \ge \tfrac{1}{4}\|\gamma_{\mathrm{f}}\|,$$

whence

$$\Phi_{\mathrm{f}}(\gamma_{\mathrm{f}}) \le e^{-\tau\|\gamma_{\mathrm{f}}\|},$$

and the induction proof of (7.55) is complete.

We turn now to part (b) of the theorem, and we prove next that, for any given $p \in (0, 1)$,

$$\min\{b_{\mathrm{w}}, b_{\mathrm{f}}\} = 0. \tag{7.62}$$

Suppose conversely that $p \in (0, 1)$ is such that $b_{\mathrm{w}}, b_{\mathrm{f}} > 0$. By (7.53) and Lemma 7.51 with $\alpha = \frac{1}{2}$, there exist $\beta_{\mathrm{w}} \in (0, b_{\mathrm{w}})$, $\beta_{\mathrm{f}} \in (0, b_{\mathrm{f}})$, and $\epsilon > 0$ such that

$$F(p, q) - \epsilon = f(\overline{\Phi}_{\mathrm{w}}^{\beta_{\mathrm{w}}}) + \beta_{\mathrm{w}} + \log r = f(\overline{\Phi}_{\mathrm{f}}^{\beta_{\mathrm{f}}}) + \beta_{\mathrm{f}} + \frac{1}{d}\log q. \tag{7.63}$$

We use this in place of (7.53) in the argument above, to obtain that $\overline{\Phi}_{\mathrm{w}}^{\beta_{\mathrm{w}}} = \Phi_{\mathrm{w}}^{\beta_{\mathrm{w}}}$ and $\overline{\Phi}_{\mathrm{w}}^{\beta_{\mathrm{f}}} = \Phi_{\mathrm{f}}^{\beta_{\mathrm{f}}}$. Equation (7.63) implies that

$$F(p, q) > f(\Phi_{\mathrm{w}}^{\beta_{\mathrm{w}}}) + \beta_{\mathrm{w}} + \log r = f(\Phi_{\mathrm{f}}^{\beta_{\mathrm{f}}}) + \beta_{\mathrm{f}} + \frac{1}{d}\log q. \tag{7.64}$$

However, by (7.43),

$$Z^1(E) = r^{|E|} q Z(E; \Phi_{\mathrm{w}}^{\beta_{\mathrm{w}}}, \beta_{\mathrm{w}}) \le (re^{\beta_{\mathrm{w}}})^{|E|} q Z(E; \Phi_{\mathrm{w}}^{\beta_{\mathrm{w}}}),$$

whence

$$F(p, q) \le \log r + \beta_{\mathrm{w}} + f(\Phi_{\mathrm{w}}^{\beta_{\mathrm{w}}})$$

in contradiction of (7.64). Therefore, (7.62) holds.

Next we show that there exists a unique p such that $b_{\mathrm{w}}(p) = b_{\mathrm{f}}(p) = 0$. The proof is deferred until later in the section.

(7.65) Lemma. *There exists $Q''' \ge Q''$ such that the following holds. For $q > Q'''$, there is a unique $p' \in (0, 1)$ such that $b_{\mathrm{w}}(p') = b_{\mathrm{f}}(p') = 0$. The ratio $r' = p'/(1 - p')$ satisfies*

$$r' = q^{1/d} \exp\{f(\Phi_{\mathrm{f}}^0) - f(\Phi_{\mathrm{w}}^0)\}. \tag{7.66}$$

Let $q > Q = Q'''$ and $\widetilde{p} = p'$, where Q''' and p' are as given in this lemma. By (7.45) and the fact that $F(p, q) \to d^{-1}\log q$ as $p \downarrow 0$, $f(\Phi_{\mathrm{f}}^{b_{\mathrm{f}}}) \to 0$ and $b_{\mathrm{f}}(p) \to 0$ as $p \downarrow 0$. Similarly, $b_{\mathrm{w}}(p) \to \infty$ as $p \downarrow 0$. By a similar argument for p close to 1, $b_{\mathrm{w}}(p) \to 0$ and $b_{\mathrm{f}}(p) \to \infty$ as $p \uparrow 1$. Statement (7.46) follows by Lemma 7.65 and the continuity of $b_{\mathrm{w}}(p)$ and $b_{\mathrm{f}}(p)$, (7.54). This completes the proof of Theorem 7.42. □

Proof of Lemma 7.51. We give the proof in the wired case, the other case being similar. Write $\Phi = \Phi_{\mathrm{w}}^{\beta}$ and let E be co-connected. For any contour $\gamma \subseteq E$ and any family Σ of compatible contours in E, we write $\Sigma \perp \gamma$ if $\gamma \notin \Sigma$ and $\Sigma \cup \{\gamma\}$ is a compatible family of contours. Since $\Phi(\gamma)$ is a smooth function of β, $\overline{\Phi}$ is piecewise-differentiable in β (see (7.48)).

Let $\alpha \in (0, \infty)$. We prove first that the function

$$z_{\mathrm{w}}^E(\beta, r) = \frac{1}{|E|} \log \mathcal{Z}(E; \overline{\Phi}_{\mathrm{w}}^{\beta})$$

satisfies

$$\left| z_{\mathrm{w}}^E(\beta, r) - z_{\mathrm{w}}^E(\beta', r) \right| \le \alpha |\beta - \beta'|, \qquad \beta, \beta' \in [0, \infty), \tag{7.67}$$

for sufficiently large q, uniformly in r and E. We fix $r \in (0, \infty)$ and shall suppress reference to r for the moment. If z_{w}^E is differentiable at β then, by (7.37),

$$\frac{d}{d\beta} z_{\mathrm{w}}^E = \frac{1}{|E| \mathcal{Z}(E; \overline{\Phi})} \sum_{\Sigma \subset E} \sum_{\gamma \in \Sigma} \frac{\overline{\Phi}(\Sigma)}{\overline{\Phi}(\gamma)} \cdot \overline{\Phi}'(\gamma),$$

where g' denotes the derivative of a function g with respect to β, and $g(\Sigma) = \prod_{\gamma \in \Sigma} g(\gamma)$. Therefore, for any given edge e,

$$\begin{aligned} \left| \frac{d}{d\beta} z_{\mathrm{w}}^E \right| &= \frac{1}{|E|} \sum_{\gamma \subseteq E} |\overline{\Phi}'(\gamma)| \frac{\mathcal{Z}_1(E \setminus \gamma; \overline{\Phi})}{\mathcal{Z}(E; \overline{\Phi})} \\ &\le \sum_{\gamma : e \in \gamma} \frac{|\overline{\Phi}'(\gamma)|}{|\gamma|}, \end{aligned} \tag{7.68}$$

where

$$\mathcal{Z}_1(E \setminus \gamma; \overline{\Phi}) = \sum_{\Sigma \subset E :\, \Sigma \perp \gamma} \overline{\Phi}(\Sigma) \le \mathcal{Z}(E; \overline{\Phi}).$$

Let $\gamma \in \mathcal{C}_{\mathrm{w}}$. We claim that

$$|\overline{\Phi}'(\gamma)| \le 2 |\overline{\gamma}| \overline{\Phi}(\gamma) \tag{7.69}$$

whenever the derivative exists. By (7.48), either the left side equals 0, or it equals $|\Phi'(\gamma)|$, and we may assume that the latter holds. Write $\mathcal{Y}(\gamma) = \mathcal{Z}(\operatorname{int} \gamma; \Phi)$. The function $\Psi = \Psi_{\mathrm{w}}^{\beta}$ satisfies $\Psi(\gamma) = \Phi(\gamma) \mathcal{Y}$ by (7.41), and also

$$\Psi'(\gamma) = -|\overline{\gamma}| \Psi(\gamma) = -|\overline{\gamma}| \Phi(\gamma) \mathcal{Y}(\gamma). \tag{7.70}$$

Hence,

$$\Phi'(\gamma) = \frac{\Psi'(\gamma) - \Phi(\gamma) \mathcal{Y}'(\gamma)}{\mathcal{Y}(\gamma)} = -\Phi(\gamma) \left(|\overline{\gamma}| + \frac{\mathcal{Y}'(\gamma)}{\mathcal{Y}(\gamma)} \right). \tag{7.71}$$

By an argument similar to that above,

$$\mathcal{Y}'(\gamma) = \frac{d}{d\beta} \sum_{\Gamma \subset \text{int}\, \gamma} \Psi(\Gamma) = \sum_{\Gamma \subset \text{int}\, \gamma} \Psi(\Gamma) \sum_{\nu \in \Gamma} \frac{\Psi'(\nu)}{\Psi(\nu)}$$
$$= - \sum_{\Gamma \subset \text{int}\, \gamma} |\overline{\Gamma}| \Psi(\Gamma),$$

whence

$$|\mathcal{Y}'(\gamma)| \le |\overline{\gamma}| \mathcal{Y}(\gamma). \tag{7.72}$$

Claim (7.69) follows from (7.71)–(7.72).

Returning to (7.68), by (7.69),

$$\left| \frac{d}{d\beta} z_{\mathrm{w}}^{E} \right| \le \sum_{\gamma : e \in \gamma} 2 \frac{|\overline{\gamma}|}{|\gamma|} \overline{\Phi}(\gamma) \le \sum_{\gamma : e \in \gamma} 2 \frac{|\overline{\gamma}|}{|\gamma|} e^{-\tau \|\gamma\|}, \tag{7.73}$$

since $\overline{\Phi}$ is a τ-functional. The Lipschitz inequality (7.67) follows by integration for $\tau = \tau(q)$ sufficiently large.

More or less the same argument may be used as follows to obtain that

$$\left| z_{\mathrm{w}}^{E}(\beta, r) - z_{\mathrm{w}}^{E}(\beta, r') \right| \le \alpha \frac{|r - r'|}{r \wedge r'}, \qquad r, r' \in (0, \infty), \tag{7.74}$$

for large q, uniformly in β and E. We now denote by g' the derivative of a function $g(r)$ with respect to r. Equation (7.68) remains valid in this new setting. Inequality (7.69) becomes

$$|\overline{\Phi}'(\gamma)| \le \frac{2}{r} |\overline{\gamma}| \overline{\Phi}(\gamma)$$

and (7.73) is replaced by

$$\left| \frac{d}{dr} z_{\mathrm{w}}^{E} \right| \le \sum_{\gamma : e \in \gamma} \frac{2|\overline{\gamma}|}{r|\gamma|} \overline{\Phi}(\gamma) \le \sum_{\gamma : e \in \gamma} \frac{2|\overline{\gamma}|}{r|\gamma|} e^{-\tau \|\gamma\|}. \tag{7.75}$$

The right side may be made small by choosing q large, and (7.74) follows by integration.

The claim of the lemma is a consequence of (7.67) and (7.74), on using the triangle inequality and passing to the limit as $E \uparrow \mathbb{E}^d$. □

Proof of Lemma 7.65. Let $p \in (0, 1)$ be such that $b_{\mathrm{w}} = b_{\mathrm{f}} = 0$, and let $r = p/(1-p)$. By (7.45), r is a root of the equation $h_1(r) = h_2(r)$ where

$$h_1(r) = f(\Phi_{\mathrm{w}}^{0}) + \log r, \quad h_2(r) = f(\Phi_{\mathrm{f}}^{0}) + \frac{1}{d} \log q,$$

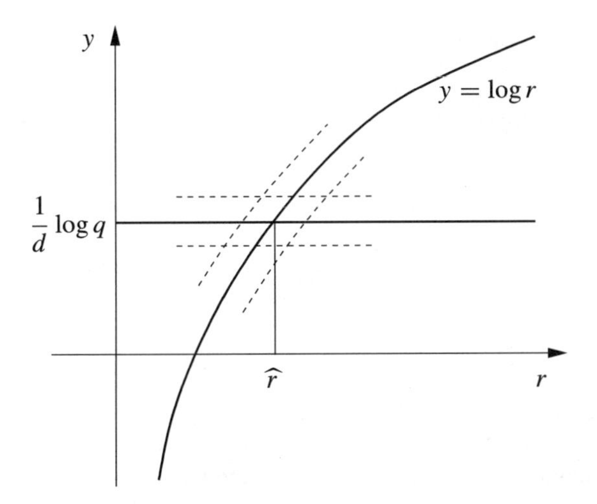

Figure 7.3. The function $y = \log r$ is plotted against r, and it intersects the constant function $y = d^{-1}\log q$ at the point $r = \widehat{r}$. The functions h_1 and h_2 are small perturbations of the two solid lines, and have Lipschitz constants which can be made as small as desired by a suitable large choice of q. Therefore, when q is large, there exists a unique intersection of h_1 and h_2, and this lies within the region delineated by dashed lines.

and thus (7.66) holds. Let $f_{\mathrm{w}}(r) = f(\overline{\Phi}^0_{\mathrm{w}})$, $f_{\mathrm{f}}(r) = f(\overline{\Phi}^0_{\mathrm{f}})$, $\widehat{r} = q^{1/d}$, and note the two following facts.

(I) Since f_{w} and f_{f} are the pressure functions of τ-functionals with $\tau > c$, we have by 'Theorem' 7.27 that $|f_{\mathrm{w}}|, |f_{\mathrm{f}}| \le e^{-c_1\tau}$.

(II) By (7.52) and Lemma 7.51 with $\beta = 0$, f_{w} and f_{f} are Lipschitz-continuous on a neighbourhood of $\widehat{r}$, with Lipschitz constants which may be made as small as desired by a suitable large choice of q.

From these facts it will follow (for sufficiently large q) that any roots of $h_1(r) = h_2(r)$ lie near $\widehat{r}$, and indeed there must be a unique such root. Some readers will accept this conclusion after looking at Figure 7.3, those wishing to check the details may read on.

Let r_1, r_2 be roots of $h_1(r) = h_2(r)$ with $0 < r_1 < r_2 < \infty$. With c_1 as in (I), we choose $Q_1 \ge Q''$ such that $e^{-c_1\tau} \le \frac{1}{8}$ for $q > Q_1$. Let $q > Q_1$. Then $r_1, r_2 \in [\widehat{r} - a, \widehat{r} + a]$ where

$$a = \big[\exp(2e^{-c_1\tau}) - 1\big]\widehat{r} \le 4e^{-c_1\tau}\widehat{r} \le \tfrac{1}{2}\widehat{r}. \tag{7.76}$$

Now, $\delta(r) = f_{\mathrm{f}}(r) - f_{\mathrm{w}}(r)$ satisfies

$$\begin{aligned} \delta(r_2) - \delta(r_1) &= \log r_2 - \log r_1 \\ &\ge \frac{r_2 - r_1}{\widehat{r} + a} \ge \frac{2}{3} \cdot \frac{r_2 - r_1}{\widehat{r}}. \end{aligned} \tag{7.77}$$

By Lemma 7.51, there exists $Q''' \geq Q_1$ such that, if $q > Q'''$,

$$|f_{\mathrm{w}}(r) - f_{\mathrm{w}}(r')| \leq \frac{|r - r'|}{8(\widehat{r} - a)}, \quad |f_{\mathrm{f}}(r) - f_{\mathrm{f}}(r')| \leq \frac{|r - r'|}{8(\widehat{r} - a)},$$

for $r, r' \geq \widehat{r} - a$. Hence, by (7.76),

$$|\delta(r_2) - \delta(r_1)| \leq \frac{r_2 - r_1}{4(\widehat{r} - a)} \leq \frac{1}{2} \cdot \frac{r_2 - r_1}{\widehat{r}}.$$

This contradicts (7.77), whence such distinct r_1, r_2 do not exist. □

Proof of Theorem 7.33. Let $p \in (0, 1)$ and $q > Q$ where Q, $\tau = \tau(q)$, $b_{\mathrm{w}} = b_{\mathrm{w}}(p, q)$, $b_{\mathrm{f}} = b_{\mathrm{f}}(p, q)$, and $\widetilde{p} = \widetilde{p}(q)$ are given as in Theorem 7.42. Let Λ be a box of $\mathbb{L}^d$, and let ϕ_Λ^1 (respectively, ϕ_Λ^0) be the wired random-cluster measure on $\mathbb{E}_\Lambda$ generated by the partition function $Z^1(\mathbb{E}_\Lambda)$ of (7.10) (respectively, the free measure generated by the partition function $Z^0(\mathbb{E}_\Lambda)$ of (7.11)).

Consider first the wired measure ϕ_Λ^1. As in (7.21), there exists a family of maximal closed wired contours Γ of $\mathbb{E}_\Lambda$ (maximal in the sense of the partial order $\gamma_1 \leq \gamma_2$ if $\overline{\gamma}_1 \subseteq \overline{\gamma}_2$) and, by (7.40)–(7.41), Γ has law

$$\kappa_{\Lambda,\mathrm{w}}^{b_{\mathrm{w}}}(\Gamma) = \frac{1}{\mathcal{Z}(\Lambda; \Phi_{\mathrm{w}}^{b_{\mathrm{w}}}, b_{\mathrm{w}})} e^{b_{\mathrm{w}}|\overline{\Gamma}|} \Psi_{\mathrm{w}}^{b_{\mathrm{w}}}(\Gamma).$$

Let $p \geq \widetilde{p}$, so that $b_{\mathrm{w}} = 0$. Then $\kappa_{\Lambda,\mathrm{w}}^{b_{\mathrm{w}}} = \kappa_{\Lambda,\mathrm{w}}^0$ is the law of the family of maximal contours in the wired contour model on Λ with weight function Φ_{w}^0.

Let $x, y \in \Lambda$, and consider the event

$$F_\Lambda(x, y) = \{x \leftrightarrow y,\ x \nleftrightarrow \partial\Lambda\}.$$

If $F_\Lambda(x, y)$ occurs, then $x, y \in V_{\mathrm{int}\,\gamma}$ for some maximal closed wired contour γ. This event has the same probability as the event that $x, y \in V_{\mathrm{int}\,\nu}$ for some contour ν of the wired contour model with weight function Φ_{w}^0. Therefore,

$$\begin{aligned} (7.78) \qquad \phi_\Lambda^1(F_\Lambda(x, y)) &\leq \kappa_{\Lambda,\mathrm{w}}^0(x, y \in V_{\mathrm{int}\,\nu} \text{ for some contour } \nu) \\ &\leq \sum_{\nu:\, x,y \in V_{\mathrm{int}\,\nu}} \Phi_{\mathrm{w}}^0(\nu) \\ &\leq \sum_{\nu:\, x,y \in V_{\mathrm{int}\,\nu}} e^{-\tau\|\nu\|}, \end{aligned}$$

by Theorem 7.32 and the fact that Φ_{w}^0 is a τ-functional. The number of such wired contours ν with $\|\nu\| = n$ grows at most exponentially in n. The leading term in the above series arises from the contour ν having smallest $\|\nu\|$, and such ν satisfies

$\|\nu\| \geq b(1+|x-y|)$ for some absolute constant $b > 0$. We may therefore find absolute constants $Q' \geq Q$ and $a > 0$ such that, for $q \geq Q'$,

$$\phi_\Lambda^1(F_\Lambda(x, y)) \leq e^{-a\tau(1+|x-y|)}. \tag{7.79}$$

Take $x = y$ in (7.79), and let $\Lambda \uparrow \mathbb{Z}^d$ to obtain by Proposition 5.11 that

$$\phi_{p,q}^1(x \not\leftrightarrow \infty) < 1$$

whence $p \geq p_c(q)$. It follows that

$$\widetilde{p} \geq p_c(q). \tag{7.80}$$

Consider next the free measure ϕ_Λ^0. Let $p \leq \widetilde{p}$, so that $b_f = 0$. By an adaptation of the argument above, there exists $Q'' \geq Q'$ and $k > 0$ such that, for $q \geq Q''$, $x, y \in \mathbb{Z}^d$, and all large Λ,

$$\phi_\Lambda^0(x \leftrightarrow y) \leq e^{-k\tau|x-y|}. \tag{7.81}$$

By Proposition 5.12 applied to $\phi_{p,q}^0$,

$$\begin{aligned} \phi_{p,q}^0(x \leftrightarrow y) &= \lim_{\Lambda \uparrow \mathbb{Z}^d} \phi_\Lambda^0(x \leftrightarrow y) \\ &\leq e^{-k\tau|x-y|}, \qquad x, y \in \mathbb{Z}^d. \end{aligned} \tag{7.82}$$

Hence $p \leq p_c(q)$, and so

$$\widetilde{p} \leq p_c(q). \tag{7.83}$$

By (7.80) and (7.83), $\widetilde{p} = p_c(q)$. By (7.82), there is exponential decay of connectivity[11] for $p \leq p_c(q)$, and a non-vanishing mass gap.

Parts (b) and (d) of the theorem have been proved for $q \geq Q''$. Part (b) implies that $\phi_{p_c(q),q}^0 \neq \phi_{p_c(q),q}^1$, and hence (a) via Theorem 4.63. The uniqueness of random-cluster measures holds generally when $p < p_c(q)$, Theorem 5.33. The proof of uniqueness when $p > p_c(q)$ has much in common with the proofs of Proposition 5.30 and Theorem 11.40, and so we present a sketch only.

Let $q \geq Q''$ and $p \in (p_c(q), 1)$. We shall show that $h^1(p, q) = \phi_{p,q}^1(e \text{ is open})$ satisfies

$$h^1(p - \epsilon, q) \uparrow h^1(p, q) \qquad \text{as } \epsilon \downarrow 0, \tag{7.84}$$

and the claim will follow by Proposition 4.28(b) and Theorem 4.63.

[11] The related issue of 'restricted complete analyticity' is considered in [110] for the case of two dimensions.

Let ϵ be such that $p_c(q) < p - \epsilon < p$, and let $\eta \in (0, 1)$. Write $\phi^1_{n,p} = \phi^1_{\Lambda_n,p,q}$ where $\Lambda_n = [-n, n]^d$. For $n > \frac{3}{2}m \geq 2$, let $E_{m,n}$ be the event that, for every $x \in \partial\Lambda_m$, if $\nu = \nu_x$ is a maximal closed wired contour of Λ_n with $x \in V_{\text{int}\,\nu}$, then $\nu \subseteq x + \mathbb{E}_{\Lambda_{m/4}}$. As in (7.78)–(7.79), there exists $\gamma = \gamma(q) > 0$ such that

$$\phi^1_{n,p-\epsilon}(E_{m,n}) \geq 1 - |\partial\Lambda_m|e^{-\gamma m},$$

and we choose $m = m(q) \geq 8$ such that

$$\phi^1_{n,p-\epsilon}(E_{m,n}) > 1 - \eta, \qquad n > \tfrac{3}{2}m. \tag{7.85}$$

Let z denote the vertex $(1, 0, 0, \ldots, 0)$. A *cutset* σ of Λ_m is defined to be a subset of $\Lambda_m \setminus \{0, z\}$ such that: every path from either 0 or z to $\partial\Lambda_m$ passes through at least one vertex in σ, and σ is minimal with this property. For any cutset σ, we write $\text{int}(\sigma)$ for the set of vertices reachable from either 0 and z along paths not intersecting σ, and $\text{out}(\sigma) = \mathbb{Z}^d \setminus \text{int}(\sigma)$. For $n > \frac{3}{2}m$ and a cutset σ, we write '$\sigma \Longrightarrow \partial\Lambda_n$ in ω' if every vertex in σ is connected to $\partial\Lambda_n$ by an ω-open path of $\text{out}(\sigma)$. We shall see below that, for $\omega \in E_{m,n}$, there exists a (random) cutset $\Sigma = \Sigma(\omega) \subseteq \Lambda_m \setminus \Lambda_{m/2}$ such that $\Sigma \Longrightarrow \partial\Lambda_n$ in ω.

Let $e = \langle 0, z\rangle$ and $n > \frac{3}{2}m$. We couple the measures $\phi^1_{n,p-\epsilon}$ and $\phi^1_{n,p}$ in such a way that the first lies beneath the second, and we do this by a sequential examination of the (paired) states of edges in Λ_n. We will follow the recipe of the proof of Theorem 3.45 (see also Proposition 5.30), but subject to a special ordering of the edges. The outcome will be a pair $\omega_0, \omega_1 \in \Omega^1_{\Lambda_n}$ such that: ω_0 has law $\phi^1_{n,p-\epsilon}$, ω_1 has law $\phi^1_{n,p}$, and $\omega_0 \leq \omega_1$. First, we determine the states $\omega_0(e), \omega_1(e)$ of edges e with both endvertices in $\Lambda_n \setminus \Lambda_{m-1}$, using some arbitrary ordering of these edges. If $\partial\Lambda_m \Longrightarrow \partial\Lambda_n$ in ω_0, we set $\Sigma = \partial\Lambda_m$ and we complete the construction of ω_0 and ω_1 according to an arbitrary ordering of the remaining edges in Λ_m.

Suppose that $\partial\Lambda_m \not\Longrightarrow \partial\Lambda_n$ in ω_0. Let A be the set of edges in $\partial\Lambda_m$ that are closed in ω_0. If $A = \varnothing$, we sample the states of the remaining edges of Λ_m in an arbitrary order as above. Suppose $A \neq \varnothing$. Pick $f \in A$, and sample the states of edges in the $(d-2)$-connected closed cluster $F_f = F_f(\omega_0)$ of f in the lower configuration ω_0. When this has been done for every $f \in A$, we complete the construction of ω_0 and ω_1 according to an arbitrary ordering of the remaining edges in Λ_m.

In examining the states of edges in F_f we will discover a set $\Delta(F_f)$ of edges, not belonging to F_f but $(d-2)$-connected to F_f, such that $\omega_0(g) = 1$ for $g \in \Delta(F_f)$. Let $\Delta_{\text{v},f}$ be the set of all vertices $v \in \Lambda_n$ lying in the infinite component of $(\mathbb{Z}^d, \mathbb{E}^d \setminus F_f)$ and such that there exists $w \in \Lambda_n$ with $\langle v, w\rangle \in \Delta(F_f) \cup F_f$. Let $\Delta_{\text{e},f}$ be the set of edges of ΔF_f joining pairs of vertices in $\Delta_{\text{v},f}$. By Theorem 7.6, the graph $(\Delta_{\text{v},f}, \Delta_{\text{e},f})$ is connected.

Suppose $\omega_0 \in E_{m,n}$. By the above, $\partial\Lambda_m \cup \left\{\bigcup_{f\in A} \Delta_{\text{v},f}\right\}$ contains a (random) cutset $\Sigma = \Sigma(\omega_0)$ such that: $\Sigma \Longrightarrow \partial\Lambda_n$ in ω_0 and, conditional on Σ and the states of edges of $\text{out}(\Sigma)$, the coupled conditional measures of $\phi^1_{n,p-\epsilon}$ and $\phi^1_{n,p}$ on the remaining edges of $\Sigma \cup \text{int}(\Sigma)$ are the appropriate wired measures.

Therefore, $h_n(p) = \phi^1_{n,p}(J_e)$ satisfies

$$\begin{aligned} h_n(p) - h_n(p-\epsilon) &\le \eta + \sum_{\sigma \in \mathcal{C}} \left| \phi^1_{\sigma,p}(J_e) - \phi^1_{\sigma,p-\epsilon}(J_e) \right| \phi^1_{n,p-\epsilon}(\Sigma = \sigma) \\ &\le \eta + \max_{\sigma \in \mathcal{C}} \left\{ \left| \phi^1_{\sigma,p}(J_e) - \phi^1_{\sigma,p-\epsilon}(J_e) \right| \right\}, \end{aligned}$$

where $\mathcal{C}$ is the set of all cutsets of Λ_m and $\phi^1_{\sigma,p}$ denotes the wired random-cluster measure on $\sigma \cup \text{int}(\sigma)$. Since m is fixed, $\mathcal{C}$ is bounded, and (7.84) follows on letting $n \to \infty$, $\epsilon \downarrow 0$, and $\eta \downarrow 0$ in that order. □

Proof of Theorem 7.34. Let q be large. Then $p_c(q) = r'/(1+r')$ where r' is given in Lemma 7.65 and satisfies (7.66). Let $p = p_c(q)$. By (7.44) and 'Theorem' 7.27, $f(\Phi^0_f), f(\Phi^0_w) \to 0$ as $q \to \infty$, and therefore $r' \sim q^{1/d}$. We sketch a derivation of the error term $O(q^{-2/d})$. The rate at which $f(\Phi^0_f) \to 0$ (respectively, $f(\Phi^0_w) \to 0$) is determined by the value $\Phi^0_f(\gamma_f)$ (respectively, $\Phi^0_w(\gamma_w)$) on the smallest free contour γ_f (respectively, smallest wired contour γ_w). The smallest free contour is the external edge-boundary γ_f of a single edge, and it is easily seen from (7.41) that $\Phi^0_f(\gamma_f) = r'q^{-1} \sim q^{-1+(1/d)}$. The shortest wired contour γ_w is a single edge, and $\Phi^0_w(\gamma_w) = 1/r' \sim q^{-1/d}$. By (7.24), as $q \to \infty$,

$$f(\Phi^0_w) = O(q^{-1/d}), \qquad f(\Phi^0_f) = O(q^{-1+(1/d)}),$$

and the claim follows by (7.66). □

7.6 Dobrushin interfaces

Until now in this chapter we have studied the critical random-cluster model for large q. We turn now to the model with $q \in [1, \infty)$ and with large p, and we prove the existence of so-called Dobrushin interfaces.

Consider for illustration the Ising model on $\mathbb{Z}^3$ with 'inverse-temperature' β and zero external-field. There is a critical value β_c marking the point at which long-range correlations cease to decay to zero. As β increases to ∞, pairs of vertices have an increasing propensity to acquire the same state, either both $+$ or both $-$. Suppose we are working on a large cube $\Lambda_L = [-L, L]^3$, to the boundary of which we give a so-called 'Dobrushin boundary condition'; that is, the upper boundary $\partial^+\Lambda_L = \{x \in \partial\Lambda_L : x_3 > 0\}$ is allocated the spin $+$, and the lower boundary $\partial^-\Lambda_L = \{x \in \partial\Lambda_L : x_3 \le 0\}$ receives spin $-$. There is a competition between the $+$ spins and the $-$ spins. There is an 'upper' domain of $+$ spins containing $\partial^+\Lambda_L$, and a 'lower' domain of $-$ spins containing $\partial^-\Lambda_L$, and these domains are separated by a (random) interface $\Delta = \Delta_L$. It is a famous result of Dobrushin, [103], that, for large β in the limit as $L \to \infty$, Δ_L deviates only locally from the horizontal plane through the centre of Λ_L. This implies in particular that there exist non-translation-invariant Gibbs measures for the three-dimensional Ising model with large β. The argument is valid in all dimensions of three or more, but not in two dimensions, for which case the interface may be thought of as a line subject to Gaussian fluctuations (see [127, 137, 187]).

Dobrushin's proof was the starting point for the study of interfaces in spin systems. His conclusions may be reformulated and generalized in the context of the random-cluster model in three or more dimensions with $q \in [1, \infty)$. This generalization of Dobrushin's theorem is achieved by defining a family of conditioned random-cluster measures, and by showing the stiffness of the ensuing interface. It is a striking fact that the conclusions hold even for the percolation model.

When cast in the more general setting of the random-cluster model on a box Λ, the correct interpretation of the boundary condition is as follows. The vertices on the upper (respectively, lower) hemisphere of Λ are wired together into a single composite vertex labelled $\partial^+\Lambda$ (respectively, $\partial^-\Lambda$). Let $\mathcal{D}$ be the event that no open path of Λ exists joining $\partial^-\Lambda$ to $\partial^+\Lambda$, and let $\overline{\phi}_{\Lambda,p,q}$ be the random-cluster measure on Λ with the above boundary condition *and conditioned on the event* $\mathcal{D}$. It is a geometrical fact that, under $\overline{\phi}_{\Lambda,p,q}$, there exists an interface separating an upper region of Λ containing $\partial^+\Lambda$ and a lower region containing $\partial^-\Lambda$, and each of these regions is in the wired phase. Dobrushin's theorem amounts to the statement that, when $q = 2$ and p is sufficiently large, this interface deviates only locally from the horizontal plane through the equator of Λ. It was proved in [139] that the same conclusion is valid for all $q \in [1, \infty)$ and all sufficiently large p, and this result is presented in the remainder of this chapter. The geometry of the interfaces for the random-cluster model is notably different from that of a spin model since the configurations are indexed by edges rather than by vertices, and this leads to difficulties not encountered in the Ising model.

Although such arguments are valid whenever $d \geq 3$, we shall assume for simplicity that $d = 3$. It is striking that the results are valid for high-density percolation on $\mathbb{Z}^d$ with $d \geq 3$, being the random-cluster model with $q = 1$. A corresponding question for supercritical percolation in two dimensions has been studied in depth in [77], where it is shown effectively that the (one-dimensional) interface converges when re-scaled to a Brownian bridge.

We have spoken above of interfaces which 'deviate only locally' from a plane, an expression made more rigorous in Section 7.11 where the principal Theorem 7.142 is presented. We include at Theorem 7.87 a weaker version of the main result which does not make use of the notation developed in later sections.

The results are proved under the assumption that $q \in [1, \infty)$ and p is sufficiently large. It is a major open question to determine whether or not such results are valid under the weaker assumption that p exceeds the critical value $p_c(q)$ of the random-cluster model. The answer may be expected to depend on the value of q and the number d of dimensions. Since the percolation measure $\overline{\phi}_{\Lambda,p,1}$ is a conditioned product measure, it may be possible to gain insight into the existence or not of such a 'roughening transition' by concentrating on the special case of percolation. The two core problems here are the following. Let $\overline{p}(q)$ be the infimum of all values of p at which the above interface is localized (a rigorous interpretation of this definition is evident after reading Theorems 7.87 and 7.142).

I. Is it the case that the interface is localized for all $p > \overline{p}(q)$?

II. For what q and d does strict inequality of critical points hold in the sense that $p_c(q) < \overline{p}(q)$?

In the case of the Ising model ($q = 2$), it is generally believed that $p_c(2) < \overline{p}(2)$ if and only if $d = 3$.

A certain amount of notation and preliminary work is required before the main theorems may be stated (in Section 7.11). In order to whet appetites, a preliminary result is included towards the end of the current section. Sections 7.7–7.8 contain some preliminary facts about random-cluster measures and interfaces. A detailed geometrical analysis of interfaces is included in Section 7.9 along the lines of Dobrushin's classification of 'walls' and 'ceilings'. This is followed in Section 7.10 by an exponential bound for the probability of finding local perturbations of a flat interface.

The *upper* and *lower boundaries* of a set Λ of vertices are defined as

$$\partial^+\Lambda = \{x \in \overline{\Lambda} : x_3 > 0,\ x \sim z \text{ for some } z \in \Lambda\},$$
$$\partial^-\Lambda = \{x \in \overline{\Lambda} : x_3 \le 0,\ x \sim z \text{ for some } z \in \Lambda\},$$

where $\overline{\Lambda} = \mathbb{Z}^d \setminus \Lambda$. For positive integers L, M, let $\Lambda_{L,M}$ denote the box $[-L, L]^2 \times [-M, M]$, and write $E_{L,M}$ for the set of edges having at least one endvertex in $\Lambda_{L,M}$. We write $\Lambda_L = \Lambda_{L,L}$, the cube of side-length $2L$, and $\Sigma_L = [-L, L]^2 \times \mathbb{Z}$, an infinite cylinder. The *equator* of the box $\Lambda_{M,N}$ is defined to be the circuit of $\Lambda_{L,M} \setminus \Lambda_{L-1,M}$ comprising all vertices x with $x_3 = \frac{1}{2}$, with a similar definition for the cylinder Σ_L.

We shall be particularly concerned with a boundary condition D corresponding to the mixed 'Dobrushin boundary' of [103]. Let $\mathrm{D} \in \Omega$ be given by

(7.86)
$$\mathrm{D}(e) = \begin{cases} 0 & \text{if } e = \langle x, y\rangle \text{ for some } x = (x_1, x_2, 0) \text{ and } y = (x_1, x_2, 1), \\ 1 & \text{otherwise.} \end{cases}$$

See Figure 7.4. Let $\Omega^{\mathrm{D}}_{L,M}$ be the set of configurations $\omega \in \Omega$ such that $\omega(f) = \mathrm{D}(f)$ if $f \notin E_{L,M}$, and let $\mathcal{I}_{L,M}$ be the event that there exists no open path connecting a vertex of $\partial^+\Lambda_{L,M}$ to a vertex of $\partial^-\Lambda_{L,M}$. The probability measure of current interest is the random-cluster measure $\phi^{\mathrm{D}}_{\Lambda_{M,N},p,q}$ conditioned on the event $\mathcal{I}_{L,M}$, which we denote by $\overline{\phi}^{\mathrm{D}}_{\Lambda_{L,M},p,q}$.

Many of the calculations concern the box $\Lambda_{L,M}$ and the measure $\overline{\phi}^{\mathrm{D}}_{\Lambda_{L,M},p,q}$. We choose however to express our conclusions in terms of the infinite cylinder $\Sigma_L = \Lambda_{L,\infty}$ and the weak limit $\overline{\phi}_{L,p,q} = \lim_{M\to\infty} \overline{\phi}^{\mathrm{D}}_{\Lambda_{L,M},p,q}$.

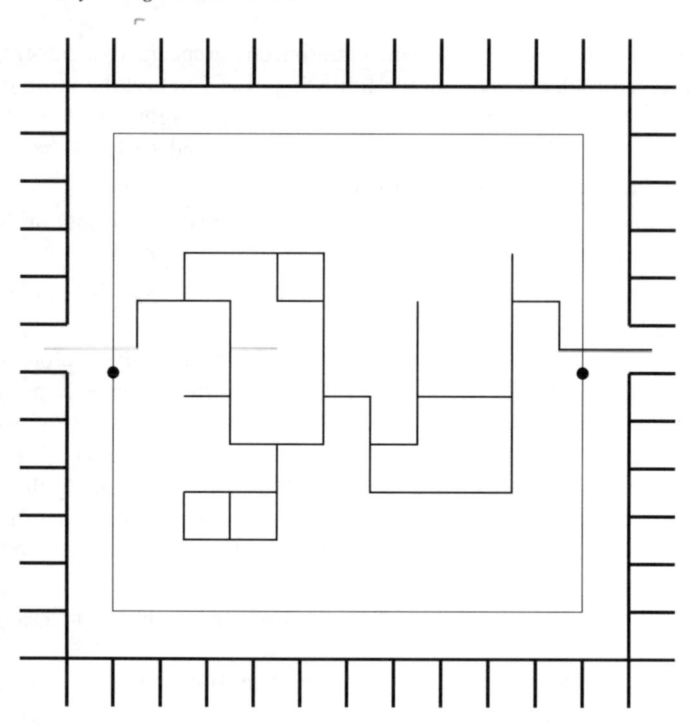

Figure 7.4. The box $\Lambda_{L,M}$. The heavy black edges are those given by the boundary condition D, and there is a two-dimensional sketch of the interface Δ.

It is shown in Lemma 7.98 that, on the event $\mathcal{I}_{L,M} \cap \Omega^{\mathrm{D}}_{L,M}$, there exists an interface spanning the equator of $\Lambda_{L,M}$. Much of the work of the subsequent sections is devoted to understanding the geometry of such an interface. We shall see in Theorem 7.142 that, in the limit as $M \to \infty$ and for sufficiently large p, this interface deviates, $\overline{\phi}_{L,p,q}$-almost-surely, only locally from the flat plane through the equator of Σ_L. Indeed, the spatial density of such deviations approaches zero as p approaches 1. The following theorem is an example of an application of the forthcoming Theorem 7.142.

(7.87) Theorem [139]. *Let $q \in [1, \infty)$. For $\epsilon > 0$, there exists $\widehat{p} = \widehat{p}(\epsilon) < 1$ such that, if $p > \widehat{p}$,*

$$\begin{aligned} \overline{\phi}_{L,p,q}(x \leftrightarrow \partial^-\Sigma_L) &> 1 - \epsilon, \\ \overline{\phi}_{L,p,q}(x + (0,0,1) \leftrightarrow \partial^+\Sigma_L) &> 1 - \epsilon, \end{aligned} \tag{7.88}$$

for all $L \geq 1$ and every $x = (x_1, x_2, 0) \in \Sigma_L$.

No proof is known of the weak convergence of $\overline{\phi}_{L,p,q}$ as $L \to \infty$, but, by the usual compactness argument[12], the sequence must possess weak limits. It is a

[12] See the proof of Theorem 4.17(a).

consequence of Theorems 7.87 and 7.142 that, for sufficiently large p, any such weak limit is non-translation-invariant.

(7.89) Theorem [139]. *Let $q \in [1, \infty)$ and $p > \widehat{p}(\frac{1}{2})$, where $\widehat{p}(\frac{1}{2})$ is given in Theorem 7.87. The family $\{\overline{\phi}_{L,p,q} : L = 1, 2, \dots\}$ possesses at least one non-translation-invariant weak limit.*

It is shown in addition at Theorem 7.144 that there exists a geometric bound, uniformly in L, on the tail of the displacement of the interface from the flat plane.

By making use of the relationship between random-cluster models and Potts models (see Sections 1.4 and 4.6), one obtains a generalization of the theorem of Dobrushin [103] to include percolation and Potts models.

The measure $\overline{\phi}_{L,p,q}$ is not a random-cluster measure in the sense of Chapter 3, even though it corresponds to a Gibbs measure when $q \in \{2, 3, \dots\}$. It may instead be termed a 'conditioned' random-cluster measure, and such measures will be encountered again in Chapter 11.

The strategy of the proofs is to follow the milestones of the paper of Dobrushin [103]. Although Dobrushin's work is a helpful indicator of the overall route to the results, a considerable amount of extra work is necessary in the context of the random-cluster model, much of which arises from the fact that the geometry of interfaces is different for the random-cluster model from that for spin systems. Heavy use is made in the remainder of this chapter of the material in [139].

7.7 Probabilistic and geometric preliminaries

We shall require two general facts about random-cluster measures, and we state these next. The first is a formula for the partition function in terms of the edge densities. For $E \subseteq \mathbb{E}^3$, let V_E denote the set of endvertices of members of E. As usual, J_e denotes the event that the edge e is open, and $Z_G^\zeta(p, q)$ is given as in (4.12). Let ζ_E^1 be the configuration obtained from $\zeta \in \Omega$ by declaring every edge in E to be open, and $k(\zeta_E^1, E)$ the number of components of ζ_E^1 that intersect V_E.

(7.90) Lemma. *Let E be a finite subset of $\mathbb{E}^3$, and $G = (V_E, E)$. Then*

$$\log Z_G^\zeta(p, q) = k(\zeta_E^1, E) \log q + \sum_{e \in E} g_{G,p,q}^\zeta(e), \qquad \zeta \in \Omega,$$

where

$$g_{G,p,q}^\zeta(e) = \int_p^1 \left[\frac{r - \phi_{G,r,q}^\zeta(J_e)}{r(1-r)} \right] dr. \tag{7.91}$$

Proof. As in the proofs of Theorems 3.73 and 4.58,

$$\frac{d}{dr} \log Z_G^\zeta(r, q) = \sum_{e \in E} \frac{\phi_{G,r,q}^\zeta(J_e) - r}{r(1-r)}.$$

This we integrate from p to 1, noting that $\log Z_G^\zeta(1,q) = k(\zeta_E^1, E)\log q$. □

Let $q \in [1,\infty)$. By Theorem 3.1,

$$\frac{r}{r+q(1-r)} \le \phi_{G,r,q}^\zeta(J_e) \le r.$$

By substitution into (7.91),

$$0 \le g_{G,p,q}^\zeta(e) \le \int_p^1 (q-1)\,dr = (1-p)(q-1), \qquad e \in E, \tag{7.92}$$

uniformly in E and ζ. The above inequalities are reversed if $q < 1$.

Let $\Lambda_n = \Lambda_{n,n}$ and write $\Lambda_n(e) = e + \Lambda_n$ for the set of translates of the endvertices of the edge e by vectors in Λ_n.

(7.93) Lemma. *Let $q \in [1,\infty)$. There exists $p^* = p^*(q) < 1$ and a constant $\alpha > 0$ such that the following holds. Let E_1 and E_2 be finite edge-sets of $\mathbb{L}^3$ such that $e \in E_1 \cap E_2$, and let $n \ge 1$ be such that $E_1 \cap \Lambda_n(e) = E_2 \cap \Lambda_n(e)$. If $p > p^*$,*

$$\left|g_{G_1,p,q}^1(e) - g_{G_2,p,q}^1(e)\right| \le e^{-\alpha n},$$

where $G_i = (V_{E_i}, E_i)$.

Proof. Let K_e be the event that the endvertices of the edge e are joined by an open path of $\mathbb{E}^d \setminus \{e\}$. By (3.3),

$$\frac{r - \phi_{G,r,q}^1(J_e)}{r(1-r)} = \frac{(q-1)(1-\phi_{G,r,q}^1(K_e))}{r+q(1-r)},$$

whence

$$\begin{aligned}\left|g_{G_1,p,q}^1(e) - g_{G_2,p,q}^1(e)\right| \\ \le \int_p^1 \frac{(q-1)}{r+q(1-r)}\left|\phi_{G_1,r,q}^1(K_e) - \phi_{G_2,r,q}^1(K_e)\right| dr.\end{aligned} \tag{7.94}$$

Let $n \ge 1$. We pursue the method of proof of Theorem 5.33(b), and shall use the notation therein. Let V be the set of vertices that are incident in $\mathbb{L}^3$ to edges of both $\Lambda_n(e)$ and its complement. We define B to be the union of V together with all vertices $x_0 \in \mathbb{Z}^3$ for which there exists a path $x_0, x_1, \dots, x_m$ of $\mathcal{L}$ such that $x_0, x_1, \dots, x_{m-1} \notin V$, $x_m \in V$, and $x_0, x_1, \dots, x_{m-1}$ are black. Let W_n be the event that there exists no $x \in B$ such that $\|x - z\| \le 10$, say, where z is the centre of e. By (5.36)–(5.37) together with estimates at the beginning of the proof of [211, Lemma (2.24)],

$$\phi_{\Lambda_n(e),r,q}^0(W_n) \ge 1 - c^n(1-\rho)^{en}, \tag{7.95}$$

where c and e are absolute positive constants, and $\rho = r/[r + q(1-r)]$. Since W_n is an increasing event,

$$\phi^1_{G_1,r,q}(W_n) \ge 1 - c^n(1-\rho)^{en}. \tag{7.96}$$

Let $H = E_1 \cap \Lambda_n(e)$. As in the proof of Theorem 5.33, and by coupling,

$$0 \le \phi^1_{H,r,q}(K_e) - \phi^1_{G_1,r,q}(K_e) \le 1 - \phi^1_{G_1,r,q}(W_n).$$

The claim follows by (7.94), (7.96), and the triangle inequality. □

As explained in Sections 7.1–7.2, the dual of the random-cluster model on $\mathbb{L}^3$ is a certain probability measure associated with the plaquettes of the dual lattice $\mathbb{L}^3_{\mathrm{d}}$. The straight line-segment joining the vertices of an edge $e = \langle x, y\rangle$ passes through the middle of exactly one plaquette, denoted by $h(e)$, which we call the *dual* plaquette of e. We declare this plaquette *open* (respectively, *closed*) if e is closed (respectively, open), see (7.9). The plaquette $h(e)$ is called *horizontal* if $y = x + (0, 0, \pm 1)$, and *vertical* otherwise.

The *regular interface* of $\mathbb{L}^3$ is the set δ_0 of plaquettes given by

$$\delta_0 = \big\{h \in \mathbb{H} : h = h(\langle x, y\rangle) \text{ for some } x = (x_1, x_2, 0) \text{ and } y = (x_1, x_2, 1)\big\}.$$

The *interface* $\Delta(\omega)$ of a configuration $\omega \in \mathcal{I}_{L,M} \cap \Omega^{\mathrm{D}}_{L,M}$ is defined to be the maximal 1-connected set of open plaquettes containing the plaquettes in the set $\delta_0 \setminus \{h(e) : e \in E_{L,M}\}$. The set of all interfaces is

$$\mathcal{D}_{L,M} = \big\{\Delta(\omega) : \omega \in \mathcal{I}_{L,M} \cap \Omega^{\mathrm{D}}_{L,M}\big\}. \tag{7.97}$$

It is tempting to think of an interface as part of a deformed plane. Interfaces may however have more complex geometry involving cavities and attachments, see Figure 7.4. The following proposition confirms that the interfaces in $\mathcal{D}_{L,M}$ separate the top of $\Lambda_{L,M}$ from its bottom.

(7.98) Lemma. *The event $\mathcal{I}_{L,M} \cap \Omega^{\mathrm{D}}_{L,M}$ comprises those configurations $\omega \in \Omega^{\mathrm{D}}_{L,M}$ for which there exists $\delta \in \mathcal{D}_{L,M}$ satisfying: $\omega(e) = 0$ whenever $h(e) \in \delta$.*

For $\delta \in \mathcal{D}_{L,M}$, we define its *extended interface* (or *closure*) $\overline{\delta}$ to be the set

$$\overline{\delta} = \delta \cup \big\{h \in \mathbb{H} : h \text{ is 1-connected to some member of } \delta\big\}. \tag{7.99}$$

See (7.4). It will be useful to introduce the 'maximal' (denoted by $\overline{\omega}_\delta$) and 'minimal' (denoted by $\underline{\omega}_\delta$) configurations in $\Omega^{\mathrm{D}}_{L,M}$ that are compatible with δ:

$$\overline{\omega}_\delta(e) = \begin{cases} 0 & \text{if } e \in \delta, \\ 1 & \text{otherwise}, \end{cases} \qquad \underline{\omega}_\delta(e) = \begin{cases} \mathrm{D}(e) & \text{if } e \notin E_{L,M}, \\ 1 & \text{if } e \in E_{L,M} \cap (\overline{\delta} \setminus \delta), \\ 0 & \text{otherwise.} \end{cases} \tag{7.100}$$

Proof of Lemma 7.98. If $\omega \in \mathcal{I}_{L,M} \cap \Omega^{\mathrm{D}}_{L,M}$, then $\omega(e) = 0$ whenever $h(e) \in \Delta(\omega)$. Suppose conversely that $\delta \in \mathcal{D}_{L,M}$, and let $\omega \in \Omega^{\mathrm{D}}_{L,M}$ satisfy $\omega(e) = 0$ whenever $h(e) \in \delta$. Since $\omega \le \overline{\omega}_\delta$, it suffices to show that $\overline{\omega}_\delta \in \mathcal{I}_{L,M}$. Since $\delta \in \mathcal{D}_{L,M}$, there exists $\xi \in \mathcal{I}_{L,M} \cap \Omega^{\mathrm{D}}_{L,M}$ such that $\delta = \Delta(\xi)$. Note that $\xi \le \overline{\omega}_\delta$. Suppose for the sake of obtaining a contradiction that $\overline{\omega}_\delta \notin \mathcal{I}_{L,M}$, and think of $\overline{\omega}_\delta$ as being obtained from ξ by declaring, in turn, a certain sequence $e_1, e_2, \dots, e_r$ with $\xi(e_i) = 0, i = 1, 2, \dots, r$, to be open. Let ξ^k be obtained from ξ by $\eta(\xi^k) = \eta(\xi) \cup \{e_1, e_2, \dots, e_k\}$. By assumption, there exists K such that $\xi^K \in \mathcal{I}_{L,M}$ but $\xi^{K+1} \notin \mathcal{I}_{L,M}$. For $\psi \in \Omega^{\mathrm{D}}_{L,M}$, let $J(\psi)$ denote the set of edges e having endvertices in $\Lambda_{L,M}$, with $\psi(e) = 1$, and both of whose endvertices are attainable from $\partial^+\Lambda_{L,M}$ by open paths of ψ. We apply Theorem 7.3 to the finite connected graph induced by $J(\xi^K)$ to find that there exists a splitting set Q of plaquettes such that: $\partial^+\Lambda_{L,M} \subseteq \mathrm{ins}([Q])$, $\partial^-\Lambda_{L,M} \subseteq \mathrm{out}([Q])$, and $\xi^K(e) = 0$ whenever $e \in E_{L,M}$ and $h(e) \in Q$. It must be the case that $h(e_{K+1}) \in Q$, since $\xi^{K+1} \notin \mathcal{I}_{L,M}$. By the 1-connectedness of Q, there exists a sequence $f_1 = e_{K+1}, f_2, f_3, \dots, f_t$ of edges such that:

(i) $h(f_i) \in Q$ for all i,

(ii) $f_i \in E_{L,M}$ for $i = 1, 2, \dots, t-1$, $f_t = h(\langle x, x - (0,0,1)\rangle)$ for some $x = (x_1, x_2, 1) \in \partial^+\Lambda_{L,M}$, and

(iii) $h(f_i) \stackrel{1}{\sim} h(f_{i+1})$ for $i = 1, 2, \dots, t-1$.

It follows that $h(f_i) \in \delta$ for $i = 1, 2, \dots, t$. In particular, $h(e_{K+1}) \in \delta$ and so $\overline{\omega}_\delta(e_{K+1}) = 0$, a contradiction. Therefore $\overline{\omega}_\delta \in \mathcal{I}_{L,M}$ as claimed. □

7.8 The law of the interface

For conciseness of notation, we abbreviate $\phi^{\mathrm{D}}_{\Lambda_{L,M},p,q}$ to $\phi_{L,M}$, and $\overline{\phi}^{\mathrm{D}}_{\Lambda_{L,M},p,q}$ to $\overline{\phi}_{L,M}$. Let $\delta \in \mathcal{D}_{L,M}$. The better to study $\phi_{L,M}(\delta) = \phi_{L,M}(\Delta = \delta)$, we develop next an expression for this probability. Consider the connected components of the graph $(\mathbb{Z}^3, \eta(\overline{\omega}_\delta))$, and denote these components by (S^i_δ, U^i_δ), $i = 1, 2, \dots, k_\delta$, where $k_\delta = k(\overline{\omega}_\delta)$. Note that U^i_δ is empty whenever S^i_δ is a singleton. Let $W(\delta)$ be the edge-set $E_{L,M} \setminus \{e \in \mathbb{E}^3 : h(e) \in \overline{\delta}\}$.

Let $\omega \in \mathcal{I}_{L,M} \cap \Omega^{\mathrm{D}}_{L,M}$ be such that $\Delta(\omega) = \delta$, so that

$$\omega(e) = \begin{cases} 0 & \text{if } h(e) \in \delta, \\ 1 & \text{if } h(e) \in \overline{\delta} \setminus \delta. \end{cases} \tag{7.101}$$

Let D be the set of edges with both endvertices in $\Lambda_{L+2,M+2}$ that either are dual to plaquettes in δ or join a vertex of $\Lambda_{L+1,M+1}$ to a vertex of $\partial\Lambda_{L+2,M+2}$. We apply Theorem 7.5 to the set D, and deduce that there are exactly k_δ components of the graph $(\mathbb{Z}^3, \eta(\omega))$ having a vertex in $V(\overline{\delta})$.

We have that

$$(7.102)\quad \phi_{L,M}(\delta) = \frac{1}{Z(E_{L,M})} p^{|\overline{\delta}\setminus\delta|}(1-p)^{|\delta|} \times \sum_{\substack{\omega\in\Omega^{\mathrm{D}}_{L,M}:\\ \Delta(\omega)=\delta}} \left\{ \prod_{e\in W(\delta)} p^{\omega(e)}(1-p)^{1-\omega(e)} \right\} q^{k(\omega)}$$

$$= \frac{Z^1(\delta)}{Z(E_{L,M})} p^{|\overline{\delta}\setminus\delta|}(1-p)^{|\delta|} q^{k_\delta - 1},$$

where $Z(E_{L,M}) = Z^{\mathrm{D}}_{\Lambda_{L,M}}(p,q)$ and $Z^1(\delta) = Z^1_{W(\delta)}(p,q)$. In this expression and later, for $H \subseteq \mathbb{H}$, $|H|$ is the cardinality of the set $H \cap \{h(e) : e \in E_{L,M}\}$. The term $q^{k_\delta - 1}$ arises since the application of '1' boundary conditions to δ has the effect of uniting the boundaries of the cavities of δ, whereby the number of clusters diminishes by $k_\delta - 1$.

For $x \in \mathbb{Z}^3$, we denote by $\tau_x : \mathbb{Z}^3 \to \mathbb{Z}^3$ the translate given by $\tau_x(y) = x + y$. The translate τ_x acts on edges and subgraphs of $\mathbb{L}^3$ in the natural way, see Section 4.3. For sets A, B of edges or vertices of $\mathbb{L}^3$, we write $A \simeq B$ if $B = \tau_x A$ for some $x \in \mathbb{Z}^3$. Note that two edges e, f satisfy $\{e\} \simeq \{f\}$ if and only if they are parallel, in which case we write $e \simeq f$.

We shall exploit properties of the partition functions $Z(\cdot)$ in order to rewrite (7.102). For $i = 1, 2$, let $L_i, M_i > 0$, $\delta_i \in \mathcal{D}_{L_i,M_i}$, and $e_i \in E(\delta_i) \cap E_{L_i,M_i}$, and

$$(7.103)\quad G(e_1, \delta_1, E_{L_1,M_1}; e_2, \delta_2, E_{L_2,M_2}) = \sup\left\{ L : \begin{array}{l} \Lambda_L(e_1) \cap E_{L_1,M_1} \simeq \Lambda_L(e_2) \cap E_{L_2,M_2} \\ \text{and } \Lambda_L(e_1) \cap E(\delta_1) \simeq \Lambda_L(e_2) \cap E(\delta_2) \end{array} \right\},$$

where $\Lambda_L(e) = e + \Lambda_L$ as before. Let $Z^1(E_{L,M}) = Z^1_{\Lambda_{L,M}}(p,q)$.

(7.104) Lemma. *Let $L, M \geq 1$ and $\delta \in \mathcal{D}_{L,M}$. We may write $\phi_{L,M}(\delta)$ as*

$$(7.105)\quad \phi_{L,M}(\delta) = \frac{Z^1(E_{L,M})}{Z(E_{L,M})} p^{|\overline{\delta}\setminus\delta|}(1-p)^{|\delta|} q^{k_\delta - 1} \exp\left(\sum_{e\in E(\delta)\cap E_{L,M}} f_p(e,\delta,L,M) \right),$$

for functions $f_p(e,\delta,L,M)$ with the following properties. For $q \in [1,\infty)$, there exist $p^* < 1$ and constants C_1, C_2, $\gamma > 0$ such that, if $p > p^*$,

$$(7.106)\quad |f_p(e,\delta,L,M)| < C_1,$$

$$(7.107)\quad \big| f_p(e_1,\delta_1,L_1,M_1) - f_p(e_2,\delta_2,L_2,M_2) \big| \leq C_2 e^{-\gamma G}, \quad e_1 \in \delta_1,\ e_2 \in \delta_2,\ e_1 \simeq e_2,$$

where $G = G(e_1,\delta_1,E_{L_1,M_1}; e_2,\delta_2,E_{L_2,M_2})$. Inequalities (7.106) *and* (7.107) *are valid for all relevant values of their arguments.*

Proof. By Lemma 7.90,

(7.108)
$$\log\left(\frac{Z^1(\delta)}{Z^1(E_{L,M})}\right) = \sum_{f\in W(\delta)} \big[g(f, W(\delta)) - g(f, E_{L,M})\big] - \sum_{f\in E(\overline{\delta})} g(f, E_{L,M}),$$

where $g(f, D) = g^1_{D,p,q}(f)$. The summations may be expressed as sums over edges $e \in E(\delta)$ in the following way. The set $\mathbb{E}^3$ may be ordered according to the lexicographic ordering of the centres of edges. Let $f \in E_{L,M}$ and $\delta \in \mathcal{D}_{L,M}$. Amongst all edges in $E(\delta) \cap E_{L,M}$ that are closest to f (in the sense that their centres are closest in the L^∞ norm), let $\nu(f, \delta)$ be the earliest edge in this ordering. By (7.108),

$$\log\left(\frac{Z^1(\delta)}{Z^1(E_{L,M})}\right) = \sum_{e\in E(\delta)\cap E_{L,M}} f_p(e, \delta, L, M) \tag{7.109}$$

where

(7.110)
$$f_p(e, \delta, L, M) = \sum_{\substack{f\in W(\delta):\\ \nu(f,\delta)=e}} \big[g(f, W(\delta)) - g(f, E_{L,M})\big] - \sum_{\substack{f\in E(\overline{\delta}):\\ \nu(f,\delta)=e}} g(f, E_{L,M}).$$

This implies (7.105) via (7.102).

It remains to show (7.106)–(7.107). Let $e = \nu(f, \delta)$ and set $r = \|e, f\|$. Then $\Lambda_{r-2}(f)$ does not intersect $\overline{\delta}$, implying by Lemma 7.93 that

$$\big|g(f, W(\delta)) - g(f, E_{L,M})\big| \le e^{-\alpha\|e,f\|+2\alpha}, \qquad p > p^*, \tag{7.111}$$

where p^* and α are given as in that lemma. Secondly, there exists an absolute constant K such that, for all e and δ, the number of edges $f \in E(\overline{\delta})$ with $e = \nu(f, \delta)$ is no greater than K. Therefore, by (7.92),

$$|f_p(e, \delta, L, M)| \le \sum_{f\in\mathbb{E}^3} e^{-\alpha\|e,f\|+2\alpha} + K(1-p)(q-1)$$

as required for (7.106).

Finally, we show (7.107) for $p > p^*$ and appropriate C_2, γ. Let $e \in \delta_1$, $e_2 \in \delta_2$, and let G be given by (7.103); we may suppose that $G > 9$. By assumption, $e_1 \simeq e_2$, whence there exists a translate τ of $\mathbb{L}^3$ such that $\tau e_1 = e_2$. For $f \in W(\delta_1) \cap \Lambda_{G/3}(e_1)$,

$$\tau[\Lambda_{G/3}(f) \cap E_{L_1,M_1}] = \Lambda_{G/3}(\tau f) \cap E_{L_2,M_2}, \tag{7.112}$$
$$\tau[\Lambda_{G/3}(f) \cap \delta_1] = \Lambda_{G/3}(\tau f) \cap \delta_2, \tag{7.113}$$

and

$$\text{for } \|f, e_1\| \le \tfrac{1}{3}G, \quad \nu(f, \delta_1) = e_1 \text{ if and only if } \nu(\tau f, \delta_2) = e_2. \tag{7.114}$$

By the definition (7.110) of the functions f_p,

$$\begin{aligned}
&\left|f_p(e_1, \delta_1, L_1, M_1) - f_p(e_2, \delta_2, L_2, M_2)\right| \\
&\le \sum_{\substack{f \in W(\delta_1) \cap \Lambda_{G/3}(e_1): \\ \nu(f,\delta_1)=e_1}} \Big\{\left|g(f, W(\delta_1)) - g(\tau f, W(\delta_2))\right| \\
&\qquad\qquad + \left|g(f, E_{L_1,M_1}) - g(\tau f, E_{L_2,M_2})\right|\Big\} \\
&\qquad + \sum_{\substack{f \subset W(\delta_1) \setminus \Lambda_{G/3}(e_1): \\ \nu(f,\delta_1)=e_1}} \left|g(f, W(\delta_1)) - g(f, E_{L_1,M_1})\right| \\
&\qquad + \sum_{\substack{f \subset W(\delta_2) \setminus \Lambda_{G/3}(e_2): \\ \nu(f,\delta_2)=e_2}} \left|g(f, W(\delta_2)) - g(f, E_{L_2,M_2})\right| + S,
\end{aligned} \tag{7.115}$$

where

$$S = \left| \sum_{\substack{f \in E(\overline{\delta}_1): \\ \nu(f,\delta_1)=e_1}} g(f, E_{L_1,M_1}) - \sum_{\substack{f \in E(\overline{\delta}_2): \\ \nu(f,\delta_2)=e_2}} g(f, E_{L_2,M_2}) \right|.$$

By (7.112)–(7.113) and Lemma 7.93, the first summation in (7.115) is bounded above by $2G^3 e^{-\frac{1}{3}\alpha G}$. By the definition of the $\nu(f, \delta_i)$, the second and third summations are bounded above, respectively, by

$$\sum_{f \notin \Lambda_{G/3}(e_i)} e^{-\alpha\|f, e_i\| + 2\alpha} \le C' e^{-\frac{1}{3}\alpha G + 2\alpha},$$

for some $C' < \infty$, as in (7.111). By (7.114),

$$S = \left| \sum_{\substack{f \in E(\overline{\delta}_1): \\ \nu(f,\delta_1)=e_1}} g(f, E_{L_1,M_1}) - g(\tau f, E_{L_2,M_2}) \right| \le K e^{-\frac{1}{3}\alpha G},$$

and (7.107) follows for an appropriate choice of γ. □

In the second part of this section, we consider measures and interfaces for the infinite cylinder $\Sigma_L = \Lambda_{L,\infty} = [-L, L]^2 \times \mathbb{Z}$. Note first by stochastic ordering that, if $q \in [1, \infty)$, then $\phi_{L,M+1} \le_{\text{st}} \phi_{L,M}$, whence the (decreasing) weak limit

$$\phi_L = \lim_{M \to \infty} \phi_{L,M} \tag{7.116}$$

exists. Let Ω_L^{D} be the set of all configurations ω such that $\omega(e) = \mathrm{D}(e)$ for $e \notin E_L = \lim_{M\to\infty} E_{L,M}$, and let $\mathcal{I}_L$ be the event that no vertex of $\partial\Sigma_L^+$ is joined by an open path to a vertex of $\partial\Sigma_L^-$. The set of interfaces on which we concentrate is $\mathcal{D}_L = \bigcup_M \mathcal{D}_{L,M} = \lim_{M\to\infty} \mathcal{D}_{L,M}$. Thus, $\mathcal{D}_L$ is the set of interfaces that span Σ_L, and every member of $\mathcal{D}_L$ is bounded in the direction of the third coordinate. It is easy to see that $\mathcal{I}_L \supseteq \lim_{M\to\infty} \mathcal{I}_{L,M}$, and it is a consequence of the next lemma that the difference between these two events has ϕ_L-probability zero.

(7.117) Lemma. *Let $q \in [1,\infty)$. The weak limit $\phi_{L,M}(\cdot \mid \mathcal{I}_{L,M}) \Rightarrow \phi_L(\cdot \mid \mathcal{I}_L)$ holds as $M \to \infty$, and*

$$\phi_L\Big(\mathcal{I}_L \setminus \lim_{M\to\infty} \mathcal{I}_{L,M}\Big) = 0.$$

For $L_i > 0$, $\delta_i \in \mathcal{D}_{L_i}$, and $e_i \in E(\delta_i) \cap E_{L_i}$, let

$$G(e_1,\delta_1,E_{L_1};e_2,\delta_2,E_{L_2}) = G(e_1,\delta_1,E_{L_1,\infty};e_2,\delta_2,E_{L_2,\infty}).$$

On the event $\mathcal{I}_L$, Δ is defined as before to be the maximal 1-connected set of open plaquettes that intersects $\delta_0 \setminus E_L$.

(7.118) Lemma.

(a) *Suppose $L > 0$, $\delta \in \mathcal{D}_L$, and $e \in E(\delta) \cap E_L$. The functions f_p given in* (7.110) *are such that the limit*

$$f_p(e,\delta,L) = \lim_{M\to\infty} f_p(e,\delta,L,M) \tag{7.119}$$

exists. Furthermore, if $p > p^*$,

$$|f_p(e,\delta,L)| < C_1, \tag{7.120}$$

and, for $L_i > 0$, $\delta_i \in \mathcal{D}_{L_i}$, and $e_i \in E(\delta_i) \cap E_{L_i}$ satisfying $e_1 \simeq e_2$,

$$\big|f_p(e_1,\delta_1,L_1) - f_p(e_2,\delta_2,L_2)\big| \le C_2 e^{-\gamma G},$$

where $G = G(e_1,\delta_1,E_{L_1};e_2,\delta_2,E_{L_2})$ and p^*, C_1, C_2, γ are as in Lemma 7.104.

(b) *For $q \in [1,\infty)$ and $\delta \in \mathcal{D}_L$, the probability $\phi_L(\delta \mid \mathcal{I}_L) = \phi_L(\Delta = \delta \mid \mathcal{I}_L)$ satisfies*

$$\phi_L(\delta \mid \mathcal{I}_L) = \frac{1}{Z_L} p^{|\overline{\delta}\setminus\delta|}(1-p)^{|\delta|} q^{k_\delta} \exp\Bigg(\sum_{e\in E(\delta)\cap E_L} f_p(e,\delta,L)\Bigg), \tag{7.121}$$

where Z_L is the appropriate normalizing constant.

Proof of Lemma 7.117. It suffices for the claim of weak convergence that

$$\phi_{L,M}(F \cap \mathcal{I}_{L,M}) \to \phi_L(F \cap \mathcal{I}_L) \qquad \text{for all cylinder events } F. \tag{7.122}$$

Let $A_{L,M} = [-L, L]^2 \times \{-M\}$ and $B_{L,M} = [-L, L]^2 \times \{M\}$, and let $T_{L,M}$ be the event that no open path exists between a vertex of $\partial\Lambda^+_{L,M} \setminus B_{L,M}$ and a vertex of $\partial\Lambda^-_{L,M} \setminus A_{L,M}$. Note that $T_{L,M} \to \mathcal{I}_L$ as $M \to \infty$. Let F be a cylinder event. Then

$$
\begin{aligned}
\text{(7.123)} \qquad \phi_{L,M}(F \cap \mathcal{I}_{L,M}) &\le \phi_{L,M}(F \cap T_{L,M'}) && \text{for } M' \le M \\
&\to \phi_L(F \cap T_{L,M'}) && \text{as } M \to \infty \\
&\to \phi_L(F \cap \mathcal{I}_L) && \text{as } M' \to \infty.
\end{aligned}
$$

In order to obtain a corresponding lower bound, we introduce the event K_r that all edges of E_L, both of whose endvertices have third coordinate equal to $\pm r$, are open. We may suppose without loss of generality that $p > 0$. By the comparison inequality (Theorem 3.21), $\phi_{L,M}$ dominates product measure with density $\pi = p/[p + q(1-p)]$, whence there exists $\beta = \beta_L < 1$ such that

$$\phi_{L,M}\left(\bigcup_{r=1}^{R} K_r\right) \ge 1 - \beta^R, \qquad R < M.$$

Now $\mathcal{I}_{L,M} \subseteq T_{L,M}$, and $T_{L,M} \setminus \mathcal{I}_{L,M} \subseteq \bigcap_{r=1}^{M-1} \overline{K_r}$, whence

$$
\begin{aligned}
\text{(7.124)} \qquad \phi_{L,M}(F \cap \mathcal{I}_{L,M}) &\ge \phi_{L,M}(F \cap T_{L,M}) - \beta^{M-1} \\
&\ge \phi_{L,M}(F \cap \mathcal{I}_L) - \beta^{M-1} \\
&\to \phi_L(F \cap \mathcal{I}_L) \qquad \text{as } M \to \infty.
\end{aligned}
$$

Equation (7.122) holds by (7.123)–(7.124). The second claim of the lemma follows by taking $F = \Omega$, the entire sample space. □

Proof of Lemma 7.118. (a) The existence of the limit follows by the monotonicity of $g(f, D_i)$ for an increasing sequence $\{D_i\}$, and the proof of (7.106). The inequalities are implied by (7.106)–(7.107).

(b) Let $\delta \in \mathcal{D}_L$, so that $\delta \in \mathcal{I}_{L,M}$ for all large M. By Lemma 7.117,

$$\phi_L(\delta \mid \mathcal{I}_L) = \lim_{M\to\infty} \phi_{L,M}(\delta \mid \mathcal{I}_{L,M}).$$

Let $M \to \infty$ in (7.105), and use part (a) to obtain the claim. □

7.9 Geometry of interfaces

A taxonomy of interfaces is required, and this is the topic of this section. Let $\delta \in \mathcal{D}_L$. While it was natural in Section 7.7 to introduce the extended interface $\overline{\delta}$, it turns out to be useful when studying the geometry of δ to work with its *semi-extended interface*

$$\delta^* = \delta \cup \{h \in \mathbb{H} : h \text{ is a horizontal plaquette that is 1-connected to } \delta\}.$$

Let $x = (x_1, x_2, x_3) \in \mathbb{Z}^3$. The *projection* $\pi(h)$ of a horizontal plaquette $h = h(\langle x, x + (0, 0, 1)\rangle)$ onto the regular interface δ_0 is defined to be the plaquette

$$\pi(h) = h\big(\langle (x_1, x_2, 0), (x_1, x_2, 1)\rangle\big).$$

The projection of the vertical plaquette $h = h(\langle x, x + (1, 0, 0)\rangle)$ is the interval

$$\pi(h) = \left[(x_1 + \tfrac{1}{2}, x_2 - \tfrac{1}{2}, \tfrac{1}{2}), (x_1 + \tfrac{1}{2}, x_2 + \tfrac{1}{2}, \tfrac{1}{2})\right],$$

and, similarly, $h = h(\langle x, x + (0, 1, 0)\rangle)$ has projection

$$\pi(h) = \left[(x_1 - \tfrac{1}{2}, x_2 + \tfrac{1}{2}, \tfrac{1}{2}), (x_1 + \tfrac{1}{2}, x_2 + \tfrac{1}{2}, \tfrac{1}{2})\right].$$

A horizontal plaquette h of the semi-extended interface δ^* is called a *c-plaquette* if h is the unique member of δ^* with projection $\pi(h)$. All other plaquettes of δ^* are called *w-plaquettes*. A *ceiling* of δ is a maximal 0-connected set of c-plaquettes. The *projection* of a ceiling C is the set $\pi(C) = \{\pi(h) : h \in C\}$. Similarly, we define a *wall* W of δ as a maximal 0-connected set of w-plaquettes, and its *projection* as

$$\pi(W) = \{\pi(h) : h \text{ is a horizontal plaquette of } W\}.$$

(7.125) Lemma. *Let $\delta \in \mathcal{D}_L$.*

(i) *The set $\delta^* \setminus \delta$ contains no c-plaquette.*

(ii) *All plaquettes of δ^* that are 1-connected to some c-plaquette are horizontal plaquettes of δ. All horizontal plaquettes that are 0-connected to some c-plaquette belong to δ^*.*

(iii) *Let C be a ceiling. There is a unique plane parallel to the regular interface that contains all the c-plaquettes of C.*

(iv) *Let C be a ceiling. Then $C = \{h \in \delta^* : \pi(h) \subseteq [\pi(C)]\}$.*

(v) *Let W be a wall. Then $W = \{h \in \delta^* : \pi(h) \subseteq [\pi(W)]\}$.*

(vi) *For each wall W, $\delta_0 \setminus \pi(W)$ has exactly one maximal infinite 0-connected component (respectively, 1-connected component).*

(vii) *Let W be a wall, and suppose that $\delta_0 \setminus \pi(W)$ comprises n maximal 0-connected sets $H_1, H_2, \dots, H_n$. The set of all plaquettes $h \in \delta^* \setminus W$*

that are 0*-connected to* W *comprises only c-plaquettes, which belong to the union of exactly* n *distinct ceilings* $C_1, C_2, \ldots, C_n$ *such that*

$$\{\pi(h) : h \text{ is a c-plaquette of } C_i\} \subseteq H_i.$$

(viii) *The projections* $\pi(W_1)$ *and* $\pi(W_2)$ *of two different walls* W_1 *and* W_2 *of* δ^* *are not* 0*-connected.*

(ix) *The projection* $\pi(W)$ *of any wall* W *contains at least one plaquette of* δ_0.

The displacement of the plane in (iii) from the regular interface, counted positive or negative, is called the *height* of the ceiling C.

Proof. (i) Let h be a c-plaquette of δ^* with $\pi(h) = h_0$. Since $\delta \in \mathcal{D}_L$, δ contains at least one plaquette with projection h_0. Yet, according to the definition of a c-plaquette, there is no such a plaquette besides h. Therefore $h \in \delta$.

(ii) Suppose h is a c-plaquette. Then h belongs to δ, and any horizontal plaquette that is 1-connected to h belongs to δ^*. It may be seen in addition that any vertical plaquette that is 1-connected to h lies in $\overline{\delta} \setminus \delta$. Suppose, on the contrary, that some such vertical plaquette h' lies in δ. Then the horizontal plaquettes that are 1-connected to h' lie in δ^*. One of these latter plaquettes has projection $\pi(h)$, in contradiction of the assumption that h is a c-plaquette.

We may now see as follows that any horizontal plaquette h'' that is 1-connected to h must lie in δ. Suppose, on the contrary, that some such plaquette h'' lies in $\overline{\delta} \setminus \delta$. We may construct a path of open edges on $(\mathbb{Z}^3, \eta(\underline{\omega}_\delta))$ connecting the vertex x just above h to the vertex $x - (0, 0, 1)$ just below h, using the open edges of $\underline{\omega}_\delta$ corresponding to the three relevant plaquettes of $\overline{\delta} \setminus \delta$. This contradicts the assumption that h is a c-plaquette of the interface δ.

The second claim of (ii) follows immediately, by the definition of δ^*.

(iii) The first part follows by the definition of ceiling, since the only horizontal plaquettes that are 0-connected with a given c-plaquette h lie in the plane containing h.

(iv) Assume that $h \in \delta^*$ and $\pi(h) \subseteq [\pi(C)]$. If h is horizontal, the conclusion holds by the definition of c-plaquette. If h is vertical, then $h \in \delta$, and all 1-connected horizontal plaquettes lie in δ^*. At least two such horizontal plaquettes project onto the same plaquette in $\pi(C)$, in contradiction of the assumption that C is a ceiling.

(v) Let C be a ceiling and let $\gamma_1, \gamma_2, \ldots, \gamma_n$ be the maximal 0-connected sets of plaquettes of $\delta_0 \setminus \pi(C)$. Let $\delta_i^* = \{h \in \delta^* : \pi(h) \subseteq [\gamma_i]\}$, and let

$$\beta_i^* = \{h \in \delta_i^* : h \text{ horizontal}, h \overset{0}{\sim} h' \text{ for some } h' \in C\}.$$

We note that[13] β_i^* is a 0-connected subset of δ_i^*.

[13] This is a consequence of [311, eqn (5.3)], see also [286, p. 40, footnote 2].

By part (iv), $\delta^* = C \cup \{\bigcup_{i=1}^n \delta_i^*\}$. We claim that each δ_i^* is 0-connected, and we prove this as follows. Let $h_1, h_2 \in \delta_i^*$. Since δ^* is 0-connected, it contains a sequence $h_1 = f_0, f_1, \dots, f_m = h_2$ of plaquettes such that $f_{i-1} \stackrel{0}{\sim} f_i$ for $i = 1, 2, \dots, m$. We need to show that such a sequence exists containing no plaquettes in C. Suppose on the contrary that the sequence (f_i) has a non-empty intersection with C. Let $k = \min\{i : f_i \in C\}$ and $l = \max\{i : f_i \in C\}$, and note that $0 < k \le l < n$.

If f_{k-1} and f_{l+1} are horizontal, then $f_{k-1}, f_{l+1} \in \beta_i^*$, whence they are 0-connected by a path of horizontal plaquettes of β_i^*, and the claim follows. A similar argument is valid if either or both of f_{k-1} and f_{l+1} is vertical. For example, if f_{k-1} is vertical, by (ii) it cannot be 1-connected to a plaquette of C. Hence, it is 1-connected to some horizontal plaquette in $\delta^* \setminus C$ that is itself 1-connected to a plaquette of C. The same conclusion is valid for f_{l+1} if vertical. In any such case, as above there exists a 0-connected sequence of w-plaquettes connecting f_{k-1} with f_{l+1}, and the claim follows.

To prove (v), we note by the above that the wall W is a subset of one of the sets δ_i^*, say δ_1^*. Next, we let C_1 be a ceiling contained in δ_1^*, if this exists, and we repeat the above procedure. Consider the 0-connected components of $\gamma_1 \setminus \pi(C_1)$, and use the fact that δ_1^* is 0-connected, to deduce that the set of plaquettes that project onto one of these components is itself 0-connected. This procedure is repeated until all ceilings have been removed, the result being a 0-connected set of w-plaquettes of which, by definition of a wall, all members belong to W.

Claim (vi) is a simple observation since walls are finite. Claim (vii) is immediate from claim (ii) and the definitions of wall and ceiling. Claim (viii) follows from (v) and (vii), and (ix) is a consequence of the definition of the semi-extended interface δ^*. □

The properties described in Lemma 7.125 allow us to describe a wall W in more detail. By (vi) and (vii), there exists a unique ceiling that is 0-connected to W and with projection in the infinite 0-connected component of $\delta_0 \setminus \pi(W)$. We call this ceiling the *base* of W. The *altitude* of W is the height of the base of W, see (iii). The *height* $D(W)$ of W is the maximum absolute value of the displacement in the third coordinate direction of $[W]$ from the horizontal plane $\{(x_1, x_2, s + \frac{1}{2}) : x_1, x_2 \in \mathbb{Z}\}$, where s is the altitude of W. The *interior* $\mathrm{int}(W)$ (of the projection $\pi(W)$) of W is the complement in δ_0 of the unique maximal infinite 0-connected component of $\delta_0 \setminus \pi(W)$, see (vi).

Let $S = (A, B)$ where A, B are sets of plaquettes. We call S a *standard wall* if there exists $\delta \in \mathcal{D}_L$ such that $A \subseteq \delta$, $B \subseteq \delta^* \setminus \delta$, and $A \cup B$ is the unique wall of δ. If $S = (A, B)$ is a standard wall, we refer to plaquettes of either A or B as plaquettes of S, and we write $\pi(S) = \pi(A \cup B)$.

(7.126) Lemma. *Let $S = (A, B)$ be a standard wall. There exists a unique $\delta \in \mathcal{D}_L$ such that*: *$A \subseteq \delta$, $B \subseteq \delta^* \setminus \delta$, and $A \cup B$ is the unique wall of δ.*

This will be proved soon. Let δ_S denote the unique such $\delta \in \mathcal{D}_L$ corresponding to the standard wall S. We shall see that standard walls are the basic building blocks for a general interface. Notice that the base of a standard wall is a subset of the regular interface. Suppose we are provided with an ordering of the plaquettes of δ_0, and let the *origin* of the standard wall S be the earliest plaquette in $\pi(S)$ that is 1-connected to some plaquette of $\delta_0 \setminus \pi(S)$. Such an origin exists by Lemma 7.125(ix), and the origin belongs to S by (ii). For $h \in \delta_0$, let $\mathcal{S}_h$ be the set of all standard walls with origin h. To $\mathcal{S}_h$ is attached the *empty wall* $\mathcal{E}_h$, interpreted as a wall with origin h but containing no plaquettes.

A family $\{S_i = (A_i, B_i) : i = 1, 2, \dots, m\}$ of standard walls is called *admissible* if:

(i) for $i \neq j$, there exists no pair $h_1 \in \pi(S_i)$ and $h_2 \in \pi(S_j)$ such that $h_1 \overset{0}{\sim} h_2$,

(ii) if, for some i, $h(e) \in S_i$ where $e \notin E_L$, then $h(e) \in A_i$ if and only if $\mathrm{D}(e) = 0$.

The members of any such family have distinct origins. For our future convenience, each S_i is labelled according to its origin $h(i)$, and we write $\{S_h : h \in \delta_0\}$ for the family, where S_h is to be interpreted as $\mathcal{E}_h$ when h is the origin of none of the S_i. We adopt the convention that, when a standard wall is denoted as S_h for some $h \in \delta_0$, then $S_h \in \mathcal{S}_h$.

We introduce next the concept of a group of walls. Let $h \in \delta_0$, $\delta \in \mathcal{D}_L$, and denote by $\rho(h, \delta)$ the number of (vertical or horizontal) plaquettes in δ whose projection is a subset of h. Two standard walls S_1, S_2 are called *close* if there exist $h_1 \in \pi(S_1)$ and $h_2 \in \pi(S_2)$ such that

$$\|h_1, h_2\| < \sqrt{\rho(h_1, \delta_{S_1})} + \sqrt{\rho(h_2, \delta_{S_2})}.$$

A family G of non-empty standard walls is called a *group of (standard) walls* if it is admissible and if, for any pair $S_1, S_2 \in G$, there exists a sequence $T_0 = S_1, T_1, T_2, \dots, T_n = S_2$ of members of G such that T_i and T_{i+1} are close for $i = 0, 1, \dots, n-1$.

The *origin* of a group of walls is defined to be the earliest of the origins of the standard walls therein. Let $\mathcal{G}_h$ denote the set of all possible groups of walls with origin $h \in \delta_0$. As before, we attach to $\mathcal{G}_h$ the *empty group* $\mathcal{E}_h$ with origin h but containing no standard wall. A family $\{G_i : i = 1, 2, \dots, m\}$ of groups of walls is called *admissible* if, for $i \neq j$, there exists no pair $S_1 \in G_i$, $S_2 \in G_j$ such that S_1 and S_2 are close.

We adopt the convention that, when a group of walls is denoted as G_h for some $h \in \delta_0$, then $G_h \in \mathcal{G}_h$. Thus, a family of groups of walls may be written as a collection $\mathbf{G} = \{G_h : h \in \delta_0\}$ where $G_h \in \mathcal{G}_h$.

(7.127) Lemma. *The set $\mathcal{D}_L$ is in one–one correspondence with both the collection of admissible families of standard walls, and with the collection of admissible families of groups of walls.*

Just as important as the existence of these one–one correspondences is their nature, as described in the proof of the lemma. Let δ_G (respectively, $\delta_{\mathbf{G}}$) denote the interface corresponding thus to an admissible family G of standard walls (respectively, an admissible family $\mathbf{G}$ of groups of walls).

Proof of Lemma 7.126. Let $\delta \in \mathcal{D}_L$ have unique wall $S = (A, B)$. By definition, every plaquette of δ^* other than those in $A \cup B$ is a c-plaquette, so that $\Sigma = \delta^* \setminus (A \cup B)$ is a union of ceilings $C_1, C_2, \dots, C_n$. Each C_i contains some plaquette h_i that is 1-connected to some $h_i' \in A$, whence, by Lemma 7.125(iii), the height of C_i is determined uniquely by knowledge of S. Hence δ is unique. □

Proof of Lemma 7.127. Let $\delta \in \mathcal{D}_L$. Let $W_1, W_2, \dots, W_n$ be the non-empty walls of δ^*, and write $W_i = (A_i, B_i)$ where $A_i = W_i \cap \delta$, $B_i = W_i \cap (\delta^* \setminus \delta)$. Let s_i be the altitude of W_i. We claim that $\tau_{(0,0,-s_i)} W_i$ is a standard wall, and we prove this as follows. Let C_{i_j}, $j = 1, 2, \dots, k$, be the ceilings that are 0-connected to W_i, and let H_{i_j} be the maximal 0-connected set of plaquettes in $\delta_0 \setminus \pi(W_i)$ onto which C_{i_j} projects. See Lemma 7.125(vii). It suffices to construct an interface $\delta(W_i)$ having $\tau_{(0,0,-s_i)} W_i$ as its unique wall. To this end, we add to $\tau_{(0,0,-s_i)} A_i$ the plaquettes in $\tau_{(0,0,-s_i)} C_{i_j}$, $j = 1, 2, \dots, k$, together with, for each j, the horizontal plaquettes in the maximal 0-connected set of horizontal plaquettes that contains $\tau_{(0,0,-s_i)} C_{i_j}$ and elements of which project onto H_{i_j}.

We now define the family $\{S_h : h \in \delta_0\}$ of standard walls by

$$S_h = \begin{cases} \tau_{(0,0,-s_i)} W_i & \text{if } h \text{ is the origin of } \tau_{(0,0,-s_i)} W_i, \\ \mathcal{E}_h & \text{if } h \text{ is the origin of no } \tau_{(0,0,-s_i)} W_i. \end{cases}$$

More precisely, in the first case, $S_h = (A_h, B_h)$ where $A_h = \tau_{(0,0,-s_i)} A_i$ and $B_h = \tau_{(0,0,-s_i)} B_i$. That this is an admissible family of standard walls follows from Lemma 7.125(viii) and from the observation that $s_i = 0$ when $E(W_i) \cap \overline{E_L} \neq \varnothing$.

Conversely, let $\{S_h = (A_h, B_h) : h \in \delta_0\}$ be an admissible family of standard walls. We shall show that there is a unique interface δ corresponding in a certain way to this family. Let $S_1, S_2 \dots, S_n$ be the non-empty walls of the family, and let δ_i be the unique interface in $\mathcal{D}_L$ having S_i as its only wall.

Consider the partial ordering on the walls given by $S_i < S_j$ if $\operatorname{int}(S_i) \subseteq \operatorname{int}(S_j)$, and re-order the non-empty walls in such a way that $S_i < S_j$ implies $i < j$. When it exists, we take the first index $k > 1$ such that $S_1 < S_k$ and we modify δ_k as follows. First, we remove the c-plaquettes that project onto $\operatorname{int}(S_1)$, and then we add translates of the plaquettes of A_1. This is done by translating these plaquettes so that the base of S_1 is raised (or lowered) to the plane containing the ceiling that is 0-connected to S_k and that projects on the maximal 0-connected set of plaquettes in $\delta_0 \setminus \pi(S_k)$ containing $\pi(S_1)$. See Lemma 7.125(viii). Let δ_k' denote the ensuing interface. We now repeat this procedure starting from the set of standard walls $S_2, S_3, \dots, S_n$ and interfaces $\delta_2, \delta_3, \dots, \delta_{k-1}, \delta_k', \delta_{k+1}, \dots, \delta_n$. If no such k exists, we continue the procedure with the reduced sequence of interfaces $\delta_2, \delta_3, \dots, \delta_{k-1}, \delta_k, \delta_{k+1}, \dots, \delta_n$.

We continue this process until we are left with interfaces δ_{i_l}'', $l = 1, 2, \dots, r$, having indices that refer to standard walls that are smaller than no other wall. The final interface δ is constructed as follows. For each l, we remove from the regular interface δ_0 all horizontal plaquettes contained in $\text{int}(S_{i_l})$, and we replace them by the plaquettes of δ''_{i_l} that project onto $\text{int}(S_{i_l})$.

The final assertion concerning admissible families of groups of walls is straightforward. □

We derive next certain combinatorial properties of walls. For $S = (A, B)$ a standard wall, let $N(S) = |A|$ and $\Pi(S) = N(S) - |\pi(S)|$. For an admissible set $F = \{S_1, S_2, \ldots, S_m\}$ of standard walls, let

$$\Pi(F) = \sum_{i=1}^{m} \Pi(S_i), \quad N(F) = \sum_{i=1}^{m} N(S_i), \quad \pi(F) = \bigcup_{i=1}^{m} \pi(S_i).$$

(7.128) Lemma. *Let $S = (A, B)$ be a standard wall, and $D(S)$ its height.*

(i) $N(S) \geq \frac{14}{13}|\pi(S)|$. *Consequently,* $\Pi(S) \geq \frac{1}{13}|\pi(S)|$ *and* $\Pi(S) \geq \frac{1}{14}N(S)$.

(ii) $N(S) \geq \frac{1}{5}|S|$.

(iii) $\Pi(S) \geq D(S)$.

Proof. (i) For each $h_0 \in \delta_0$, let $U(h_0) = \{h \in \delta_0 : h = h_0 \text{ or } h \overset{1}{\sim} h_0\}$. We call two plaquettes $h_1, h_2 \in \delta_0$ *separated* if $U(h_1) \cap U(h_2) = \varnothing$. Denote by $H_{\text{sep}} = H_{\text{sep}}(S) \subseteq \pi(S)$ a set of pairwise-separated plaquettes in $\pi(S)$ having maximum cardinality, and let $H = \bigcup_{h_1 \in H_{\text{sep}}} [U(h_1) \cap \pi(S)]$. Note that

$$|H_{\text{sep}}| \geq \tfrac{1}{13}|\pi(S)|. \tag{7.129}$$

For every $h_0 \in \pi(S)$, there exists a horizontal plaquette $h_1 \in \delta_S$ such that $\pi(h_1) = h_0$. Since $A \cup B$ contains no c-plaquette of δ_S, h_1 is a w-plaquette, whence $h_1 \in A$. In particular, $N(S) \geq |\pi(S)|$.

For $h_0 = \pi(h_1) \in H_{\text{sep}}$ where $h_1 \in A$, we claim that

$$\Big|\big\{h \in A : \text{ either } \pi(h) \subseteq [h_0] \text{ or } \pi(h) \in U(h_0)\big\}\Big| \geq |U(h_0) \cap \pi(S)| + 1. \tag{7.130}$$

By (7.129)–(7.130),

$$\begin{aligned} N(S) &\geq \sum_{h_0 \in H_{\text{sep}}} \big\{|U(h_0) \cap \pi(S)| + 1\big\} + |\pi(S) \setminus H| \\ &= |H| + |H_{\text{sep}}| + |\pi(S)| - |H| \geq \tfrac{14}{13}|\pi(S)|. \end{aligned}$$

In order to prove (7.130), we argue first that $U(h_0) \cap \pi(S)$ contains at least one (horizontal) plaquette besides h_0. Suppose that this is not true. Then $U(h_0) \setminus h_0$ contains the projections of c-plaquettes of δ_S^* only. By Lemma 7.125(ii, iii), these c-plaquettes belong to the same ceiling C and therefore lie in the same plane. Since h_1 is by assumption a w-plaquette, there must be at least one other horizontal plaquette of δ_S^* projecting onto h_0. Only one such plaquette, however, is 1-connected with the c-plaquettes. Since δ_S^* is 1-connected, the other plaquettes projecting onto h_0 must be 1-connected with at least one other plaquette of δ_S^*. Each of these further plaquettes projects into $\pi(C)$, in contradiction of Lemma 7.125(iv).

We now prove (7.130) as follows. Since h_1 is a w-plaquette, there exists $h_2 \in A \cup B$, $h_2 \neq h_1$, such that $\pi(h_2) = h_0$. If there exists such h_2 belonging to A, then (7.130) holds. Suppose the contrary, and let h_2 be such a plaquette with $h_2 \subset B$. Since $h_1 \in A$, for every $\eta \in U(h_0) \cap \pi(S)$, $\eta \neq h_0$, there exists $\eta' \in A$ such that $\pi(\eta') \subseteq [\eta]$ and $\eta' \stackrel{1}{\sim} h_1$. [If this were false for some η then, as in the proof of Lemma 7.125(ii), in any configuration with interface δ_S, there would exist a path of open edges joining the vertex just above h_1 to the vertex just beneath h_1. Since, by assumption, all plaquettes of $A \cup B$ other than h_1, having projection h_0, lie in B, this would contradict the fact that δ_S is an interface.] If any such η' is vertical, then (7.130) follows. Assume that all such η' are horizontal. Since $h_2 \in B$, there exists $h_3 \in A$ such that $h_3 \stackrel{1}{\sim} h_2$, and (7.130) holds in this case also.

(ii) The second part of the lemma follows from the observation that each of the plaquettes in A is 1-connected to no more than four horizontal plaquettes of B.

(iii) Recall from the remark after (7.129) that A contains at least $|\pi(S)|$ horizontal plaquettes. Furthermore, A must contain at least $D(S)$ vertical plaquettes, and the claim follows. □

Finally in this section, we derive an exponential bound for the number of groups of walls satisfying certain constraints.

(7.131) Lemma. *Let $h \in \delta_0$. There exists a constant K such that: for $k \geq 1$, the number of groups of walls $G \in \mathcal{G}_h$ satisfying $\Pi(G) = k$ is no greater than K^k.*

Proof. Let $G = \{S_1, S_2, \dots, S_n\} \in \mathcal{G}_h$ where the $S_i = (A_i, B_i)$ are non-empty standard walls and $S_1 \in \mathcal{S}_h$. For $j \in \delta_0$, let

$$R_j = \left\{ h' \in \delta_0 : \|j, h'\| \leq \sqrt{\rho(j, \delta_G)} \right\} \setminus \pi(G),$$

$$\widetilde{G} = \left\{ \bigcup_{i=1}^{n} (A_i \cup B_i) \right\} \cup \left\{ \bigcup_{j \in \pi(G)} R_j \right\}.$$

There exist constants C' and C'' such that, by Lemma 7.128,

$$|\widetilde{G}| \leq |G| + C' \sum_{j \in \pi(G)} \rho(j, \delta_G) \leq C''|G| \leq 5 \cdot 14 C'' \Pi(G),$$

where $|G| = \left| \bigcup_i (A_i \cup B_i) \right|$.

It may be seen that $\widetilde{G}$ is a 0-connected set of plaquettes containing h. Moreover, the 0-connected sets obtained by removing all the horizontal plaquettes $h' \in \widetilde{G}$, for which there exists no other plaquette $h'' \in \widetilde{G}$ with $\pi(h'') = \pi(h')$, are the standard walls of G. Hence, the number of such groups of walls with $\Pi(G) = k$ is no greater than the number of 0-connected sets of plaquettes containing no more than $70C''k$ elements including h. It is proved in [103, Lemma 2] that there exists $\nu < \infty$ such that the number of 0-connected sets of size n containing h is no larger than ν^n. Given any such set, there are at most 2^n ways of partitioning the plaquettes between the A_i and the B_i. The claim of the lemma follows. □

7.10 Exponential bounds for group probabilities

The probabilistic expressions of Section 7.8 may be combined with the classification of Section 7.9 to obtain an estimate concerning the geometry of the interface. Let $\mathbf{G} = \{G_h : h \in \delta_0\}$ be a family of groups of walls. If $\mathbf{G}$ is admissible, there exists by Lemma 7.127 a unique corresponding interface $\delta_{\mathbf{G}}$. We may pick a random family $\zeta = \{\zeta_h : h \in \delta_0\}$ of groups of walls according to the probability measure $\mathbb{P}_L$ induced by $\overline{\phi}_L$ thus:

$$\mathbb{P}_L(\zeta = \mathbf{G}) = \begin{cases} \overline{\phi}_L(\Delta = \delta_{\mathbf{G}}) & \text{if } \mathbf{G} \text{ is admissible,} \\ 0 & \text{otherwise.} \end{cases}$$

(7.132) Lemma. *Let $q \in [1, \infty)$, and let p^* be as in Lemma 7.104. There exist constants C_3, C_4 such that*

$$\mathbb{P}_L\big(\zeta_{h'} = G_{h'} \,\big|\, \zeta_h = G_h \text{ for } h \in \delta_0,\ h \neq h'\big) \leq C_3[C_4(1-p)]^{\Pi(G_{h'})},$$

for $p > p^*$, and for all $h' \in \delta_0$, $G_{h'} \in \mathcal{G}_{h'}$, $L > 0$, and for any admissible family $\{G_h : h \in \delta_0,\ h \neq h'\}$ of groups of walls.

Proof. The claim is trivial if $\mathbf{G} = \{G_h : h \in \delta_0\}$ is not admissible, and therefore we may assume it to be admissible. Let $h' \in \delta_0$, and let $\mathbf{G}'$ agree with $\mathbf{G}$ except at h', where $G_{h'}$ is replaced by the empty group $\mathcal{E}_{h'}$. Then

$$\mathbb{P}_L\big(\zeta_{h'} = G_{h'} \,\big|\, \zeta_h = G_h \text{ for } h \in \delta_0,\ h \neq h'\big) \leq \frac{\overline{\phi}_L(\delta)}{\overline{\phi}_L(\delta')}, \tag{7.133}$$

where $\delta = \delta_{\mathbf{G}}$ and $\delta' = \delta_{\mathbf{G}'}$.

In using (7.121) to bound the right side of this expression, we shall require bounds for $|\delta| - |\delta'|$, $|\overline{\delta} \setminus \delta| - |\overline{\delta'} \setminus \delta'|$, $k_\delta - k_{\delta'}$, and

$$\sum_{e \in E(\delta) \cap E_L} f_p(e, \delta, L) - \sum_{e \in E(\delta') \cap E_L} f_p(e, \delta', L). \tag{7.134}$$

It is easy to see from the definition of δ that

$$|\delta| = |\delta_0| + \sum_{h \in \delta_0} \big[N(G_h) - |\pi(G_h)|\big],$$

and therefore,

$$(7.135) \qquad |\delta| - |\delta'| = N(G_{h'}) - |\pi(G_{h'})| = \Pi(G_{h'}).$$

A little thought leads to the inequality

$$(7.136) \qquad |\overline{\delta} \setminus \delta| - |\overline{\delta}' \setminus \delta'| \geq 0,$$

and the reader may be prepared to omit the explanation that follows. We claim that (7.136) follows from the inequality

$$(7.137) \qquad |P(\overline{\delta})| - |P(\overline{\delta}')| \geq 0,$$

where $P(\overline{\delta})$ (respectively, $P(\overline{\delta}')$) is the set of plaquettes in $\overline{\delta} \setminus \delta$ (respectively, $\overline{\delta}' \setminus \delta'$) that project into $[\pi(G_{h'})]$. In order to see that (7.137) implies (7.136), we argue as follows. The extended interface $\overline{\delta}$ may be constructed from $\overline{\delta}'$ in the following manner. First, we remove all the plaquettes from $\overline{\delta}'$ that project into $[\pi(G_{h'})]$, and we fill the gaps by introducing the walls of $G_{h'}$ one by one along the lines of the proof of Lemma 7.127. Then we add the plaquettes of $\overline{\delta} \setminus \delta$ that project into $[\pi(G_{h'})]$. During this operation on interfaces, we remove $P(\overline{\delta}')$ and add $P(\overline{\delta})$, and the claim follows.

By Lemma 7.125(viii), there exists no vertical plaquette of $\overline{\delta}' \setminus \delta'$ that projects into $[\pi(G_{h'})]$ and is in addition 1-connected to some wall not belonging to $G_{h'}$. Moreover, since all the horizontal plaquettes of $\overline{\delta}'$ belong to the semi-extended interface δ'^*, those that project onto $[\pi(G_{h'})]$ are c-plaquettes of δ'^*; hence, such plaquettes lie in δ'. It follows that $P(\overline{\delta}')$ comprises the vertical plaquettes that are 1-connected with $\pi(G_{h'})$.

It is therefore sufficient to construct an injective map T that maps each vertical plaquette, 1-connected with $\pi(G_{h'})$, to a different vertical plaquette in $P(\overline{\delta})$. We noted in the proof of Lemma 7.128(i) that, for every $h_0 \in \pi(G_h')$, there exists a horizontal plaquette $h_1 \in \delta$ with $\pi(h_1) = h_0$. For every vertical plaquette $h^{\mathrm{v}} \stackrel{1}{\sim} h_0$, there exists a translate $h_1^{\mathrm{v}} \stackrel{1}{\sim} h_1$. Suppose h^{v} lies above δ_0. If $h_1^{\mathrm{v}} \in \overline{\delta} \setminus \delta$, we set $T(h^{\mathrm{v}}) = h_1^{\mathrm{v}}$. If $h_1^{\mathrm{v}} \in \delta$, we consider the (unique) vertical plaquette 'above' it, which we denote by h_2^{v}. We repeat this procedure up to the first n for which we meet a plaquette $h_n^{\mathrm{v}} \in \overline{\delta} \setminus \delta$, and we set $T(h^{\mathrm{v}}) = h_n^{\mathrm{v}}$. When h^{v} lies below δ_0, we act similarly to find a plaquette $T(h^{\mathrm{v}})$ of $\overline{\delta} \setminus \delta$ beneath h^{v}. The resulting T is as required.

We turn now to the quantity $k_\delta - k_{\delta'}$, and we shall use the notation around (7.101). Note that exactly two of the components (S_δ^i, U_δ^i) are infinite, and these may be taken as those with indices 1 and 2. For $i = 3, 4, \dots, k_\delta$, let $H(S_\delta^i)$ be the set of plaquettes that are dual to edges having exactly one endvertex in S_δ^i. The finite component (S_δ^i, U_δ^i) is in a natural way surrounded by a particular wall, namely that to which all the plaquettes of $H(S_\delta^i)$ belong. This follows from Lemma 7.125(v, viii) and the facts that

$$P_i = \{\pi\big(h(\langle x, x+(0,0,1)\rangle)\big) : x \in S_\delta^i\}$$

is a 1-connected subset of δ_0, and that $[\pi(H(S_\delta^i))] = [P_i]$.

Therefore,

$$k_\delta - k_{\delta'} = k_{\delta''} - 2, \tag{7.138}$$

where $\delta'' = \delta_{G_{h'}}$. It is elementary by Lemma 7.128(i) that

$$k_{\delta''} \le 2N(G_{h'}) \le 28\Pi(G_{h'}). \tag{7.139}$$

Finally, we estimate (7.134). Let $H_1, H_2, \dots, H_r$ be the maximal 0-connected sets of plaquettes in $\delta_0 \setminus \pi(G_{h'})$, and let δ_i (respectively, δ_i') be the set of plaquettes of δ (respectively, δ') that project into $[H_i]$. Recalling the construction of an interface from its standard walls in the proof of Lemma 7.127, there is a natural one–one correspondence between the plaquettes of δ_i and those of δ_i', and hence between the plaquettes in $U = \bigcup_{i=1}^r \delta_i$ and those in $U' = \bigcup_{i=1}^r \delta_i'$. We denote by T the corresponding bijection mapping an edge e with $h(e) \in \bigcup_{i=1}^r \delta_i$ to the edge $T(e)$ with corresponding dual plaquette in $\bigcup_{i=1}^r \delta_i'$. Note that $T(e)$ is a vertical translate of e.

If e is such that $h(e) \in U$,

$$G(e, \delta, E_L; T(e), \delta', E_L) \ge \|\pi'(h(e)), \pi(G_{h'})\| - 1,$$

where $\pi'(h)$ is the earliest plaquette h'' of δ_0 such that $\pi(h) \subseteq [h'']$, and

$$\|h_1, H\| = \min\{\|h_1, h_2\| : h_2 \in H\}.$$

Let $p > p^*$. In the notation of Lemmas 7.104 and 7.118,

(7.140)

$$\begin{aligned}
&\Big| \sum_{e \in E(\delta) \cap E_L} f_p(e, \delta, L) - \sum_{e \in E(\delta') \cap E_L} f_p(e, \delta', L) \Big| \\
&\le \sum_{e \in E(U) \cap E_L} \big| f_p(e, \delta, L) - f_p(T(e), \delta', L) \big| \\
&\qquad + \sum_{e \in E(\delta \setminus U) \cap E_L} f_p(e, \delta, L) + \sum_{e \in E(\delta' \setminus U') \cap E_L} f_p(e, \delta', L) \\
&\le C_2 e^\gamma \sum_{e \in E(U) \cap E_L} \exp\big(-\gamma \|\pi'(h(e)), \pi(G_{h'})\|\big) + C_1\big[N(G_{h'}) + |\pi(G_{h'})|\big].
\end{aligned}$$

By Lemma 7.128, the second term of the last line is no greater than $C_5\Pi(G_{h'})$ for some constant C_5. Using the same lemma and the definition of a group of walls, the first term is no larger than

$$
\begin{aligned}
(7.141)\quad C_2e^{\gamma} &\sum_{h\in\delta_0\setminus\pi(G_{h'})} \rho(h,\delta)\exp\big(-\gamma\|h,\pi(G_{h'})\|\big) \\
&\le C_2e^{\gamma} \sum_{h\in\delta_0\setminus\pi(G_{h'})} \|h,\pi(G_{h'})\|^2 \exp\big(-\gamma\|h,\pi(G_{h'})\|\big) \\
&\le C_2e^{\gamma} \sum_{h''\in\pi(G_{h'})} \sum_{h\in\delta_0\setminus\pi(G_{h'})} \|h,h''\|^2 \exp\big(-\gamma\|h,h''\|\big) \\
&\le C_6|\pi(G_{h'})| \le 13C_6\Pi(G_{h'}),
\end{aligned}
$$

for some constant C_6.

The required conditional probability is, by (7.121) and (7.133),

$$
p^{|\overline{\delta}\setminus\delta|-|\overline{\delta}'\setminus\delta'|}(1-p)^{|\delta|-|\delta'|}q^{k_\delta-k_{\delta'}} \times \exp\Bigg(\sum_{e\in E(\delta)\cap E_L} f_P(e,\delta,L) - \sum_{e\in E(\delta')\cap E_L} f_P(e,\delta',L)\Bigg),
$$

which, by (7.135)–(7.141), is bounded as required. □

7.11 Localization of interface

The principal theorem states in rough terms the following. Let $q\in[1,\infty)$ and let p be sufficiently large. With $\overline{\phi}_L$-probability close to 1, the interface $\Delta(\omega)$ deviates from the flat plane δ_0 only through local perturbations. An ant living on $\Delta(\omega)$ is able, with large probability, to visit a positive density of the interface via horizontal meanderings only.

Let $h\in\delta_0$. For $\omega\in\Omega_L^{\mathrm{D}}$, we write $h\leftrightarrow\infty$ if there exists a sequence $h=h_0,h_1,\dots,h_r$ of plaquettes in δ_0 such that:

(a) $h_i \stackrel{1}{\sim} h_{i+1}$ for $i=0,1,\dots,r-1$,
(b) each h_i is a c-plaquette of $\Delta(\omega)$, and
(c) $h_r=h(e)$ for some $e\notin E_L$.

(7.142) Theorem [139]. *Let $q\in[1,\infty)$. For $\epsilon>0$, there exists $\widehat{p}=\widehat{p}(\epsilon)<1$ such that, if $p>\widehat{p}$,*

$$
(7.143)\qquad \overline{\phi}_L(h\leftrightarrow\infty) > 1-\epsilon, \qquad h\in\delta_0,\ L\ge 1.
$$

Since, following Theorem 7.142, $h\in\delta_0$ is a c-plaquette with high probability, the vertex of $\mathbb{Z}^3$ immediately beneath (respectively, above) the centre of h is joined to $\partial^- \Sigma_L$ (respectively, $\partial^+ \Sigma_L$) with high probability. Theorem 7.87 follows. Furthermore, since $h \nleftrightarrow \infty$ with high probability, such connections may be found within the plane of $\mathbb{Z}^3$ comprising vertices x with $x_3 = 0$ (respectively, $x_3 = 1$).

The existence of non-translation-invariant (conditioned) random-cluster measures follows from Theorem 7.142, as in the following sketch argument. For $e \in \mathbb{E}^3$, let $e^\pm = e \pm (0,0,1)$, and let $\omega \in \Omega$. If $h = h(e) \in \delta_0$ is a c-plaquette of $\Delta = \Delta(\omega)$, then e is closed, and $h(e^\pm) \notin \overline{\Delta}$. The configurations in the two regions above and below Δ are governed by wired random-cluster measures[14]. Therefore, under (7.143),

$$\overline{\phi}_L(e \text{ is open}) \le \epsilon, \quad \overline{\phi}_L(e^\pm \text{ is open}) \ge \frac{(1-\epsilon)p}{p + q(1-p)},$$

by stochastic ordering. Note that these inequalities concern the probabilities of cylinder events. This implies Theorem 7.89.

Our second main result concerns the vertical displacement of the interface, and asserts the existence of a geometric bound on the tail of the displacement, uniformly in L. Let $\delta \in \mathcal{D}_L$, $(x_1, x_2) \in \mathbb{Z}^2$, and $x = (x_1, x_2, \frac{1}{2})$. We define the *displacement* of δ at x by

$$D(x, \delta) = \sup\bigl\{|z - \tfrac{1}{2}| : (x_1, x_2, z) \in [\delta]\bigr\}.$$

(7.144) Theorem [139]. *Let $q \in [1, \infty)$. There exists $\widehat{p} < 1$ and $\alpha(p)$ satisfying $\alpha(p) > 0$ when $p > \widehat{p}$ such that*

$$\overline{\phi}_L(D(x, \Delta) \ge z) \le e^{-z\alpha(p)}, \qquad z \ge 1,\ (x_1, x_2) \in \mathbb{Z}^2,\ L \ge 1.$$

Proof of Theorem 7.142. Let $h \in \delta_0$. We have not so far specified the ordering of plaquettes in δ_0 used to identify the origin of a standard wall or of a group of walls. We assume henceforth that this ordering is such that: for $h_1, h_2 \in \delta_0$, $h_1 > h_2$ implies $\|h, h_1\| \ge \|h, h_2\|$.

For any standard wall S there exists, by Lemma 7.125(vi), a unique maximal infinite 1-connected component $I(S)$ of $\delta_0 \setminus \pi(S)$. Let $\omega \in \Omega_L^{\mathrm{D}}$. The interface $\Delta(\omega)$ gives rise to a family of standard walls, and $h \nleftrightarrow \infty$ if and only if[15], for each such wall S, h belongs to $I(S)$. Suppose on the contrary that $h \notin I(S_j)$ for some such standard wall S_j, for some $j \in \delta_0$, belonging in turn to some maximal admissible group $G_{h'} \in \mathcal{G}_{h'}$ of walls of Δ, for some $h' \in \delta_0$. By Lemma 7.128 and the above ordering of δ_0,

$$13\Pi(G_{h'}) \ge |\pi(G_{h'})| \ge |\pi(S_j)| \ge \|h, j\| + 1 \ge \|h, h'\| + 1.$$

[14] We have used Lemma 7.117 here.

[15] This is a consequence of a standard property of $\mathbb{Z}^2$, see [210, Appendix].

Let K be as in Lemma 7.131, and p^*, C_4 as in Lemma 7.132. Let $\widetilde{p}$ be sufficiently large that $\widetilde{p} > p^*$ and that

$$\lambda = \lambda(p) = -\tfrac{1}{13}\log[KC_4(1-p)]$$

satisfies $\lambda(\widetilde{p}) > 0$. By the last lemma, when $p > \widetilde{p}$,

$$\begin{aligned}
1 - \overline{\phi}_L(h \leftrightarrow \infty) &\le \sum_{h' \in \delta_0} \mathbb{P}_L\big(\Pi(\zeta_{h'}) \ge \tfrac{1}{13}[\|h, h'\| + 1]\big) \\
&\le \sum_{h' \in \delta_0} \sum_{n \ge (\|h,h'\|+1)/13} \sum_{\substack{G \in \mathcal{G}_{h'}: \\ \Pi(G) = n}} \mathbb{P}_L(\zeta_{h'} = G) \\
&\le \sum_{h' \in \delta_0} \sum_{n \ge (\|h,h'\|+1)/13} K^n C_3[C_4(1-p)]^n \\
&\le C_3 \sum_{h' \in \delta_0} \exp\big(-\lambda(\|h, h'\| + 1)\big) \le C_7 e^{-\lambda},
\end{aligned}$$

for appropriate constants C_i. The claim follows on choosing p sufficiently close to 1. □

Proof of Theorem 7.144. If $D(x, \Delta) \ge z$, there exists r satisfying $1 \le r \le z$ such that the following statement holds. There exist distinct plaquettes $h_1, h_2, \dots, h_r \in \delta_0$, and maximal admissible groups G_{h_i}, $i = 1, 2, \dots, r$, of walls of Δ such that: $x = (x_1, x_2, \frac{1}{2})$ lies in the interior of one or more standard wall of each G_{h_i}, and

$$\sum_{i=1}^{r} \Pi(G_{h_i}) \ge z.$$

Recall Lemma 7.128(iii). Let $m_i = \lfloor \frac{1}{13}(\|x, h_i\| + 1) \rfloor$ where $\|x, h\| = \|x - y\|$ and y is the centre of h. By Lemma 7.132, and as in the previous proof,

$$\begin{aligned}
&\overline{\phi}_L(D(x,\Delta) \ge z) \\
&\quad \le \sum_{\substack{h_1, h_2, \dots, h_r \\ 1 \le r \le z}} \mathbb{P}_L\Big(\sum_i \Pi(\zeta_{h_i}) \ge z,\ \Pi(\zeta_{h_i}) \ge m_i \vee 1\Big) \\
&\quad = \sum_{\substack{h_1, h_2, \dots, h_r \\ 1 \le r \le z}} \sum_{s=z}^{\infty} \sum_{\substack{z_1, z_2, \dots, z_r: \\ z_1 + z_2 + \cdots + z_r = s \\ z_i \ge m_i \vee 1}} \mathbb{P}_L\big(\Pi(\zeta_{h_i}) = z_i \text{ for } i = 1, 2, \dots, r\big) \\
&\quad \le \sum_{h_i} \sum_{s \ge z} C_8[KC_4(1-p)]^s \sum_{\substack{z_1, z_2, \dots, z_r: \\ z_1 + z_2 + \cdots + z_r = s \\ z_i \ge m_i \vee 1}} 1,
\end{aligned}$$

for some constant C_8. The last summation is the number of ordered partitions of the integer s into r parts, the ith of which is at least $m_i \vee 1$. By adapting the classical solution to this enumeration problem when $m_i = 1$ for all i, we find that

$$\sum_{\substack{z_1, z_2, \ldots, z_r: \\ z_1 + z_2 + \cdots + z_r = s \\ z_i \geq m_i \vee 1}} 1 \leq \binom{s - 1 - \sum_i (m_i \vee 1)}{r - 1} \leq 2^{s-1-\sum_i (m_i \vee 1)} \leq 2^{s-1-\sum_i m_i},$$

whence, for some C_9,

$$\overline{\phi}_L(D(x, \Delta) \geq z) \leq C_9 \sum_{s=z}^{\infty} [2KC_4(1-p)]^s \left(\sum_{h \in \delta_0} 2^{-\lfloor \|x,h\|/13 \rfloor} \right)^z, \qquad z \geq 1.$$

The right side decays exponentially as $z \to \infty$ when $2KC_4(1-p)$ is sufficiently small. □

Chapter 8

Dynamics of Random-Cluster Models

Summary. One may associate time-dynamics with the random-cluster model in a variety of natural ways. Amongst Glauber-type processes, the Gibbs sampler is especially useful and is well suited to the construction of a 'coupling from the past' algorithm resulting in a sample with the random-cluster measure as its (exact) law. In the Swendsen–Wang algorithm, one interleaves transitions of the random-cluster model and the associated Potts model. The random-cluster model for different values of p may be coupled together via a certain Markov process on a more general state space. This provides a mechanism for studying the 'equilibrium' model.

8.1 Time-evolution of the random-cluster model

The random-cluster model as studied so far is random in space but not in time. There are a variety of ways of introducing time-dynamics into the model, and some good reasons for so doing. The principal reason is that, in our $3 + 1$ dimensional universe, the time-evolution of processes is fundamental. It entails the concepts of equilibrium and convergence, of metastability, and of chaos. A rigorous theory of time-evolution in statistical mechanics is one of the major achievements of modern probability theory with which the names Dobrushin, Spitzer, and Liggett are easily associated.

There is an interplay between the time-dynamics of an ergodic system and its equilibrium measure. The equilibrium is determined by the dynamics, and thus, in models where the equilibrium may itself be hard of access, the dynamics may allow an entrance. Such difficulties arise commonly in applications of Bayesian statistics, in situations where one wishes to sample from a posterior distribution μ having complex structure. One way of doing this is to construct a Markov chain with invariant measure μ, and to follow the evolution of this chain as it approaches equilibrium. The consequent field of 'Monte Carlo Markov chains' is now established as a key area of modern statistics. Similarly, the dynamical theory of the random-cluster model allows an insight into the equilibrium random-cluster measures. It provides in addition a mechanism for studying the way in which the system 'relaxes' to its equilibrium. We note that the simulation of a Markov chain can, after some time, result in samples whose distribution is close to the invariant measure μ. Such samples will in general have laws which differ from μ, and it can be a difficult theoretical problem to obtain a useful estimate of the distance between the actual sample and μ.

Consider first the random-cluster model on a finite graph G with given values of p and q. Perhaps the most obvious type of dynamic is a so-called Glauber process in which single edges change their states at rates chosen in such a way that the equilibrium measure is the random-cluster measure on G. These are the spin-flip processes which, in the context of the Ising model and related systems, have been studied in many works including Liggett's book [235]. There is a difficulty in constructing such a process on an *infinite* graph, since the natural speed functions are not continuous in the product topology.

There is a special Glauber process, termed the 'Gibbs sampler' or 'heat-bath algorithm', which we describe in Section 8.3 in discrete time. This is particularly suited to the exposition in Section 8.4 of the method of 'coupling from the past'. This beautiful approach to simulation results in a sample having the exact target distribution, unlike the approximate samples produced by Monte Carlo Markov chains. The random-cluster model is a natural application for the method when $q \in [1, \infty)$, since $\phi_{G,p,q}$ is monotonic: the model has 'smallest' and 'largest' configurations, and the target measure is attained at the moment of coalescence of the two trajectories beginning respectively at these extremes.

The speed of convergence of Glauber processes has been studied in detail for Ising and related models, and it turns out that the rate of convergence to the unique invariant measure can be very slow. This occurs for example if the graph is a large box of a lattice with, say, $+$ boundary conditions, the initial configuration has $-$ everywhere in the interior, and the temperature is low. The process remains for a long time close to the $-$ state; then it senses the boundary, and converges duly to the $+$ state. There is an alternative dynamic for the Ising (and Potts) model, termed Swendsen–Wang dynamics, which converges rather faster to the unique equilibrium so long as the temperature is different from its critical value. This method proceeds by a progressive coupling of the Ising/Potts system with the random-cluster model, and by interleaving a Markovian transition for these two systems in turn. It is described in Section 8.5.

The remaining sections of this chapter are devoted to an exposition of Glauber dynamics on finite and infinite graphs, implemented in such a way as to highlight the effect of varying the parameter p. We begin in Section 8.6 with the case of a finite graph, and proceed in Sections 8.7–8.8 to a process on the infinite lattice which incorporates in a monotone manner a time-evolving random-cluster process for every value of $p \in (0, 1)$. The unique invariant measure of this composite Markov process may be viewed as a coupling of random-cluster measures on the lattice for different values of p. One consequence of this approach is a proof of the left-continuity of the percolation probability for random-cluster models with

$q \in [1, \infty)$, see Theorem 5.16. It leads in Section 8.9 to an open question of 'simultaneous uniqueness' of infinite open clusters.

8.2 Glauber dynamics

Let $G = (V, E)$ be a finite graph, with $\Omega = \{0, 1\}^E$ as usual. Let $p \in (0, 1)$ and $q \in (0, \infty)$. We shall construct a reversible Markov chain in continuous time having as unique invariant measure the random-cluster measure $\phi_{p,q}$ on Ω. A feature of the Glauber dynamics of this section is that the set of permissible jumps comprises exactly those in which the state of a single edge, e say, changes. To this end, we recall first the notation of (1.25). For $\omega \in \Omega$ and $e \in E$, let ω^e and ω_e be the configurations obtained by 'switching e on' and 'switching e off', respectively.

Let $X = (X_t : t \geq 0)$ be a continuous-time Markov chain, [164, Chapter 6], on the state space Ω with generator $Q = (q_{\omega,\omega'} : \omega, \omega' \in \Omega)$ satisfying

$$(8.1) \qquad q_{\omega_e,\omega^e} = p, \quad q_{\omega^e,\omega_e} = (1-p)q^{D(e,\omega_e)}, \qquad \omega \in \Omega,\ e \in E,$$

where $D(e, \xi)$ is the indicator function of the event that the endvertices of e are joined by no open path of ξ. Equations (8.1) specify the rates at which single edges are acquired or lost by the present configuration. We set $q_{\omega,\xi} = 0$ if ω and ξ differ on two or more edges, and we choose the diagonal elements $q_{\omega,\omega}$ in such a way that Q, when viewed as a matrix, has row-sums zero, that is,

$$q_{\omega,\omega} = -\sum_{\xi : \xi \neq \omega} q_{\omega,\xi}, \qquad \omega \in \Omega.$$

Note that X proceeds by transitions in which single edges change their states, it is not permissible for two or more edge-states to change simultaneously. We say in this regard that X proceeds by 'local moves'.

It is elementary that the so-called 'detailed balance equations'

$$(8.2) \qquad \phi_{p,q}(\omega)q_{\omega,\omega'} = \phi_{p,q}(\omega')q_{\omega',\omega}, \qquad \omega, \omega' \in \Omega,$$

hold, whence X is reversible with respect to the random-cluster measure $\phi_{p,q}$. It is easily seen that the chain is irreducible, and therefore $\phi_{p,q}$ is the unique invariant measure of the chain and, in particular, $X_t \Rightarrow \phi_{p,q}$ as $t \to \infty$, where '$\Rightarrow$' denotes weak convergence. There are of course many Markov chains with generators satisfying the detailed balance equations (8.2). It is important only that the ratio $q_{\omega,\omega'}/q_{\omega',\omega}$ satisfies

$$(8.3) \qquad \frac{q_{\omega,\omega'}}{q_{\omega',\omega}} = \frac{\phi_{p,q}(\omega')}{\phi_{p,q}(\omega)}, \qquad \omega, \omega' \in \Omega.$$

We call a Markov chain on Ω a *Glauber process* if it proceeds by local moves and has a generator Q satisfying (8.3), see [235, p. 191]. We have concentrated here on continuous-time processes, but Glauber processes may be constructed in discrete time also.

Two extensions of this dynamical structure which have proved useful are as follows. The evolution may be specified in terms of a so-called graphical representation, constructed via a family of independent Poisson processes. This allows a natural coupling of the measures $\phi_{p,q}$ for different p and q. Such couplings are monotone in p when $q \in [1, \infty)$. One may similarly couple the unconditional measure $\phi_{p,q}(\cdot)$ and the conditioned measure $\phi_{p,q}(\cdot \mid A)$. Such couplings permit probabilistic interpretations of differences of the form $\phi_{p',q}(B \mid A) - \phi_{p,q}(B)$ when $q \in [1, \infty)$, $p \le p'$, and A and B are increasing, and this can be useful in particular calculations, see [39, 151, 152].

One needs to be more careful when G is an *infinite* graph. In this case, one may construct a Glauber process on a finite subgraph H of G, and then pass to the thermodynamic limit as $H \uparrow G$. Such a limit may be justified when $q \in [1, \infty)$ using the positive association of random-cluster measures, [152]. We shall discuss such limits in Section 8.8 in the more general context of 'coupled dynamics'. For a reason which will emerge later, we will give the details for the Gibbs sampler of Section 8.3, rather than for the Glauber process of (8.1). The latter case may however be treated in an essentially identical manner.

Note that the generator (8.1) of the Markov chain given above depends on the random variable $D(e, \omega_e)$, and that this random variable is 'non-local' in the sense that it is not everywhere continuous in ω. It is this feature of non-locality which leads to an interesting complication when the graph is infinite, linked in part to the 0/1-infinite-cluster property introduced before Theorem 4.31. Further discussion may be found in [152, 272].

8.3 Gibbs sampler

Once again we take $G = (V, E)$ to be a finite graph, and we let $p \in (0, 1)$ and $q \in (0, \infty)$. We consider in this section a special Glauber process termed the *Gibbs sampler* (or *heat-bath algorithm*). This is a Markov chain X on the state space $\Omega = \{0, 1\}^E$ which proceeds by local moves. Its basic rule is as follows. We choose an edge e at random, and we set the state of e according to the conditional measure of $\omega(e)$ given the current states of the other edges. This may be done in either discrete or continuous time, we give the details for continuous time here and shall return to the case of discrete time in Section 8.4.

Let $X = (X_t : t \ge 0)$ be the Markov chain on the state space Ω with generator $Q = (q_{\omega,\omega'}; \omega, \omega' \in \Omega)$ given by

$$
\begin{aligned}
q_{\omega_e,\omega^e} &= \frac{\phi_{p,q}(\omega^e)}{\phi_{p,q}(\omega^e) + \phi_{p,q}(\omega_e)},\\
q_{\omega^e,\omega_e} &= \frac{\phi_{p,q}(\omega_e)}{\phi_{p,q}(\omega^e) + \phi_{p,q}(\omega_e)},
\end{aligned}
\tag{8.4}
$$

for $\omega \in \Omega$ and $e \in E$. Thus, each edge is selected at rate 1, and the state of that edge is changed according to the correct conditional measure. It is evident that the detailed balance equations (8.2) hold as before, whence X is reversible with respect to $\phi_{p,q}$. By irreducibility, $\phi_{p,q}$ is the unique invariant measure of the chain and thus, in particular, $X_t \Rightarrow \phi_{p,q}$ as $t \to \infty$.

There is a useful way of formulating the transition rules (8.4). With each edge e is associated an 'exponential alarm clock' that rings at the times of a Poisson process with intensity 1. Suppose that the alarm clock at e rings at time T, and let U be a random variable with the uniform distribution on the interval $[0, 1]$. Let $X_{T-} = \omega$ denote the current state of the process. The state of e jumps to the value $X_T(e)$ given as follows:

$$\begin{aligned} &\text{when } \omega(e) = 1, \quad \text{we set } X_T(e) = 0 \quad \text{if} \quad U \le \frac{\phi_{p,q}(\omega_e)}{\phi_{p,q}(\omega^e) + \phi_{p,q}(\omega_e)}, \\ &\text{when } \omega(e) = 0, \quad \text{we set } X_T(e) = 1 \quad \text{if} \quad U > \frac{\phi_{p,q}(\omega_e)}{\phi_{p,q}(\omega^e) + \phi_{p,q}(\omega_e)}. \end{aligned} \tag{8.5}$$

The state of e is unchanged if the appropriate inequality is false. It is easily checked that this rule generates a Markov chain which satisfies (8.4) and proceeds by local moves. This version of such a chain has two attractive properties. First, it is a neat way of implementing the Gibbs sampler in practice since it requires only two random mechanisms: one that samples edges at random, and a second that produces uniformly distributed random variables.

A second benefit is that it provides a coupling of a variety of such Markov chains with different values of p and q, and with different initial states. We explain this next. Suppose that $0 < p_1 < p_2 < 1$ and $q_1 > q_2 \ge 1$. It is easily checked, as in Section 3.4, that

$$\frac{\phi_{p_1,q_1}(\omega_e)}{\phi_{p_1,q_1}(\omega^e) + \phi_{p_1,q_1}(\omega_e)} \ge \frac{\phi_{p_2,q_2}(\xi_e)}{\phi_{p_2,q_2}(\xi^e) + \phi_{p_2,q_2}(\xi_e)}, \qquad \omega \le \xi. \tag{8.6}$$

Let $U(e) = (U_j(e) : j = 1, 2, \dots)$, $e \in E$, be independent families of independent random variables each having the uniform distribution on $[0, 1]$. Let $X^i = (X^i_t : t \ge 0)$, $i = 1, 2$, be Markov processes on Ω constructed as follows. The process X^i evolves according to the above rules, with parameters p_i, q_i, and using the value $U_j(e)$ at the jth ring of the alarm clock at the edge e. By (8.5)–(8.6), if $X^1_0 \le X^2_0$, then $X^1_t \le X^2_t$ for all $t \ge 0$. We have therefore constructed a coupling which preserves ordering between processes with different parameters p, q, and with different initial configurations. The key to this ordering is the fact that the coupled processes utilize the same variables $U_j(e)$. This discussion will be developed in the next section.

8.4 Coupling from the past

When performing simulations of the random-cluster model, one is required to sample from the probability measure $\phi_{p,q}$. The Glauber processes of the last two sections certainly converge weakly to $\phi_{p,q}$ as $t \to \infty$, but this is not as good as having a sample with the *exact* distribution. The Propp–Wilson approach to sampling termed 'coupling from the past', [282], provides a mechanism for obtaining samples with the correct distribution, and is in addition especially well suited to the random-cluster model when $q \in [1, \infty)$. We describe this here. Some illustrations of the method in practice may be found in [173, 195, 243].

Let $G = (V, E)$ be a finite graph and let $p \in (0, 1)$ and $q \in (0, \infty)$. We shall later restrict ourselves to the case $q \in [1, \infty)$, since this will be important in the subsequent analysis of the algorithm. We provide ourselves first with a discrete-time reversible Markov chain $Z = (Z_n : n = 0, 1, 2, \dots)$ with state space Ω and having unique invariant measure $\phi_{p,q}$. The discrete-time Gibbs sampler provides a suitable example of such a chain, and proceeds as follows, see Section 8.3 and [175]. At each stage, we pick a random edge e, chosen uniformly from E and independently of all earlier choices, and we make e open with the correct conditional probability, given the configuration on the other edges. This Markov chain proceeds by local moves, and has transition matrix $\Pi = (\pi_{\omega,\omega'} : \omega, \omega' \in \Omega)$ satisfying

$$\begin{aligned}
\pi_{\omega_e,\omega^e} &= \frac{1}{|E|} \cdot \frac{\phi_{p,q}(\omega^e)}{\phi_{p,q}(\omega^e) + \phi_{p,q}(\omega_e)},\\
\pi_{\omega^e,\omega_e} &= \frac{1}{|E|} \cdot \frac{\phi_{p,q}(\omega_e)}{\phi_{p,q}(\omega^e) + \phi_{p,q}(\omega_e)},
\end{aligned}$$

for $\omega \in \Omega$ and $e \in E$. A neat way to implement this is to follow the recipe of the last section. Suppose that $Z_n = \omega$. Let e_n be a random edge of E, and let U_n be uniformly distributed on the interval $[0, 1]$, these variables being chosen independently of all earlier choices. We obtain Z_{n+1} by retaining the states of all edges except possibly that of e_n, and by setting

$$Z_{n+1}(e_n) = 0 \quad \text{if and only if} \quad U_n \le \frac{\phi_{p,q}(\omega_{e_n})}{\phi_{p,q}(\omega^{e_n}) + \phi_{p,q}(\omega_{e_n})}. \tag{8.7}$$

The evolution of the chain is determined by the sequences e_n, U_n, and the initial state Z_0. One may make this construction explicit by writing

$$Z_{n+1} = \psi(Z_n, e_n, U_n)$$

for some deterministic function $\psi : \Omega \times E \times [0, 1] \to \Omega$.

We highlight a certain monotonicity of ψ, valid when $q \in [1, \infty)$. Fix $e \in E$ and $u \in [0, 1]$. The configuration $\Psi_\omega = \psi(\omega, e, u)$, viewed as a function of ω, is constant on edges $f \ne e$, and takes values 0, 1 on e with

$$\Psi_\omega(e) = 0 \quad \text{if and only if} \quad u \le \frac{\phi_{p,q}(\omega_e)}{\phi_{p,q}(\omega^e) + \phi_{p,q}(\omega_e)}, \qquad \omega \in \Omega.$$

As in (8.6), when $q \in [1, \infty)$,

$$\frac{\phi_{p,q}(\omega_e)}{\phi_{p,q}(\omega^e) + \phi_{p,q}(\omega_e)} \geq \frac{\phi_{p,q}(\xi_e)}{\phi_{p,q}(\xi^e) + \phi_{p,q}(\xi_e)}, \qquad \omega \leq \xi,$$

implying that $\Psi_\omega(e) \leq \Psi_\xi(e)$, and hence

$$\psi(\omega, e, u) \leq \psi(\xi, e, u), \qquad \omega \leq \xi. \tag{8.8}$$

Let $Z^\nu = (Z_n^\nu : n = 0, 1, 2, \dots)$ be the Markov chain constructed via (8.7) with initial state $Z_0 = \nu$. By (8.8),

$$Z_n^\omega \leq Z_n^\xi \quad \text{for all } n, \quad \text{if} \quad \omega \leq \xi \text{ and } q \in [1, \infty), \tag{8.9}$$

which is to say that the coupling is monotone in the initial state: if one such chain starts below another, then it remains below for all time.

Instead of running the chain Z 'forwards' in time in order to approximate the invariant measure $\phi_{p,q}$, we shall run it 'backwards' in time in a certain special manner which results in a sample with the exact target distribution. Let $W = (W(\omega) : \omega \in \Omega)$ be a vector of random variables such that each $W(\omega)$ has the law of Z_1 conditional on $Z_0 = \omega$,

$$\mathbb{P}(W(\omega) = \xi) = \pi_{\omega,\xi}, \qquad \omega, \xi \in \Omega.$$

Following the scheme described above, we may take $W(\omega) = \psi(\omega, e, U)$ where e and U are chosen uniformly at random. Let W_{-m}, $m = 1, 2, \dots$, be independent random vectors distributed as W, that is, $W_{-m}(\cdot) = \psi(\cdot, e_m, U_m)$ where the set $\{(e_m, U_m) : m = 1, 2, \dots\}$ comprises independent pairs of independent uniformly-distributed random variables. We construct a sequence Y_{-n}, $n = 1, 2, \dots$, of random maps from Ω to Ω by the following inductive procedure. First, we define $Y_0 : \Omega \to \Omega$ to be the identity mapping. Having found $Y_0, Y_{-1}, Y_{-2}, \dots, Y_{-m}$ for $m = 0, 1, 2, \dots$, we define

$$Y_{-m-1}(\omega) = Y_{-m}(W_{-m-1}(\omega)).$$

That is, $Y_{-m-1}(\omega)$ is obtained from ω by passing in one step to $W_{-m-1}(\omega)$, and then applying Y_{-m} to this new state. The exact dependence structure of this scheme is an important ingredient of its analysis.

We terminate the process Y at the earliest time M of coalescence,

$$M = \min\big\{m : Y_{-m}(\cdot) \text{ is a constant function}\big\}. \tag{8.10}$$

By the definition of M, the value $Y_{-M} = Y_{-M}(\omega)$ does not depend on the choice of ω. The process of coalescence is illustrated in Figure 8.1. We prove next that Y_{-M} has law $\phi_{p,q}$.

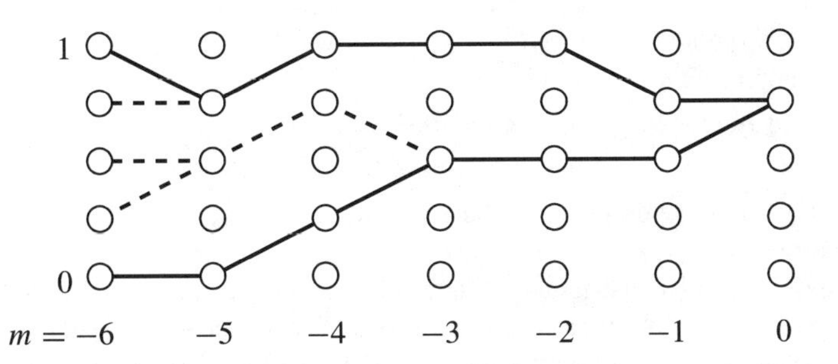

Figure 8.1. An illustration of 'coupling from the past' in a situation where the intermediate states are sandwiched by the extreme states. There are five states in this illustration. The heavy lines map the evolution of the smallest (0) and largest (1) configurations, and the dotted lines show the evolution of states that are sandwiched between these extremes.

(8.11) Theorem [282]. *The random variable M is almost-surely finite, and*

$$\mathbb{P}(Y_{-M} = \omega) = \phi_{p,q}(\omega), \qquad \omega \in \Omega.$$

The above procedure may seem unwieldy in practice, since Ω will often be large, and it appears necessary to keep track in (8.10) of the $Y_{-m}(\omega)$ *for every* $\omega \in \Omega$. The reality is simpler at least when $q \in [1, \infty)$, which we henceforth assume. By the monotonicity (8.9) of the coupling when $q \in [1, \infty)$, it suffices to follow the trajectories of the 'smallest' and 'largest' configurations, namely those beginning, respectively, with *every edge closed* and with *every edge open*. The processes starting at intermediate configurations remain sandwiched between these extremal processes at all future times. Thus one may define M instead by

$$M = \min\{m : Y_{-m}(0) = Y_{-m}(1)\}, \tag{8.12}$$

where 0 and 1 denote the vectors of zeros and ones as before. This brings a substantial computational advantage, since one is required to calculate only the $Y_{-m}(b)$ for $b = 0, 1$, and to find the earliest m at which they are equal.

We make two notes prior to the proof. In classical Monte Carlo experiments, the time-n measure converges to the target measure as $n \to \infty$. An estimate of the rate of convergence is necessary in order to know when to cease the process. Such estimates are not central to coupling-from-the-past, since this method results, after a finite (random) time, in a sample having the target measure as its exact law. That said, the method of proof implies a geometric rate of convergence. Secondly, the implementation of the method is greatly simplified by the monotonicity[1].

Proof of Theorem 8.11. We follow [282]. Let $q \in (0, \infty)$. By elementary properties of the Gibbs sampler (8.7), we may choose L such that

$$\mathbb{P}(Y_{-L} \text{ is a constant function}) > 0.$$

[1] The method may be implemented successfully in some situations where there is no such monotonicity, see [243, Chapter 32] for example.

We extend the notation prior to (8.10) as follows. Let $(Y_{-s,-t} : 0 \le t \le s)$ be functions mapping Ω to Ω given by:

(i) $Y_{-t,-t}$ is the identity map, for $t = 0, 1, 2, \dots$,

(ii) $Y_{-s,-t}(\omega) = Y_{-s+1,-t}(W_{-s}(\omega))$, for $t = 0, 1, \dots, s-1$.

The map $Y_{-s,-t}$ depends only on the set $\{(e_m, U_m, W_{-m}) : t < m \le s\}$ of random variables. Therefore, the maps $Y_{-kL,-(k-1)L}$, $k = 1, 2, \dots$, are independent and identically distributed. Since each is a constant function with some fixed positive probability, there exists almost surely a (random) integer K such that $Y_{-KL,-(K-1)L}$ is a constant function. It follows that $M \le KL$, whence $\mathbb{P}(M < \infty) = 1$.

Let C be chosen randomly from Ω with law $\phi_{p,q}$, and write $C_m = Y_{-m}(C)$. Since the law of C is the unique invariant measure $\phi_{p,q}$ of the Gibbs sampler, C_m has law $\phi_{p,q}$ for all $m = 0, 1, 2, \dots$. By the definition of M,

$$Y_{-M} = C_m \qquad \text{on the event } \{M \le m\}.$$

For $\omega \in \Omega$ and $m = 0, 1, 2, \dots$,

$$\begin{aligned}\mathbb{P}(Y_{-M} = \omega) &= \mathbb{P}(Y_{-M} = \omega,\ M \le m) + \mathbb{P}(Y_{-M} = \omega,\ M > m)\\ &= \mathbb{P}(C_m = \omega,\ M \le m) + \mathbb{P}(Y_{-M} = \omega,\ M > m)\\ &\le \phi_{p,q}(\omega) + \mathbb{P}(M > m),\end{aligned}$$

and similarly,

$$\phi_{p,q}(\omega) = \mathbb{P}(C_m = \omega) \le \mathbb{P}(Y_{-M} = \omega) + \mathbb{P}(M > m).$$

We combine these two inequalities to obtain that

$$\left|\mathbb{P}(Y_{-M} = \omega) - \phi_{p,q}(\omega)\right| \le \mathbb{P}(M > m), \qquad \omega \in \Omega,$$

and we let $m \to \infty$ to obtain the result. □

8.5 Swendsen–Wang dynamics

It is a major target of statistical physics to understand the time-evolution of disordered systems, and a prime example lies in the study of the Ising model. A multiplicity of types of dynamics have been proposed. The majority of these share a property of 'locality' in the sense that the evolution involves changes to the states of vertices in close proximity to one another, perhaps single spin-flips or spin-exchanges. The state space is generally large, of size 2^N where N is the number of vertices, and the Hamiltonian may have complicated structure. When subjected to 'local dynamics', the process may approach equilibrium quite slowly[2]. Other forms of dynamics are 'non-local' in that they permit large moves around the state space relatively unconstrained by neighbourly relations, and such processes can approach equilibrium faster. The random-cluster model has played a role in the development of a simple but attractive system of non-local dynamics proposed by Swendsen and Wang [310] and described as follows for the Potts model with q states.

As usual, $G = (V, E)$ is a finite graph, typically a large box in $\mathbb{Z}^d$, and we let $q \in \{2, 3, \dots\}$. Consider a q-state Potts model on G, with state space $\Sigma = \{1, 2, \dots, q\}^V$ and parameter $\beta \in (0, \infty)$. The corresponding random-cluster model has state space $\Omega = \{0, 1\}^E$ and parameter $p = 1 - e^{-\beta}$. The Swendsen–Wang evolution for the Potts model is as follows.

Suppose that, at some time n, we have obtained a configuration $\sigma_n \in \Sigma$. We construct σ_{n+1} as follows.

I. Let $\omega_n \in \Omega$ be given by: for all $e = \langle x, y \rangle \in E$,

$$\text{if } \sigma_n(x) \neq \sigma_n(y), \quad \text{let } \omega_n(e) = 0,$$
$$\text{if } \sigma_n(x) = \sigma_n(y), \quad \text{let } \omega_n(e) = \begin{cases} 1 & \text{with probability } p, \\ 0 & \text{otherwise,} \end{cases}$$

different edges receiving independent states. The edge-configuration ω_n is carried forward to the next stage.

II. To each cluster C of the graph $(V, \eta(\omega_n))$ we assign an integer chosen uniformly at random from the set $\{1, 2, \dots, q\}$, different clusters receiving independent labels. Let $\sigma_{n+1}(x)$ be the value thus assigned to the cluster containing the vertex x.

(8.13) Theorem [310]. *The Markov chain $\sigma = (\sigma_n : n = 0, 1, 2, \dots)$ has as unique invariant measure the q-state Potts measure on Σ with parameter β.*

Proof. There is a strictly positive probability that $\omega_n(e) = 0$ for all $e \in E$. Therefore, $\mathbb{P}(\sigma_{n+1} = \sigma \mid \sigma_n = \sigma') > 0$ for all $\sigma, \sigma' \in \Sigma$, so that the chain is irreducible. The invariance of $\phi_{p,q}$ is a consequence of Theorem 1.13. □

The Swendsen–Wang algorithm generates a Markov chain $(\sigma_n : n = 0, 1, \dots)$. It is generally the case that this chain converges to the equilibrium Potts measure faster than time-evolutions defined via local dynamics. This is especially evident in the 'high β' (or 'low temperature') phase, for the following reason. Consider for example the simulation of an Ising model on a finite box with free boundary conditions, and suppose that the initial state is $+1$ at all vertices. If β is large, local dynamics result in samples that remain close to the '$+$ phase' for a very long time. Only after a long delay will the process achieve an average magnetization close to 0. Swendsen–Wang dynamics, on the other hand, can achieve large jumps in average magnetization even in a single step, since the spin allocated to a given large

[2]See [249, 292] for accounts of recent work of relevance.

cluster of the corresponding random-cluster model is equally likely to be either of the two possibilities. A rigorous analysis of rates of convergence is however incomplete. It turns out that, *at* the critical point, Swendsen–Wang dynamics approach equilibrium only slowly, [64]. A further discussion may be found in [136].

Algorithms of Swendsen–Wang type have been described for other statistical mechanical models with graphical representations of random-cluster-type, see [93, 94]. Related work may be found in [322].

8.6 Coupled dynamics on a finite graph

Let $G = (V, E)$ be a graph, possible infinite. Associated with G there is a family $\phi_{G,p,q}$ of random-cluster measures indexed by the parameters $p \in [0, 1]$ and $q \in (0, \infty)$; we defer a discussion of boundary conditions to the next section. It has proved fruitful to couple these measures, for fixed q, by finding a family $(Z_q(e) : e \in E)$ of random variables taking values in $[0, 1]$ whose 'level-sets' are governed by the $\phi_{G,p,q}$. It might be the case for example that, for any given $p \in (0, 1)$, the configuration $(Z_{p,q}(e) : e \in E)$ given by

$$Z_{p,q}(e) = \begin{cases} 1 & \text{if } Z_q(e) \le p, \\ 0 & \text{otherwise}, \end{cases}$$

has law $\phi_{G,p,q}$. Such a coupling has been valuable in the study of percolation theory (that is, when $q = 1$), where one may simply take a family of independent random variables $Z(e)$ with the uniform distribution on the interval $[0, 1]$, see [154, 178]. The picture for random-cluster measures is more complex owing to the dependence structure of the process. Such a coupling has been explored in detail in [152] but we choose here to follow a minor variant which might be termed a 'coupled Gibbs sampler'. We shall assume for the moment that G is finite, returning in the next two sections to the case of an infinite graph G.

Let $G = (V, E)$ be finite, and let $q \in [1, \infty)$. Let $X = [0, 1]^E$, and let $\mathcal{B}$ be the Borel σ-field of subsets of X, that is, the σ-field generated by the open subsets. We shall construct a Markov process $Z = (Z_t : t \ge 0)$ on the state space X, and we do this via a so-called graphical construction. We shall consider the states of edges chosen at random as time passes, and to this end we provide ourselves with a family of independent Poisson processes termed 'alarm clocks'. For each arrival-time of these processes, we shall require a uniformly distributed random variable.

(a) For each edge $e \in E$, let $A(e) = (A_n(e) : n = 1, 2, \dots)$ be the (increasing) sequence of arrival times of a Poisson process with intensity 1.

(b) Let $(\alpha_n(e) : e \in E,\ n = 1, 2, \dots)$ be a family of independent random variables each of which is uniformly distributed on the interval $[0, 1]$.

We write $\mathbb{P}$ for the appropriate probability measure[3].

Let $e = \langle x, y \rangle$, and let $\mathcal{P}_e$ be the set of paths in G having endvertices x and y but not using the edge e. Let $F : E \times X \to [0, 1]$ be given by

$$F(e, \nu) = \inf_{\pi \in \mathcal{P}_e} \max_{f \in \pi} \nu(f), \qquad e \in E, \ \nu \in X, \tag{8.14}$$

where the infimum of the empty set is taken to be 1. The maximum is taken over all edges f in the path π, and the infimum is taken over the finite set $\mathcal{P}_e$; we shall later consider situations in which $\mathcal{P}_e$ is infinite.

The state of the edge e may jump only at the times $A_n(e)$. When it jumps, it takes a new value which depends on the states of the other edges, and also on the value of the corresponding $\alpha_n(e)$. We describe next the value to which it will jump.

Suppose that the Poisson alarm clock at edge e rings at time $T = A_N(e)$, with corresponding uniform random variable $\alpha = \alpha_N(e)$. Let $\nu \in X$, let the current state of the process be $Z_{T-} = \nu$, and write $F = F(e, \nu)$. We define Z_T by

$$Z_T(f) = \begin{cases} \nu(f) & \text{if } f \neq e, \\ \rho(e) & \text{if } f = e, \end{cases} \tag{8.15}$$

where the new value $\rho(e)$ is given by

$$\rho(e) = \begin{cases} \alpha & \text{if } \alpha > F, \\ F & \text{if } \dfrac{F}{F + q(1-F)} < \alpha \leq F, \\ \dfrac{q\alpha}{1 - \alpha + q\alpha} & \text{if } \alpha \leq \dfrac{F}{F + q(1-F)}. \end{cases} \tag{8.16}$$

Since $q \in [1, \infty)$,

$$\frac{F}{F + q(1-F)} \leq F.$$

Here is a more formal definition of the process. Let $C(X)$ be the space of continuous real-valued functions on X. The generator S of the process is the mapping $S : C(X) \to C(X)$ given by

$$Sg(\nu) = \sum_{e \in E} \int_0^1 [g(\nu_e^u) - g(\nu)] \, dH_{e,\nu}(u), \qquad g \in C(X), \ \nu \in X. \tag{8.17}$$

Here, the configuration $\nu_e^u \in X$ is given by

$$\nu_e^u(f) = \begin{cases} \nu(f) & \text{if } f \neq e, \\ u & \text{if } f = e, \end{cases}$$

[3] In order to avoid certain standard difficulties later, we shall assume that the $A_n(e)$ are distinct, and that, for each $e \in E$, the set $\{A_n(e) : n = 1, 2, \ldots\}$ has no accumulation points. We adjust the probability space accordingly.

and, by (8.16), the distribution function $H_{e,\nu}(\cdot) : [0,1] \to [0,1]$ satisfies

$$H_{e,\nu}(u) = \begin{cases} u & \text{if } F \le u \le 1, \\ \dfrac{u}{u + q(1-u)} & \text{if } 0 \le u < F. \end{cases} \tag{8.18}$$

Equation (8.17) is to be interpreted as follows. Suppose the alarm clock at e rings at time T, and the current state of the process is ν. The local state at e jumps to a new value $\rho(e)$ which (conditional on ν) does not depend on its previous value and has distribution function $H_{e,\nu}$.

There follows the main theorem of this section. The proof is based on that to be found in [152] and is deferred until later in the section.

(8.19) Theorem. *The Markov process $Z = (Z_t : t \ge 0)$ has a unique invariant measure μ and, in particular, $Z_t \Rightarrow \mu$ as $t \to \infty$.*

The purpose of the above construction is to achieve a level-set representation of evolving random-cluster processes on G. Let $p \in [0,1]$, and recall that $\Omega = \{0,1\}^E$. We define two 'projection operators' $\Pi^p, \Pi_p : X \to \Omega$ by

$$\Pi^p \nu(e) = \begin{cases} 1 & \text{if } \nu(e) \le p, \\ 0 & \text{if } \nu(e) > p, \end{cases} \qquad e \in E,\ \nu \in X, \tag{8.20}$$

$$\Pi_p \nu(e) = \begin{cases} 1 & \text{if } \nu(e) < p, \\ 0 & \text{if } \nu(e) \ge p, \end{cases} \qquad e \in E,\ \nu \in X, \tag{8.21}$$

and point out that

$$\Pi_p \nu \le \Pi^p \nu, \qquad p \in [0,1],\ \nu \in X, \tag{8.22}$$

$$\Pi_{p_1} \nu_1 \le \Pi_{p_2} \nu_2, \quad \Pi^{p_1} \nu_1 \le \Pi^{p_2} \nu_2, \qquad p_1 \le p_2,\ \nu_1 \ge \nu_2. \tag{8.23}$$

In writing $\nu_1 \ge \nu_2$, we are using the partial order $\ge$ on X given by: $\nu_1 \ge \nu_2$ if and only if $\nu_1(e) \ge \nu_2(e)$ for all $e \in E$. A source of possible confusion later is that fact that $\Pi^p \nu$ and $\Pi_p \nu$ are *decreasing* functions of ν.

We concentrate next on the projected processes $\Pi^p Z = (\Pi^p Z_t : t \ge 0)$ and $\Pi_p Z = (\Pi_p Z_t : t \ge 0)$. An important difference between these two processes will become clear in the next section when we introduce boundary conditions.

(8.24) Theorem [152]. *Let $p \in (0,1)$.*

(a) *The process $\Pi^p Z = (\Pi^p Z_t : t \ge 0)$ is a Markov chain on the state space Ω with unique invariant distribution $\phi_{p,q}$, and it is reversible with respect to $\phi_{p,q}$. Furthermore,*

$$\Pi^{p_1} Z_t \le \Pi^{p_2} Z_t \quad \textit{for all } t, \qquad \textit{if } p_1 \le p_2. \tag{8.25}$$

(b) *Statement* (a) *is valid with the operator Π^p replaced throughout by Π_p.*

This theorem provides a coupling of the random-cluster measures $\phi_{p,q}$ for fixed $q \in [1,\infty)$ and varying p. We make two notes concerning the parameter q. First, the above construction may be extended in order to couple together random-cluster processes with different values of p *and* different values of $q \in [1, \infty)$. Secondly, some of the arguments of this section may be re-cast in the 'non-FKG' case when $q \in (0, 1)$.

It is noted that the level-set processes are reversible, unlike the process Z.

Proof of Theorem 8.24. (a) We begin with a calculation involving the function F defined in (8.14). Let $e \in E$, $\nu \in X$, and let $\gamma = \Pi^p \nu \in \Omega$. We claim that

$$F(e, \nu) > p \quad \text{if and only if} \quad \gamma = \Pi^p \nu \in D_e, \tag{8.26}$$

where $D_e \subseteq \Omega$ is the set of configurations in which the endvertices of e are joined by no open path of $E \setminus \{e\}$. This may be seen from (8.14) by noting that: $F(e, \nu) > p$ if and only if, for every $\pi \in \mathcal{P}_e$, there exists an edge $f \in \pi$ such that $\nu(f) > p$.

The projected process $\Pi^p Z$ changes its value only when Z changes its value. Assume that $Z_t = \nu$ and $\Pi^p Z_t = \Pi^p \nu = \gamma$. Let $\gamma' \in \Omega$. By the discussion around (8.16)–(8.18), the rate at which $\Pi^p Z$ jumps subsequently to the new state γ' depends only on:

(i) the arrival-times of the Poisson processes $A(e)$ subsequent to t,

(ii) the associated values of the random variables α, and

(iii) the set $F_\nu = \{e \in E : F(e, \nu) > p\}$ of edges.

By (8.26), $F_\nu = \{e \in E : \gamma \in D_e\}$, which depends on γ only and not further on ν. It follows that $\Pi^p Z = (\Pi^p Z_t : t \geq 0)$ is a time-homogeneous Markov chain on Ω. This argument is expanded in the following computation of the jump rates.

Let $Q = (q_{\gamma,\omega} : \gamma, \omega \in \Omega)$ denote the generator of the process $\Pi^p Z$. Since Z proceeds by local moves,

$$q_{\gamma,\omega} = 0 \quad \text{if} \quad H(\gamma, \omega) \geq 2,$$

where H denotes Hamming distance. It remains to calculate the terms q_{γ_e,γ^e} and q_{γ^e,γ_e} for $\gamma \in \Omega$ and $e \in E$. Consider first q_{γ_e,γ^e}. By (8.17),

$$\mathbb{P}\big(\Pi^p Z_{t+h} = \gamma^e \,\big|\, \Pi^p Z_t = \gamma_e\big) = h H_{e,\nu}(p) + \mathrm{o}(h) \qquad \text{as } h \downarrow 0,$$

whence, by (8.18) and (8.26),

$$q_{\gamma_e,\gamma^e} = H_{e,\nu}(p) = \begin{cases} \dfrac{p}{p + q(1-p)} & \text{if } \gamma \in D_e, \\ p & \text{if } \gamma \notin D_e. \end{cases} \tag{8.27}$$

By a similar argument,

$$q_{\gamma^e,\gamma_e} = 1 - H_{e,\nu}(p) = \begin{cases} \dfrac{q(1-p)}{p + q(1-p)} & \text{if } \gamma \in D_e, \\ 1 - p & \text{if } \gamma \notin D_e. \end{cases} \tag{8.28}$$

Therefore,

(8.29)
$$\frac{q_{\gamma_e,\gamma^e}}{q_{\gamma^e,\gamma_e}} = \frac{H_{e,\nu}(p)}{1-H_{e,\nu}(p)} = \begin{cases} \dfrac{p}{q(1-p)} & \text{if } \gamma \in D_e, \\ \dfrac{p}{1-p} & \text{if } \gamma \notin D_e, \end{cases}$$
$$= \frac{\phi_{p,q}(\gamma^e)}{\phi_{p,q}(\gamma_e)}.$$

It follows as in (8.3) that the detailed balance equations hold, and the process $\Pi^p Z$ is reversible with respect to $\phi_{p,q}$. That $\phi_{p,q}$ is the unique invariant measure is a consequence of the irreducibility of the chain. Inequality (8.25) follows by (8.23).
(b) A similar argument is valid with Π^p replaced by Π_p, and (8.26) by

(8.30)
$$F(e,\nu) \geq p \quad \text{if and only if} \quad \gamma = \Pi_p \nu \in D_e,$$

and on replacing $H_{e,\nu}(p)$ by

$$H_{e,\nu}(p-) = \lim_{u \uparrow p} H_{e,\nu}(u)$$

in the calculations (8.27)–(8.29). □

We turn now to the proof of Theorem 8.19, which is preceeded by a lemma. The product space $X = [0,1]^E$ is equipped with the Borel σ-field $\mathcal{B}$. An event $A \in \mathcal{B}$ is called *increasing* if it has the property that $\nu' \in A$ whenever there exists $\nu \in A$ such that $\nu \leq \nu'$, and it is called *decreasing* if its complement is increasing. For $\zeta \in X$, let $Z^\zeta = (Z_t^\zeta : t \geq 0)$ denote the above Markov process with initial state $Z_0 = \zeta$.

(8.31) Lemma.

(a) *If $\zeta \leq \nu$ then $Z_t^\zeta \leq Z_t^\nu$ for all t.*

(b) *Let E be an increasing event in $\mathcal{B}$. The function $g^b(t) = \mathbb{P}(Z_t^b \in E)$ is non-decreasing in t if $b = 0$, and is non-increasing if $b = 1$.*

Proof of Lemma 8.31. (a) This follows from the transition rules (8.15)–(8.16) together with the fact that $F(e,\nu)$ is non-decreasing in ν.
(b) Using conditional expectation,

$$g^b(s+t) = \mathbb{P}\big\{\mathbb{P}(Z_{s+t}^b \in E \mid Z_s^b)\big\}, \qquad b = 0, 1.$$

By the time-homogeneity of the processes (A, α), the fact that $0 \leq Z_s^b \leq 1$, and part (a),

$$g^b(s+t) \begin{cases} \geq g^b(t) & \text{if } b = 0, \\ \leq g^b(t) & \text{if } b = 1. \end{cases}$$
□

Proof of Theorem 8.19. In order to prove the existence of a unique invariant probability measure μ, we shall prove that Z_t converges weakly as $t \to \infty$, and we shall write μ for the weak limit. By Lemma 8.31(a),

$$Z_t^0 \le Z_t^\nu \le Z_t^1, \qquad t \ge 0,\ \nu \in X. \tag{8.32}$$

By Lemma 8.31(b), Z_t^b is stochastically increasing in t if $b = 0$, and stochastically decreasing if $b = 1$. It therefore suffices to show that

$$Z_t^1 - Z_t^0 \Rightarrow 0 \qquad \text{as } t \to \infty. \tag{8.33}$$

Let $\epsilon > 0$, and write $\mathcal{E} = \{k/N : k = 1, 2, \dots, N-1\}$ where N is a positive integer satisfying $N^{-1} < \epsilon$. Then

$$\mathbb{P}\big(|Z_t^1(e) - Z_t^0(e)| > \epsilon \text{ for some } e \in E\big) \le \sum_{e\in E}\sum_{p\in\mathcal{E}} \mathbb{P}\big(Z_t^0(e) < p < Z_t^1(e)\big).$$

Now,

$$\begin{aligned}\mathbb{P}\big(Z_t^0(e) < p < Z_t^1(e)\big) &\le \mathbb{P}(\Pi_p Z_t^0(e) = 1) - \mathbb{P}(\Pi_p Z_t^1(e) = 1)\\ &\to 0 \qquad \text{as } t \to \infty\end{aligned}$$

by the ergodicity of the Markov chain $\Pi_p Z$, see Theorem 8.24. □

8.7 Box dynamics with boundary conditions

In the last section, we constructed a Markov process Z on the state space $X = [0,1]^E$ for a *finite* edge-set E. In moving to an infinite graph, we shall require a discussion of boundary conditions. Let $d \ge 1$ and $X = [0,1]^{\mathbb{E}^d}$, a compact metric space when equipped with the Borel σ-field $\mathcal{B}$ generated by the open sets.

Since our target is to study processes on the lattice $\mathbb{L}^d = (\mathbb{Z}^d, \mathbb{E}^d)$, we shall assume for convenience that our finite graphs are boxes in this lattice. Let Λ be such a box. For $\zeta \in X$, let

$$X_\Lambda^\zeta = \big\{\nu \in X : \nu(e) = \zeta(e) \text{ for } e \notin \mathbb{E}_\Lambda\big\}. \tag{8.34}$$

As in (8.14), we define $F : \mathbb{E}^d \times X \to \mathbb{R}$ by

$$F(e,\nu) = \inf_{\pi\in\mathcal{P}_e} \max_{f\in\pi} \nu(f), \qquad e = \langle x, y\rangle \in \mathbb{E}^d,\ \nu \in X, \tag{8.35}$$

where $\mathcal{P}_e$ is the (infinite) set of all (finite) paths of $\mathbb{E}^d \setminus \{e\}$ that join x to y.

Let $q \in [1, \infty)$. We provide ourselves[4] with a family of independent Poisson processes $A(e) = (A_n(e) : n = 1, 2, \dots)$, $e \in \mathbb{E}^d$, with intensity 1, and an associated collection $(\alpha_n(e) : e \in \mathbb{E}^d,\ n = 1, 2, \dots)$ of independent random variables with the uniform distribution on $[0, 1]$. Let $\zeta \in X$. The above variables may be used as in the last section to construct a family of coupled Markov processes $Z_\Lambda^\zeta = (Z_{\Lambda,t}^\zeta : t \geq 0)$ taking values in X_Λ^ζ and indexed by the pair Λ, ζ. The process Z_Λ^ζ has generator S_Λ^ζ given by (8.17)–(8.18) for $\nu \in X_\Lambda^\zeta$ and with $F = F(e, \nu)$ given in (8.35).

As in Lemma 8.31(a),

$$Z_{\Lambda,t}^\zeta \leq Z_{\Lambda,t}^\nu, \qquad \zeta \leq \nu, \quad t \geq 0. \tag{8.36}$$

For $\nu, \zeta \in X$ and a box Λ, we denote by (ν, ζ) $[= (\nu, \zeta)_\Lambda \in X]$ the composite configuration that agrees with ν on $\mathbb{E}_\Lambda$ and with ζ off $\mathbb{E}_\Lambda$. We sometimes suppress the subscript Λ when using this notation. For example, the expression $Z_{\Delta,t}^{(\nu,\zeta)}$ denotes the value of the process on the box Δ at time t, with initial value $(\nu, \zeta)_\Delta$. Finally, with Π^p, Π_p given as in (8.20)–(8.21), we write Υ_Λ^p for the set of all $\zeta \in X$ with the property that $\Pi^p[(1, \zeta)_\Lambda]$ has at most one infinite cluster.

(8.37) Theorem. *Let $\zeta \in X$ and let Λ be a box of $\mathbb{L}^d$. The Markov process $Z_\Lambda^{(\nu,\zeta)} = (Z_{\Lambda,t}^{(\nu,\zeta)} : t \geq 0)$, viewed as a process on $(X, \mathcal{B})$, has a unique invariant measure μ_Λ^ζ and, in particular, $Z_{\Lambda,t}^{(\nu,\zeta)} \Rightarrow \mu_\Lambda^\zeta$ as $t \to \infty$.*

We turn as before to the projected processes $\Pi^p Z_\Lambda^\zeta$ and $\Pi_p Z_\Lambda^\zeta$. A complication arises in the case of the first of these, depending on whether or not $\zeta \in \Upsilon_\Lambda^p$.

(8.38) Theorem. *Let $p \in (0, 1)$, $\zeta \in X$, and let Λ be a box of $\mathbb{L}^d$.*

(a) *The process $\Pi_p Z_\Lambda^\zeta = (\Pi_p Z_{\Lambda,t}^\zeta : t \geq 0)$ is a Markov chain on the state space $\Pi_p X_\Lambda^\zeta$ with unique invariant measure $\phi_{\Lambda,p,q}^{\Pi_p\zeta}$, and it is reversible with respect to this measure. Furthermore,*

$$\Pi_{p_1} Z_{\Lambda,t}^\zeta \leq \Pi_{p_2} Z_{\Lambda,t}^\zeta \quad \textit{for } t \geq 0, \qquad \textit{if } p_1 \leq p_2. \tag{8.39}$$

(b) *Assume that $\zeta \in \Upsilon_\Lambda^p$. Statement* (a) *is valid with the operator Π_p replaced throughout by Π^p.*

We note two further facts for future use. First, there is a sample-path monotonicity of the graphical representation which will enable us to pass to the limit of the *processes* Z_Λ^ζ as $\Lambda \uparrow \mathbb{Z}^d$. Secondly, if ν and ζ are members of X that are close to one another, then so are $Z_{\Lambda,t}^{(\nu,b)}$ and $Z_{\Lambda,t}^{(\zeta,b)}$, for $b \in \{0, 1\}$. These observations are made formal as follows.

---

[4] We make the same assumption as in the footnote on page 233.

(8.40) Lemma.

(a) *Let* Λ *and* Δ *be boxes satisfying* $\Lambda \subseteq \Delta$. *Then*:

$$Z_{\Lambda,t}^{(\zeta,0)} \le Z_{\Delta,t}^{(\zeta,0)}, \qquad \zeta \in X,\ t \ge 0, \tag{8.41}$$

$$Z_{\Lambda,t}^{(\zeta,1)} \ge Z_{\Delta,t}^{(\zeta,1)}, \qquad \zeta \in X,\ t \ge 0. \tag{8.42}$$

(b) *Let* Λ *be a box*, $b \in \{0,1\}$, *and* $\nu, \zeta \in X$. *Then*:

$$\left|Z_{\Lambda,t}^{(\nu,b)}(e) - Z_{\Lambda,t}^{(\zeta,b)}(e)\right| \le \max_{f \in \mathbb{E}_\Lambda} \left\{|\nu(f) - \zeta(f)|\right\}, \qquad t \ge 0,\ e \in \mathbb{E}_\Lambda. \tag{8.43}$$

Proof of Theorem 8.38. (a) The projected process $(\Pi_p Z_{\Lambda,t}^{\zeta} : t \ge 0)$ takes values in the state space $\Omega_\Lambda^{\Pi_p \zeta} = \Pi_p X_\Lambda^{\zeta}$. The proof now follows that of Theorem 8.24(b), the key observation being that (8.30) remains valid with D_e the set of all configurations in $\Omega = \{0,1\}^{\mathbb{E}^d}$ such that the endvertices of e are joined by no open path of $\mathbb{E}^d \setminus \{e\}$.

(b) The claim will follow as in Theorem 8.24(a) once we have proved (8.26) for $\nu \in \Upsilon_\Lambda^p$. We are thus required to show that:

(8.44) $\qquad$ for $\nu \in \Upsilon_\Lambda^p$, $\qquad F(e,\nu) > p$ if and only if $\gamma = \Pi^p \nu \in D_e$.

Let $e \in \mathbb{E}_\Lambda$ and $\nu \in X$. If $F(e,\nu) > p$, then $\Pi^p \nu \in D_e$. Suppose conversely that $\nu \in \Upsilon_\Lambda^p$ and $\Pi^p \nu \in D_e$. By the definition (8.20) of $\Pi^p \nu$, the function $\mu : \mathcal{P}_e \to [0,1]$ given by

$$\mu(\pi) = \max_{f \in \pi} \nu(f), \qquad \pi \in \mathcal{P}_e,$$

satisfies

$$\mu(\pi) > p, \qquad \pi \in \mathcal{P}_e.$$

By (8.35), $F(e,\nu) \ge p$. Suppose $F(e,\nu) = p$. There exists an infinite sequence $(\pi_n : n = 1, 2, \dots)$ of distinct paths in $\mathcal{P}_e$ such that $\mu(\pi_n) > p$ and $\mu(\pi_n) \to p$ as $n \to \infty$. Let $\mathcal{E}$ be the set of edges belonging to infinitely many of the paths π_n. Now,

$$\nu(f) \le \lim_{n \to \infty} \mu(\pi_n) = p, \qquad f \in \mathcal{E},$$

so that $\Pi^p \nu(f) = 1$ for $f \in \mathcal{E}$.

Write $e = \langle x, y \rangle$, and let $C(x)$ (respectively, $C(y)$) denote the set of vertices of $\mathbb{L}^d$ joined to x (respectively, y) by paths comprising edges f with $\Pi^p \nu(f) = 1$. By a counting argument, we have that x (respectively, y) lies in some infinite path of $\mathcal{E}$, and therefore $|C(x)| = |C(y)| = \infty$. Since $\nu \in \Upsilon_\Lambda^p$, $\Pi^p \nu$ has at most one infinite cluster. Therefore, $C(x) = C(y)$, whence $\Pi^p \nu \notin D_e$, a contradiction. This proves that $F(e,\nu) > p$, as required for (8.44). $\square$

Proof of Theorem 8.37. This follows the proof of Theorem 8.19, with Theorem 8.38 used in place of Theorem 8.24. □

Proof of Lemma 8.40. (a) We shall consider (8.41), inequality (8.42) being exactly analogous. Certainly,

$$0 = Z_{\Lambda,t}^{(\zeta,0)}(e) \le Z_{\Delta,t}^{(\zeta,0)}(e), \qquad e \in \mathbb{E}^d \setminus \mathbb{E}_\Lambda.$$

Let $e \in \mathbb{E}_\Lambda$, and note that $Z_{\Lambda,0}^{(\zeta,0)}(e) = Z_{\Delta,0}^{(\zeta,0)}(e)$, since $\Lambda \subseteq \Delta$. It suffices to check that, at each ring of the alarm clock on the edge e, the process $Z_{\Lambda,\cdot}^{(\zeta,0)}(e)$ cannot jump above $Z_{\Delta,\cdot}^{(\zeta,0)}(e)$. As in Lemma 8.31(a), this is a consequence of the transition rules (8.15)–(8.16) on noting that $F(e,\nu)$ is non-decreasing in ν.
(b) Let $b \in \{0,1\}$ and $\nu, \zeta \in X$. It suffices to show that

$$M_t = \max_{f \in \mathbb{E}_\Lambda} \left\{ \left| Z_{\Lambda,t}^{(\nu,b)}(f) - Z_{\Lambda,t}^{(\zeta,b)}(f) \right| \right\} \tag{8.45}$$

is a non-increasing function of t. Now, M_t is constant except when an alarm clock rings. Suppose that $A_N(e) = T$ for some $N \ge 1$ and $e \in \mathbb{E}_\Lambda$. It is enough to show that

$$\left| Z_{\Lambda,T}^{(\nu,b)}(e) - Z_{\Lambda,T}^{(\zeta,b)}(e) \right| \le M_{T-}. \tag{8.46}$$

By (8.35),

$$\left| F(e,\xi) - F(e,\xi') \right| \le \max_{f \in \mathbb{E}^d} \left\{ |\xi(f) - \xi'(f)| \right\}, \qquad \xi, \xi' \in X,$$

and (8.46) follows by (8.16). □

8.8 Coupled dynamics on the infinite lattice

The reader is reminded of the assumption that $q \in [1,\infty)$. We have constructed two Markov processes $Z_\Lambda^b = (Z_{\Lambda,t}^b : t \ge 0)$ on the state space $X = [0,1]^{\mathbb{E}^d}$, indexed by the finite box Λ and the boundary condition $b \in \{0,1\}$. Similar processes may be constructed on the infinite lattice $\mathbb{L}^d$ by passing to limits 'pathwise', and exploiting the monotonicity in Λ of the processes Z_Λ^b.

The following (monotone) limits exist by Lemma 8.40,

$$Z_t^{(\zeta,0)} = \lim_{\Lambda \uparrow \mathbb{Z}^d} Z_{\Lambda,t}^{(\zeta,0)}, \quad Z_t^{(\zeta,1)} = \lim_{\Lambda \uparrow \mathbb{Z}^d} Z_{\Lambda,t}^{(\zeta,1)}, \tag{8.47}$$

and satisfy

$$Z_t^{(\zeta,0)} \le Z_t^{(\zeta,1)}, \qquad \zeta \in X,\ t \ge 0. \tag{8.48}$$

We write in particular

$$Z_t^0 = Z_t^{(0,0)}, \quad Z_t^1 = Z_t^{(1,1)}. \tag{8.49}$$

It is proved in this section that the processes $Z^b = (Z_t^b : t \geq 0)$, $b = 0, 1$, are Markovian, and that their level-set invariant measures are the free and wired random-cluster measures $\phi_{p,q}^b$. The arguments of this section are those of [152], where closely related results are obtained.

The state space $X = [0,1]^{\mathbb{E}^d}$ is a compact metric space equipped with the Borel σ-field $\mathcal{B}$ generated by the open sets. Let $B(X)$ denote the space of bounded measurable functions from X to $\mathbb{R}$, and $C(X)$ the space of continuous functions.

We now introduce two transition functions and semigroups, as follows[5]. For $b \in \{0,1\}$ and $t \geq 0$, let

$$P_t^b(\zeta, A) = \mathbb{P}(Z_t^{(\zeta,b)} \in A), \qquad \zeta \in X,\ A \in \mathcal{B}, \tag{8.50}$$

and let $S_t^b : B(X) \to B(X)$ be given by

$$S_t^b g(\zeta) = \mathbb{P}(g(Z_t^{(\zeta,b)})), \qquad \zeta \in X,\ g \in B(X). \tag{8.51}$$

(8.52) Theorem. *Let $b \in \{0,1\}$. The process $Z^b = (Z_t^b : t \geq 0)$ is a Markov process with Markov transition functions $(P_t^b : t \geq 0)$.*

(8.53) Theorem. *There exists a translation-invariant probability measure μ on $(X, \mathcal{B})$ that is the unique invariant measure of each of the two processes Z^0, Z^1. In particular, $Z_t^0, Z_t^1 \Rightarrow \mu$ as $t \to \infty$.*

By the last theorem and monotonicity (see (8.36) and (8.47)),

$$Z_t^{(\zeta,b)} \Rightarrow \mu \quad \text{as } t \to \infty, \qquad \zeta \in X,\ b = 0, 1. \tag{8.54}$$

The 'level-set processes' of Z_t^0 and Z_t^1 are given as follows. Let $p \in (0,1)$, and write

$$L_{p,t}^0 = \Pi_p Z_t^1, \quad L_{p,t}^1 = \Pi^p Z_t^0, \qquad t \geq 0, \tag{8.55}$$

where the projections Π^p and Π_p are defined in (8.20)–(8.21). Note the apparent reversal of boundary conditions in (8.55).

---

[5] A possible alternative to the methodology of this section might be the 'martingale method' described in [186, 235]. For general accounts of the theory of Markov processes, the reader may consult the books [51, 113, 235, 299].

(8.56) Theorem.

(a) *Let $b \in \{0, 1\}$ and $p \in (0, 1)$. The process L_p^b is a Markov process on the state space $\Omega = \{0, 1\}^{\mathbb{E}^d}$, with as unique invariant measure the random-cluster measure $\phi_{p,q}^b$ on $\mathbb{L}^d$. The process L_p^b is reversible with respect to $\phi_{p,q}^b$.*

(b) *The measures $\phi_{p,q}^b$, $b = 0, 1$, are 'level-set' measures of the invariant measure μ of Theorem 8.53 in the sense that, for $A \in \mathcal{F}$,*

$$\phi_{p,q}^0(A) = \mu\big(\{\zeta : \Pi_p \zeta \in A\}\big), \quad \phi_{p,q}^1(A) = \mu\big(\{\zeta : \Pi^p \zeta \in A\}\big). \quad (8.57)$$

We make several remarks before proving the above theorems. First, the invariant measures $\phi_{p,q}^0$ and $\phi_{p,q}^1$ of Theorem 8.56 are identical if and only if $p \notin \mathcal{D}_q$, where $\mathcal{D}_q$ is that of Theorem 4.63.

Secondly, with μ as in Theorem 8.53, and $e \in \mathbb{E}^d$, let $J : [0, 1] \to [0, 1]$ be given by

$$J(x) = \mu\big(\{\zeta \in X : \zeta(e) = x\}\big), \qquad x \in [0, 1]. \quad (8.58)$$

Thus, J is the atomic component of the marginal measure of μ at the edge e and, by translation-invariance, it does not depend on the choice of e. We recall from (4.61) the edge-densities

$$h^b(p, q) = \phi_{p,q}^b(e \text{ is open}), \qquad b = 0, 1.$$

(8.59) Proposition. *It is the case that*

$$J(p) = h^1(p, q) - h^0(p, q), \qquad p \in (0, 1).$$

We deduce by Theorem 4.63 that $p \in \mathcal{D}_q$ if and only if $J(p) \neq 0$, thereby providing a representation of $\mathcal{D}_q$ in terms of atoms of the weak limit μ. This may be used to prove the left-continuity of the percolation probability $\theta^0(\cdot, q)$. See Theorem 5.16(a), the proof of which is included at the end of the current section.

As discussed after Theorem 4.63, it is believed that there exists $Q = Q(d)$ such that

$$\mathcal{D}_q = \begin{cases} \varnothing & \text{if } q < Q, \\ \{p_c(q)\} & \text{if } q > Q, \end{cases}$$

and it is a first-rate challenge to prove this. The above results provide some incomplete probabilistic justification for such a claim, as follows. The set $\mathcal{D}_q$ is the set of atoms of the one-dimensional marginal measure of μ. Such atoms arise presumably through an accumulation of edges e having the same value $Z_t^b(e)$. Two edges e and f acquire the same state in the process Z by way of transitions at some time T for which, say, the alarm clock at e rings and $F = F(e, Z_{T-}) = Z_{T-}(f)$. Discounting events with probability zero, this can occur only when the new state at e is at the (unique) atom of the function $H_{e,\nu}$ in (8.18), where $\nu = Z_{T-}$. The size of this atom is

$$F - \frac{F}{F + q(1 - F)}$$

which is an increasing function of q. This is evidence that the number of pairs e, f of edges having the same state increases with q.

Finally, we describe the transition rules of the projected processes L_p^0 and L_p^1. It turns out that the transition mechanisms of these two chains differ in an interesting but ultimately unimportant regard. It is convenient to summarize the following discussion by writing down the two infinitesimal generators.

Let $e = \langle x, y \rangle \in \mathbb{E}^d$. As in (8.35), let $\mathcal{P}_e$ be the set of all paths of $\mathbb{E}^d \setminus \{e\}$ that join x to y. Let $\mathcal{Q}_e$ be the set of all pairs $\alpha = (\alpha_1, \alpha_2, \dots)$, $\beta = (\beta_1, \beta_2, \dots)$ of vertex-disjoint semi-infinite paths (where α_i and β_j are the vertices of these paths) with $\alpha_1 = x$ and $\beta_1 = y$; we require $\alpha_i \neq \beta_j$ for all i, j. Thus $\mathcal{Q}_e$ comprises pairs (α, β) of paths and, for $\omega \in \Omega$, we call an element (α, β) of $\mathcal{Q}_e$ *open* if all the edges of both α and β are open.

For $b = 0, 1$, let G^b be the linear operator, with domain a suitable subset of $C(\Omega)$, given by

(8.60)
$$G^b g(\omega) = \sum_{e \in \mathbb{E}^d} \left[q^b_{\omega,\omega^e} \{g(\omega^e) - g(\omega)\} + q^b_{\omega,\omega_e} \{g(\omega_e) - g(\omega)\} \right], \qquad \omega \in \Omega,$$

where

$$q^b_{\omega,\omega^e} = p(1 - 1_{D_e^b}) + \frac{p}{p + q(1-p)} 1_{D_e^b}, \tag{8.61}$$

$$q^b_{\omega,\omega_e} = 1 - q^b_{\omega,\omega^e} = (1-p)(1 - 1_{D_e^b}) + \frac{q(1-p)}{p + q(1-p)} 1_{D_e^b}, \tag{8.62}$$

with

$$D_e^0 = \{\text{no path in } \mathcal{P}_e \text{ is open}\}, \tag{8.63}$$

$$D_e^1 = \{\text{no element in } \mathcal{P}_e \cup \mathcal{Q}_e \text{ is open}\}. \tag{8.64}$$

Note that $G^b g$ is well defined for all cylinder functions g, since the infinite sum in (8.60) may then be written as a finite sum. However, $G^b g$ is not generally continuous when $q \in (1, \infty)$, even for cylinder functions g. For example, let $q \in (1, \infty)$, let g be the indicator function of the event that a given edge e is open, and let ω be a configuration satisfying:

(a) $\omega(e) = 1$,

(b) no path in $\mathcal{P}_e$ is open in ω,

(c) some pair (α, β) in $\mathcal{Q}_e$ is open in ω.

Then $G^b g(\omega) = -q^b_{\omega,\omega_e}$. However, q^b_{ω,ω_e} is discontinuous at ω for $b = 0, 1$ since, for every finite box Λ, there exists $\rho \in \Omega$ agreeing with ω on $\mathbb{E}_\Lambda$ such that

$q^b_{\rho,\rho_e} \neq q^b_{\omega,\omega_e}$. However, the set of configurations satisfying (a), (b), (c) has zero $\phi^b_{p,q}$-probability since such configurations have two or more infinite open clusters (see the remark following Theorem 4.34). One may see that the Markov transition functions of L^0_p and L^1_p are not Feller, see Proposition 8.90.

We shall make use of the following two lemmas in describing the transition rules of the processes L^0_p and L^1_p. Let $G : \mathbb{E}^d \times X \to [0,1]$ be given by

$$G(e,\nu) = \inf_{\pi \in \mathcal{P}_e \cup \mathcal{Q}_e} \sup_{f \in \pi} \nu(f), \qquad e \in \mathbb{E}^d,\ \nu \in X. \tag{8.65}$$

Here, $\mathcal{P}_e$ contains certain paths π, and $\mathcal{Q}_e$ contains certain pairs $\pi = (\alpha, \beta)$ of paths; for $\pi = (\alpha, \beta) \in \mathcal{Q}_e$, the infimum in (8.65) is over all edges f lying in the union of α and β.

(8.66) Lemma. *Let $e \in \mathbb{E}^d$, $\nu \in X$, and let $(\nu_\Lambda : \Lambda \subseteq \mathbb{Z}^d)$ be a family of elements of X indexed by boxes Λ.*

(a) *If $\nu_\Lambda \downarrow \nu$ as $\Lambda \uparrow \mathbb{Z}^d$, then*

$$F(e,\nu_\Lambda) \downarrow F(e,\nu) \qquad \text{as } \Lambda \uparrow \mathbb{Z}^d. \tag{8.67}$$

(b) *If $\nu_\Lambda \in X^0_\Lambda$ and $\nu_\Lambda \uparrow \nu$ as $\Lambda \uparrow \mathbb{Z}^d$, then*

$$F(e,\nu_\Lambda) \uparrow G(e,\nu) \qquad \text{as } \Lambda \uparrow \mathbb{Z}^d. \tag{8.68}$$

(8.69) Lemma. *Let $e \in \mathbb{E}^d$ and $\nu \in X$. Then:*

(i) *$p \le F(e,\nu)$ if and only if $\Pi_p \nu \in D^0_e$,*

(ii) *$p < G(e,\nu)$ if and only if $\Pi^p \nu \in D^1_e$.*

Consider the process $L^0_p = \Pi_p Z^1_t$. Since Z^1_t is the decreasing limit of $Z^{(1,1)}_{\Lambda,t}$,

$$L^0_{p,t} = \lim_{\Lambda \uparrow \mathbb{Z}^d} \Pi_p Z^{(1,1)}_{\Lambda,t}. \tag{8.70}$$

Fix $t \ge 0$, and write $\zeta_\Lambda = Z^{(1,1)}_{\Lambda,t}$ and $\zeta = \lim_{\Lambda \uparrow \mathbb{Z}^d} \zeta_\Lambda$, so that

$$L^0_{p,t} = \Pi_p \zeta = \lim_{\Lambda \uparrow \mathbb{Z}^d} \Pi_p \zeta_\Lambda. \tag{8.71}$$

Let $e \in \mathbb{E}^d$, and assume first that ζ is such that $\Pi_p \zeta(e) = 0$. At what rate does the state of e change from 0 to 1 in the process L^0_p? Since $\zeta(e) \ge p$, we have that $\zeta_\Lambda(e) \ge p$ for all Λ. The process $\Pi_p Z^{(1,1)}_{\Lambda,\cdot}$ acquires[6] the edge e at rate $H_{e,\zeta_\Lambda}(p-)$ given by (8.18) with $F = F(e,\zeta_\Lambda)$. Now,

$$H_{e,\zeta_\Lambda}(p-) = \begin{cases} p & \text{if } p > F(e,\zeta_\Lambda), \\ \dfrac{p}{p+q(1-p)} & \text{if } p \le F(e,\zeta_\Lambda), \end{cases}$$

[6]We speak of a process 'acquiring' (respectively, 'losing') the edge e when the state of e changes from closed to open (respectively, open to closed).

which by Lemma 8.66(a) converges as $\Lambda \uparrow \mathbb{Z}^d$ to

$$H_{e,\zeta}(p-) = \begin{cases} p & \text{if } p > F(e,\zeta), \\ \dfrac{p}{p+q(1-p)} & \text{if } p \le F(e,\zeta). \end{cases}$$

Thus, by Lemma 8.69(i), L_p^0 acquires the edge e at rate

$$\begin{cases} p & \text{if } \Pi_p\zeta \notin D_e^0, \\ \dfrac{p}{p+q(1-p)} & \text{if } \Pi_p\zeta \in D_e^0. \end{cases} \tag{8.72}$$

Assume next that ζ is such that $\Pi_p\zeta(e) = 1$, and consider the rate at which L_p^0 loses the edge e. Since $\zeta(e) < p$, we have that $\zeta_\Lambda(e) < p$ for all large Λ. As above, $\Pi_p Z_{\Lambda,\cdot}^{(1,1)}$ loses e at rate $1 - H_{e,\zeta_\Lambda}(p-)$, whence L_p^0 loses e at rate

$$\begin{cases} 1-p & \text{if } \Pi_p\zeta \notin D_e^0, \\ \dfrac{q(1-p)}{p+q(1-p)} & \text{if } \Pi_p\zeta \in D_e^0. \end{cases} \tag{8.73}$$

These calculations are in agreement with (8.60) with $b = 0$.

We turn next to the process $L_{p,t}^1 = \Pi^p Z_t^0$. This time, Z_t^0 is the increasing limit of $Z_{\Lambda,t}^0$ as $\Lambda \uparrow \mathbb{Z}^d$, and

$$L_{p,t}^1 = \lim_{\Lambda \uparrow \mathbb{Z}^d} \Pi^p Z_{\Lambda,t}^{(0,0)}. \tag{8.74}$$

The above argument is followed, noting that decreasing limits are replaced by increasing limits, Π_p by Π^p, $F(e,\nu)$ by $G(e,\nu)$, and D_e^0 by D_e^1. The conclusion is in agreement with (8.60) with $b = 1$.

Proof of Lemma 8.66. (a) Let $e \in \mathbb{E}^d$ and $\nu_\Lambda \downarrow \nu$ as $\Lambda \uparrow \mathbb{Z}^d$. Certainly $F(e,\nu_\Lambda)$ is non-increasing in Λ, whence the limit

$$\lambda = \lim_{\Lambda \uparrow \mathbb{Z}^d} F(e,\nu_\Lambda)$$

exists and satisfies $\lambda \ge F(e,\nu)$. We prove next that

$$\lambda \le F(e,\nu). \tag{8.75}$$

Since $F(e,\nu_\Lambda) \ge \lambda$ for all Λ, by (8.35),

$$\forall \pi \in \mathcal{P}_e,\ \forall \Lambda,\ \exists f \in \pi \text{ with } \nu_\Lambda(f) \ge \lambda.$$

Since all paths in $\mathcal{P}_e$ are finite, this implies

$$\forall \pi \in \mathcal{P}_e,\ \exists f \in \pi \text{ with } \nu(f) \ge \lambda,$$

which implies (8.75). We deduce as required that $\lambda = F(e, \nu)$.

(b) Let $e = \langle x, y \rangle \in \mathbb{E}^d$. Suppose $\nu_\Lambda \in X_\Lambda^0$ and $\nu_\Lambda \uparrow \nu$ as $\Lambda \uparrow \mathbb{Z}^d$. We prove first that the increasing limit

$$\lambda = \lim_{\Lambda \uparrow \mathbb{Z}^d} F(e, \nu_\Lambda) \tag{8.76}$$

satisfies

$$\lambda \geq G(e, \nu). \tag{8.77}$$

Let $\delta \in (0, 1)$, and suppose $G(e, \nu) > \delta$; we shall deduce that $\lambda > \delta$, thus obtaining (8.77).

A finite set S of edges of $\mathbb{L}^d$ is called a *cutset* (for e) if:

(i) $e \notin S$,

(ii) every path in $\mathcal{P}_e$ contains at least one edge of S,

(iii) S is minimal with the two properties above, in the sense that no strict subset of S satisfies (i) and (ii).

We claim that:

$$\text{there exists a cutset } S \text{ with } \nu(f) > \delta \text{ for all } f \in S, \tag{8.78}$$

and we prove this as follows. First, we write $G(e, \nu) = \min\{A, B\}$ where

$$A = F(e, \nu) = \inf_{\pi \in \mathcal{P}_e} \max_{f \in \pi} \nu(f), \quad B = \inf_{\pi \in \mathcal{Q}_e} \sup_{f \in \pi} \nu(f). \tag{8.79}$$

Since $G(e, \nu) > \delta$, we have that $A, B > \delta$. For $w \in \mathbb{Z}^d$, let $C_w(\nu)$ denote the set of vertices of $\mathbb{L}^d$ that are connected to w by paths π of $\mathbb{L}^d$ satisfying:

(a) π does not contain the edge e, and

(b) every edge f of π satisfies $\nu(f) \leq \delta$.

If $x \in C_y(\nu)$, then there exists $\pi \in \mathcal{P}_e$ with $\nu(f) \leq \delta$ for all $f \in \pi$, which contradicts the fact that $A > \delta$. Therefore $x \notin C_y(\nu)$. Furthermore, either $C_x(\nu)$ or $C_y(\nu)$ (or both) is finite, since if both were infinite, then there would exist $\pi = (\alpha, \beta) \in \mathcal{Q}_e$ with $\nu(f) \leq \delta$ for all f in α and β, thereby contradicting the fact that $B > \delta$. We may suppose without loss of generality that $C_x(\nu)$ is finite, and we let R be the subset of $\mathbb{E}^d \setminus \{e\}$ containing all edges g with exactly one endvertex in $C_x(\nu)$. Certainly $\nu(g) > \delta$ for all $g \in R$, and additionally every path in $\mathcal{P}_e$ contains some edge of R. However, R may fail to be minimal with the last property, in which case we replace R by a subset $S \subseteq R$ that is minimal. The set S is the required cutset, and (8.78) is proved.

Since S is finite and $\nu(f) > \delta$ for all $f \in S$,

$$\text{for all large } \Lambda \text{ and all } f \in S, \quad \nu_\Lambda(f) > \delta,$$

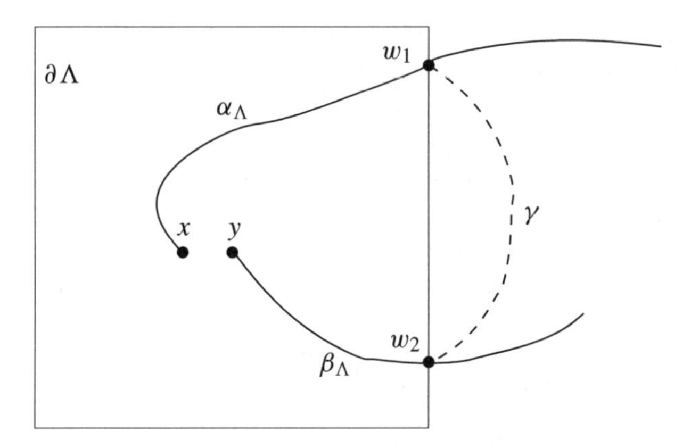

Figure 8.2. A path from x to y may be constructed from two disjoint infinite paths.

and therefore (using the finiteness of S again)

$$\text{for all large } \Lambda, \quad F(e, \nu_\Lambda) > \delta,$$

implying that $\lambda > \delta$ as required for (8.77).

We prove secondly that

$$\lambda \leq G(e, \nu), \tag{8.80}$$

by proving in turn that $\lambda \leq A$ and $\lambda \leq B$. That $\lambda \leq A$ is an immediate consequence of the assumption $\nu_\Lambda \leq \nu$, so we concentrate on the inequality $\lambda \leq B$. For $\pi = (\alpha, \beta) \in \mathcal{Q}_e$, where α has endvertex x, and β has endvertex y, let α_Λ (respectively, β_Λ) denote the initial segment of α (respectively, β) joining x (respectively, y) to the earliest vertex w_1 of α (respectively, w_2, of β) lying in $\partial\Lambda$. Since $w_1, w_2 \in \partial\Lambda$ and $w_1 \neq w_2$, there exists a path γ joining w_1 to w_2 and using no other vertex of Λ. We denote by π' the path comprising α_Λ, followed by γ, followed by β_Λ taken in reverse order; note that $\pi' \in \mathcal{P}_e$, and denote by $\mathcal{P}_{e,\Lambda}$ the set of all $\pi' \in \mathcal{P}_e$ obtainable in this way from any $\pi = (\alpha, \beta) \in \mathcal{Q}_e$. See Figure 8.2. Now,

$$\begin{aligned}
F(e, \nu_\Lambda) &\leq \inf_{\pi' \in \mathcal{P}_{e,\Lambda}} \max_{f \in \pi'} \nu_\Lambda(f) && \text{since } \mathcal{P}_{e,\Lambda} \subseteq \mathcal{P}_e \\
&= \inf_{\pi' \in \mathcal{P}_{e,\Lambda}} \max_{f \in \pi' \cap \mathbb{E}_\Lambda} \nu_\Lambda(f) && \text{since } \nu_\Lambda(f) = 0 \text{ for } f \notin \mathbb{E}_\Lambda \\
&\leq \inf_{\pi' \in \mathcal{P}_{e,\Lambda}} \max_{f \in \pi' \cap \mathbb{E}_\Lambda} \nu(f) && \text{since } \nu_\Lambda \leq \nu \\
&\leq \inf_{\pi \in \mathcal{Q}_e} \max_{f \in \pi \cap \mathbb{E}_\Lambda} \nu(f) \\
&\leq \inf_{\pi \in \mathcal{Q}_e} \sup_{f \in \pi} \nu(f) = B,
\end{aligned}$$

where we have used the fact that every $\pi' \in \mathcal{P}_{e,\Lambda}$ arises in the above manner from some $\pi \in \mathcal{Q}_e$. Inequality (8.80) follows. □

Proof of Lemma 8.69. (i) By (8.35), $p \le F(e, \nu)$ if and only if every $\pi \in \mathcal{P}_e$ contains some edge f with $\nu(f) \ge p$, which is to say that $\Pi_p \nu \in D_e^0$.
(ii) Suppose that $p < G(e, \nu)$. For $\pi \in \mathcal{P}_e \cup \mathcal{Q}_e$, there exists an edge $f \in \pi$ such that $\nu(f) > p$. Therefore, $\Pi^p \nu \in D_e^1$.

Suppose conversely that $\Pi^p \nu \in D_e^1$. It is elementary that $p \le G(e, \nu)$. Suppose in addition that $p = G(e, \nu)$, and we shall derive a contradiction. Let $e = \langle x, y \rangle$, and let $C_x(\nu)$ (respectively, $C_y(\nu)$) be the set of vertices attainable from x (respectively, y) along open paths of $\Pi^p \nu$ not using e. Since $\Pi^p \nu \in D_e^1$, $C_x(\nu)$ and $C_y(\nu)$ are disjoint. We shall prove that $C_x(\nu)$ (and similarly $C_y(\nu)$) is infinite. Since $p = G(e, \nu)$, there exists an infinite sequence $(\alpha_n : n = 1, 2, \dots)$ of distinct (finite or infinite) paths of $\mathbb{E}^d \setminus \{e\}$ with endvertex x such that

$$\sup_{f \in \alpha_n} \nu(f) \downarrow p \qquad \text{as } n \to \infty. \tag{8.81}$$

If $|C_x(\nu)| < \infty$, there exists some edge $g \ne e$, having exactly one endvertex in $C_x(\nu)$, and belonging to infinitely many of the paths α_n. By (8.81), any such g has $\nu(g) \le p$, in contradiction of the definition of $C_x(\nu)$. Therefore $C_x(\nu)$ (and similarly $C_y(\nu)$) is infinite.

Since $C_x(\nu)$ and $C_y(\nu)$ are disjoint and infinite, there exists $\pi = (\alpha, \beta) \in \mathcal{Q}_e$ such that $\nu(f) \le p$ for $f \in \alpha \cup \beta$, in contradiction of the assumption $\Pi^p \nu \in D_e^1$. The proof is complete. □

Proof of Theorem 8.52. Let $b \in \{0, 1\}$. The transitions of the process $(Z_t^b : t \ge 0)$ are given in terms of families of independent doubly-stochastic Poisson processes. In order that Z_t^b be a Markov process, it suffices therefore to prove that the conditional distribution of $(Z_{s+t}^b : t \ge 0)$, given $(Z_u^b : 0 \le u \le s)$, depends only on Z_s^b.

Here is an informal proof. We have that $Z_{s+t}^b = \lim_{\Lambda \uparrow \mathbb{Z}^d} Z_{\Lambda, s+t}^b$, where the processes $Z_{\Lambda, s+t}^b$ are given in terms of a graphical representation of compound Poisson processes. It follows that, given $(Z_{\Lambda,u}^b,\ Z_u^b : 0 \le u \le s,\ \Lambda \subseteq \mathbb{Z}^d)$, $(Z_{s+t}^b : t \ge 0)$ has law depending only on the family $(Z_{\Lambda,s}^b : \Lambda \subseteq \mathbb{Z}^d)$. Write $\zeta_\Lambda = Z_{\Lambda,s}^b$ and $\zeta = \lim_{\Lambda \uparrow \mathbb{Z}^d} \zeta_\Lambda = Z_s^b$. We need to show that the (conditional) law of $(Z_{s+t}^b : t \ge 0)$ does not depend on the family $(\zeta_\Lambda : \Lambda \subseteq \mathbb{Z}^d)$ further than on its limit ζ. Lemma 8.40(b) is used for this.

Let $s, t \ge 0$ and $\nu \in X$. Denote by $Y_{\Lambda, s+t}^{(\nu,b)}$ the state (in X_Λ^b) at time $s + t$ obtained from the evolution rules given prior to (8.36), starting at time s in state $(\nu, b) = (\nu, b)_\Lambda$.

Suppose that $b = 0$, so that $\zeta_\Lambda \uparrow \zeta$ as $\Lambda \to \mathbb{Z}^d$. Let $\epsilon > 0$ and let Δ be a finite box. There exists a box Λ' such that $\Lambda' \supseteq \Delta$ and

$$\zeta(e) - \epsilon \le \zeta_\Lambda(e) \le \zeta(e), \qquad e \in \mathbb{E}_\Delta,\ \Lambda \supseteq \Lambda'.$$

By Lemma 8.40(b),

$$Y_{\Delta, s+t}^{(\zeta,b)} - \epsilon \le Y_{\Delta, s+t}^{(\zeta_\Lambda,b)} \le Y_{\Lambda, s+t}^{(\zeta_\Lambda,b)} \le Y_{\Lambda, s+t}^{(\zeta,b)}, \qquad \Lambda \supseteq \Lambda'.$$

Now, $Y^{(\zeta_\Lambda,b)}_{\Lambda,s+t} = Z^b_{\Lambda,s+t}$, and we pass to the limits as $\Lambda \uparrow \mathbb{Z}^d$, $\Delta \uparrow \mathbb{Z}^d$, $\epsilon \downarrow 0$, to obtain that

(8.82)
$$\lim_{\Lambda\uparrow\mathbb{Z}^d} Y^{(\zeta,b)}_{\Lambda,s+t} = Z^b_{s+t},$$

implying as required that Z^b_{s+t} depends on ζ but not further on the family $(\zeta_\Lambda : \Lambda \subseteq \mathbb{Z}^d)$. The same argument is valid when $b = 1$, with the above inequalities reversed and the sign of ϵ changed.

The Markov transition function of Z^b_t is the family $(Q^b_{s,t} : 0 \le s \le t)$ given by

$$Q^b_{s,t}(\zeta, A) = \mathbb{P}(Z^b_t \in A \mid Z^b_s = \zeta), \qquad \zeta \in X,\ A \in \mathcal{B}.$$

In the light of the remarks above and particularly (8.82),

$$\begin{aligned} Q^b_{s,t}(\zeta, A) &= Q^b_{0,t-s}(\zeta, A) \\ &= \mathbb{P}(Z^{(\zeta,b)}_{t-s} \in A) = P^b_{t-s}(\zeta, A). \end{aligned}$$ □

Proof of Theorem 8.53. As in Lemma 8.31, the limits

$$\psi^b(A) = \lim_{t\to\infty} \mathbb{P}(Z^b_t \in A), \qquad b = 0, 1,$$

exist for any increasing event $A \in \mathcal{B}$. The space X is compact, and the increasing events are convergence-determining, and therefore Z^0_t and Z^1_t converge weakly as $t \to \infty$. It suffices to show that

$$Z^1_t - Z^0_t \Rightarrow 0 \qquad \text{as } t \to \infty.$$

Since we are working with the product topology on X, it will be enough to show that, for $\epsilon > 0$ and $f \in \mathbb{E}^d$,

(8.83)
$$\mathbb{P}\big(|Z^1_t(f) - Z^0_t(f)| > \epsilon\big) \to 0 \qquad \text{as } t \to \infty.$$

Let $\mathcal{D} = \mathcal{D}_q$ be as in Theorem 4.63, and let $\epsilon > 0$. Pick a finite subset $\mathcal{E}$ of $\overline{\mathcal{D}} = (0, 1) \setminus \mathcal{D}$ such that every interval of the form $(\delta, \delta + \epsilon)$ contains some point of $\mathcal{E}$, as δ ranges over $[0, 1 - \epsilon)$. By Theorem 4.63,

(8.84)
$$\phi^0_{p,q} = \phi^1_{p,q}, \qquad p \in \mathcal{E}.$$

For $f \in \mathbb{E}^d$,

$$
\begin{aligned}
\mathbb{P}\big(|Z_t^1(f) - Z_t^0(f)| > \epsilon\big)
&\le \sum_{p\in\mathcal{E}} \mathbb{P}\big(Z_t^0(f) \le p \le Z_t^1(f)\big) \\
&\le \sum_{p\in\mathcal{E}} \mathbb{P}\big(Z_{\Lambda,t}^0(f) \le p \le Z_{\Lambda,t}^1(f)\big) && \text{for all boxes } \Lambda \\
&= \sum_{p\in\mathcal{E}} \mathbb{P}\big(\Pi^p Z_{\Lambda,t}^0(f) = 1,\ \Pi_p Z_{\Lambda,t}^1(f) = 0\big) \\
&\to \sum_{p\in\mathcal{E}} \big[\phi_{\Lambda,p,q}^1(J_f) - \phi_{\Lambda,p,q}^0(J_f)\big] && \text{as } t \to \infty \\
&\to \sum_{p\in\mathcal{E}} \big[\phi_{p,q}^1(J_f) - \phi_{p,q}^0(J_f)\big] && \text{as } \Lambda \uparrow \mathbb{Z}^d \\
&= 0 && \text{by (8.84),}
\end{aligned}
$$

where J_f is the event that f is open.

The translation-invariance of the limit measure μ is a consequence of the fact that the limits in (8.47)–(8.48) do not depend on the way in which the increasing limit $\Lambda \uparrow \mathbb{Z}^d$ is taken. □

Proof of Theorem 8.56. (a) That the projected processes $(L_{p,t}^b : t \ge 0)$, $b = 0, 1$, are Markovian follows from Theorem 8.52 and the discussion after Lemma 8.69.

Let $A \in \mathcal{F}$ be increasing. As in Lemma 8.31, the limits

$$\psi_p^b(A) = \lim_{t\to\infty} \mathbb{P}(L_{p,t}^b \in A)$$

exist for $b = 0, 1$. Since $L_{p,t}^0 \le L_{p,t}^1$,

$$\psi_p^0(A) \le \psi_p^1(A) \qquad \text{for increasing } A \in \mathcal{F}. \tag{8.85}$$

Let $A \in \mathcal{F}$ be an increasing cylinder event. Then

$$
\begin{aligned}
\psi_p^0(A) &= \lim_{t\to\infty} \mathbb{P}(L_{p,t}^0 \in A) \\
&\ge \lim_{t\to\infty} \mathbb{P}(\Pi_p Z_{\Lambda,t}^1 \in A) && \text{since } L_{p,t}^0 \ge \Pi_p Z_{\Lambda,t}^1 \\
&= \phi_{\Lambda,p,q}^0(A) && \text{by Theorem 8.38} \\
&\to \phi_{p,q}^0(A) && \text{as } \Lambda \to \mathbb{Z}^d,
\end{aligned}
$$

and similarly

$$\psi_p^1(A) \le \phi_{p,q}^1(A). \tag{8.86}$$

Let $\mathcal{D}_q$ be given as in Theorem 4.63. Since $\phi^0_{p,q} = \phi^1_{p,q}$ for $p \notin \mathcal{D}_q$, we have by (8.85)–(8.86) that

$$\phi^0_{p,q}(A) = \psi^0_p(A) = \psi^1_p(A) = \phi^1_{p,q}(A), \qquad p \notin \mathcal{D}_q.$$

Since $\mathcal{F}$ is generated by the increasing cylinder events, $\phi^b_{p,q}$ is the unique invariant measure of L^b_p whenever $p \notin \mathcal{D}_q$.

In order to show that

$$\phi^0_{p,q}(A) = \psi^0_p(A), \quad \phi^1_{p,q}(A) = \psi^1_p(A),$$

for *all* p and any increasing cylinder event A, it suffices to show that $\psi^0_p(A)$ is left-continuous in p, and $\psi^1_p(A)$ is right-continuous (the conclusion will then follow by Proposition 4.28). We confine ourselves to the case of $\psi^0_p(A)$, since the other case is exactly similar.

Let $p \in (0,1)$, and let $A \in \mathcal{F}$ be an increasing cylinder event. Let

$$B_p = \{\zeta \in X : \Pi_p\zeta \in A\}, \quad C_p = \{\zeta \in X : \Pi^p\zeta \in A\},$$

be the corresponding events in $\mathcal{B}$, and note from the definitions of Π_p and Π^p that B_p is decreasing and open, and that C_p is decreasing and closed. Furthermore, $C_{p-\epsilon} \subseteq B_p$ for $\epsilon > 0$, and

$$B_p \setminus C_{p-\epsilon} \to \varnothing \qquad \text{as } \epsilon \downarrow 0. \tag{8.87}$$

By stochastic monotonicity, the limit $\lim_{t\to\infty} \mathbb{P}(Z^1_t \in B_p)$ exists and, by weak convergence (see Theorem 8.53),

$$\lim_{t\to\infty} \mathbb{P}(Z^1_t \in B_p) \geq \mu(B_p).$$

We claim further that $\mathbb{P}(Z^1_t \in B_p) \leq \mu(B_p)$ for all t, whence

$$\mathbb{P}(Z^1_t \in B_p) \to \mu(B_p) \qquad \text{as } t \to \infty. \tag{8.88}$$

Suppose on the contrary that

$$\mathbb{P}(Z^1_T \in B_p) > \mu(B_p) + \eta \qquad \text{for some } T \text{ and } \eta > 0.$$

Now $Z^1_t \leq_{\text{st}} Z^1_T$ for $t \geq T$, and hence

$$\mathbb{P}(Z^1_t \in C_{p-\epsilon}) > \mu(C_{p-\epsilon}) + \tfrac{1}{2}\eta \qquad \text{for some } \epsilon > 0 \text{ and all } t \geq T,$$

by (8.87). Since $C_{p-\epsilon}$ is closed, this contradicts the fact that $Z^1_t \Rightarrow \mu$.

For $h > 0$,

$$\begin{aligned}\psi_p^0(A) - \psi_{p-h}^0(A) &= \lim_{t\to\infty} \big[\mathbb{P}(Z_t^1 \in B_p) - \mathbb{P}(Z_t^1 \in B_{p-h})\big] \\ &= \mu(B_p \setminus B_{p-h}) \qquad \text{by (8.88).}\end{aligned}$$

The sets B_p and B_{p-h} are open, and $B_p \setminus B_{p-h} \to \varnothing$ as $h \downarrow 0$. Hence $\psi_{p-h}^0(A) \to \psi_p^0(A)$ as $h \downarrow 0$.

In the corresponding argument for $\psi_p^1(A)$, the set B_p is replaced by the decreasing closed event C_p, and the difference $B_p \setminus B_{p-h}$ is replaced by $C_{p+h} \setminus C_p$.

We prove finally that $L_{p,t}^0$ is reversible with respect to $\phi_{p,q}^0$; the argument is similar for $L_{p,t}^1$. Let f and g be increasing non-negative cylinder functions mapping Ω to $\mathbb{R}$, and let $U_{\Lambda,t}^0$ (respectively, U_t^0) be the transition semigroup of the process $\Pi_p Z_{\Lambda,t}^1$ (respectively, $L_{p,t}^0 = \Pi_p Z_t^1$). For $\Lambda \subseteq \Delta$,

$$f(\eta)U_{\Lambda,t}^0 g(\eta) \le f(\eta)U_{\Delta,t}^0 g(\eta) \le f(\eta)U_t^0 g(\eta), \qquad \eta \in \Omega,$$

by Lemmas 8.31 and 8.40. Therefore,

$$\begin{aligned}\phi_{\Delta,p,q}^0\big(f(\eta)U_{\Lambda,t}^0 g(\eta)\big) &\le \phi_{\Delta,p,q}^0\big(f(\eta)U_{\Delta,t}^0 g(\eta)\big) \\ &\le \phi_{p,q}^0\big(f(\eta)U_t^0 g(\eta)\big), \qquad \Lambda \subseteq \Delta,\end{aligned}$$

since $\phi_{\Delta,p,q}^0 \le_{\text{st}} \phi_{p,q}^0$. Let $\Delta \uparrow \mathbb{Z}^d$ and $\Lambda \uparrow \mathbb{Z}^d$, and deduce by the monotone convergence theorem that

$$\phi_{\Delta,p,q}^0\big(f(\eta)U_{\Delta,t}^0 g(\eta)\big) \to \phi_{p,q}^0\big(f(\eta)U_t^0 g(\eta)\big) \qquad \text{as } \Delta \uparrow \mathbb{Z}^d. \tag{8.89}$$

The left side of (8.89) is unchanged when f and g are exchanged, by the reversibility of $\Pi_p Z_{\Delta,t}^1$, see Theorem 8.38. Therefore, the right side is unchanged by this exchange, implying the required reversibility (see [235, p. 91]).

(b) It suffices to prove (8.57) for increasing cylinder events A, since such events generate $\mathcal{F}$. For such A, (8.57) follows from (8.88) in the case of $\phi_{p,q}^0$, and similarly for $\phi_{p,q}^1$. □

Proof of Proposition 8.59. This is a consequence of Theorem 8.56(b). □

(8.90) Proposition. *Let $q \in (1,\infty)$ and $p \in (0,1)$. The Markov processes L_p^0 and L_p^1 are not Feller processes.*

Proof. For simplicity we take $d = 2$ and $b = 0$; a similar argument is valid for $d > 2$ and/or $b = 1$. Let e be the edge with endvertices $(0,0)$ and $(1,0)$, and let J_e be the indicator function of the event that e is open. We shall show that the function $U_s^0 J_e : \Omega \to \mathbb{R}$ is not continuous for sufficiently small positive values

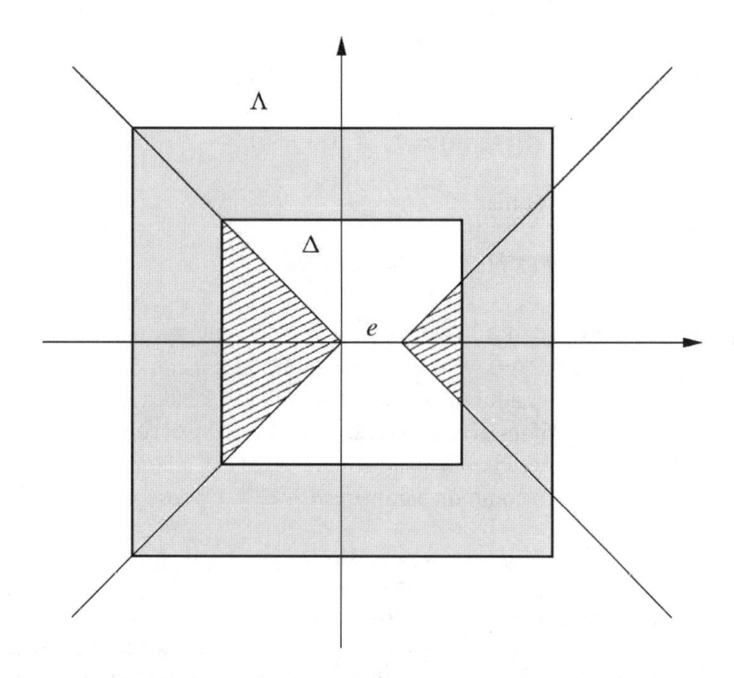

Figure 8.3. In the inner square, edges other than e are open if and only if they are in the shaded area. In the grey area $\Lambda \setminus \Delta$, all edges have the same state, namely 0 for ω^0 and 1 for ω^1. The endvertices of e are joined by an open path of $\mathbb{E}_\Lambda \setminus \{e\}$ in ω^1 but not in ω^0.

of s, where $(U_s^0 : s \geq 0)$ is the transition semigroup of L_p^0. Let V be the set of vertices $x = (x_1, x_2) \in \mathbb{Z}^2$ satisfying

$$\text{either} \quad x_1 \geq |x_2| + 1 \quad \text{or} \quad -x_1 \geq |x_2|,$$

and let $\mathbb{E}_V$ ($\ni e$) be the set of edges with both endvertices in V. See Figure 8.3. Let n be a positive integer, and Δ the box $[-n, n]^2$. Let ω^0, $\omega^1 \in \Omega$ be the configurations given by

$$\omega^b(f) = \begin{cases} 1 & \text{if } f \in \mathbb{E}_\Delta \cap \mathbb{E}_V, \\ 0 & \text{if } f \in \mathbb{E}_\Delta \setminus \mathbb{E}_V, \\ b & \text{otherwise.} \end{cases}$$

Note that ω^0 and ω^1 depend on n, and also that $\omega^1 \notin D_e$ but $\omega^0 \in D_e$, where D_e is the event that there exists no open path of $\mathbb{E}^2 \setminus \{e\}$ joining the endvertices of e. We shall couple together two processes, with respective initial configurations ω^0, ω^1, and we claim that there exists a non-zero time interval during which, with a strictly positive probability, the lower of these two processes remains in D_e and the upper process remains in its complement.

For $b = 0, 1$, let $K_{\Lambda,t}^b$ be the process $\Pi_p Z_{\Lambda,t}^{(\zeta^b,1)}$ for some $\zeta^b \in X$ satisfying $\omega^b = \Pi_p \zeta^b$; the value of ζ^b is otherwise immaterial. We write $K_t^b =$

$\lim_{\Lambda\uparrow\mathbb{Z}^d} K^b_{\Lambda,t}$, a limit which exists by the usual monotonicity. We claim that there exist $\epsilon, \eta > 0$, independent of the value of n, such that

$$\mathbb{P}\big(K^1_\eta(e) = 1,\ K^0_\eta(e) = 0\big) > \epsilon. \tag{8.91}$$

Inequality (8.91) implies that

$$\mathbb{P}(K^1_\eta(e) = 1) - \mathbb{P}(K^0_\eta(e) = 1) > \epsilon,$$

irrespective of the value of n, and therefore that the semigroup U^0_s is not Feller.

In order to prove (8.91), we use a percolation argument. Let $\eta > 0$. As in Section 8.6, we consider a family of rate-1 alarm clocks indexed by $\mathbb{E}^2$. For each edge f, we set $B_f = 0$ if the alarm clock at f does not ring during the time-interval $[0, \eta]$, and $B_f = 1$ otherwise. Thus, $(B_f : f \in \mathbb{E}^2)$ is a family of independent Bernoulli variables with common parameter $1 - e^{-\eta}$. Choose η sufficiently small such that

$$1 - e^{-\eta} < \tfrac{1}{4},$$

noting that $\frac{1}{4}$ is less than the critical probability of bond percolation on the square lattice (see Chapter 6 and [154]). Routine percolation arguments may now be used to obtain the existence of $\epsilon' > 0$ such that, for all boxes Λ containing $[-2n, 2n]^2$,

$$\mathbb{P}\big(K^1_{\Lambda,t} \notin D_e,\ K^0_{\Lambda,t} \in D_e,\ \text{for all } t \in [0,\eta] \,\big|\, \mathcal{G}_\eta\big) > \epsilon' \qquad \mathbb{P}\text{-a.s.},$$

where $\mathcal{G}_\eta$ is the σ-field generated by the ringing times of the alarm clock at e up to time t, together with the associated values of α (in the language of Section 8.6).

Suppose that the alarm clock at e rings once only during the time-interval $[0, \eta]$, at the random time T, say. By (8.72)–(8.73), there exists $\epsilon'' = \epsilon''(p, q) > 0$ such that: there is (conditional) probability at least ϵ'' that, for all $\Lambda \supseteq [-2n, 2n]^2$, the edge e is declared closed at time T in the lower process $K^0_{\Lambda,T}$ but not in the upper process $K^1_{\Lambda,T}$. The conditioning here is over all values of the doubly-stochastic Poisson processes indexed by edges other than e. Therefore,

$$\mathbb{P}\big(K^1_{\Lambda,\eta}(e) = 1,\ K^0_{\Lambda,\eta}(e) = 0\big) > \epsilon'\epsilon''\eta e^{-\eta},$$

for all Λ containing $[-2n, 2n]^2$. Let $\Lambda \uparrow \mathbb{Z}^d$ to obtain (8.91) with an appropriate value of ϵ. □

Proof of Theorem 5.16(a). This was deferred from Section 5.2. We follow the argument of [36] as reported in [154]. For $p \in (0, 1]$ and $\zeta \in X$, we say that an edge e is *p-open* if $\Pi_p\zeta(e) = 1$, which is to say that $\zeta(e) < p$. Let $C_p = C_p(\zeta)$ be the p-open cluster of $\mathbb{L}^d$ containing the origin, and note that $C_{p'} \subseteq C_p$ if $p' \le p$.

By Theorem 8.56(b),

$$\theta^0(p, q) = \mu(|C_p| = \infty),$$

where μ is given in Theorem 8.53. Therefore,

(8.92)
$$\theta^0(p,q) - \theta^0(p-,q) = \lim_{p' \uparrow p} \mu\big(|C_p| = \infty,\ |C_{p'}| < \infty\big)$$
$$= \mu\big(|C_p| = \infty,\ |C_{p'}| < \infty \text{ for all } p' < p\big).$$

Assume that $p > p_c(q)$, and suppose $|C_p| = \infty$. If $p_c(q) < \alpha < p$, there exists (almost surely) an α-open infinite cluster I_α, and furthermore I_α is (almost surely) a subgraph of C_p, by the 0/1-infinite-cluster property of the 0-boundary-condition random-cluster measures. Therefore, there exists a p-open path π joining the origin to some vertex of I_α. Such a path π has finite length and each edge e in π satisfies $\zeta(e) < p$, whence $\beta = \max\{\zeta(e) : e \in \pi\}$ satisfies $\beta < p$. If p' satisfies $p' \geq \alpha$ and $\beta < p' < p$ then there exists a p'-open path joining the origin to some vertex of I_α, so that $|C_{p'}| = \infty$. However, $p' < p$, implying that the event on the right side of (8.92) has probability zero. □

8.9 Simultaneous uniqueness

One of the key facts for supercritical percolation is the (almost-sure) uniqueness of the infinite open cluster, which may be stated in the following form. Let ϕ_p be the percolation (product) measure on $\Omega = \{0,1\}^{\mathbb{E}^d}$ where $d \geq 2$. We have that:

(8.93) $\qquad$ for all $p \in [0,1]$, $\quad \phi_p$ has the 0/1-infinite-cluster property.

It has been asked whether or not there exists a unique infinite cluster *simultaneously for all values of* p. This question may be formulated as follows. First, we couple together the percolation processes for different values of p by defining

$$\eta_p(e) = \begin{cases} 1 & \text{if } U(e) < p, \\ 0 & \text{otherwise,} \end{cases}$$

where the $U(e)$, $e \in \mathbb{E}^d$, are independent and uniformly distributed on the interval $[0,1]$. Let $I(\omega)$ be the number of infinite open clusters in a configuration $\omega \in \Omega$. It is proved in [13] that there exists a deterministic non-decreasing function $i : [0,1] \to \{0,1\}$ such that

(8.94) $\qquad \mathbb{P}\big(I(\eta_p) = i(p) \text{ for all } p \in [0,1]\big) = 1,$

a statement to which we refer as 'simultaneous uniqueness'. By (8.93) and the definition of the critical probability p_c,

$$i(p) = \begin{cases} 0 & \text{if } p < p_c, \\ 1 & \text{if } p > p_c. \end{cases}$$

It is an open question to prove the conjecture that $i(p_c) = 0$ for $d \geq 2$. See the discussion in [154, Section 8.2].

Simultaneous uniqueness may be conjectured for the random-cluster model also, using the coupling of the last section.

(8.95) Conjecture (Simultaneous uniqueness). *Let $q \in [1, \infty)$, and consider the coupling μ of the random-cluster measures $\phi^b_{p,q}$ on $\mathbb{L}^d$ with parameter q. There exist non-decreasing functions $i_q, i'_q : [0, 1] \to \{0, 1\}$ such that*

$$\mu\big(I(\Pi_p\zeta) = i_q(p) \text{ and } I(\Pi^p\zeta) = i'_q(p), \text{ for all } p \in [0, 1]\big) = 1.$$

It must be the case that $i_q(p) = i'_q(p)$ for $p \neq p_c(q)$.

Here is a sufficient condition for simultaneous uniqueness. For $r \in (0, 1)$ and a box Λ, let $E_\Lambda(r)$ be the subset of the configuration space X containing all ν with $\nu(e) < r$ for all $e \in \mathbb{E}_\Lambda$. Thus, $E_\Lambda(r)$ is the event that every edge in $\mathbb{E}_\Lambda$ is open in the configuration $\Pi_r\nu$. By [13, Thm 1.8], it suffices to show that μ has a property termed 'positive finite energy'. This is in turn implied by:

$$\mu(E_\Lambda(r) \mid \mathcal{T}_\Lambda) > 0, \qquad \mu\text{-a.s.} \tag{8.96}$$

for all $r \in (0, 1)$ and boxes Λ. Here as earlier, $\mathcal{T}_\Lambda$ is the σ-field generated by the states of edges not belonging to $\mathbb{E}_\Lambda$. It seems reasonable in the light of Theorem 4.17(b) to conjecture the stronger inequality

$$\mu(E_\Lambda(r) \mid \mathcal{T}_\Lambda) \geq \left(\frac{r}{r + q(1-r)}\right)^{|\mathbb{E}_\Lambda|}, \qquad \mu\text{-a.s.}$$

Chapter 9

Flows in Poisson Graphs

Summary. The random-cluster partition function with integer q on a graph G may be transformed into the mean flow-polynomial of a 'Poissonian' random graph obtained from G by randomizing the numbers of edges between neighbouring pairs. This leads to a flow representation for the two-point Potts correlation function, and extends to general q the so-called 'random-current expansion' of the Ising model. In the last case, one may derive the Simon–Lieb inequality together with largely complete solutions to the problems of exponential decay and the continuity of the phase transition. It is an open problem to adapt such methods to general Potts and random-cluster models.

9.1 Potts models and flows

The Tutte polynomial is a function of two variables (see Section 3.6). For suitable values of these variables, one obtains counts of colourings, forests, and flows, together with other combinatorial quantities, in addition to the random-cluster and Potts partition functions. The algebra of the Tutte polynomial may be used to obtain representations of the Potts correlation functions, which have in turn the potential to explain the decay of correlations in the two phases of an infinite-volume Potts measure. It is thus that many beautiful results have been derived for the Ising model (when $q = 2$), see [3, 5, 9]. The cases $q \in \{3, 4, \dots\}$, and more generally $q \in (1, \infty)$, remain largely unexplained. We summarize this methodology in this chapter, beginning with the definition of a flow on a directed graph.

Let $H = (W, F)$ be a finite graph with vertex-set W and edge-set F, and let $q \in \{2, 3, \dots\}$. We permit H to have multiple edges and loops. To each edge $e \in F$ we allocate a direction, turning H thus into a directed graph denoted by $\vec{H} = (W, \vec{F})$. When the edge $e = \langle u, v \rangle \in F$ is directed from u to v, we write

$\vec{e} = [u, v\rangle$ for the corresponding directed edge[1], and we speak of u as the *tail* and v as the *head* of $\vec{e}$. It will turn out that the choices of directions are immaterial to the principal conclusions that follow. A function $f : \vec{F} \to \{0, 1, 2, \dots, q-1\}$ is called a *mod-q flow* on $\vec{H}$ if

$$\text{(9.1)} \qquad \sum_{\substack{\vec{e}\in\vec{F}:\\ \vec{e} \text{ has head } w}} f(\vec{e}) - \sum_{\substack{\vec{e}\in\vec{F}:\\ \vec{e} \text{ has tail } w}} f(\vec{e}) = 0 \mod q, \qquad \text{for all } w \in W,$$

which is to say that flow is conserved (modulo q) at every vertex. A mod-q flow f is called *non-zero* if $f(\vec{e}) \neq 0$ for all $\vec{e} \in \vec{F}$. We write $C_H(q)$ for the number of non-zero mod-q flows on $\vec{H}$. It is standard (and an easy exercise) that $C_H(q)$ does not depend on the directions allocated to the edges of H, [313]. The function $C_H(q)$, viewed as a function of q, is called the *flow polynomial* of H.

The flow polynomial of H is an evaluation of its Tutte polynomial. Recall from Section 3.6 the (Whitney) rank-generating function and the Tutte polynomial,

$$\text{(9.2)} \qquad W_H(u, v) = \sum_{F' \subseteq F} u^{r(H')} v^{c(H')}, \qquad u, v \in \mathbb{R},$$

$$\text{(9.3)} \qquad T_H(u, v) = (u-1)^{|W|-1} W_H\big((u-1)^{-1}, v-1\big),$$

where $r(H') = |W| - k(H')$ is the *rank* of the subgraph $H' = (W, F')$, $c(H') = |F'| - |W| + k(H')$ is its *co-rank*, and $k(H')$ is the number of its connected components (including isolated vertices). Note that

$$\text{(9.4)} \qquad W_H(u, v) = (u/v)^{|W|} \sum_{F' \subseteq F} v^{|F'|} (v/u)^{k(H')}, \qquad u, v \neq 0.$$

The flow polynomial of H satisfies

$$\text{(9.5)} \qquad \begin{aligned} C_H(q) &= (-1)^{|F|} W_H(-1, -q) \\ &= (-1)^{|F|-|W|+1} T_H(0, 1-q), \qquad q \in \{2, 3, \dots\}. \end{aligned}$$

See [40, 313]. When the need for a different notation arises, we shall write $C(H; q)$ for $C_H(q)$, and similarly for other polynomials.

We return now to the random-cluster and Potts models on the finite graph $G = (V, E)$. It is convenient to allow a separate parameter for each edge of G, and thus we let $\mathbf{J} = (J_e : e \in E)$ be a vector of non-negative numbers, and we take $\beta \in (0, \infty)$. For $q \in \{2, 3, \dots\}$, the q-state Potts measure on the configuration space $\Sigma = \{1, 2, \dots, q\}^V$ is written in this chapter as

$$\text{(9.6)} \qquad \pi_{\beta\mathbf{J},q}(\sigma) = \frac{1}{Z_{\mathrm{P}}} \exp\left\{ \sum_{e\in E} \beta J_e (q\delta_e(\sigma) - 1) \right\}, \qquad \sigma \in \Sigma,$$

[1] This is not a good notation since H may have multiple edges.

where, for $e = \langle x, y\rangle \in E$,

$$\delta_e(\sigma) = \delta_{\sigma_x,\sigma_y} = \begin{cases} 1 & \text{if } \sigma_x = \sigma_y, \\ 0 & \text{otherwise,} \end{cases}$$

and Z_{P} is the partition function

$$Z_{\mathrm{P}} = \sum_{\sigma\in\Sigma} \exp\left\{\sum_{e\in E} \beta J_e(q\delta_e(\sigma) - 1)\right\}. \tag{9.7}$$

This differs slightly from (1.5)–(1.6) in that different edges e may have different interactions J_e, and these interactions have been 're-parametrized' by the factor q. The reason for defining $\pi_{\beta\mathbf{J},q}$ thus will emerge in the calculations that follow.

The corresponding two-point correlation function is given as in (1.14) by

$$\tau_{\beta\mathbf{J},q}(x, y) = \pi_{\beta\mathbf{J},q}(\sigma_x = \sigma_y) - \frac{1}{q}, \qquad x, y \in V. \tag{9.8}$$

We shall work often with the quantity $q\tau_{\beta\mathbf{J},q}(x, y) = \pi_{\beta\mathbf{J},q}(q\delta_{\sigma_x,\sigma_y} - 1)$ and, for ease of notation in the following, we write

$$\sigma(x, y) = q\tau_{\beta\mathbf{J},q}(x, y), \qquad x, y \in V, \tag{9.9}$$

thereby suppressing reference to the parameters $\beta\mathbf{J}$ and q. Note that, for the Ising case with $q = 2$, $\sigma(x, y)$ is simply the mean of the product $\sigma_x\sigma_y$ of the Ising spins at x and at y, see (1.7).

From the graph $G = (V, E)$ we construct next a certain random graph. For any vector $m = (m(e) : e \in E)$ of non-negative integers, let $G_m = (V, E_m)$ be the graph with vertex set V and, for each $e \in E$, with exactly $m(e)$ edges in parallel joining the endvertices of the edge e [the original edge e is itself removed]. Note that

$$|E_m| = \sum_{e\in E} m(e). \tag{9.10}$$

Let $\boldsymbol{\lambda} = (\lambda_e : e \in E)$ be a family of non-negative reals, and let $P = (P(e) : e \in E)$ be a family of independent random variables such that $P(e)$ has the Poisson distribution with parameter λ_e. We now consider the random graph $G_P = (V, E_P)$, which we call a *Poisson graph with intensity* $\boldsymbol{\lambda}$. Write $\mathbb{P}_{\boldsymbol{\lambda}}$ and $\mathbb{E}_{\boldsymbol{\lambda}}$ for the corresponding probability measure and expectation operator.

For $x, y \in V$, $x \neq y$, we denote by $G_P^{x,y}$ the graph obtained from G_P by adding an edge with endvertices x, y. If x and y are already adjacent in G_P, we add exactly one further edge between them. Potts-correlations and flows are related by the following theorem[2].

[2] The relationship between flows and correlation functions has been explored also in [112, 246, 247].

(9.11) Theorem [146, 157]. *Let $q \in \{2, 3, \dots\}$ and $\lambda_e = \beta J_e$. Then*

(9.12) $$\sigma(x, y) = \frac{\mathbb{E}_\lambda(C(G_P^{x,y}; q))}{\mathbb{E}_\lambda(C(G_P; q))}, \qquad x, y \in V.$$

This formula takes an especially simple form when $q = 2$, since non-zero mod-2 flows necessarily take the value 1 only. A finite graph $H = (W, F)$ is called *even* if the degree of every vertex $w \in W$ is even. It is elementary that $C_H(2) = 1$ (respectively, $C_H(2) = 0$) if H is even (respectively, not even), and therefore

(9.13) $$\mathbb{E}_\lambda(C_H(2)) = \mathbb{P}_\lambda(H \text{ is even}).$$

By (9.12), for any graph G,

(9.14) $$\sigma(x, y) = \frac{\mathbb{P}_\lambda(G_P^{x,y} \text{ is even})}{\mathbb{P}_\lambda(G_P \text{ is even})},$$

when $q = 2$. Observations of this sort have led to the so-called 'random-current' expansion for Ising models, thereby after some work [3, 5, 9] yielding proofs amongst other things of the exponential decay of correlations in the high-temperature regime. We return to the case $q = 2$ in Sections 9.2–9.4.

Whereas Theorem 9.11 concerns Potts models only, there is a random-cluster generalization. We restrict ourselves here to the situation in which every edge has the same parameter p, but we note that the result is easily generalized to allowing different parameters for each edge. Recall that $\phi_{G,p}$ denotes product measure on $\Omega = \{0, 1\}^E$ with density p.

(9.15) Theorem [146, 157]. *Let $p \in [0, 1)$ and $q \in (0, \infty)$. Let $\lambda_e = \lambda$ for all $e \in E$, where $p = 1 - e^{-\lambda q}$.*

(a) *For $x, y \in V$,*

$$(q-1)\phi_{G,p,q}(x \leftrightarrow y) = \frac{\mathbb{E}_\lambda\big((-1)^{1+|E_P|}T(G_P^{x,y}; 0, 1-q)\big)}{\mathbb{E}_\lambda\big((-1)^{|E_P|}T(G_P; 0, 1-q)\big)}, \tag{9.16}$$

and, in particular,

$$(q-1)\phi_{G,p,q}(x \leftrightarrow y) = \frac{\mathbb{E}_\lambda(C(G_P^{x,y}; q))}{\mathbb{E}_\lambda(C(G_P; q))}, \qquad q \in \{2, 3, \dots\}. \tag{9.17}$$

(b) *For $q \in \{2, 3, \dots\}$,*

$$\phi_{G,p}(q^{k(\omega)}) = (1-p)^{|E|(q-2)/q} q^{|V|} \mathbb{E}_\lambda(C(G_P; q)). \tag{9.18}$$

When $q = 2$, equation (9.18) reduces by (9.13) to

(9.19) $$\phi_{G,p}(2^{k(\omega)}) = 2^{|V|}\mathbb{P}_\lambda(G_P \text{ is even}).$$

This may be simplified further. Let $\zeta(e) = P(e)$ modulo 2. It is easily seen that G_P is even if and only if G_ζ is even, and that the $\zeta(e)$, $e \in E$, are independent Bernoulli variables with

$$\mathbb{P}_\lambda(\zeta(e) = 1) = \tfrac{1}{2}(1 - e^{-2\lambda}) = \tfrac{1}{2}p.$$

Equation (9.18) may therefore be written as

$$\phi_{G,p}(2^{k(\omega)}) = 2^{|V|}\phi_{G,p/2}(\text{the open subgraph of } G \text{ is even}). \tag{9.20}$$

Proof of Theorem 9.11. Since the parameter β appears always with the multiplicative factor J_e, we may without loss of generality take $\beta = 1$.

We begin with a calculation involving the Potts partition function Z_{P} given in (9.7). Let $\mathbb{Z}_+ = \{0, 1, 2, \dots\}$ and consider vectors $m = (m_e : e \in E) \in \mathbb{Z}_+^E$. By a Taylor expansion in the variables J_e,

$$\begin{aligned}
\exp\Big\{-\sum_{e\in E} J_e\Big\} Z_{\mathrm{P}} &= \sum_{m\in\mathbb{Z}_+^E} \Big(\prod_{e\in E} \frac{J_e^{m_e}}{m_e!} e^{-J_e}\Big) \partial^m Z_{\mathrm{P}}\Big|_{\mathbf{J}=0} \\
&= \mathbb{E}_\lambda\Big(\partial^P Z_{\mathrm{P}}\Big|_{\mathbf{J}=0}\Big)
\end{aligned} \tag{9.21}$$

where

$$\partial^m Z_{\mathrm{P}} = \Big(\prod_{e\in E} \frac{\partial^{m_e}}{\partial J_e^{m_e}}\Big) Z_{\mathrm{P}}, \qquad m \in \mathbb{Z}_+^E.$$

By (9.7) with $\beta = 1$, and similarly to the proof of Theorem 1.10(a),

$$\begin{aligned}
\partial^m Z_{\mathrm{P}}\Big|_{\mathbf{J}=0} &= \sum_{\sigma\in\Sigma} \prod_{e\in E} (q\delta_e(\sigma) - 1)^{m_e} \\
&= \sum_{\sigma\in\Sigma} \prod_{e\in E_m} (q\delta_e(\sigma) - 1) \\
&= \sum_{\sigma\in\Sigma} \prod_{e\in E_m} \sum_{n_e\in\{0,1\}} [-\delta_{n_e,0} + \delta_{n_e,1} q\delta_e(\sigma)] \\
&= \sum_{n\in\{0,1\}^{E_m}} \sum_{\sigma\in\Sigma} (-1)^{|\{e:\, n_e=0\}|} q^{|\{e:\, n_e=1\}|} \Big(\prod_{e\in E_m} \delta_e(\sigma)^{n_e}\Big) \\
&= \sum_{n\in\{0,1\}^{E_m}} (-1)^{|\{e:\, n_e=0\}|} q^{|\{e:\, n_e=1\}|} q^{k(m,n)},
\end{aligned} \tag{9.22}$$

where $k(m, n)$ is the number of connected components of the graph obtained from G_m after deletion of all edges e with $n_e = 0$. Therefore, by (9.4)–(9.5),

$$\begin{aligned}
\partial^m Z_{\mathrm{P}}\Big|_{\mathbf{J}=0} &= (-1)^{|E_m|} \sum_{n\in\{0,1\}^{E_m}} (-q)^{|\{e:\, n_e=1\}|} q^{k(m,n)} \\
&= (-1)^{|E_m|} q^{|V|} W_{G_m}(-1, -q) \\
&= q^{|V|} C(G_m; q).
\end{aligned} \tag{9.23}$$

Combining (9.21)–(9.23), we conclude that

$$\exp\Big\{-\sum_{e\in E} J_e\Big\} Z_{\mathrm{P}} = q^{|V|}\mathbb{E}_\lambda(C(G_P;q)). \tag{9.24}$$

Note in passing that equation (9.18) follows as in (1.12).

Let $x, y \in V$. We define the unordered pair $f = (x, y)$, and write $\delta_f(\sigma) = \delta_{\sigma_x,\sigma_y}$ for $\sigma \in \Sigma$. Then

$$\begin{aligned} \sigma(x,y) &= \pi_{\beta\mathbf{J},q}(q\delta_f(\sigma)-1) \\ &= \frac{1}{Z_{\mathrm{P}}}\sum_{\sigma\in\Sigma}(q\delta_f(\sigma)-1)\exp\Big\{\sum_{e\in E}\beta J_e(q\delta_e(\sigma)-1)\Big\}. \end{aligned} \tag{9.25}$$

By an analysis parallel to (9.21)–(9.24),

$$\begin{aligned} \exp\Big\{-\sum_{e\in E} J_e\Big\}\sum_{\sigma\in\Sigma}(q\delta_f(\sigma)-1)\exp\Big\{\sum_{e\in E}\beta J_e(q\delta_e(\sigma)-1)\Big\} \\ = q^{|V|}\mathbb{E}_\lambda(C(G_P^{x,y};q)), \end{aligned} \tag{9.26}$$

and (9.12) follows by (9.24) and (9.25). □

Proof of Theorem 9.15. This theorem may be proved directly, but we shall derive it from Theorem 9.11.

(a) Equation (9.17) holds by Theorems 1.16 and 9.11. By (9.5), equation (9.16) holds for $q \in \{2, 3, \dots\}$. Since both sides are the ratios of polynomials in q of finite order, (9.16) is an identity in $q \in (0, \infty)$.

(b) This was noted after (9.24) above. □

9.2 Flows for the Ising model

Henceforth in this chapter we assume that $q = 2$, and we begin with a reminder. Let $H = (W, F)$ be a finite graph, and let $\deg_F(w)$ denote the degree of the vertex w. We call H an *even* graph if $\deg_F(w)$ is even for every $w \in W$. Let $\vec{H} = (V, \vec{F})$ be a directed graph obtained from H by assigning a direction to each edge in F. Since a non-zero mod-2 flow on $\vec{H}$ may by definition take only the value 1,

$$C_H(2) = \begin{cases} 1 & \text{if } H \text{ is even,} \\ 0 & \text{otherwise.} \end{cases} \tag{9.27}$$

Consider the Ising model on a finite graph $G = (V, E)$ with parameters $\lambda_e = \beta J_e$, $e \in E$. As in (9.14),

$$\sigma(x,y) = 2\tau_{\lambda,2}(x,y) = \frac{\mathbb{P}_\lambda(G_P^{x,y} \text{ is even})}{\mathbb{P}_\lambda(G_P \text{ is even})}. \tag{9.28}$$

The value of such a representation will become clear in the following discussion, which is based on material in [3, 234, 300]. In advance of this, we make a remark concerning (9.28). In deciding whether G_P or $G_P^{x,y}$ is an even graph, we need only know the numbers $P(e)$ when reduced modulo 2. That is, we can work with $\zeta \in \Omega = \{0,1\}^E$ given by $\zeta(e) = P(e) \bmod 2$. Since $P(e)$ has the Poisson distribution with parameter λ_e, $\zeta(e)$ has the Bernoulli distribution with parameter

$$p'_e = \mathbb{P}_\lambda(P(e) \text{ is odd}) = \tfrac{1}{2}(1 - e^{-2\lambda_e}).$$

We obtain thus from (9.28) that

$$\sigma(x,y) = \frac{\phi_{\mathbf{p}'}(\partial\zeta = \{x,y\})}{\phi_{\mathbf{p}'}(\partial\zeta = \varnothing)}, \tag{9.29}$$

where $\mathbf{p}' = (p'_e : e \in E)$, $\phi_{\mathbf{p}'}$ denotes product measure on Ω with edge-densities p'_e, and

$$\partial\zeta = \Big\{v \in V : \sum_{e:\, e\sim v} \zeta(e) \text{ is odd}\Big\}, \qquad \zeta \in \Omega,$$

where the sum is over all edges e incident to v.

Let $M = (M_e : e \in E)$ be a sequence of disjoint finite sets indexed by E, and let $m_e = |M_e|$. As noted in the last section, the vector M may be used to construct a multigraph $G_m = (V, E_m)$ in which each $e \in E$ is replaced by m_e edges in parallel; we may take M_e to be the set of such edges. For $x, y \in V$, we write '$x \leftrightarrow y$ in m' if x and y lie in the same component of G_m. We define the set ∂M of *sources* of M by

$$\partial M = \Big\{v \in V : \sum_{e:\, e\sim v} m_e \text{ is odd}\Big\}. \tag{9.30}$$

For example, G_m is even if and only if $\partial M = \varnothing$. From the vector M we construct a vector $N = (N_e : e \in E)$ by deleting each member of each M_e with probability $\frac{1}{2}$, independently of all other elements. That is, we let B_i, $i \in \bigcup_e M_e$, be independent Bernoulli random variables with parameter $\frac{1}{2}$, and we set

$$N_e = \{i \in M_e : B_i = 1\}, \qquad e \in E.$$

Let $\mathbb{P}^M$ denote the appropriate probability measure.

The following technical lemma is pivotal for the computations that follow.

(9.31) Theorem. *Let M and m be as above. If $x, y \in V$ are such that $x \ne y$ and $x \leftrightarrow y$ in m then, for $A \subseteq V$,*

$$\mathbb{P}^M\big(\partial N = \{x,y\},\ \partial(M \setminus N) = A\big) = \mathbb{P}^M\big(\partial N = \varnothing,\ \partial(M \setminus N) = A \triangle \{x,y\}\big).$$

Proof. Take M_e to be the set of edges of G_m parallel to e, and assume that $x \leftrightarrow y$ in m. Let $A \subseteq V$. Let $\mathcal{M}$ be the set of all vectors $n = (n_e : e \in E)$ with $n_e \subseteq M_e$ for $e \in E$. Let α be a fixed path of G_m with endvertices x, y, viewed as a set of edges, and consider the map $\rho : \mathcal{M} \to \mathcal{M}$ given by

$$\rho(n) = n \bigtriangleup \alpha, \qquad n \in \mathcal{M}.$$

The map ρ is one–one, and maps $\{n \in \mathcal{M} : \partial n = \{x, y\},\ \partial(M \setminus n) = A\}$ to $\{n' \in \mathcal{M} : \partial n' = \varnothing,\ \partial(M \setminus n') = A \bigtriangleup \{x, y\}\}$. Each member of $\mathcal{M}$ is equiprobable under $\mathbb{P}^M$, and the claim follows. □

Let $\boldsymbol{\lambda} = (\lambda_e : e \in E)$ be a vector of non-negative reals, and recall the Poisson graph with parameter $\boldsymbol{\lambda}$. The following is a fairly immediate corollary of the last theorem. Let $M = (M_e : e \in E)$ and $M' = (M'_e : e \in E)$ be vectors of disjoint finite sets satisfying $M_e \cap M'_f = \varnothing$ for all $e, f \in E$, and let $m_e = |M_e|, m'_e = |M'_e|$, $e \in E$, be independent random variables such that each m_e and m'_e have the Poisson distribution with parameter λ_e. Let $M \cup M' = (M_e \cup M'_e : e \in E)$, and write $\mathbb{P}$ for the appropriate probability measure. The following lemma is based on the so-called switching lemma of [3].

(9.32) Corollary (Switching lemma). *If $x, y \in V$ are such that $x \neq y$ and $x \leftrightarrow y$ in $m + m'$ then, for $A \subseteq V$,*

$$\begin{aligned}\mathbb{P}\big(\partial M = \{x, y\},\ \partial M' = A \,\big|\, M \cup M'\big)& \\ = \mathbb{P}\big(\partial M = \varnothing,\ \partial M' = A \bigtriangleup \{x, y\} \,\big|\, M \cup M'\big),& \qquad \mathbb{P}\text{-a.s.}\end{aligned}$$

Proof. Conditional on the sets $M_e \cup M'_e$, $e \in E$, the sets M_e are selected by the independent removal of each element with probability $\frac{1}{2}$. The claim follows from Theorem 9.31. □

We present two applications of Theorem 9.32 to the Ising model, as in [3]. For $m = (m_e : e \in E) \in \mathbb{Z}_+^E$, let

$$\partial m = \Big\{v \in V : \sum_{e:\, e \sim v} m_e \text{ is odd}\Big\}, \tag{9.33}$$

as in (9.30). In our study of the correlation functions $\tau_{\boldsymbol{\lambda},2}(x, y)$, we shall as before write

$$\sigma(x, y) = 2\tau_{\boldsymbol{\lambda},2}(x, y) = \pi_{\boldsymbol{\lambda},2}(2\delta_{\sigma_x,\sigma_y} - 1), \qquad x, y \in V.$$

By (9.29),

$$\sigma(x, y) = \frac{\mathbb{P}_{\boldsymbol{\lambda}}(\partial P = \{x, y\})}{\mathbb{P}_{\boldsymbol{\lambda}}(\partial P = \varnothing)}. \tag{9.34}$$

Let $\mathbb{Q}_A$ denote the law of P *conditional* on the event $\{\partial P = A\}$, that is,

$$\mathbb{Q}_A(E) = \mathbb{P}_{\boldsymbol{\lambda}}(P \in E \mid \partial P = A).$$

We shall require two independent copies P_1, P_2 of P with potentially different conditionings, and thus we write $\mathbb{Q}_{A;B} = \mathbb{Q}_A \times \mathbb{Q}_B$.

(9.35) Theorem [3]. *Let $x, y, z \in V$ be distinct vertices. Then:*

$$\sigma(x,y)^2 = \mathbb{Q}_{\varnothing;\varnothing}(x \leftrightarrow y \text{ in } P_1 + P_2),$$
$$\sigma(x,y)\sigma(y,z) = \sigma(x,z)\mathbb{Q}_{\{x,z\};\varnothing}(x \leftrightarrow y \text{ in } P_1 + P_2).$$

Proof. By (9.34) and Theorem 9.32,

$$\begin{aligned}
\sigma(x,y)^2 &= \frac{\mathbb{P}_\lambda \times \mathbb{P}_\lambda(\partial P_1 = \{x,y\},\ \partial P_2 = \{x,y\})}{\mathbb{P}_\lambda(\partial P = \varnothing)^2} \\
&= \frac{\mathbb{P}_\lambda \times \mathbb{P}_\lambda(\partial P_1 = \{x,y\},\ \partial P_2 = \{x,y\},\ x \leftrightarrow y \text{ in } P_1 + P_2)}{\mathbb{P}_\lambda(\partial P = \varnothing)^2} \\
&= \frac{\mathbb{P}_\lambda \times \mathbb{P}_\lambda(\partial P_1 = \partial P_2 = \varnothing,\ x \leftrightarrow y \text{ in } P_1 + P_2)}{\mathbb{P}_\lambda(\partial P = \varnothing)^2} \\
&= \mathbb{Q}_{\varnothing;\varnothing}(x \leftrightarrow y \text{ in } P_1 + P_2).
\end{aligned}$$

Similarly,

$$\begin{aligned}
&\sigma(x,y)\sigma(y,z) \\
&\quad = \frac{\mathbb{P}_\lambda \times \mathbb{P}_\lambda(\partial P_1 = \{x,y\},\ \partial P_2 = \{y,z\})}{\mathbb{P}_\lambda(\partial P = \varnothing)^2} \\
&\quad = \frac{\mathbb{P}_\lambda \times \mathbb{P}_\lambda(\partial P_1 = \varnothing,\ \partial P_2 = \{x,z\},\ x \leftrightarrow y \text{ in } P_1 + P_2)}{\mathbb{P}_\lambda(\partial P = \varnothing)^2} \\
&\quad = \frac{\mathbb{P}_\lambda(\partial P_2 = \{x,z\})}{\mathbb{P}_\lambda(\partial P = \varnothing)} \cdot \mathbb{P}_\lambda \times \mathbb{P}_\lambda\big(x \leftrightarrow y \text{ in } P_1 + P_2 \,\big|\, \partial P_1 = \varnothing,\ \partial P_2 = \{x,z\}\big) \\
&\quad = \sigma(x,z)\mathbb{Q}_{\{x,z\};\varnothing}(x \leftrightarrow y \text{ in } P_1 + P_2).
\end{aligned}$$

□

Theorem 9.35 leads to an important correlation inequality known as the 'Simon inequality'. Let $x, z \in V$ be distinct vertices. A subset $W \subseteq V$ is said to *separate* x and z if $x, z \notin W$ and every path from x to z contains some vertex of W.

(9.36) Corollary (Simon inequality) [300]. *Let $x, z \in V$ be distinct vertices, and let W separate x and z. Then*

$$\sigma(x,z) \le \sum_{y \in W} \sigma(x,y)\sigma(y,z).$$

Proof. By Theorem 9.35,

$$\begin{aligned}
\sum_{y \in W} \frac{\sigma(x,y)\sigma(y,z)}{\sigma(x,z)} &= \sum_{y \in W} \mathbb{Q}_{\{x,z\};\varnothing}(x \leftrightarrow y \text{ in } P_1 + P_2) \\
&= \mathbb{Q}_{\{x,z\};\varnothing}\big(|\{y \in W : x \leftrightarrow y \text{ in } P_1 + P_2\}|\big).
\end{aligned}$$

Assume that the event $\partial P_1 = \{x, z\}$ occurs. On this event, $x \leftrightarrow z$ in $P_1 + P_2$. Since W separates x and z, the set $\{y \in W : x \leftrightarrow y \text{ in } P_1 + P_2\}$ is non-empty on this event. Therefore, its mean cardinality is at least one under the measure $\mathbb{Q}_{\{x,z\};\varnothing}$, and the claim follows. □

The Ising model on $G = (V, E)$ corresponds as described in Chapter 1 to a random-cluster measure $\phi_{G,p,q}$ with $q = 2$. By Theorem 1.10, if $\lambda_e = \lambda$ for all e,

$$\sigma(x, y) = 2\tau_{\lambda,2}(x, y) = \phi_{G,p,q}(x \leftrightarrow y),$$

where $p = 1 - e^{-\lambda q}$ and $q = 2$. Therefore, the Simon inequality[3] may be written in the form

$$\phi_{G,p,q}(x \leftrightarrow z) \le \sum_{y \in W} \phi_{G,p,q}(x \leftrightarrow y)\phi_{G,p,q}(y \leftrightarrow z) \tag{9.37}$$

whenever W separates x and z. It is a curious fact that this inequality holds also when $q = 1$, as noticed by Hammersley [177]; see [154, Chapter 6]. It may be conjectured that it holds whenever $q \in [1, 2]$.

The Simon inequality has an important consequence for the random-cluster model with $q = 2$ on an infinite lattice, namely that the two-point correlation function decays exponentially whenever it is summable. Let $\phi_{p,q}$ be the random-cluster measure on $\mathbb{L}^d$ where $d \ge 2$. We shall consider only the case $p < p_c(q)$, and it is therefore unnecessary to mention boundary conditions.

(9.38) Corollary [300]. *Let $d \ge 2$, $q = 2$, and let p be such that*

$$\sum_{x \in \mathbb{Z}^d} \phi_{p,q}(0 \leftrightarrow x) < \infty. \tag{9.39}$$

There exists $\gamma = \gamma(p, q) \in (0, \infty)$ such that

$$\phi_{p,q}(0 \leftrightarrow z) \le e^{-\|z\|\gamma(p,q)}, \qquad z \in \mathbb{Z}^d.$$

By the corollary, condition (9.39) is both necessary and sufficient for exponential decay. Related results for exponential decay appear in Section 5.4–5.6.

Proof. We use the Simon inequality in the form (9.37) as in [177, 300]. Let $\Lambda_n = [-n, n]^d$ and $\partial\Lambda_n = \Lambda_n \setminus \Lambda_{n-1}$, and take $q = 2$. By (9.37) with $G = \Lambda_n$, and Proposition 5.12,

$$\phi_{p,q}(x \leftrightarrow z) \le \sum_{y \in W} \phi_{p,q}(x \leftrightarrow y)\phi_{p,q}(y \leftrightarrow z), \tag{9.40}$$

[3] In association with related inequalities of Hammersley [177] and Lieb [234], see Theorem 9.44(b), this is an example of what is sometimes called the Hammersley–Simon–Lieb inequality. The Simon inequality is a special case of the Boel–Kasteleyn inequalities, [56, 57].

for $x, z \in \mathbb{Z}^d$ and any finite separating set W.

By (9.39), there exists $c \in (0,1)$ and $N \geq 1$ such that

$$\sum_{x \in \partial\Lambda_N} \phi_{p,q}(0 \leftrightarrow x) < c.$$

For any integer $k > 1$ and any vertex $z \in \partial\Lambda_{kN}$, we have by progressive use of (9.40) and the translation-invariance of $\phi_{p,q}$ (see Theorem 4.19(b)) that

$$\begin{aligned}
&\phi_{p,q}(0 \leftrightarrow z) \\
&\quad \leq \sum_{\substack{x_1: \\ \|x_1\|=N}} \phi_{p,q}(0 \leftrightarrow x_1)\phi_{p,q}(x_1 \leftrightarrow z) \\
&\quad \leq \sum_{\substack{x_1: \\ \|x_1\|=N}} \sum_{\substack{x_2: \\ \|x_2-x_1\|=N}} \phi_{p,q}(0 \leftrightarrow x_1)\phi_{p,q}(x_1 \leftrightarrow x_2)\phi_{p,q}(x_2 \leftrightarrow z) \\
&\quad \leq \sum_{\substack{x_1: \\ \|x_1\|=N}} \cdots \sum_{\substack{x_k: \\ \|x_k-x_{k-1}\|=N}} \phi_{p,q}(0 \leftrightarrow x_1)\cdots\phi_{p,q}(x_{k-1} \leftrightarrow x_k)\phi_{p,q}(x_k \leftrightarrow z) \\
&\quad \leq c^k.
\end{aligned}$$

Therefore, there exists $g > 0$ such that

$$\phi_{p,q}(0 \leftrightarrow z) \leq e^{-\|z\|g} \qquad \text{if } \|z\| \text{ is a multiple of } N.$$

More generally, let $z \in \mathbb{Z}^d$ and write $\|z\| = kN + l$ where $0 \leq l < N$. By (9.40),

$$\phi_{p,q}(0 \leftrightarrow z) \leq \sum_{\substack{x: \\ \|x\|=kN}} \phi_{p,q}(0 \leftrightarrow x)\phi_{p,q}(x \leftrightarrow z) \leq e^{-kNg}.$$

Furthermore, $\phi_{p,q}(0 \leftrightarrow z) < 1$ for $z \neq 0$, and the claim follows. □

We close this section with an improvement of the Simon inequality due to Lieb [234]. This improvement may seem at first sight to be slender, but it leads to a significant conclusion termed the 'vanishing of the mass gap'.

We first re-visit Theorem 9.31. As usual, $G = (V, E)$ is a finite graph, and we partition E as $E = F \cup H$, where $F \cap H = \varnothing$. Let $M = (M_e : e \in E)$ be a vector of disjoint finite sets with cardinalities $m_e = |M_e|$. We write $M^F = (M_e : e \in F)$ and define the vector m^F by

$$m_e^F = \begin{cases} m_e & \text{if } e \in F, \\ 0 & \text{otherwise,} \end{cases}$$

and similarly for M^H and m^H. It is elementary that $m = m^F + m^H$, and that the sets of sources of M^F and M^H are related by

$$\partial M^F \,\triangle\, \partial M^H = \partial M. \tag{9.41}$$

As before Theorem 9.31, we select subsets N_e from the M_e by deleting each member independently at random with probability $\frac{1}{2}$. For given M, the associated probability measure is denoted by $\mathbb{P}^M$.

(9.42) Theorem. *Let F, H, M, and m be as above. If $x, y \in V$ are such that $x \neq y$ and $x \leftrightarrow y$ in m^F then, for $A \subseteq V$,*

$$\begin{aligned}\mathbb{P}^M\big(\partial N^F = \{x, y\},\ \partial N^H = \varnothing,\ \partial(M \setminus N) = A\big)\\ = \mathbb{P}^M\big(\partial N^F = \varnothing,\ \partial N^H = \varnothing,\ \partial(M \setminus N) = A \triangle \{x, y\}\big).\end{aligned}$$

Proof. This follows that of Theorem 9.31. Let α be a fixed path of G_{m^F} with endvertices x and y, and consider the map $\rho(n) = n \triangle \alpha$, $n \in \mathcal{M}$. This map is a one–one correspondence between the two subsets of $\mathcal{M}$ corresponding to the two events in question. □

We obtain as in the switching lemma, Theorem 9.32, the following corollary involving the two independent random vectors M and M', each being such that $m_e = |M_e|$ and $m'_e = |M'_e|$ have the Poisson distribution with parameter $\lambda \in [0, \infty)$. The proof follows that of Theorem 9.32.

(9.43) Corollary. *Let E be partitioned as $E = F \cup H$. If $x, y \in V$ are such that $x \neq y$ and $x \leftrightarrow y$ in $m^F + m'^F$ then, for $A \subseteq V$,*

$$\begin{aligned}\mathbb{P}\big(\partial M^F = \{x, y\},\ \partial M^H = \varnothing,\ \partial M' = A \,\big|\, M \cup M'\big)\\ = \mathbb{P}\big(\partial M^F = \varnothing,\ \partial M^H = \varnothing,\ \partial M' = A \triangle \{x, y\} \,\big|\, M \cup M'\big), \qquad \mathbb{P}\text{-a.s.}\end{aligned}$$

Let P_1 and P_2 be independent copies of the Poisson field P, with intensity $\lambda \in [0, \infty)$, and let E be partitioned as $E = F \cup H$. We write $\mathbb{Q}_{A,B;C}$ for the probability measure governing the pair P_1, P_2 *conditional on the event* $\{\partial P_1^F = A\} \cap \{\partial P_1^H = B\} \cap \{\partial P_2 = C\}$. We recall from (9.28) that $\sigma(x, y)$ denotes a certain correlation function associated with the graph $G = (V, E)$, and we write $\sigma^F(x, y)$ for the quantity defined similarly on the smaller graph (V, F).

(9.44) Theorem. *Let $x, y, z \in V$ be distinct vertices, and let $F \subseteq E$.*

(a) *We have that*

$$\sigma^F(x, y)\sigma(y, z) = \sigma(x, z)\mathbb{Q}_{\varnothing,\varnothing;\{x,z\}}(x \leftrightarrow z \text{ in } P_1^F + P_2^F).$$

(b) Lieb inequality [234]. *Let W separate x and z, and let F be the set of edges with at least one endvertex not separated by W from x. Then*

$$\sigma(x, z) \leq \sum_{y \in W} \sigma^F(x, y)\sigma(y, z).$$

The sets W and F of part (b) are illustrated in Figure 9.1. By the random-cluster representation of Theorem 1.16 and positive association,

$$\begin{aligned}\sigma^F(x, y) &= \phi_{G,p,q}(x \leftrightarrow y \mid \text{all edges in } E \setminus F \text{ are closed})\\ &\leq \phi_{G,p,q}(x \leftrightarrow y) = \sigma(x, y),\end{aligned}$$

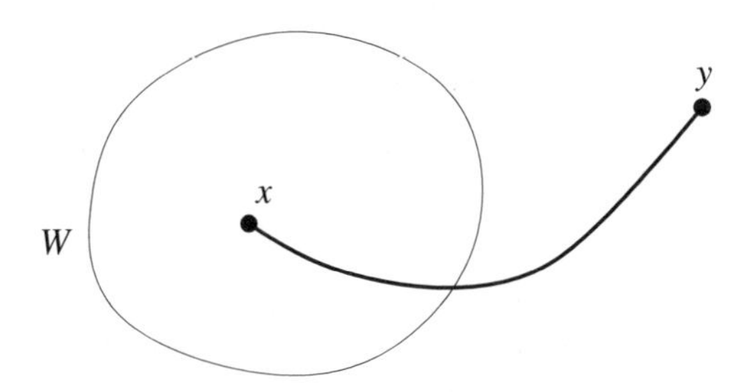

Figure 9.1. Every path from x to y passes through W. The edges inside W comprise F.

where $q = 2$ and $p = 1 - e^{-2\lambda}$ as before. Therefore, the Lieb inequality is a strengthening of the Simon inequality.

Proof. (a) Since F and H are disjoint, P_1^F and P_1^H are independent random vectors. As in the proof of Theorem 9.35, by Corollary 9.43,

$$\begin{aligned}
&\sigma^F(x, y)\sigma(y, z) \\
&\quad = \frac{\mathbb{P}_\lambda \times \mathbb{P}_\lambda(\partial P_1^F = \{x, y\},\ \partial P_1^H = \varnothing,\ \partial P_2 = \{y, z\})}{\mathbb{P}_\lambda(\partial P_1^F = \varnothing)\mathbb{P}_\lambda(\partial P_1^H = \varnothing)\mathbb{P}_\lambda(\partial P_2 = \varnothing)} \\
&\quad = \frac{\mathbb{P}_\lambda \times \mathbb{P}_\lambda(\partial P_1^F = \partial P_1^H = \varnothing,\ \partial P_2 = \{x, z\},\ x \leftrightarrow y \text{ in } P_1^F + P_2^F)}{\mathbb{P}_\lambda(\partial P_1^F = \varnothing)\mathbb{P}_\lambda(\partial P_1^H = \varnothing)\mathbb{P}_\lambda(\partial P_2 = \varnothing)} \\
&\quad = \sigma(x, z) \cdot \frac{\mathbb{P}_\lambda \times \mathbb{P}_\lambda(\partial P_1^F = \partial P_1^H = \varnothing,\ \partial P_2 = \{x, z\},\ x \leftrightarrow y \text{ in } P_1^F + P_2^F)}{\mathbb{P}_\lambda(\partial P_1^F = \varnothing)\mathbb{P}_\lambda(\partial P_1^H = \varnothing)\mathbb{P}_\lambda(\partial P_2 = \{x, z\})} \\
&\quad = \sigma(x, z)\mathbb{Q}_{\varnothing,\varnothing;\{x,z\}}(x \leftrightarrow y \text{ in } P_1^F + P_2^F),
\end{aligned}$$

(b) Evidently,

$$\begin{aligned}
\sum_{y \in W} \frac{\sigma^F(x, y)\sigma(y, z)}{\sigma(x, z)} &= \sum_{y \in W} \mathbb{Q}_{\varnothing,\varnothing;\{x,z\}}(x \leftrightarrow y \text{ in } P_1^F + P_2^F) \\
&= \mathbb{Q}_{\varnothing,\varnothing;\{x,z\}}\big(|\{y \in W : x \leftrightarrow y \text{ in } P_1^F + P_2^F\}|\big) \\
&\geq 1,
\end{aligned}$$

since, conditional on $\partial P_2 = \{x, z\}$, P_2 contains (almost surely) a path from x to z, and any such path necessarily intersects W. □

We return now to the question of exponential decay, which we formulate in the context of the random-cluster model on $\mathbb{Z}^d$ with $q = 2$. By Theorem 9.44(b) with $W = \partial\Lambda_k$ as before, $\phi_{p,q}(0 \leftrightarrow z)$ decays exponentially as $\|z\| \to \infty$ if and only if

$$\sum_{x \in \partial\Lambda_k} \phi_{\Lambda_k,p,q}(0 \leftrightarrow x) < 1 \qquad \text{for some } k \geq 1, \tag{9.45}$$

where $q = 2$. Condition (9.45) is a 'finite-volume condition' in that it uses probability measures on *finite* graphs only. We arrive thus at the following result, sometimes termed the 'vanishing of the mass gap'. Let $q = 2$ and

$$\psi(p,q) = \lim_{n\to\infty}\left\{-\frac{1}{n}\log\phi^0_{p,q}(0\leftrightarrow\partial\Lambda_n)\right\}$$

as in Theorem 5.45. It is clear that $\psi(p,q)$ is non-increasing in p, and $\psi(p,q) = 0$ if $p > p_c(q)$. One of the characteristics of a first-order phase transition is the (strict) exponential decay of free-boundary-condition connectivity probabilities *at* the critical point, see Theorems 6.35(c) and 7.33.

(9.46) Theorem (Vanishing mass gap) [234]. *Let* $q = 2$. *Then* $\psi(p,q)$ *decreases to* 0 *as* $p \uparrow p_c(q)$. *In particular,* $\psi(p_c(q),q) = 0$.

Proof. We consider only values of p satisfying $\epsilon < p < 1-\epsilon$ where $\epsilon > 0$ is fixed and small. Let $k \ge 1$, and let $\eta(\omega)$ be the set of open edges of a configuration ω. By Theorem 3.12 and the Cauchy–Schwarz inequality, with $q = 2$ throughout,

$$\begin{aligned} 0 &\le \frac{d}{dp}\sum_{x\in\partial\Lambda_k}\phi_{\Lambda_k,p,q}(0\leftrightarrow x)\\ &\le \sum_{x\in\partial\Lambda_k}\frac{1}{\epsilon(1-\epsilon)}\mathrm{cov}_k(|\eta|, 1_{\{0\leftrightarrow x\}})\\ &\le \sum_{x\in\partial\Lambda_k}\frac{1}{\epsilon(1-\epsilon)}\sqrt{\phi_{\Lambda_k,p,q}(|\eta|^2)}\\ &\le C_1 k^{2d-1}, \end{aligned}$$

for some constant $C_1 = C_1(\epsilon)$, where cov_k denotes covariance with respect to $\phi_{\Lambda_k,p,q}$. Therefore, for $\epsilon < p < p' < 1-\epsilon$,

$$\sum_{x\in\partial\Lambda_k}\phi_{\Lambda_k,p',q}(0\leftrightarrow x) \le C_1 k^{2d-1}(p'-p) + \sum_{x\in\partial\Lambda_k}\phi_{\Lambda_k,p,q}(0\leftrightarrow x).$$

It follows that, if (9.45) holds for some $p \in (\epsilon, 1-\epsilon)$, then it holds for some $p' > p$. That is, if $\phi_{p,q}(0\leftrightarrow z)$ decays exponentially as $\|z\|\to\infty$, then the same holds for some p' satisfying $p' > p$. The set $\{p \in (0,1) : \psi(p,q) > 0\}$ is therefore open. Since $\psi(p,q) = 0$ for $p > p_c(q)$, we deduce that $\psi(p_c(q),q) = 0$.

By Theorem 4.28(c) and the second inequality of (5.46), $\psi(p,q)$ is the limit from above of upper-semicontinuous functions of p. Therefore, $\psi(p,q)$ is itself upper-semicontinuous, and hence left-continuous. □

Could some of the results of this section be valid for more general values of q than simply $q = 2$? It is known that the mass gap vanishes when $q = 1$, [154, Thm 6.14], and does not vanish for sufficiently large values of q (and any $d \ge 2$), see [224] and Section 7.5. Therefore, Theorem 9.44(b) is not generally true for large q. It seems possible that the conclusions hold for sufficiently small q, but this is unproven.

One may ask whether the weaker Simon inequality, Corollary 9.36, might hold for more general values of q. The following example would need to be assimilated in any such result.

(9.47) Example[4]. Let $G = (V, E)$ be a cycle of length m, illustrated in Figure 9.2. We work with the partition function

$$Y = \sum_{\omega \in \Omega} \left(\frac{p}{1-p} \right)^{|\eta(\omega)|} q^{k(\omega)}, \tag{9.48}$$

where $\Omega = \{0, 1\}^E$ as usual. Since

$$k(\omega) = \begin{cases} 1 & \text{if } \eta(\omega) = E, \\ m - |\eta(\omega)| & \text{otherwise,} \end{cases}$$

we have that

$$\begin{aligned} Y &= \sum_{j=0}^{m-1} \binom{m}{j} \alpha^j q^{m-j} + \alpha^m q = (\alpha + q)^m - \alpha^m + \alpha^m q \\ &= Q^m + (q-1)\alpha^m. \end{aligned} \tag{9.49}$$

where

$$\alpha = \frac{p}{1-p}, \qquad Q = q + \alpha.$$

Let $x, y \in V$, let P_1, P_2 be the two paths joining x and y, and let k and l be their respective lengths. Configurations which contain P_1 but not P_2 contribute

$$Y_1 = \sum_{j=0}^{l-1} \binom{l}{j} \alpha^{k+j} q^{m-k-j} = \alpha^k \sum_{j=0}^{l-1} \binom{l}{j} \alpha^j q^{l-j} = \alpha^k (Q^l - \alpha^l)$$

to the summation of (9.48), with a similar contribution Y_2 from configurations containing P_2 but not P_1. The single configuration containing both P_1 and P_2 contributes $Y_{12} = q\alpha^m$ to the summation. Therefore,

$$\begin{aligned} \phi_{p,q}(x \leftrightarrow y) &= \frac{Y_1 + Y_2 + Y_{12}}{Y} \\ &= \frac{(\alpha/Q)^k + (\alpha/Q)^l + (q-2)(\alpha/Q)^m}{1 + (q-1)(\alpha/Q)^m}. \end{aligned} \tag{9.50}$$

[4] Calculations by S. Janson, on 11 March 2003 at Melbourn.

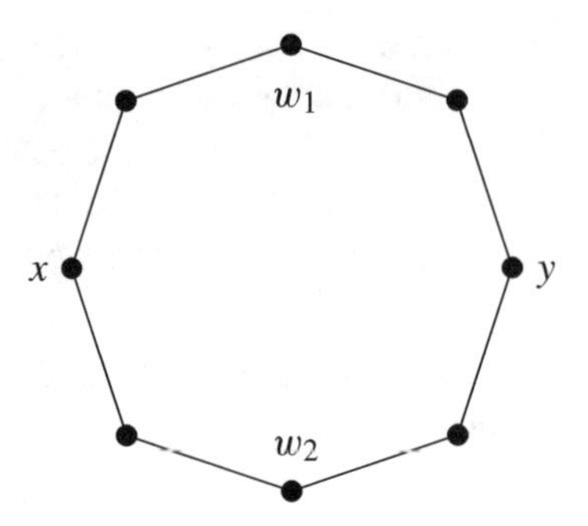

Figure 9.2. A cycle of length 8 with four marked vertices.

Consider now the specific example illustrated in Figure 9.2. Take $m = 8$, let x, y be opposite one another, and let w_1, w_2 be the intermediate vertices indicated in the figure. For fixed q and small α, by (9.50),

$$\phi_{p,q}(x \leftrightarrow y) = 2\left(\frac{\alpha}{Q}\right)^4 + (q-2)\left(\frac{\alpha}{Q}\right)^8 + \mathrm{O}(\alpha^{12}). \tag{9.51}$$

[A corresponding expression is valid for fixed α and large q.] Similarly,

$$\phi_{p,q}(x \leftrightarrow w_j) = \phi_{p,q}(w_j \leftrightarrow y) = \left(\frac{\alpha}{Q}\right)^2 + \left(\frac{\alpha}{Q}\right)^6 + \mathrm{O}(\alpha^8), \qquad j = 1, 2.$$

Hence,

$$\begin{aligned}\sum_{j=1}^{2} \phi_{p,q}(x \leftrightarrow w_j)\phi_{p,q}(w_j \leftrightarrow y) &= 2\left\{\left(\frac{\alpha}{Q}\right)^2 + \left(\frac{\alpha}{Q}\right)^6 + \mathrm{O}(\alpha^8)\right\}^2 \\ &= 2\left(\frac{\alpha}{Q}\right)^4 + 4\left(\frac{\alpha}{Q}\right)^8 + \mathrm{O}(\alpha^{10}).\end{aligned} \tag{9.52}$$

Comparing (9.51)–(9.52), we see that

$$\phi_{p,q}(x \leftrightarrow y) > \sum_{j=1}^{2} \phi_{p,q}(x \leftrightarrow w_j)\phi_{p,q}(w_j \leftrightarrow y)$$

if $q > 6$ and α is sufficiently small. This may be compared with (9.37).

9.3 Exponential decay for the Ising model

In the remaining two sections of this chapter, we review certain aspects of the mathematics of the Ising model in two and more dimensions. Several of the outstanding problems for Potts and random-cluster models have rigorous solutions in the Ising case, when $q = 2$, and it is a challenge of substance to extend such results (where valid) to the case of general $q \in [1, \infty)$. Our account of the Ising model will be restricted to the work of Aizenman, Barsky, and Fernández as reported in two major papers [5, 9], of which we begin in this section with the first. The principal technique of these papers is the so-called 'random-current representation', that is, the representation of the Ising random field in terms of non-zero mod-2 flows. See, for example, the representation (9.28) for the two-point correlation function. Without more ado, we state the main theorem in the language of the random-cluster model.

(9.53) Theorem (Finite susceptibility for $q = 2$ random-cluster model) [5]. *Let $p \in [0, 1]$, $q = 2$, $d \geq 2$, and let $\phi^1_{p,q}$ be the wired random-cluster measure on $\mathbb{L}^d$. The open cluster C at the origin satisfies*

$$\phi^1_{p,q}(|C|) < \infty, \qquad p < p_c(q).$$

This implies exponential decay, by Theorem 9.38: if $p < p_c(q)$, the connectivity function $\phi^1_{p,q}(0 \leftrightarrow z)$ decays exponentially to zero as $\|z\| \to \infty$. When $d = 2$, it implies that $p_c(2) = \sqrt{2}/(1 + \sqrt{2})$, see Theorem 6.18.

(9.54) Theorem (Mean-field bound) [5]. *Under the conditions stated in Theorem 9.53, there exists a constant $c = c(d) > 0$ such that the percolation probability $\theta^1(p, q) = \phi^1_{p,q}(0 \leftrightarrow \infty)$ satisfies*

$$\theta^1(p, 2) \geq c(p - p_c)^{\frac{1}{2}}, \qquad p > p_c = p_c(2). \tag{9.55}$$

Through the use of scaling theory (see [154, Chapter 9]), one is led to predictions concerning the existence of critical exponents for quantities exhibiting singularities at the critical point $p_c(q)$. It is believed in particular that the function $\theta(\cdot, 2)$ possesses a critical exponent[5] in that there exists $b \in (0, \infty)$ satisfying

$$\theta^1(p, 2) = |p - p_c|^{b(1+o(1))} \qquad \text{as } p \downarrow p_c = p_c(2). \tag{9.56}$$

If this is true, then $b \leq \frac{1}{2}$ by Theorem 9.54. It turns out that the latter inequality is sharp in the sense that, when $d \geq 4$, it is satisfied with equality; see Theorem 9.58. The value $b = \frac{1}{2}$ is in addition the 'mean-field' value of the critical exponent, as we shall see in Section 10.7 in the context of the random-cluster model on a complete graph.

[5] We write b rather than the more usual β for the critical exponent associated with the percolation probability, in order to avoid duplication with the inverse-temperature of the Ising model.

Proofs of the above theorems may be found in [5], and are omitted from the current work since they are Ising-specific and have not (yet) been generalized to the random-cluster setting for general q. The key ingredient is the random-current representation of the last section, utilized with ingenuity.

Nevertheless, included here is the briefest sketch of the approach; there is a striking similarity to, but also striking differences from, that used to prove corresponding results for percolation, see [4], [154, Section 5.3]. First, one introduces an external field h into the ferromagnetic Ising model with inverse-temperature β. This amounts in the context of the random-cluster model to the inclusion of a special vertex called by some the 'ghost', to which every vertex is joined by an edge with parameter $\gamma = 1 - e^{-\beta h}$. One works on a finite box Λ with 'toroidal' boundary conditions. An important step in the proof is the following differential inequality for the mean spin-value $M_\Lambda(\beta, h)$ at the origin:

$$M_\Lambda \le \tanh(\beta h)\frac{\partial M_\Lambda}{\partial(\beta h)} + M_\Lambda^2\left(\beta\frac{\partial M_\Lambda}{\partial \beta} + M_\Lambda\right). \tag{9.57}$$

The proof of this uses the random-current representation.

Equation (9.57) is complemented by two further differential inequalities:

$$\frac{\partial M_\Lambda}{\partial \beta} \le \beta J \frac{\partial M_\Lambda}{\partial(\beta h)}, \qquad \frac{\partial M_\Lambda}{\partial(\beta h)} \le \frac{M_\Lambda}{\beta h}.$$

Using an analysis presented in [4] for percolation, the three inequalities above imply Theorem 9.54.

9.4 The Ising model in four and more dimensions

Just as two-dimensional systems have special properties, so there are special arguments valid when the number d of dimensions is sufficiently large. For example, percolation in 19 and more dimensions is rather well understood through the work of Hara and Slade and others, [23], [154, Section 10.3], [179, 303], using the so-called 'lace expansion'. One expects that results for percolation in high dimensions will be extended in due course to $d > 6$, and even in part to $d \ge 6$. Key to this work is the so-called 'triangle condition', namely that $T(p_c) < \infty$ where $p_c = p_c(1)$ and

$$T(p) = \sum_{x,y \in \mathbb{Z}^d} \phi_p(0 \leftrightarrow x)\phi_p(x \leftrightarrow y)\phi_p(y \leftrightarrow 0).$$

The situation for the Ising model, and therefore for the $q = 2$ random-cluster model, is also well understood, but this time under the considerably less restrictive assumption that $d \ge 4$. The counterpart of the triangle condition is the 'bubble condition', namely that $B(\beta_c) < \infty$ where, in the usual notation of the Ising model without external field,

$$B(\beta) = \sum_{x \in \mathbb{Z}^d} \langle \sigma_0 \sigma_x \rangle^2.$$

In the language of the random-cluster model with $p = 1 - e^{-\beta}$, the corresponding quantity is

$$B(\beta) = \sum_{x \in \mathbb{Z}^d} \phi^0_{p,2}(0 \leftrightarrow x)^2.$$

Once again, one introduces an external field and then establishes a differential inequality via the random-current representation. We state the main result in the language of the random-cluster model.

(9.58) Theorem (Critical exponent for $q = 2$ random-cluster model) [9]. *Let $q = 2$ and $d \geq 4$. We have that*

$$\theta^1(p, q) = (p - p_c)^{\frac{1}{2}(1+o(1))} \qquad \textit{as } p \downarrow p_c = p_c(2).$$

Thus, the critical exponent b exists when $d \geq 4$, and it takes its 'mean-field' value $b = \frac{1}{2}$. This implies in particular that the percolation probability $\theta^1(p, 2)$ is a continuous function of p at the critical value $p_c(2)$. Continuity has been proved by classical methods in two dimensions[6], and there remains only the $d = 3$ case for which the continuity of $\theta^1(\cdot, 2)$ is as yet unproved. In summary, it is proved when $d \neq 3$ that the phase transition is of second order, and this is believed to be so when $d = 3$ also.

Similarly to the results of the last section, Theorem 9.58 is proved by an analysis of the model parametrized by the two variables β, h. This yields several further facts including an exact critical exponent for the behaviour of the Ising magnetization $M(\beta, h)$ with $\beta = \beta_c$ and $h \downarrow 0$, namely

$$M(\beta_c, h) = h^{\frac{1}{3}(1+o(1))} \qquad \text{as } h \downarrow 0.$$

We refer the reader to [5, 9] for details of the random-current representation in practice, for proofs of the above results and of more detailed asymptotics, and for a more extensive bibliography. The random-cluster representation is a key ingredient in the derivation of a lace expansion for the Ising model with either nearest-neighbour or spread-out interactions, [288]. This has led to asymptotic formulae for the two-point correlation function when $d > 4$. A broader perspective on phase transitions may be found in [118].

---

[6] See, for example, [252]. A probabilistic proof would be worthwhile.

Chapter 10

On Other Graphs

Summary. Exact solutions are known for the random-cluster models on complete graphs and on regular trees, and these provide theories of mean-field-type. There is a special argument for the complete graph which utilizes the theory of Erdős–Rényi random graphs, and leads to exact calculations valid for all values of $q \in (0, \infty)$. The transition is of first order if and only if $q \in (2, \infty)$. The (non-)uniqueness of random-cluster measures on a tree, when subject to a variety of boundary conditions, may be studied via an iterative formula permitting exact calculations of the critical value and the percolation probability. There is a discussion of the random-cluster model on a general non-amenable graph.

10.1 Mean-field theory

The theory of phase transitions addresses primarily singularities associated with spaces of finite dimension. There are two reasons for considering a 'mean-field' theory in which the number d of dimensions may be considered to take the value ∞. Firstly, the major problems confronting the mathematics lie in the geometrical constraints imposed by *finite-dimensional* Euclidean space; a solution for 'infinite dimension' can cast light on the case of finite dimension. The second reason is the desire to understand better the d-dimensional process in the limit of large d. One is led thus to the problems of establishing the theory of a process viewed as ∞-dimensional, and to proving that this is the limit in an appropriate sense of the d-dimensional process. Progress is well advanced on these two problems for percolation (see [154, Chapter 10]) but there remains much to be done for the random-cluster model.

Being informed by progress for percolation, it is natural to consider as mean-field models the random-cluster models on complete graphs and on an infinite tree. In the former case, we consider the model on the complete graph K_n on n vertices, and we pass to the limit as $n \to \infty$. The vertex-degrees tend to ∞ as $n \to \infty$, and some re-scaling is done in order to establish a non-trivial limit. The correct way to do this is to set $p = \lambda/n$ for fixed $\lambda > 0$. The consequent theory may be regarded as an extension of the usual Erdős–Rényi theory of random graphs, [61, 194]. This model is expounded in Section 10.2. The main results are described in Section 10.3, and are proved in Sections 10.4–10.6. The nature of the phase transition is discussed in Section 10.7, and the consequences for large deviations of cluster-counts are presented in Section 10.8. The principal reference[1] is [62], of which heavy use is made in this chapter.

The random-cluster model on a finite tree is essentially trivial. Owing to the absence of circuits, a random-cluster measure thereon is simply a product measure. The tree is a more interesting setting when it is infinite and subjected to boundary conditions. There is a continuum of random-cluster measures indexed by the set of possible boundary conditions. The present state of knowledge is summarized in Sections 10.9–10.11. The relevant references are [160, 167, 196] but the current treatment is fundamentally different.

Trees are examples of graphs whose boxes have surface/volume ratios bounded away from 0. Such graphs are termed 'non-amenable' and, subject to further conditions, they may have three phases rather than the more usual two. A brief account of this phenomenon may be found in Section 10.12.

10.2 On complete graphs

Let $n \ge 1$, and let $K_n = (V, V^{(2)})$ be the complete graph on the vertex set $V = V_n = \{1, 2, \dots, n\}$, with edge-set the set $V^{(2)}$ of all $\binom{n}{2}$ pairs of unordered elements of V. We shall consider the random-cluster measure on K_n with parameters $p \in (0, 1)$ and $q \in (0, \infty)$. We define the 'weight function'

$$\widetilde{P}_{n,p,q}(F) = p^{|F|}(1-p)^{\binom{n}{2}-|F|}q^{k(V,F)}, \qquad F \subseteq V^{(2)}, \tag{10.1}$$

where $k(V, F)$ denotes the number of components of the graph (V, F). The partition function is

$$Z_{n,p,q} = \sum_{F \subseteq V^{(2)}} \widetilde{P}_{n,p,q}(F) \tag{10.2}$$

and the random-cluster measure on subsets of $V^{(2)}$ is then given by

$$\phi_{n,p,q}(F) = \frac{\widetilde{P}_{n,p,q}(F)}{Z_{n,p,q}}, \qquad F \subseteq V^{(2)}. \tag{10.3}$$

Thus, for any given n, p, q, the measure $\phi_{n,p,q}$ is the law of a random graph with n vertices which we denote by $G_{n,p,q}$. We sometimes write $\phi_{n,p,q}(F)$ as $\phi_{V,p,q}(F)$.

[1]The random-cluster model on the complete graph is related to the 'first-shell' model of Whittle, [317, 318].

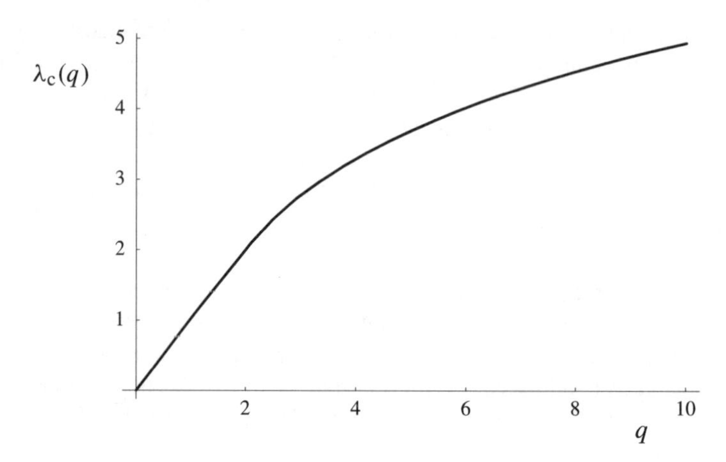

Figure 10.1. The critical value $\lambda_c(q)$. There is a discontinuity in the second derivative at the value $q = 2$.

In order to facilitate the notation later, we have chosen to take as sample space the set of subsets of $V^{(2)}$ rather than the vector space $\{0, 1\}^{V^{(2)}}$.

The random-cluster measure of (10.3) has two parameters, p and q. When $q = 1$, we recover the usual Erdős–Rényi model usually denoted by $G_{n,p}$, see [61, 194]. When $q \in \{2, 3, \dots\}$, the random-cluster model corresponds in the usual way to a Potts model on the complete graph K_n with q states and with inverse-temperature $\beta = -\log(1 - p)$.

The principal technique for analysing the mean-field Potts model relies heavily upon the assumption that q is an integer, see [324]. This technique is invalid for general real values of q, and one needs a new method in order to understand the full model. The principal extra technique, described in Section 10.3, is a method whereby properties of $G_{n,p,q}$ may be studied via corresponding properties of the usual random graph $G_{n,p}$. Unlike the case of lattice systems, this allows an essentially complete analysis of the asymptotic properties of random-cluster measures on K_n for all values of $p \in (0, 1)$ and $q \in (0, \infty)$. Results for $\phi_{n,p,q}$ are obtained using combinatorial estimates, and no use is made of the FKG inequality.

As in the Erdős–Rényi theory of the giant component when $q = 1$, we set $p = \lambda/n$ where λ is a positive constant, and we study the size of the largest component of the ensuing graph $G_{n,\lambda/n,q}$ in the limit as $n \to \infty$. It turns out that there is a critical value of λ, depending on the value of q, that marks the arrival of a 'giant component' of the graph. This critical value is given by

$$(10.4) \qquad \lambda_c(q) = \begin{cases} q & \text{if } q \in (0, 2], \\ 2\left(\dfrac{q-1}{q-2}\right)\log(q-1) & \text{if } q \in (2, \infty), \end{cases}$$

and is plotted in Figure 10.1

It will turn out that the proportion of vertices in the largest component is roughly constant, namely $\theta(\lambda, q)$, for large n. It is convenient to introduce a definition of θ immediately, namely

$$\theta(\lambda, q) = \begin{cases} 0 & \text{if } \lambda < \lambda_c(q), \\ \theta_{\max} & \text{if } \lambda \geq \lambda_c(q), \end{cases} \tag{10.5}$$

where $\theta_{\max}$ is the largest root of the equation

$$e^{-\lambda\theta} = \frac{1-\theta}{1+(q-1)\theta}. \tag{10.6}$$

The roots of (10.6) are illustrated in Figure 10.2.

We note some of the properties of $\theta(\lambda, q)$. Firstly, $\theta(\lambda, q) > 0$ if and only if

$$\begin{aligned} \text{either:} \quad & \lambda > \lambda_c(q), \\ \text{or:} \quad & \lambda = \lambda_c(q) \text{ and } q > 2, \end{aligned}$$

see Lemma 10.12. Secondly, for all $q \in (0, \infty)$, $\theta(\lambda, q)$ is non-decreasing in λ, and it follows that $\theta(\cdot, q)$ is continuous if $q \in (0, 2]$, and has a unique (jump) discontinuity at $\lambda = \lambda_c(q)$ if $q \in (2, \infty)$. This jump discontinuity corresponds to a phase transition of first order.

We say that 'almost every (a.e.) $G_{n,p,q}$ satisfies property Π', for a given sequence $p = p_n$ and a fixed q, if

$$\phi_{n,p,q}(G_{n,p,q} \text{ has } \Pi) \to 1 \qquad \text{as } n \to \infty.$$

We summarize the main results of the following sections as follows.

(a) If $0 < \lambda < \lambda_c(q)$ and $q \in (0, \infty)$, then almost every $G_{n,\lambda/n,q}$ has largest component of order $\log n$.

(b) If $\lambda > \lambda_c(q)$ and $q \in (0, \infty)$, then almost every $G_{n,\lambda/n,q}$ consists of a 'giant component' of order $\theta(\lambda, q)n$, together with other components of order $\log n$ or smaller.

(c) If $\lambda = \lambda_c(q)$ and $q \in (0, 2]$, then almost every $G_{n,\lambda/n,q}$ has largest component of order $n^{2/3}$.

The behaviour of $G_{n,\lambda/n,q}$ with $q \in (2, \infty)$ and $\lambda = \lambda_n \to \lambda_c(q)$ has been studied further in the combinatorial analysis of [238].

There are two main steps in establishing the above facts. The first is to establish the relation (10.6) by studying the size of the largest component of $G_{n,\lambda/n,q}$. When $q \in (2, \infty)$, (10.6) has three solutions for large λ, see Figure 10.2. In order to decide which of these is the density of the largest component, we shall study the number of edges in $G_{n,\lambda/n,q}$. That is to say, we shall find the function $\psi(\lambda, q)$ such that almost every $G_{n,\lambda/n,q}$ has (order) $\psi(\lambda, q)n$ edges. It will turn out that the function $\psi(\cdot, q)$ is discontinuous at the critical point of a first-order phase transition.

The material presented here for the random-cluster model on K_n is taken from [62]. See also [238].

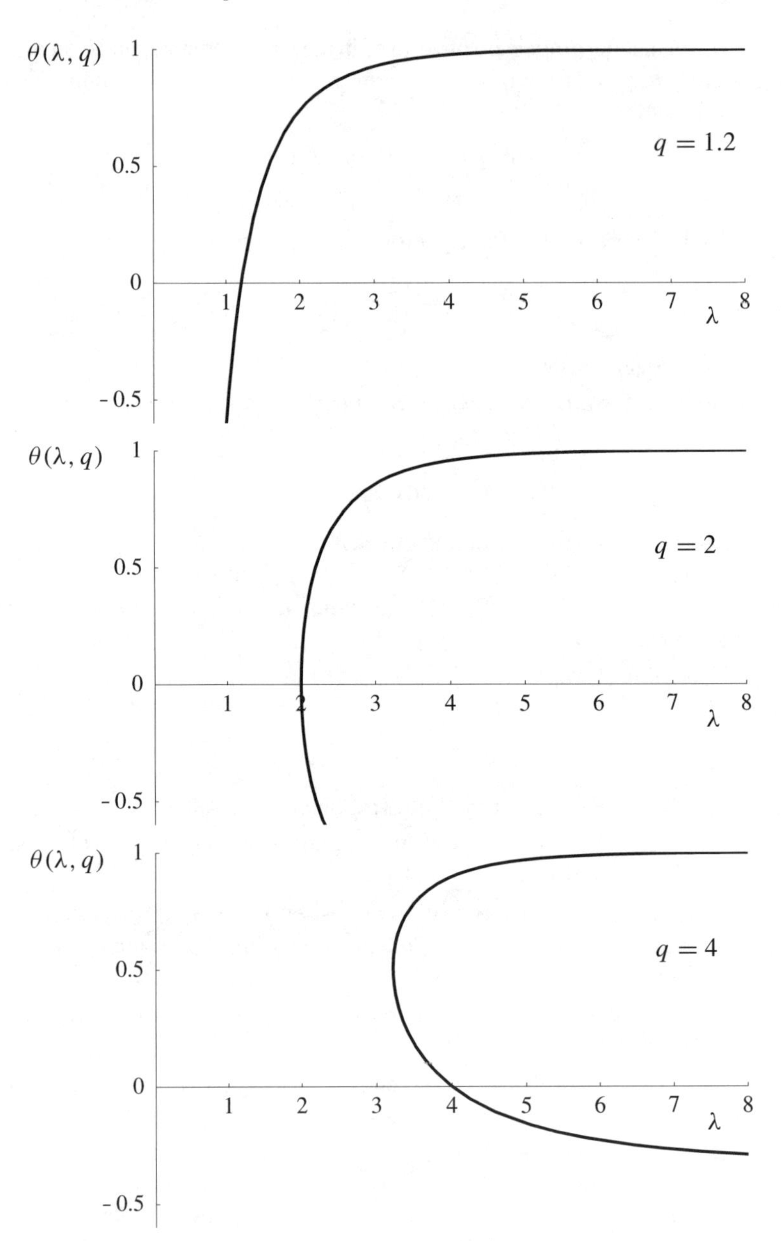

Figure 10.2. The roots of equation (10.6) are plotted against λ in the three cases $q = 1.2$, $q = 2$, $q = 4$. There is always a root $\theta = 0$, and there is a further root which is drawn here. The latter has been extended into the lower half-plane (of negative θ), although this region has no apparent probabilistic significance.

10.3 Main results for the complete graph

Let $q \in (0, \infty)$ and $p = \lambda/n$ where λ is a positive constant. For ease of notation, we shall sometimes suppress explicit reference to q. We shall make heavy use of the critical value $\lambda_c(q)$ given in (10.4), and the function $\theta(\lambda) = \theta(\lambda, q)$ defined in (10.5)–(10.6). The properties of roots of (10.6) will be used in some detail, but these are deferred until Lemma 10.12. For the moment we note only that $\theta(\lambda) = 0$ if and only if: either $\lambda < \lambda_c(q)$, or $\lambda = \lambda_c(q)$ and $q \le 2$.

There are three principal theorems dealing respectively with the subcritical case $\lambda < \lambda_c(q)$, the supercritical case $\lambda > \lambda_c(q)$, and the critical case $\lambda = \lambda_c(q)$. In the matter of notation, for a sequence $(X_n : n = 1, 2, \dots)$ of random variables, we write $X_n = \mathrm{O_p}(f(n))$ if $X_n/f(n)$ is bounded in probability:

$$\mathbb{P}\big(|X_n| \le f(n)\omega(n)\big) \to 1 \qquad \text{as } n \to \infty$$

for any sequence $\omega(n)$ satisfying $\omega(n) \to \infty$ as $n \to \infty$. Similarly, we write $X_n = \mathrm{o_p}(f(n))$ if $X_n/f(n) \to 0$ in probability as $n \to \infty$:

$$\mathbb{P}\big(|X_n| \le f(n)/\omega(n)\big) \to 1 \qquad \text{as } n \to \infty$$

for some sequence $\omega(n)$ satisfying $\omega(n) \to \infty$. Convergence in probability is denoted by the symbol $\xrightarrow{\mathrm{P}}$.

(10.7) Theorem (Subcritical case) [62]. *Let $q \in (0, \infty)$ and $\lambda < \lambda_c(q)$.*

(a) *Almost every $G_{n,\lambda/n,q}$ comprises trees and unicyclic components only.*

(b) *There are $\mathrm{O_p}(1)$ unicyclic components with a total number $\mathrm{O_p}(1)$ of vertices.*

(c) *The largest component of almost every $G_{n,\lambda/n,q}$ is a tree with order $\alpha \log n + \mathrm{O_p}(\log\log n)$, where*

$$\frac{1}{\alpha} = -\log(\lambda/q) + \frac{\lambda}{q} - 1 > 0.$$

(d) *The number of edges in $G_{n,\lambda/n,q}$ is $\lambda n/(2q) + \mathrm{o_p}(n)$.*

(10.8) Theorem (Supercritical case) [62]. *Let $q \in (0, \infty)$ and $\lambda > \lambda_c(q)$.*

(a) *Almost every $G_{n,\lambda/n,q}$ consists of a giant component, trees, and unicyclic components.*

(b) *The number of vertices in the giant component is $\theta(\lambda)n + \mathrm{o_p}(n)$, and the number of edges is*

$$\lambda\theta(\lambda)\left\{\frac{1}{q} + \left(\frac{1}{2} - \frac{1}{q}\right)\theta(\lambda)\right\} n + \mathrm{o_p}(n).$$

(c) *The largest tree in almost every $G_{n,\lambda/n,q}$ has order $\alpha \log n + \mathrm{o_p}(\log n)$, where*

$$\frac{1}{\alpha} = -\log\beta + \beta - 1 > 0, \quad \beta = \frac{\lambda}{q}(1 - \theta(\lambda)).$$

(d) *There are* $O_p(1)$ *unicyclic components with a total number* $O_p(1)$ *of vertices.*

(e) *The number of edges in* $G_{n,\lambda/n,q}$ *is*

$$\frac{\lambda}{2q}\left[1+(q-1)\theta(\lambda)^2\right]n+o_p(n).$$

(10.9) Theorem (Critical case) [62]. *Let* $q \in [1,2]$ *and* $\lambda = \lambda_c(q)$.

(a) *Almost every* $G_{n,\lambda/n,q}$ *consists of trees, unicyclic components, and* $O_p(1)$ *components with more than one cycle.*

(b) *The largest component has order* $o_p(n)$.

(c) *The total number of vertices in unicyclic components is* $O_p(n^{2/3})$.

(d) *The largest tree has order* $O_p(n^{2/3})$.

More detailed asymptotics are available for $G_{n,\lambda/n,q}$ by looking deeper into the proofs. The last theorem has been extended to the cases $q \in (0,1)$ and $q \in (2,\infty)$ in [238], where a detailed combinatorial analysis has been performed.

The giant component, when it exists, has order approximately $\theta(\lambda)n$, with $\theta(\lambda)$ given by (10.5)–(10.6). We study next the roots of (10.6). Note first that $\theta = 0$ satisfies (10.6) for all λ and q, and that all strictly positive roots satisfy $0 < \theta < 1$. Let

$$f(\theta) = \frac{1}{\theta}\left[\log\{1+(q-1)\theta\} - \log(1-\theta)\right], \qquad \theta \in (0,1), \tag{10.10}$$

and note that $\theta \in (0,1)$ satisfies (10.6) if and only if $f(\theta) = \lambda$. Here are two elementary lemmas concerning the function f.

(10.11) Lemma. *The function* f *is strictly convex on* $(0,1)$, *and satisfies* $f(0+) = q$ *and* $f(1-) = \infty$.

(a) *If* $q \in (0,2]$, *the function* f *is strictly increasing.*

(b) *If* $q \in (2,\infty)$, *there exists* $\theta_{\min} \in (0,1)$ *such that* f *is strictly decreasing on* $(0,\theta_{\min})$ *and strictly increasing on* $(\theta_{\min},1)$.

Proof. If $t > -1$ then $(1+t\theta)^{-1}$ is a strictly convex function of θ on $(0,1)$. Hence, the function

$$f(\theta) = \int_{-1}^{q-1} \frac{dt}{1+t\theta}$$

is strictly convex. Furthermore,

$$\lim_{\theta\uparrow 1} \log\left(\frac{1+(q-1)\theta}{1-\theta}\right) = \lim_{\epsilon\downarrow 0} \log\left(\frac{q-(q-1)\epsilon}{\epsilon}\right) = \infty,$$

implying that $f(1-) = \infty$. Applying Taylor's theorem about the point $\theta = 0$, we find that

$$\begin{aligned} f(\theta) &= \frac{1}{\theta}\left[(q-1)\theta - \tfrac{1}{2}(q-1)^2\theta^2 + \theta + \tfrac{1}{2}\theta^2 + O(\theta^3)\right] \\ &= q + \tfrac{1}{2}q(2-q)\theta + O(\theta^2), \end{aligned}$$

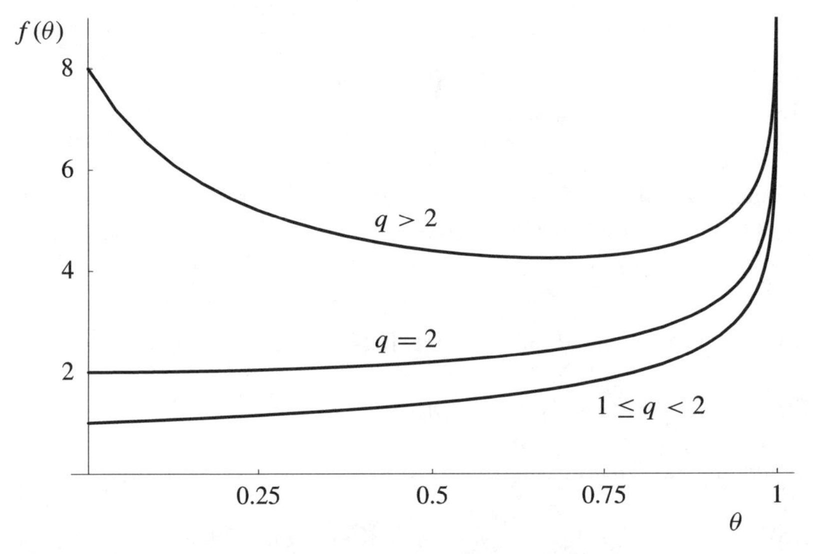

Figure 10.3. Sketches of the function $f(\theta)$ in the three cases $1 \le q < 2$, $q = 2$, and $q > 2$. The respective values of q can be read off from the y-axis, since $f(0+) = q$. Note that f is strictly increasing if and only if $q \le 2$, and $f'(0) = 0$ when $q = 2$. Recall that the positive roots of (10.6) are obtained by intersecting the graph of f by the horizontal line $f(\theta) = \lambda$.

whence $f(0+) = q$ and $f'(0+) = \frac{1}{2}q(2-q)$. These facts imply parts (a) and (b) of the lemma. □

In Figure 10.3 is plotted f against θ in the three cases $q \in [1, 2)$, $q = 2$, and $q \in (2, \infty)$. Since $\theta \in (0, 1)$ is a root of (10.6) if and only if $f(\theta) = \lambda$, Lemma 10.11 has the following consequence.

(10.12) Lemma. *The non-negative roots of equation* (10.6) *are given as follows, in addition to the root* $\theta = 0$.

(a) *Let* $q \in (0, 2]$.

  (i) *If* $0 < \lambda \le \lambda_c(q) = q$, *there exists a unique root* $\theta = 0$.

  (ii) *If* $q < \lambda$, *there exists a unique positive root* $\theta_{\max}(\lambda)$, *which satisfies* $\theta_{\max}(q+) = 0$.

(b) *Let* $q \in (2, \infty)$, *and let* $\lambda_{\min} = f(\theta_{\min})$ *where* $\theta_{\min}$ *is given in Lemma 10.11.*

  (i) *If* $0 < \lambda < \lambda_{\min}$, *there exists a unique root* $\theta = 0$.

  (ii) *If* $\lambda = \lambda_{\min}$, *then* $\theta_{\min}$ *is the unique positive root.*

  (iii) *If* $\lambda_{\min} < \lambda < q$, *there exist exactly two positive roots,* $\theta_1(\lambda)$ *and* $\theta_{\max}(\lambda)$.

  (iv) *If* $\lambda \ge q$, *there exists a unique positive root* $\theta_{\max}(\lambda)$.

We shall see later that $\lambda_{\min} < \lambda_c(q) < q$ when $q > 2$, and that the function

$\theta(\lambda)$ of Theorems 10.7–10.9 satisfies

$$\theta(\lambda_c(q)) = \begin{cases} 0 & \text{if } q \le 2, \\ \dfrac{q-2}{q-1} & \text{if } q > 2. \end{cases} \tag{10.13}$$

Furthermore, we shall obtain in Section 10.6 the following result for the asymptotic behaviour of the partition function $Z_{n,\lambda/n,q}$ as $n \to \infty$. This will find application in Section 10.8 to large deviations for the cluster-count of the random-cluster measure.

(10.14) Theorem (Existence of pressure) [62]. *If $q \in (0,\infty)$ and $\lambda \in (0,\infty)$,*

$$\frac{1}{n} \log Z_{n,\lambda/n,q} \to \eta(\lambda) \qquad \text{as } n \to \infty,$$

where the 'pressure' $\eta(\lambda) = \eta(\lambda, q)$ is given by

$$\eta(\lambda) = \frac{g(\theta(\lambda))}{2q} - \frac{q-1}{2q}\lambda + \log q \tag{10.15}$$

and $g(\theta)$ is given as in (10.46) *by*

$$g(\theta) = -(q-1)(2-\theta)\log(1-\theta) - [2 + (q-1)\theta] \log[1 + (q-1)\theta].$$

The proofs of Theorems 10.7–10.9 and 10.14 are given in Section 10.6 for $q \in (1,\infty)$. For proofs in the case $q \in (0,1)$, the reader is referred to [62].

10.4 The fundamental proposition

There is a fundamental technique which allows the study of $G_{n,p,q}$ via the properties of the usual random graph $G_{N,p,1}$. Let $r \in [0,1]$ be fixed. Given a random graph $G_{n,p,q}$, we colour each component either *red* (with probability r) or *green* (with probability $1-r$); different components are coloured independently of one another. The union of the red components is called the *red subgraph* of $G_{n,p,q}$, and the green components form the *green subgraph*. Let R be the set of *red* vertices, that is, the (random) vertex-set of the red subgraph. We see in the next lemma that the (conditional) distribution of the red subgraph is a random-cluster measure.

(10.16) Proposition. *Let V_1 be a subset of $V = \{1,2,\dots,n\}$ with cardinality $|V_1| = n_1$. Conditional on the event $\{R = V_1\}$, the red subgraph of $G_{n,p,q}$ is distributed as $G_{V_1,p,rq}$, and the green subgraph is distributed as $G_{V\setminus V_1,p,(1-r)q}$. Furthermore, the red subgraph is conditionally independent of the green subgraph.*

Proof. Set $V_2 = V \setminus V_1$, $n_2 = |V_2| = n - n_1$, and let $E_i \subseteq V_i^{(2)}$ for $i = 1,2$. With $k(U,F)$ the number of components of the graph (U,F),

$$k(V, E_1 \cup E_2) = k(V_1, E_1) + k(V_2, E_2).$$

Therefore, the probability that the red graph is (V_1, E_1) and the green graph is (V_2, E_2) equals

$$\left\{\frac{p^{|E_1\cup E_2|}(1-p)^{\binom{n}{2}-|E_1\cup E_2|}q^{k(V,E_1\cup E_2)}}{Z_{n,p,q}}\right\} r^{k(V_1,E_1)}(1-r)^{k(V_2,E_2)}$$
$$= c\phi_{V_1,p,rq}(E_1)\phi_{V_2,p,(1-r)q}(E_2),$$

for some positive constant $c = c(n, p, q, n_1)$. Hence, conditional on $R = V_1$ and the green subgraph being (V_2, E_2), the probability that the red subgraph is (V_1, E_1) is precisely $\phi_{V_1,p,rq}(E_1)$. □

In this context, we shall write N rather than n_1 for the (random) number of red vertices. Thus N is a random variable, and $G_{N,p,rq}$ is a random graph on a random number of vertices.

If $q \in [1, \infty)$ and $r = q^{-1}$, the red subgraph is distributed as $G_{N,p}$. Much is known about such a random graph, see [61, 194]. By studying the distribution of N and using known facts about $G_{N,p}$, one may deduce much about the structure of $G_{n,p,q}$. Similarly, in order to study the random-cluster model with $q \in (0, 1)$, one applies Proposition 10.16 to $G_{n,p}$ with $r = q$, obtaining that the red subgraph is distributed as $G_{N,p,q}$. By using known facts about $G_{n,p}$, together with some distributional properties of N, we may derive results for $G_{m,p,q}$ with m large. The details of the $q \in (0, 1)$ case are omitted but may be found in [62].

Here is a corollary which will be of use later.

(10.17) Lemma. *Let $q \in [1, \infty)$. For any sequence $p = p_n$, almost every $G_{n,p,q}$ has at most one component with order at least $n^{3/4}$.*

Proof. Let $L = L(G)$ be the number of components of a random graph G having order at least $n^{3/4}$. Suppose $L \ge 2$, and pick two of these in some arbitrary way. With probability r^2 both of these are coloured red. Setting $r = q^{-1}$, we find by [61, Thm VI.9] that

$$\begin{aligned} r^2\phi_{n,p,q}(L \ge 2) &\le \sum_{n^{3/4}\le m\le n} \phi_{m,p,1}(L \ge 2)\phi_{n,p,q}(|R| = m) \\ &\le \max_{n^{3/4}\le m\le n} \phi_{m,p,1}(L \ge 2) \\ &\to 0 \qquad \text{as } n \to \infty. \end{aligned}$$

□

10.5 The size of the largest component

We assume henceforth that $q \in [1, \infty)$. Let $\Theta_n n$ denote the number of vertices in the largest component of $G_{n,\lambda/n,q}$, and note that $0 < \Theta_n \le 1$. If two or more 'largest components' exist, we pick one of these at random. All other components are called 'small' and, by Lemma 10.17, all small components of almost every $G_{n,\lambda/n,q}$ have orders less than $n^{3/4}$.

Consider the colouring scheme of Proposition 10.16 with $r = q^{-1}$, and suppose that $G_{n,\lambda/n,q}$ has components of order $\Theta_n n, \nu_2, \nu_3, \dots, \nu_k$ where k is the total number of components and we shall assume that $\nu_i \le n^{3/4}$ for $i \ge 2$. The number of red vertices in the small components has conditional expectation

$$\sum_{i=2}^{k} \nu_i r = r(1 - \Theta_n)n$$

and variance

$$\sum_{i=2}^{k} \nu_i^2 r(1-r) \le \sum_{i=2}^{k} \nu_i^2 \le n \max_{i \ge 2} \nu_i \le n^{7/4}.$$

Hence, there is a total of $r(1-\Theta_n)n + \mathrm{o_p}(n)$ red vertices in the small components.

Since the largest component may or may not be coloured red, there are two possibilities for the red graph:

(i) with probability r, it has

$$\Theta_n n + r(1 - \Theta_n)n + \mathrm{o_p}(n) = [r + (1-r)\Theta_n]n + \mathrm{o_p}(n)$$

vertices, of which $\Theta_n n$ belong to the largest component,

(ii) with probability $1 - r$, it has $r(1 - \Theta_n)n + \mathrm{o_p}(n)$ vertices, and the largest component has order less than $n^{3/4}$.

In the first case, the red graph is distributed as a supercritical $G_{n',\lambda'/n'}$ graph, and in the second case as a subcritical $G_{n'',\lambda''/n''}$ graph. Here, n' and n'' are random integers and, with probability tending to 1, $\lambda' = n'p > 1 > \lambda'' = n''p$. This leads to the next lemma.

(10.18) Lemma. *If $\lambda > q \ge 1$, there exists $\theta_0 > 0$ such that $\Theta_n \ge \theta_0$ for almost every $G_{n,\lambda/n,q}$.*

Proof. The assertion is well known when $q = 1$, see for example [61, Thm VI.11]. Therefore, we may assume $q > 1$ and thus $r < 1$.

Let $\theta_0 = (\lambda - q)/(2\lambda)$, $\pi_n = \phi_{n,p,q}(\Theta_n < \theta_0)$, and $\epsilon > 0$. By considering the event that the largest component is not coloured red, we find that, with probability at least $(1 - r)\pi_n + \mathrm{o}(1)$, the number N of red vertices satisfies

$$N \ge r(1 - \theta_0)n - \epsilon n,$$

and there are no red components of order at least $n^{3/4}$. When this happens,

$$Np \geq \lambda[r(1-\theta_0)-\epsilon] = \frac{1}{2} + \frac{\lambda}{2q} - \epsilon\lambda > 1 \tag{10.19}$$

for ϵ sufficiently small, and we pick ϵ accordingly. Conditional on the value N, almost every $G_{N,p}$ has a component of order at least δN ($\geq \delta n/\lambda$ by (10.19)) for some $\delta > 0$. Therefore, $(1-r)\pi_n \to 0$ as $n \to \infty$. □

(10.20) Lemma. *If $q \in [1,\infty)$ then, for any sequence $\lambda = \lambda_n$,*

$$e^{-\lambda_n\Theta_n} - \frac{1-\Theta_n}{1+(q-1)\Theta_n} \xrightarrow{\text{P}} 0 \qquad \text{as } n \to \infty.$$

Proof. For $q = 1$ and constant $\lambda = \lambda_n$, this follows from the well known fact that $\Theta_n \xrightarrow{\text{P}} \theta$ where $e^{-\lambda\theta} = 1-\theta$, see [61, Thm VI.11] and the remark after [61, Thm V.7]. The case of varying λ_n is not hard to deduce by looking down convergent subsequences. We may express this by writing

$$e^{-p_n\Theta_n n} + \frac{\Theta_n n}{n} - 1 \xrightarrow{\text{P}} 0 \qquad \text{when } q = 1,$$

for the random graph G_{n,p_n} and any sequence (p_n). Applying this to the red subgraph, on the event that it contains the largest component of $G_{n,\lambda/n,q}$, we obtain for general $q \in [1,\infty)$ that

$$\begin{aligned} e^{-\lambda\Theta_n} + \frac{\Theta_n}{r+(1-r)\Theta_n} - 1 &= e^{-p\Theta_n n} + \frac{\Theta_n n}{N} - 1 + o_{\text{P}}(1) \\ &\xrightarrow{\text{P}} 0 \qquad \text{as } n \to \infty, \end{aligned}$$

where N is the number of red vertices. The claim follows. □

Combining these lemmas, we arrive at the following theorem.

(10.21) Theorem [62].

(a) *If $q \in [1,2]$ and $\lambda \leq q$, or if $q \in (2,\infty)$ and $\lambda < \lambda_{\min}$ where $\lambda_{\min}$ is given in Lemma 10.12*(b)*, then $\Theta_n \xrightarrow{\text{P}} 0$ as $n \to \infty$.*

(b) *If $q \in [1,\infty)$ and $\lambda > q$, then $\Theta_n \xrightarrow{\text{P}} \theta(\lambda)$ where $\theta(\lambda)$ is the unique (strictly) positive solution of* (10.6).

This goes some way towards proving Theorems 10.7–10.8. Overlooking for the moment the more detailed asymptotical claims of those theorems, we note that the major remaining gap is when $q \in (2,\infty)$ and $\lambda_{\min} \leq \lambda \leq q$. In this case, by Lemma 10.20, Θ_n is approximately equal to one of the three roots of (10.6) (including the trivial root $\theta = 0$). Only after the analysis of the next two sections shall we see which root is the correct one for given λ.

Proof. The function

$$\phi(\theta) = e^{-\lambda\theta} - \frac{1-\theta}{1+(q-1)\theta}$$

is continuous on $[0, 1]$, and the set Z of zeros of ϕ is described in Lemma 10.12. Since $\phi(\Theta_n) \xrightarrow{\text{P}} 0$, by Lemma 10.20, it follows that, for all $\epsilon > 0$,

$$\phi_{n,p,q}\big(\Theta_n \in Z + (-\epsilon, \epsilon)\big) \to 1 \qquad \text{as } n \to \infty.$$

Under the assumption of (a), Z contains the singleton 0, and the claim follows. Under (b), Z contains a unique strictly positive number $\theta(\lambda)$, and the claim follows by Lemma 10.18. □

We turn now to the number of edges in the largest component. Let $\Psi_n n$ denote the number of edges of $G_{n,p,q}$. We pick one of its largest components at random, and write $\Xi_n n$ for the number of its edges. Let $q \in (1, \infty)$. Arguing as in Sections 10.4–10.5 with $r = q^{-1}$, almost every $G_{n,p,q}$ has at most $n^{3/4}$ edges in each small component (a 'small' component is any component except the largest, picked above)[2]. Furthermore, the total number of red edges in the small components is $r(\Psi_n - \Xi_n)n + \text{o}_\text{P}(n)$. Hence, the red subgraph has either:

(i) *with probability* r, $[\Theta_n + r(1-\Theta_n)]n + \text{o}_\text{P}(n)$ vertices and $[\Xi_n + r(\Psi_n - \Xi_n)]n + \text{o}_\text{P}(n)$ edges, or

(ii) *otherwise*, $r(1-\Theta_n)n + \text{o}_\text{P}(n)$ vertices and $r(\Psi_n - \Xi_n)n + \text{o}_\text{P}(n)$ edges.

Assume that $p = \text{O}(n^{-1})$. Since almost every $G_{N,p}$ has

$$\binom{N}{2}p + \text{O}_\text{P}(Np^{1/2}) = \tfrac{1}{2}N^2 p + \text{o}_\text{P}(N)$$

edges, the following two equations follow from the two cases above,

$$\text{(10.22)} \qquad \big[\Xi_n + r(\Psi_n - \Xi_n)\big]n = \tfrac{1}{2}\big[\Theta_n + r(1-\Theta_n)\big]^2 n^2 p + \text{o}_\text{P}(n),$$

$$\text{(10.23)} \qquad r(\Psi_n - \Xi_n)n = \tfrac{1}{2}[r(1-\Theta_n)]^2 n^2 p + \text{o}_\text{P}(n),$$

yielding when $p = \lambda/n$ that

$$\text{(10.24)} \qquad \Xi_n + r(\Psi_n - \Xi_n) = \tfrac{1}{2}\lambda\big[\Theta_n + r(1-\Theta_n)\big]^2 + \text{o}_\text{P}(1),$$

$$\text{(10.25)} \qquad r(\Psi_n - \Xi_n) = \tfrac{1}{2}\lambda[r(1-\Theta_n)]^2 + \text{o}_\text{P}(1).$$

We solve for Ξ_n and Ψ_n, and let $n \to \infty$ to obtain the next theorem.

[2]One needs here the corresponding result for $q = 1$, which follows easily from the corresponding result for the number of vertices used above, together with results on the components having more edges than vertices given in [61, 192, 193].

(10.26) Theorem [62]. *If $q \in (1, \infty)$ and $\lambda \in (0, \infty)$ then, as $n \to \infty$,*

$$\Psi_n - \frac{\lambda}{2q}\big[1 + (q-1)\Theta_n^2\big] \xrightarrow{\mathrm{P}} 0, \tag{10.27}$$

$$\Xi_n - \frac{\lambda}{q}\Theta_n\big[1 + (\tfrac{1}{2}q - 1)\Theta_n\big] \xrightarrow{\mathrm{P}} 0. \tag{10.28}$$

Whereas we proved this theorem under the assumption that $q > 1$, its conclusions are valid for $q = 1$ also, by [61, Thms VI.11, VI.12].

10.6 Proofs of main results for complete graphs

The results derived so far are combined next with a new argument in order to prove Theorems 10.7–10.9 for $q \in [1, \infty)$. The results are well known when $q = 1$ (see [61, Chapters V, VI] and [239]), and we assume henceforth that $q \in (1, \infty)$. The *acyclic* part of a graph is the union of all components that are trees, and the *cyclic* part is the union of the remaining components. A graph is called *cyclic* if its acyclic part is empty. We begin by showing that the cyclic part of almost every $G_{n,\lambda/n,q}$ consists principally of the largest component only (when this component is cyclic).

(10.29) Lemma. *The numbers of vertices and edges in the small cyclic components of $G_{n,\lambda/n,q}$ are $\mathrm{o_p}(n)$.*

Proof. Let k be an integer satisfying $k \ge q$. In the colouring scheme of Section 10.4 with $r = q^{-1}$, we introduce the refinement that each component is coloured *dark red* with probability k^{-1} and *light red* with probability $r - k^{-1}$. Let M be the number of edges in the small cyclic components of $G_{n,\lambda/n,q}$.

By a symmetry argument, with probability at least k^{-1}, at least M/k of these edges are coloured dark red. To see this, let M_i be the number of such edges coloured χ_i when each component is coloured by a random colour from the set $\{\chi_1, \chi_2, \dots, \chi_k\}$, each such colour having equal probability. If

$$\phi_{n,p,q}(M_i \ge M/k) < \frac{1}{k}, \qquad i = 1, 2, \dots, k,$$

then

$$\phi_{n,p,q}(M_i \ge M/k \text{ for some } i) < 1,$$

in contradiction of the equality $\sum_{i=1}^{k} M_i = M$.

Therefore, with probability at least r/k, the red subgraph contains the largest component together with small cyclic components having at least M/k edges. The result now follows from the known case $q = 1$, see [60], [61, Thm VI.11]. □

Let $\widetilde{P}_{n,p,q}(m, j, k, l)$ be the sum of $\widetilde{P}_{n,p,q}(F)$ over edge-sets F that define a graph with $|F| = m$ edges and a cyclic part with j components, k vertices, and l

edges. Since such graphs have an acyclic part with $n-k$ vertices and $m-l$ edges, and therefore $n-k-m+l$ components, we obtain

(10.30)
$$\widetilde{P}_{n,p,q}(m,j,k,l) = \binom{n}{k} c(j,k,l) f(n-k,m-l) p^m (1-p)^{\binom{n}{2}-m} q^{n-k-m+l+j}$$

where $c(j,k,l)$ is the number of cyclic graphs with j components, k labelled vertices, and l edges, and $f(n,m)$ is the number of forests with n labelled vertices and m edges.

Assume now that $n \to \infty$, that $\lambda = np > 0$ and $q \in [1,\infty)$ are fixed, and that

$$m/n \to \psi, \quad k/n \to \theta, \quad l/n \to \xi, \quad j/n \to 0, \tag{10.31}$$

where $\theta \geq 0$ satisfies (10.6), and

$$\xi = \frac{\lambda}{q}\theta\big[1 + (\tfrac{1}{2}q - 1)\theta\big], \tag{10.32}$$

$$\psi = \xi + \frac{\lambda}{2q}(1-\theta)^2. \tag{10.33}$$

See (10.27) and (10.28). If $\lambda > q$, we assume also that $\theta > 0$, see Lemma 10.18 and Theorem 10.21(b).

Since

$$f(n,m) \leq \binom{\binom{n}{2}}{m},$$

the total number of graphs with m edges on n vertices,

(10.34)
$$\begin{aligned}
&\widetilde{P}_{n,p,q}(m,j,k,l)\\
&\quad\leq \binom{n}{k} c(j,k,l) \binom{\binom{n-k}{2}}{m-l} p^m (1-p)^{\binom{n}{2}-m} q^{n-k-m+l+j}\\
&\quad= \binom{n}{k} c(j,k,l) \binom{n-k}{2}^{m-l} \left(\frac{e}{m-l}\right)^{m-l} p^m q^{n-k-m+l} e^{-\frac{1}{2}\lambda n + o(n)}\\
&\quad= c(j,k,l)\binom{n}{k}\left(\frac{(n-k)^2 e\lambda}{2(m-l)n}\right)^{m-l} p^l q^{n-k-m+l} e^{-\frac{1}{2}\lambda n + o(n)}\\
&\quad= p^l c(j,k,l)\binom{n}{k}\left(\frac{(1-\theta)^2\lambda}{2(\psi-\xi)}\right)^{m-l} q^{n-k-m+l} \exp\big(m - l - \tfrac{1}{2}\lambda n + o(n)\big)\\
&\quad= p^l e^{-l} c(j,k,l)\binom{n}{k} q^{n-k} \exp\big(m - \tfrac{1}{2}\lambda n + o(n)\big),
\end{aligned}$$

where we used (10.33) in the last step.

We shall be interested only in values of λ and roots θ of (10.6) satisfying

$$\text{either } \theta > 0, \quad \text{or } \theta = 0 \text{ and } \lambda \le q. \tag{10.35}$$

We claim that, under these assumptions, (10.34) is an equality in that

$$\widetilde{P}_{n,p,q}(m,j,k,l) = p^l e^{-l} c(j,k,l)\binom{n}{k} q^{n-k} \exp\big(m - \tfrac{1}{2}\lambda n + \mathrm{o}(n)\big). \tag{10.36}$$

To see this when either $\theta > 0$, or $\theta = 0$ and $\lambda < q$, set $n_0 = n-k$ and $m_0 = m-l$, and observe that

$$\frac{m_0}{n_0} = \frac{m-l}{n-k} \to \frac{\psi - \xi}{1-\theta} = \frac{\lambda}{2q}(1-\theta) < \frac{1}{2},$$

where we have used the fact that, by (10.6),

$$\lambda\theta < e^{\lambda\theta} - 1 = \frac{q\theta}{1-\theta}, \qquad \theta \in (0,\infty).$$

Hence in this case, the 'fixed edge-number' random graph G_{n_0,m_0} has average vertex-degree not exceeding $1-\epsilon$ for some positive constant ϵ independent of n, m, k, l. Therefore, there exists $\delta > 0$ such that

$$\mathbb{P}(G_{n_0,m_0} \text{ is a forest}) > \delta,$$

and hence

$$f(n_0, m_0) > \delta \binom{\binom{n_0}{2}}{m_0}.$$

This implies (10.36), via (10.30) and (10.34). When $\theta = 0$ and $\lambda = q$, we have that $m_0/n_0 \to \psi = \frac{1}{2}$, and hence

$$f(n_0, m_0) = \binom{\binom{n_0}{2}}{m_0} e^{\mathrm{o}(n)}, \tag{10.37}$$

implying (10.36). To see (10.37) note that, with $0 < \epsilon < \frac{1}{4}$ and $s \asymp \epsilon n$,

$$\begin{aligned} f(n_0, m_0) &\ge (n_0 - 1)_{s-1} f(n_0 - s, m_0 - s + 1) \\ &\ge e^{-\epsilon n} n_0^s \binom{\binom{n_0 - s}{2}}{m_0 - s + 1} \\ &\ge e^{2\log(1-\epsilon)n} \binom{\binom{n_0}{2}}{m_0}, \end{aligned}$$

for large n, by counting only forests where vertex 1 is an endvertex of an isolated path of length $s-1$.

We estimate $c(j,k,l)$ next. Suppose first that $\theta = 0$. Then $c(j,k,l)$ is no greater than the total number of graphs with k vertices and l edges, that is,

$$p^l e^{-l} c(j,k,l) \le \left(\frac{p}{e}\right)^l \left(\frac{k^2}{2}\right)^l \frac{1}{l\,!} \le \left(\frac{\lambda n}{l}\right)^l$$
$$= \exp\bigl[l\{\log\lambda - \log(l/n)\}\bigr] = e^{o(n)}.$$

Equality holds here for some suitablc triplc j, k, l: just set $j = k = l = 0$, for which $p^l e^{-l} c(j,k,l) = 1$. It is easily checked that $\binom{n}{k} = e^{o(n)}$ when $\theta = 0$, and therefore,

$$(10.38)\qquad p^l e^{-l} c(j,k,l) \le \binom{n}{k}^{-1} e^{o(n)}$$

with equality for some suitable j, k, l.

Our estimate of $c(j,k,l)$ when $\theta > 0$ uses the fact that $\widetilde{P}_{n,p,q}(\cdot)$ is a probability measure when $q = 1$. Suppose $\theta > 0$, define $n_1 = n_1(\theta) = \lfloor \theta n + r(1-\theta)n \rfloor$ where $r = q^{-1}$ as usual, and set

$$m_1 = l + r(m-l) + o(n) = [\xi + r(\psi - \xi)]n + o(n),$$
$$\lambda_1 = [\theta + r(1-\theta)]\lambda.$$

Then,

$$(10.39)\qquad \frac{m_1}{n_1} \to \psi_1 = \frac{\xi + r(\psi - \xi)}{\theta + r(1-\theta)} = \frac{1}{2}\lambda_1,$$

$$(10.40)\qquad \frac{k}{n_1} \to \theta_1 = \frac{\theta}{\theta + r(1-\theta)},$$

$$(10.41)\qquad \frac{l}{n_1} \to \xi_1 = \frac{\xi}{\theta + r(1-\theta)}, \quad \frac{j}{n_1} \to 0.$$

It is easy to check the analogues of (10.6) and (10.32)–(10.33), namely,

$$(10.42)\quad e^{-\lambda_1\theta_1} = 1 - \theta_1, \quad \xi_1 = \lambda_1\theta_1(1 - \tfrac{1}{2}\theta_1), \quad \psi_1 = \xi_1 + \tfrac{1}{2}\lambda_1(1-\theta_1)^2.$$

Now, (10.36) is valid with $q = 1$, since $\theta > 0$. Hence,

$$(10.43)\qquad 1 \ge \widetilde{P}_{n_1,p_1,1}(m_1, j, k, l)$$
$$= p_1^l e^{-l} c(j,k,l) \binom{n_1}{k} \exp\bigl(m_1 - \tfrac{1}{2}\lambda_1 n_1 + o(n)\bigr)$$
$$= p^l e^{-l} c(j,k,l) \binom{n_1}{k} e^{o(n)}$$

by (10.39), where $p_1 = \lambda_1/n_1 = p(1+\mathrm{O}(n^{-1}))$. Therefore,

$$p^l e^{-l} c(j,k,l) \le \binom{n_1}{k}^{-1} e^{\mathrm{o}(n)}. \tag{10.44}$$

We claim that there exist suitable j, k, l such that equality holds in (10.44). To see this, note that G_{n_1,p_1} has $\binom{n_1}{2}p_1 + \mathrm{o_p}(n_1)$ edges, a giant component with $\theta_1 n_1 + \mathrm{o_p}(n_1)$ vertices and $\xi_1 n_1 + \mathrm{o_p}(n_1)$ edges, $\mathrm{O_p}(1)$ unicyclic components with a total of $\mathrm{O_p}(1)$ vertices and edges, and no other cyclic components, see [61, Thm VI.11]. By considering the number of possible combinations of values of m_1, j, k, l satisfying the above constraints, there exist m_1, j, k, l such that

$$\widetilde{P}_{n_1,p_1,1}(m_1, j, k, l) \ge n^{-4}$$

for all large n. Combining this with (10.43), equality follows in (10.44) for some suitable j, k, l.

In conclusion, whatever the root θ of (10.6) (subject to (10.35)), inequality (10.44) holds with equality for some suitable j, k, l, and where $n_1 = n_1(0)$ is interpreted as n (that is, when $\theta = 0$). We substitute (10.44) into (10.34) to obtain

(10.45)
$$\begin{aligned}
&\widetilde{P}_{n,p,q}(m, j, k, l) \\
&\quad \le \binom{n_1}{k}^{-1}\binom{n}{k} q^{n-k} \exp\left(m - \tfrac{1}{2}\lambda n + \mathrm{o}(n)\right) \\
&\quad = \frac{n^n}{(n-k)^{n-k}} \frac{(n_1-k)^{n_1-k}}{n_1^{n_1}} q^{n-k} \exp\left(\frac{\lambda}{2q}[1+(q-1)\theta^2]n - \tfrac{1}{2}\lambda n + \mathrm{o}(n)\right) \\
&\quad = \left[\frac{[r(1-\theta)]^{r(1-\theta)}}{(1-\theta)^{1-\theta}[\theta + r(1-\theta)]^{\theta + r(1-\theta)}} \right. \\
&\qquad\qquad \left. \times\, q^{1-\theta} \exp\left(\frac{\lambda}{2q}[1+(q-1)\theta^2] - \tfrac{1}{2}\lambda + \mathrm{o}(1)\right)\right]^n \\
&\quad = \exp\left(n\left[\frac{g(\theta)}{2q} - \frac{q-1}{2q}\lambda + \log q + \mathrm{o}(1)\right]\right),
\end{aligned}$$

where

$$g(\theta) = -(q-1)(2-\theta)\log(1-\theta) - [2+(q-1)\theta]\log[1+(q-1)\theta]. \tag{10.46}$$

We have used (10.32)–(10.33) in order to obtain the second line of (10.45). To pass to the last line, we used the fact that θ is a root of (10.6), thus enabling the substitution

$$\exp\left(\frac{\lambda}{2q}[1+(q-1)\theta^2]\right) = e^{\lambda/(2q)}\left\{\frac{1+(q-1)\theta}{1-\theta}\right\}^{(q-1)\theta/(2q)}.$$

In addition, equality holds in (10.45) for at least one suitable choice of j, k, l. Let $\theta^* = \theta^*(\lambda)$ be the root[3] of (10.6) that maximizes $g(\theta)$ and satisfies (10.35). By (10.45) and the equality observed above,

$$Z_{n,p,q} = \sum_{m,j,k,l} \widetilde{P}_{n,p,q}(m, j, k, l) \geq \exp\left\{n\left[\frac{g(\theta^*)}{2q} - \frac{q-1}{2q}\lambda + \log q + o(1)\right]\right\}, \tag{10.47}$$

whence

$$\liminf_{n\to\infty}\left\{\frac{1}{n}\log Z_{n,p,q}\right\} \geq \frac{g(\theta^*)}{2q} - \frac{q-1}{2q}\lambda + \log q. \tag{10.48}$$

On the other hand, by Lemmas 10.18 and 10.20, there exists a root θ of (10.6) satisfying (10.35), and a function $\omega(n)$ satisfying $\omega(n) \to \infty$, such that

$$\liminf_{n\to\infty} \phi_{n,p,q}\big(|\Theta_n - \theta| < \omega(n)^{-1}\big) > 0. \tag{10.49}$$

For such θ there exist, by Lemma 10.29 and Theorem 10.26, sequences m, j, k, l satisfying (10.31)–(10.33) such that

$$1 \geq \frac{\widetilde{P}_{n,p,q}(m, j, k, l)}{Z_{n,p,q}} \geq n^{-4}$$

for all large n (this is shown by considering the number of possible combinations of m, j, k, l satisfying (10.31)–(10.33) and the above-mentioned results). By (10.45),

$$\limsup_{n\to\infty}\left\{\frac{1}{n}\log Z_{n,p,q}\right\} \leq \frac{g(\theta)}{2q} - \frac{q-1}{2q}\lambda + \log q, \tag{10.50}$$

which, by (10.48), implies $g(\theta) \geq g(\theta^*)$, and therefore $\theta = \theta^*$. Theorem 10.14 follows by (10.48) and (10.50). Furthermore, θ^* is the only root of (10.6) satisfying (10.35) such that (10.49) holds for some $\omega(n)$. Therefore,

$$\Theta_n \xrightarrow{\text{P}} \theta^* \qquad \text{as } n \to \infty. \tag{10.51}$$

Next we calculate $\theta^*(\lambda)$. As in Theorem 10.21, when $q \in [1, 2]$, $\theta^*(\lambda)$ is the largest non-negative root of (10.6). Assume that $q \in (2, \infty)$. By a straightforward computation,

$$g(0) = g\left(\frac{q-2}{q-1}\right) = 0, \quad g'(0) = 0,$$

$$g''(\theta) = -\frac{q(q-1)[q-2-2(q-1)\theta]\theta}{(1-\theta)^2[1+(q-1)\theta]^2}.$$

---

[3]We shall see that there is a unique such θ^*, except possibly when $\lambda = \lambda_c(q)$ and $q > 2$.

Therefore, $g''(\theta)$ has a unique zero in $(0, 1)$, at the point $\theta = \frac{1}{2}(q-2)/(q-1)$. At this point, $g'(\theta)$ has a negative minimum. It follows that $g(\theta) < 0$ on $(0, \theta_0)$, and $g(\theta) > 0$ on $(\theta_0, 1)$ where $\theta_0 = (q-2)/(q-1)$.

Substituting θ_0 into (10.6), we find that θ_0 satisfies (10.6) if

$$\lambda = \lambda_c(q) = 2\left(\frac{q-1}{q-2}\right)\log(q-1),$$

and, for this value of λ, the three roots of (10.6) are 0, $\frac{1}{2}\theta_0$, θ_0. Therefore, $\lambda_{\min} < \lambda_c(q) < q$, and

$$\theta^* = \begin{cases} 0 & \text{if } \lambda < \lambda_c(q), \\ \theta_{\max}(\lambda) & \text{if } \lambda > \lambda_c(q). \end{cases}$$

This completes the proof of the assertions concerning the order of the largest component. The claims concerning the numbers of edges in $G_{n,p,q}$ and in the largest component follow by Theorem 10.26. Proofs of the remaining assertions about the structure of $G_{n,p,q}$ are omitted, but may be obtained easily using the colouring argument and known facts for $G_{n,p}$, see [61, 239].

10.7 The nature of the singularity

It is an important problem of statistical physics to understand the nature of the singularity at a point of phase transition. For the mean-field random-cluster model on a complete graph, the necessary calculations may be performed explicitly, and the conclusions are as follows.

Let $q \in [1, \infty)$ be fixed, and consider the functions $\theta(\lambda)$, given in (10.5), and $\psi(\lambda)$, $\xi(\lambda)$ defined by

$$\psi(\lambda) = \frac{\lambda}{2q}\big[1 + (q-1)\theta(\lambda)^2\big], \quad \xi(\lambda) = \frac{\lambda}{q}\big[\theta(\lambda) + (\tfrac{1}{2}q - 1)\theta(\lambda)^2\big],$$

describing the order of the giant component, and the numbers of edges in the graph and in its giant component, respectively. All three functions are non-decreasing on $(0, \infty)$. In addition, ψ is strictly increasing, while $\theta(\lambda)$ and $\xi(\lambda)$ equal 0 for $\lambda < \lambda_c$ and are strictly increasing on $[\lambda_c, \infty)$.

A fourth function of interest is the pressure $\eta(\lambda)$ given in Theorem 10.14. These four functions are real-analytic on $(0, \infty) \setminus \{\lambda_c\}$. At the singularity λ_c, the following may be verified with reasonable ease.

(a) Let $q \in [1, 2)$. Then θ, ψ, ξ, and η are continuous at the point $\lambda_c(q) = q$. The functions θ and ξ have discontinuous first derivatives at λ_c, with

$$\theta'(\lambda_c-) = \xi'(\lambda_c-) = 0, \quad \theta'(\lambda_c+) = \xi'(\lambda_c+) = \frac{2}{q(2-q)}.$$

In particular,

$$\theta(\lambda) \sim \frac{2(\lambda - \lambda_c)}{q(2-q)} \qquad \text{as } \lambda \downarrow \lambda_c.$$

Similarly, ψ' and η'' are continuous, but ψ'' and η''' have discontinuities at λ_c, except when $q = 1$.

(b) Let $q = 2$. Once again, θ, ψ, ξ, and η are continuous at the point λ_c. In this case,

$$\theta(\lambda) \sim \xi(\lambda) \sim \left[\tfrac{3}{2}(\lambda - \lambda_c)\right]^{\frac{1}{2}} \qquad \text{as } \lambda \downarrow \lambda_c.$$

Thus, $\theta'(\lambda_c+) = \xi'(\lambda_c+) = \infty$. The function ψ' has a jump at λ_c in that $\psi'(\lambda_c-) = \frac{1}{4}$, $\psi'(\lambda_c+) = 1$. Also, η' is continuous, but η'' has a jump at λ_c in that $\eta''(\lambda_c-) = 0$, $\eta''(\lambda_c+) = \frac{3}{8}$. The functions ψ and η are real-analytic on $(0, \lambda_c]$ and on $[\lambda_c, \infty)$.

(c) Let $q \in (2, \infty)$. Then θ, ψ, and ξ have jumps at λ_c, and it may be checked that $\psi(\lambda_c-) = \lambda_c/(2q) < \frac{1}{2} < \psi(\lambda_c+)$. The pressure η is continuous at λ_c, but its derivative η' has a jump at λ_c,

$$\eta'(\lambda_c-) = -\frac{q-1}{2q}, \qquad \eta'(\lambda_c+) = -\frac{2q-3}{2q(q-1)}.$$

10.8 Large deviations

The partition function $Z_{n,p,q}$ of (10.2) may be written[4] as the exponential expectation

$$Z_{n,p,q} = \phi_{n,p,1}(q^{k(\omega)}).$$

This suggests a link, via a Legendre transform, to the theory of large deviations of the cluster-count $k(\omega)$ in a random-cluster model. We summarize the consequent theory in this section, and we refer the reader to [62] for the proofs. Related arguments concerning the random-cluster model on a lattice may be found in [298].

Let[5] $q \in [1, \infty)$, $\lambda \in (0, \infty)$, and let C_n be the number of components of the graph $G_{n,\lambda/n,q}$. Our target is to show how the exact calculation of pressure in Theorem 10.14 may be used to estimate probabilities of the form $\phi_{n,p,q}(C_n \le \alpha n)$ and $\phi_{n,p,q}(C_n \ge \beta n)$ for given constants α, β. When $q = 1$, this gives information about the probabilities of large deviations of C_n in an Erdős–Rényi random graph.

As in the language of large-deviation theory, [99, 164], let

$$\Lambda_{n,\lambda,q}(\nu) = \log \phi_{n,p,q}(e^{\nu C_n/n}), \qquad \nu \in \mathbb{R},$$

[4] See (3.59) also.

[5] The conclusions of this section are valid when $q \in (0, 1)$ also, see [62].

and note that

$$\Lambda_{n,\lambda,q}(\nu) = \log\left\{\frac{Z_{n,\lambda/n,qe^{\nu/n}}}{Z_{n,\lambda/n,q}}\right\},$$

whence

$$\text{(10.52)} \qquad \frac{1}{n}\Lambda_{n,\lambda,q}(n\nu) \to \Lambda_{\lambda,q}(\nu) = \eta(\lambda, qe^{\nu}) - \eta(\lambda, q) \qquad \text{as } n \to \infty,$$

where $\eta(\lambda, q)$ denotes the pressure function of Theorem 10.14. The Legendre transform $\Lambda^*_{\lambda,q}$ of $\Lambda_{\lambda,q}$ is given by

$$\text{(10.53)} \qquad \Lambda^*_{\lambda,q}(x) = \sup_{\nu\in\mathbb{R}}\{\nu x - \Lambda_{\lambda,q}(\nu)\}, \qquad x \in \mathbb{R}.$$

It may be proved directly, or see [99, Lemma 2.3.9], that $\Lambda_{\lambda,q}$ and $\Lambda^*_{\lambda,q}$ are convex functions, and that

$$\text{(10.54)} \qquad \Lambda^*_{\lambda,q}(x) = \delta x - \Lambda_{\lambda,q}(\delta) \qquad \text{if } \Lambda'_{\lambda,q}(\delta) = x.$$

Since we have an exact formula for $\Lambda_{\lambda,q}$, see (10.15) and (10.52), we may compute its derivative whenever it exists. Consequently,

$$\Lambda^*_{\lambda,q}(x)\begin{cases} < \infty & \text{if } x \in [0,1], \\ = \infty & \text{otherwise.}\end{cases}$$

A large-deviation principle (LDP) may be established for $n^{-1}C_n$ in terms of the 'rate function' $\Lambda^*_{\lambda,q}$. The details of the LDP depend on the set of points x at which $\Lambda^*_{\lambda,q}$ is strictly convex, and we investigate this next. There is a slight complication arising from the discontinuity of the phase transition when $q \in (2, \infty)$. The function

$$\text{(10.55)} \qquad \kappa(\lambda, q) = q\frac{\partial \eta}{\partial q},$$

turns out to play a central role. This derivative exists except when $\lambda = \lambda_c(q)$ and $q \in (2, \infty)$, and satisfies

$$\text{(10.56)} \qquad \begin{aligned}\kappa(\lambda, q) &= \lim_{n\to\infty}\left\{\frac{1}{n}\phi_{n,\lambda/n,q}(C_n)\right\} \\ &= 1 - \theta(\lambda) - [1-\theta(\lambda)]^2\frac{\lambda}{2q}.\end{aligned}$$

When $\lambda = \lambda_c(q)$ and $q \in (2, \infty)$, the limits

$$\kappa^{\pm}(\lambda, q) = q\frac{\partial \eta}{\partial q}(\lambda, q\pm)$$

exist with $\kappa^-(\lambda, q) < \kappa^+(\lambda, q)$. Also, $\kappa^-(\lambda, q)$ is given by (10.56), and

$$\kappa^+(\lambda, q) = 1 - \frac{\lambda}{2q}.$$

Details of the above calculations may be found in [62].

We write $\mathcal{F}_{\lambda,q}$ for the set of 'exposed points' of $\Lambda^*_{\lambda,q}$, and one may see after some work that

$$\mathcal{F}_{\lambda,q} = \begin{cases} (0,1) & \text{if } \lambda \le 2, \\ (0,1) \setminus [\kappa^-(\lambda, Q), \kappa^+(\lambda, Q)] & \text{if } \lambda > 2, \end{cases} \tag{10.57}$$

where Q is chosen to satisfy $\lambda = \lambda_c(Q)$. The following LDP is a consequence of the Gärtner–Ellis theorem, [99, Thm 2.3.6].

(10.58) Theorem (Large deviations) [62]. *Let $q \in [1, \infty)$ and $\lambda \in (0, \infty)$.*

(a) *For any closed subset F of $\mathbb{R}$,*

$$\limsup_{n\to\infty} \left\{ \frac{1}{n} \log \phi_{n,p,q}(n^{-1}C_n \in F) \right\} \le - \inf_{x \in F} \Lambda^*_{\lambda,q}(x).$$

(b) *For any open subset G of $\mathbb{R}$,*

$$\liminf_{n\to\infty} \left\{ \frac{1}{n} \log \phi_{n,p,q}(n^{-1}C_n \in G) \right\} \ge - \inf_{x \in G \cap \mathcal{F}_{\lambda,q}} \Lambda^*_{\lambda,q}(x).$$

Of especial interest are the cases when F takes the form $[0, \alpha]$ or $[\beta, 1]$, analysed as follows using Theorem 10.58.

(i) Let $q \in [1, 2]$. Then, as $n \to \infty$,

$$\frac{1}{n} \log \phi_{n,p,q}(C_n \le \alpha n) \to -\Lambda^*_{\lambda,q}(\alpha), \tag{10.59}$$

$$\frac{1}{n} \log \phi_{n,p,q}(C_n \ge \beta n) \to -\Lambda^*_{\lambda,q}(\beta), \tag{10.60}$$

whenever $0 < \alpha \le \kappa(\lambda, q) \le \beta < 1$.

(ii) Let $q \in (2, \infty)$ and $\lambda = \lambda_c(q)$. Then (10.59)–(10.60) hold for α, β satisfying

$$0 < \alpha \le \kappa^-(\lambda, q) < \kappa^+(\lambda, q) \le \beta < 1.$$

(iii) Let $q \in (2, \infty)$ and $\lambda \ne \lambda_c(q)$. Let Q be such that $\lambda = \lambda_c(Q)$. Then (10.59)–(10.60) hold for any α, β satisfying $0 < \alpha \le \kappa(\lambda, q) \le \beta < 1$ except possibly when

$$\kappa^-(\lambda, Q) < \alpha \le \kappa^+(\lambda, Q) \quad \text{or} \quad \kappa^-(\lambda, Q) \le \beta < \kappa^+(\lambda, Q).$$

We note that $\kappa^+(\lambda, Q) < \kappa(\lambda, q)$ if $Q < q$, and $\kappa^-(\lambda, Q) > \kappa(\lambda, q)$ if $Q > q$, so that only one of these two cases can occur for any given q.

We summarize the above facts as follows. Excepting the special case when $\lambda = \lambda_c(q)$ and $q \in (2, \infty)$, the limit

$$\kappa = \lim_{n\to\infty} \left\{ \frac{1}{n} \phi_{n,p,q}(C_n) \right\}$$

exists, and the probabilities $\phi_{n,p,q}(C_n \le \alpha n)$, $\phi_{n,p,q}(C_n \ge \beta n)$ decay at least as fast as exponentially when $\alpha < \kappa < \beta$. The exact (exponential) rate of decay can be determined except when the levels αn and βn lie within the interval of discontinuity of a first-order phase transition. In the exceptional case with $\lambda = \lambda_c(q)$ and $q \in (2, \infty)$, a similar conclusion holds when $\alpha < \kappa^-$ and $\beta > \kappa^+$.

Since first-order transitions occur only when $q \in (2, \infty)$, and since the critical λ-values of such q fill the interval $(2, \infty)$, there is a weak sense in which the value $\lambda = 2$ marks a singularity of the asymptotics of the random graph $G_{n,\lambda/n,q}$. This holds for any value of q, including $q = 1$. That is, the Erdős–Rényi random graph senses the existence of a first-order phase transition in the random-cluster model, but only through its large deviations. It is well known that the Erdős–Rényi random graph undergoes a type of phase transition at $\lambda = 1$, and it follows from the above that it has a (weak) singularity at $\lambda = 2$ also.

10.9 On a tree

A random-cluster measure on a finite tree is simply a product measure — it is the circuits of a graph which cause dependence between the states of different edges and, when there are no circuits, there is no dependence. This may be seen explicitly as follows. Let $p \in [0, 1]$ and $q \in (0, \infty)$, and let $T = (V, E)$ be a finite tree. For $\omega \in \Omega = \{0, 1\}^E$, the number of open clusters is $k(\omega) = |V| - |\eta(\omega)|$, so that the corresponding random-cluster measure $\phi_{p,q}$ satisfies

$$\phi_{p,q}(\omega) \propto \left(\frac{p}{q(1-p)} \right)^{|\eta(\omega)|} = \left(\frac{\pi}{1-\pi} \right)^{|\eta(\omega)|}, \qquad \omega \in \Omega, \tag{10.61}$$

where

$$\pi = \pi(p, q) = \frac{p}{p + q(1-p)}. \tag{10.62}$$

Therefore, $\phi_{p,q}$ is the product measure on Ω with density π. The situation becomes more interesting when we introduce boundary conditions.

Let T be an infinite labelled tree with root 0, and let $\mathcal{R} = \mathcal{R}(T)$ be the set of all infinite (self-avoiding) paths of T beginning at 0, termed 0-*rays*. We may think of a boundary condition on T as being an equivalence relation $\sim$ on $\mathcal{R}$, the

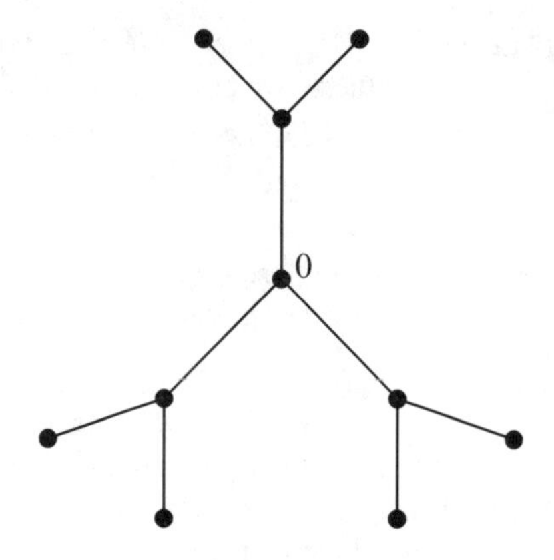

Figure 10.4. Part of the infinite binary tree T_2.

'physical' meaning of which is that two rays ρ, ρ' are considered to be 'connected at infinity' whenever $\rho \sim \rho'$. Such connections affect the counts of connected components of subgraphs. The two extremal boundary conditions are usually termed 'free' (meaning that there exist no connections at infinity) and 'wired' (meaning that all rays are equivalent). The wired boundary condition on T has been studied in [167, 196], and general boundary conditions in [160]. There has been a similar development for Ising models on trees with boundary conditions, see for example [48, 49, 188] in the statistical-physics literature and [114, 248, 256] in the probability literature under the title 'broadcasting on trees'.

We restrict ourselves to the so-called binary tree $T = T_2$, the calculations are easily extended to a regular m-ary tree T_m with $m \in \{2, 3, \dots\}$. Thus $T = (V, E)$ is taken henceforth to be a regular labelled tree, with a distinguished *root* labelled 0, and such that every vertex has degree 3. See Figure 10.4.

We turn T into a directed tree by directing every edge away from 0. There follows some notation concerning the paths of T. Let x be a vertex. An x*-ray* is defined to be an infinite directed path of T with (unique) endvertex x. We denote by $\mathcal{R}_x$ the set of all x-rays of T, and we abbreviate $\mathcal{R}_0$ to $\mathcal{R}$. We shall use the term *ray* to mean a member of some $\mathcal{R}_x$. The edge of T joining vertices x and y is denoted by $\langle x, y\rangle$ when undirected, and by $[x, y\rangle$ when directed from x to y. For any vertex x, we write $\mathcal{R}'_x$ for the subset of $\mathcal{R}$ comprising all rays that pass through x. Any ray $\rho_x \in \mathcal{R}_x$ is a sub-ray of a unique ray $\rho'_x \in \mathcal{R}$, and thus there is a natural one–one correspondence $\rho_x \leftrightarrow \rho'_x$ between $\mathcal{R}_x$ and $\mathcal{R}'_x$.

Let $\mathcal{E}$ be the set of equivalence relations on the set $\mathcal{R}$. Any equivalence relation $\sim$ on $\mathcal{R}$ may be extended to an equivalence relation on $\bigcup_{v\in V} \mathcal{R}_v$ by: for $\rho_u \in \mathcal{R}_u$, $\rho_v \in \mathcal{R}_v$, we have $\rho_u \sim \rho_v$ if and only if $\rho'_u \sim \rho'_v$.

One may define the random-cluster measure corresponding to any given member $\sim$ of a fairly large sub-class of $\mathcal{E}$, but for the sake of simplicity we shall concentrate in the main on the two extremal equivalence relations, as follows.

There is a partial order $\le$ on $\mathcal{E}$ given by:

(10.63) $\quad \sim_1 \le \sim_2$ if: for all $\rho, \rho' \in \mathcal{R}$, $\rho \sim_2 \rho'$ whenever $\rho \sim_1 \rho'$.

There is a minimal (respectively, maximal) partial order which we denote by $\sim^0$ (respectively, $\sim^1$). The equivalence classes of $\sim^0$ are singletons, whereas $\sim^1$ has the single equivalence class $\mathcal{R}$. We refer to $\sim^0$ (respectively, $\sim^1$) as the 'free' (respectively, 'wired') boundary condition.

Let Λ be a finite subset of V, and let E_Λ be the set of edges of T having both endvertices in Λ. For $\xi \in \Omega = \{0, 1\}^E$, we write Ω_Λ^ξ for the (finite) subset of Ω containing all configurations ω satisfying $\omega(e) = \xi(e)$ for $e \in E \setminus E_\Lambda$; these are the configurations that agree with ξ off Λ. For simplicity, we shall restrict ourselves to sets Λ of a certain form. A subset C of V is called a *cutset* if every infinite path from 0 intersects C, and C is minimal with this property. It may be seen by an elementary argument that every cutset is finite. Let C be a cutset, and write $\text{out}(C)$ for the set of all vertices x such that: $x \notin C$ and the (unique) path from 0 to x intersects C. A *box* Λ is a set of the form $V \setminus \text{out}(C)$ for some cutset C, and we write $\partial\Lambda$ for the corresponding C.

Let Λ be a box, and let $\sim \in \mathcal{E}$, $\xi \in \Omega$, and $\omega \in \Omega_\Lambda^\xi$. The configuration ω gives rise to an 'open graph' on Λ, namely $G(\Lambda, \omega) = (\Lambda, \eta(\omega) \cap E_\Lambda)$. We augment this graph by adding certain new edges representing the action of the equivalence relation $\sim$ in the presence of the external configuration ξ. Specifically, for distinct $u, v \in \partial\Lambda$, we add a new edge between the pair u, v if there exist ξ-open rays $\rho_u \in \mathcal{R}_u$, $\rho_v \in \mathcal{R}_v$ satisfying $\rho_u \sim \rho_v$. We write $G^{\xi,\sim}(\Lambda, \omega)$ for the resulting augmented graph, and we let $k^{\xi,\sim}(\Lambda, \omega)$ be the number of connected components of $G^{\xi,\sim}(\Lambda, \omega)$. These definitions are motivated by the idea that each equivalence class of rays leads to a common 'point at infinity' through which vertices may be connected by open paths.

We define next a random-cluster measure corresponding to a given equivalence relation $\sim$. Let $\xi \in \Omega$, and let $p \in [0, 1]$ and $q \in (0, \infty)$. We define $\phi_{\Lambda,p,q}^{\xi,\sim}$ as the random-cluster measure on the box (Λ, E_Λ) with boundary condition $(\xi, \sim)$. More precisely, $\phi_{\Lambda,p,q}^{\xi,\sim}$ is the probability measure on the pair $(\Omega, \mathcal{F})$ given by

(10.64)
$$\phi_{\Lambda,p,q}^{\xi,\sim}(\omega) = \begin{cases} \dfrac{1}{Z_{\Lambda,p,q}^{\xi,\sim}} \left\{ \displaystyle\prod_{e \in E_\Lambda} p^{\omega(e)}(1-p)^{1-\omega(e)} \right\} q^{k^{\xi,\sim}(\Lambda,\omega)} & \text{if } \omega \in \Omega_\Lambda^\xi, \\ 0 & \text{otherwise,} \end{cases}$$

where $Z_{\Lambda,p,q}^{\xi,\sim}$ is the appropriate normalizing constant,

(10.65)
$$Z_{\Lambda,p,q}^{\xi,\sim} = \sum_{\omega \in \Omega_\Lambda^\xi} \left\{ \prod_{e \in E_\Lambda} p^{\omega(e)}(1-p)^{1-\omega(e)} \right\} q^{k^{\xi,\sim}(\Lambda,\omega)}.$$

In the special case when $\xi = 1$ and $\sim\, = \,\sim^1$, we write $\phi_{\Lambda,p,q}^1$ for $\phi_{\Lambda,p,q}^{\xi,\sim}$. This measure will be referred to as the random-cluster measure on Λ with 'wired' boundary conditions, and it has been studied in a slightly disguised form in [167, 196].

For any finite subset $\Lambda \subseteq V$, let $\mathcal{T}_\Lambda$ denote the σ-field generated by the set $\{\omega(e) : e \in E \setminus E_\Lambda\}$ of states of edges having at least one endvertex outside Λ. For $e \in E$, $\mathcal{T}_e$ denotes the σ-field generated by the states of edges other than e.

Let $p \in [0, 1]$, $q \in (0, \infty)$, and let $\sim$ be an equivalence relation that satisfies a certain measurability condition to be stated soon. A probability measure ϕ on $(\Omega, \mathcal{F})$ is called a ($\sim$)*DLR-random-cluster measure* with parameters p and q if: for all $A \in \mathcal{F}$ and all boxes Λ,

$$\phi(A \mid \mathcal{T}_\Lambda)(\xi) = \phi^{\xi,\sim}_{\Lambda,p,q}(A) \qquad \text{for } \phi\text{-a.e. } \xi. \tag{10.66}$$

The set of such measures is denoted by $\mathcal{R}^{\sim}_{p,q}$. The set $\mathcal{R}^{\sim}_{p,q}$ is convex whenever it is non-empty (as in Theorem 4.34).

We introduce next the relevant measurability assumption on the equivalence relation $\sim$. Since the left side of (10.66) is a measurable function of ξ, the right side must be measurable also. For a box Λ and distinct vertices $u, v \in \partial\Lambda$, let $K^{\sim}_{u,v,\Lambda}$ denote the set of $\omega \in \Omega$ such that there exist ω-open rays $\rho_u \in \mathcal{R}_u$, $\rho_v \in \mathcal{R}_v$ satisfying $\rho_u \sim \rho_v$. We call the equivalence relation $\sim$ *measurable* if $K^{\sim}_{u,v,\Lambda} \in \mathcal{F}$ for all such u, v, Λ. It is an easy exercise to deduce, if $\sim$ is measurable, that $\phi^{\xi,\sim}_{\Lambda,p,q}(A)$ is a measurable function of ξ, thus permitting condition (10.66). We write $\mathcal{E}_{\mathrm{m}}$ for the set of all measurable elements of $\mathcal{E}$. It is easily seen that the extremal equivalence relations $\sim^0$, $\sim^1$ are measurable.

For simplicity of notation we write $\mathcal{R}^{\sim^0}_{p,q} = \mathcal{R}^0_{p,q}$ and similarly $\mathcal{R}^{\sim^1}_{p,q} = \mathcal{R}^1_{p,q}$. Members of $\mathcal{R}^0_{p,q}$ (respectively, $\mathcal{R}^1_{p,q}$) are called 'free' random-cluster measures (respectively, 'wired' random-cluster measures). There follows an existence theorem. Any probability measure μ on $(\Omega, \mathcal{F})$ is called *automorphism-invariant* if the vectors $(\omega(e) : e \in E)$ and $(\omega(\tau e) : e \in E)$ have the same laws under μ, for any automorphism τ of the tree T.

(10.67) Theorem [167]. *Let $p \in [0, 1]$ and $q \in (0, \infty)$.*

(a) *The set $\mathcal{R}^0_{p,q}$ of free random-cluster measures comprises the singleton ϕ_π only, where $\pi = \pi(p, q)$ is given in (10.62). The product measure ϕ_π belongs to $\mathcal{R}^1_{p,q}$ if and only if $\pi \le \frac{1}{2}$.*

(b) *The set $\mathcal{R}^1_{p,q}$ of wired random-cluster measures is non-empty.*

(c) *If $q \in [1, \infty)$, the weak limit*

$$\phi^1_{p,q} = \lim_{\Lambda \uparrow V} \phi^1_{\Lambda,p,q} \tag{10.68}$$

exists and belongs to $\mathcal{R}^1_{p,q}$. Furthermore, $\phi^1_{p,q}$ is an extremal element of the convex set $\mathcal{R}^1_{p,q}$ and is automorphism-invariant.

Here are some comments on this theorem. Part (b) will be proved at Theorem 10.82(c). Parts (a) and (c) are proved later in the current section, and we anticipate this with a brief discussion of the condition $\pi \le \frac{1}{2}$. This will be recognized as the condition for the almost-sure extinction of a branching process whose family-sizes have the binomial bin$(2, \pi)$ distribution. That is, $\pi \le \frac{1}{2}$ if and only if

$$\phi_\pi(0 \leftrightarrow \infty) = 0, \tag{10.69}$$

see [164, Thm 5.4.5]. It turns out that the product measure ϕ_π lies in $\mathcal{R}^1_{p,q}$ if and only if it does not 'feel' the wired boundary condition $\sim^1$, that is to say, if there exist (ϕ_π-almost-surely) no infinite clusters[6].

We turn briefly to more general boundary conditions than merely the free and wired, see [160] for further details. The set $\mathcal{R}$ of rays may be viewed as a compact topological space with the product topology. Let $\sim$ be an equivalence relation on $\mathcal{R}$. We call $\sim$ *closed* if the set $\{(\rho_1, \rho_2) \in \mathcal{R}^2 : \rho_1 \sim \rho_2\}$ is a closed subset of $\mathcal{R}^2$. It turns out that closed equivalence relations are necessarily measurable. For $q \in [1, \infty)$ and a closed relation $\sim$, the existence of the weak limit $\phi^{1,\sim}_{p,q} = \lim_{\Lambda \uparrow V} \phi^{1,\sim}_{\Lambda,p,q}$ follows by stochastic ordering, and it may be shown that $\phi^{1,\sim}_{p,q}$ is a $(\sim)$DLR-random-cluster measure.

Theorem 10.67 leaves open the questions of deciding when $\phi_\pi = \phi^1_{p,q}$, and when $\mathcal{R}^1_{p,q}$ comprises a singleton only. We return to these questions in Sections 10.10–10.11.

Proof of Theorem 10.67. (a) Consider the free boundary condition $\sim^0$, and let A be a cylinder event. By (10.61),

$$\phi^{\xi,\sim^0}_{\Lambda,p,q}(A) = \phi_\pi(A)$$

for all boxes Λ that are sufficiently large that A is defined on the edge-set E_Λ. For $\phi \in \mathcal{R}^0_{p,q}$, by (10.66),

$$\phi(A \mid \mathcal{T}_\Lambda) = \phi_\pi(A), \qquad \phi\text{-almost-surely,}$$

for all sufficiently large Λ, and therefore

$$\phi(A) = \phi(\phi(A \mid \mathcal{T}_\Lambda)) = \phi_\pi(A)$$

as required. The second part of (a) is proved after the proof of (c).

(c) The existence of the weak limit in (10.68) follows by positive assocation as in the proof of Theorem 4.19(a). In order to show that the limit measure lies in $\mathcal{R}^1_{p,q}$, we shall make use of the characterization of random-cluster measures provided by Proposition 4.37; this was proved with the lattice $\mathbb{L}^d$ in mind but is valid also in the present setting with the same proof.

For $v \in V$, let Π_v be the set of infinite undirected paths of T with endvertex v. Let $e = \langle x, y\rangle$, and let K^1_e be the event that there exist open vertex-disjoint paths

[6]See also [168].

$\nu_x \in \Pi_x$ and $\nu_y \in \Pi_y$. For any box Λ and $\omega \in \Omega$, let ω_Λ^1 denote the configuration that agrees with ω on E_Λ and equals 1 elsewhere, which is to say that

$$\omega_\Lambda^1(e) = \begin{cases} \omega(e) & \text{if } e \in E_\Lambda, \\ 1 & \text{otherwise.} \end{cases}$$

We define the event

$$K_{e,\Lambda}^1 = \{\omega \in \Omega : \omega_\Lambda^1 \in K_e^1\}.$$

Note that $K_{e,\Lambda}^1$ is a cylinder event, and is decreasing in Λ. It is easily checked that

$$K_{e,\Lambda}^1 \downarrow K_e^1 \qquad \text{as } \Lambda \uparrow V. \tag{10.70}$$

We may now state the relevant conclusion of Proposition 4.37 in the current context, namely that $\phi \in \mathcal{R}_{p,q}^1$ if and only if, for all $e \in E$,

$$\phi(J_e \mid \mathcal{T}_e) = \pi + (p - \pi)1_{K_e^1} \qquad \phi\text{-almost-surely}, \tag{10.71}$$

where $J_e = \{e \text{ is open}\}$.

For $\xi \in \Omega$ and $W \subseteq V$, write $[\xi]_W$ for the set of all configurations that agree with ξ on E_W. For $e \in E_W$, let $[\xi]_{W\setminus e}$ be an abbreviation for $[\xi]_{E_W\setminus\{e\}}$. We shall omit explicit reference to the values of p and q in the rest of this proof. Thus, for example, $\phi^1 = \phi_{p,q}^1$.

By the martingale convergence theorem (see [164, Ex. 12.3.9]), for $e = \langle x, y\rangle \in E$ and ϕ^1-almost-every ξ,

$$\begin{aligned} \phi^1(J_e \mid \mathcal{T}_e)(\xi) &= \lim_{\Lambda\uparrow V} \frac{\phi^1(J_e,\ [\xi]_{\Lambda\setminus e})}{\phi^1([\xi]_{\Lambda\setminus e})} \\ &= \lim_{\Lambda\uparrow V}\lim_{\Delta\uparrow V} \frac{\phi_\Delta^1(J_e,\ [\xi]_{\Lambda\setminus e})}{\phi_\Delta^1([\xi]_{\Lambda\setminus e})} \\ &= \lim_{\Lambda\uparrow V}\lim_{\Delta\uparrow V} \phi_\Delta^1\big(\phi_\Delta^1(J_e \mid [\xi]_{\Delta\setminus e}) \bigm| [\xi]_{\Lambda\setminus e}\big) \\ &= \lim_{\Lambda\uparrow V}\lim_{\Delta\uparrow V} \phi_\Delta^1(g_\Delta \mid [\xi]_{\Lambda\setminus e}), \end{aligned} \tag{10.72}$$

by Theorem 3.1, where

$$g_\Delta(\xi) = \pi + (p - \pi)1_{K_{e,\Delta}^1}(\xi).$$

By (10.70), $g_\Delta \downarrow g$ as $\Delta \uparrow V$, where $g = \pi + (p - \pi)1_{K_e^1}$.

We claim that

$$\phi_\Lambda^1(g_\Delta \mid [\xi]_{\Lambda\setminus e}) \to \phi^1(g \mid [\xi]_{\Lambda\setminus e}) \qquad \text{as } \Delta \uparrow V, \tag{10.73}$$

and we prove this as follows. Let Δ', Δ'' be boxes satisfying $\Lambda \subseteq \Delta' \subseteq \Delta \subseteq \Delta''$. Since $\psi_\Delta(\cdot) = \phi_\Delta^1(\cdot \mid [\xi]_{\Lambda\setminus e})$ is a random-cluster measure on an altered graph (see Theorem 3.1(a)) and since g_Δ is increasing on Ω and non-increasing in Δ, we have by positive association that

$$\psi_{\Delta''}(g_\Delta) \le \psi_\Delta(g_\Delta) \le \psi_\Delta(g_{\Delta'}).$$

Let $\Delta'' \uparrow V$, $\Delta \uparrow V$, and $\Delta' \uparrow V$, in that order, to conclude (10.73) by monotone convergence.

By the martingale convergence theorem again,

$$\phi^1(g \mid [\xi]_{\Lambda \setminus e}) \to g(\xi) \qquad \text{as } \Lambda \uparrow V, \text{ for } \phi^1\text{-a.e. } \xi,$$

and (10.71) follows by (10.72)–(10.73).

The extremality of $\phi^1_{p,q}$ is a consequence of positive association, on noting that $\phi^1_{p,q} \ge_{\text{st}} \phi$ for all $\phi \in \mathcal{R}^1_{p,q}$. Let τ be an automorphism of the graph T. In the notation of Section 4.3, for any increasing cylinder event A and all boxes Λ,

$$\phi^1_{\Lambda,p,q}(A) = \phi^1_{\tau\Lambda,p,q}(\tau^{-1}A),$$

and, by positive association,

$$\phi^1_{\tau\Lambda,p,q}(\tau^{-1}A) \ge \phi^1_{\Delta,p,q}(\tau^{-1}A) \qquad \text{if } \Delta \supseteq \tau\Lambda.$$

Letting $\Delta \uparrow V$, we obtain that

$$\phi^1_{\Lambda,p,q}(A) \ge \phi^1_{p,q}(\tau^{-1}A),$$

so that $\phi^1_{p,q}(A) \ge \phi^1_{p,q}(\tau^{-1}A)$. Equality must hold here, and the claim of automorphism-invariance follows.

Turning to the final statement of part (a), by the discussion around (10.71), $\phi_\pi \in \mathcal{R}^1_{p,q}$ if and only if $\phi_\pi(K^1_e) = 0$ for all $e \in E$. Since ϕ_π is a product measure, this condition is equivalent to (10.69). □

10.10 The critical point for a tree

We concentrate henceforth on the binary tree $T = T_2 = (V, E)$ and the wired equivalence relation $\sim^1$. It is shown in this section how the series/parallel laws may be used to study random-cluster measures on T. Corresponding results are valid for the m-ary tree with $m \ge 2$.

The results of this section are valid for all $q \in (0, \infty)$, and we begin by proving the existence of the wired weak-limit for all p and q, thereby extending part of Theorem 10.67(c). The limit as $\Lambda \uparrow V$ is taken along an arbitrary increasing sequence of boxes.

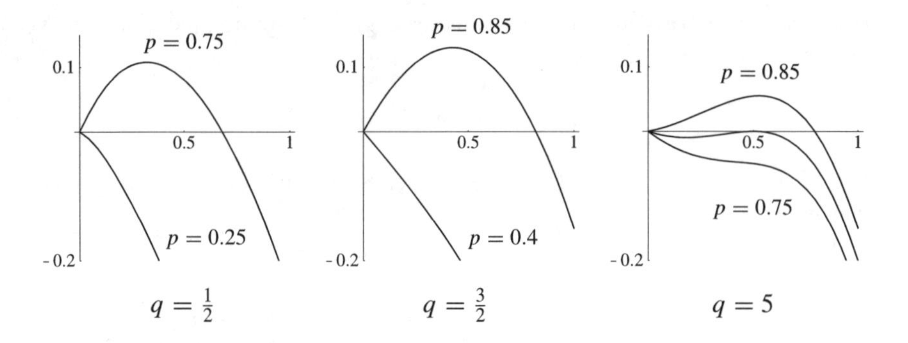

Figure 10.5. The function $f_{p,q}(x) - x$ is plotted in the three cases $q < 1$, $1 < q \le 2$, $q > 2$. The maximal solution ρ of $f_{p,q}(x) = x$ satisfies $\rho > 0$ if and only if $p > \kappa_q$ (respectively, $p \ge \kappa_q$) when $0 < q \le 2$ (respectively, $q > 2$). The intermediate curve in the third graph corresponds to the critical case with $p = \kappa_q$ and $q > 2$. Note in this case that $\rho = \rho(p, q)$ is a discontinuous function of p.

(10.74) Theorem. *The weak limit*

$$\phi^1_{p,q} = \lim_{\Lambda \uparrow V} \phi^1_{\Lambda,p,q} \tag{10.75}$$

exists for all $p \in [0, 1]$ *and* $q \in (0, \infty)$.

Consider now the existence (or not) of infinite open clusters under the weak limit $\phi^1_{p,q}$. Let

$$\theta(p, q) = \phi^1_{p,q}(0 \leftrightarrow \infty), \tag{10.76}$$

and define the critical value of p by

$$p_c(q) = \sup\{p : \theta(p, q) = 0\}. \tag{10.77}$$

The calculation of $\theta(p, q)$ and $p_c(q)$ makes use of certain quantities which we introduce next.

Let κ_q be defined by

$$\kappa_q = \begin{cases} \dfrac{q}{q+1} & \text{if } 0 < q \le 2, \\[2ex] \dfrac{2\sqrt{q-1}}{1+2\sqrt{q-1}} & \text{if } q > 2. \end{cases} \tag{10.78}$$

Let $F_{p,q} : [0, 1]^2 \to [0, 1]$ be given by

$$F_{p,q}(x, y) = \frac{p[1 - (1-x)(1-y)]}{1 + (q-1)(1-p)(1-x)(1-y)},$$

and let $f_{p,q}, g_q : [0,1] \to [0,1]$ be given by

$$f_{p,q}(x) = F_{p,q}(x,x), \tag{10.79}$$

$$g_q(y) = \frac{1-(1-y)^3}{1+(q-1)(1-y)^3}. \tag{10.80}$$

An important quantity is the maximal root $\rho = \rho(p,q)$ in $[0,1]$ of the equation $f_{p,q}(x) = x$. In particular, we will need to know under what conditions $\rho(p,q)$ is strictly positive.

(10.81) Proposition. *Let $p \in [0,1]$ and $q \in (0,\infty)$. Let $\rho = \rho(p,q)$ be the maximal solution in the interval $[0,1]$ of the equation $f_{p,q}(x) = x$. Then:*

$$\rho > 0 \quad \textit{if and only if} \quad p \begin{cases} > \kappa_q & \textit{when } 0 < q \le 2, \\ \ge \kappa_q & \textit{when } q > 2. \end{cases}$$

The proof of this proposition is elementary and is omitted. Illustrations of the three cases $q \in (0,1)$, $q \in [1,2]$, $q \in (2,\infty)$ appear in Figure 10.5. We now state the main theorem of this section.

(10.82) Theorem. *Let $p \in [0,1]$ and $q \in (0,\infty)$. Then:*

(a) *$\theta(p,q) = g_q(\rho)$ where ρ is the maximal root in $[0,1]$ of the equation $f_{p,q}(x) = x$,*

(b) *$p_c(q) = \kappa_q$ where κ_q is given in (10.78),*

(c) *$\phi^1_{p,q} \in \mathcal{R}^1_{p,q}$,*

(d) *$\mathcal{R}^1_{p,q} = \{\phi_\pi\}$ whenever $\theta(p,q) = 0$.*

This theorem may be found in essence in [167] but with different proofs. In contrast to the direct calculations[7] of this section, the proofs in [167] proceed via a representation of random-cluster measures on T in terms of a certain class of multi-type branching processes.

Proof of Theorem 10.74. We use the series/parallel laws of Theorem 3.91. The basic fact is that three edges in the configuration on the left side of Figure 10.6, with parameter-values as given there, may be replaced as indicated by a single edge with parameter $F_{p,q}(x,y)$. This is easy to check: the two lower edges in parallel may be replaced by a single edge with parameter $1-(1-x)(1-y)$, and the latter may then be combined with the upper edge in series.

Let $\Lambda_n = \{x \in V : |x| \le n\}$, where $|x|$ denotes the number of edges in the path from 0 to x. We consider first the measures $\phi^1_{\Lambda_n,p,q}$, in the limit as $n \to \infty$.

Let H_r be the graph obtained from the finite tree $(\Lambda_r, E_{\Lambda_r})$ by adding two new edges $[x,x'\rangle$, $[x,x''\rangle$ to each terminal vertex $x \in \partial\Lambda_r$. We colour these new

[7] The current method was mentioned in passing in [160].

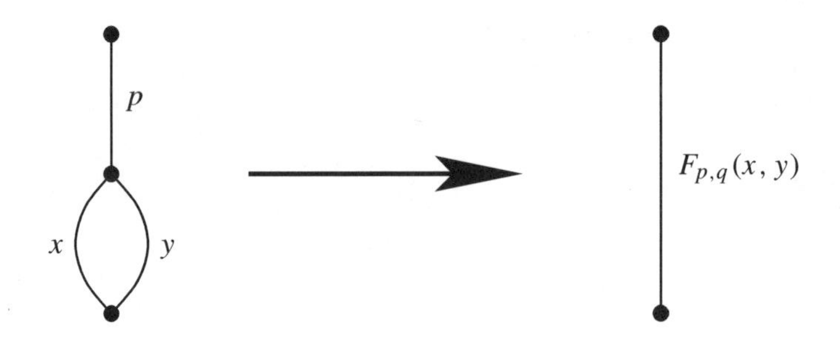

Figure 10.6. The parallel and series laws are used to replace the left graph by a single edge. The parameter values are as marked.

edges *green*, together with all new endvertices x', x''. We write H_r^1 for the graph obtained from H_r by an identification of all green vertices. See Figure 10.7.

Let $p \in [0, 1]$, $q \in (0, \infty)$ and $1 \le r \le s - 2$. The wired measure on Λ_s may be viewed as the random-cluster measure on the graph obtained from Λ_s by identifying the set $\partial\Lambda_s$. We write Λ_s^1 for the graph obtained thus, and we will not dwell on the changes of notation required in order to do this properly. We propose to use the series/parallel laws in order to replace edges belonging to Λ_s^1 but not Λ_r^1. Edges in Λ_s^1 incident to the composite vertex $\partial\Lambda_s$ come in pairs, and each such pair e_1, e_2 has an immediate ancestor edge e_3. The trio e_1, e_2, e_3 may be replaced by a single edge with parameter $f_{p,q}(p)$. When all such trios have been replaced, the resulting graph is Λ_{s-1}^1. This process is iterated until Λ_s^1 has been reduced to H_r^1. The green edges of H_r^1 have acquired parameter-value $f_{p,q}^{(s-r-1)}(p)$, where $f_{p,q}^{(k)}$ denotes the kth iterate of $f_{p,q}$. Note that $f_{p,q}(1) = p$, and hence $f_{p,q}^{(s-r-1)}(p) = f_{p,q}^{(s-r)}(1)$.

The function $f_{p,q}$ is increasing on $[0, 1]$ with $f_{p,q}(0) = 0$ and $f_{p,q}(1) = p$. Therefore,

$$f_{p,q}^{(n)}(1) \to \rho \qquad \text{as } n \to \infty, \tag{10.83}$$

where ρ is the maximal root in $[0, 1]$ of the equation $f_{p,q}(x) = x$.

Let $E \in \mathcal{F}_{\Lambda_r}$. Let $\phi_{r,s}^1$ be the random-cluster measure on H_r^1 with edge-parameter $f^{(s-r)}(1)$ (respectively, p) for the green (respectively, non-green) edges. By the above,

$$\phi_{\Lambda_s,p,q}^1(E) = \phi_{r,s}^1(E). \tag{10.84}$$

A (random-cluster) probability $\phi_{G,\mathbf{p},q}(E)$ is a continuous function of the edge-parameters $\mathbf{p}$. Therefore, by (10.83),

$$\phi_{\Lambda_s,p,q}^1(E) \to \phi_{r,\infty}^1(E) \qquad \text{as } s \to \infty, \tag{10.85}$$

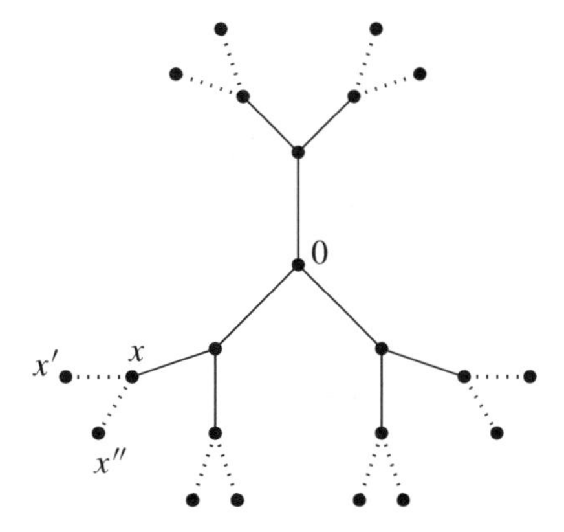

Figure 10.7. To each boundary vertex x of the box Λ_2 is attached two new (green) edges $[x, x'\rangle$, $[x, x''\rangle$. The resulting graph is denoted by H_2.

where $\phi^1_{r,\infty}$ is the wired random-cluster measure on H_r in which the green edges have parameter ρ.

When $q \in [1, \infty)$, the random-cluster measure is positively associated, and (10.85) implies (10.75) for general Λ. When $q \in (0, 1)$, a separate argument is needed in order to extend the limit in (10.85) to a general increasing sequence of boxes. Let Λ be a box with $\Lambda \supseteq \Lambda_{r+1}$, and let

$$a = a(\Lambda) = \max\{n : \Lambda_n \subseteq \Lambda\}, \quad b = b(\Lambda) = \min\{n : \Lambda \subseteq \Lambda_n\}.$$

The measure $\phi^1_{\Lambda,p,q}$ may be viewed as the random-cluster measure on Λ^1_b in which edges of $E_{\Lambda_b} \setminus E_\Lambda$ have parameter 1. We may reduce Λ^1_b to H^1_r via the series/parallel laws as above. Since $F_{p,q}(x, y)$ is increasing in p, x, y, the green edges of H^1_r acquire parameter values lying between $f^{(b-r)}_{p,q}(1)$ and $f^{(a-r)}_{p,q}(1)$. Now $a, b \to \infty$ as $\Lambda \to V$, and

$$f^{(b-r)}_{p,q}(1) \to \rho, \quad f^{(a-r)}_{p,q}(1) \to \rho.$$

It follows as above that

$$\phi^1_{\Lambda,p,q}(E) \to \phi^1_{r,\infty}(E) \qquad \text{as } \Lambda \uparrow V. \tag{10.86}$$

There remains a detail. Each $\phi^1_{\Lambda,p,q}$ is a probability measure on the compact state space Ω. Therefore, the family of such $\phi^1_{\Lambda,p,q}$, as Λ ranges over boxes, is tight. By Prohorov's theorem, [42], every subsequence contains a convergent sub(sub)sequence. The limiting probability of any cylinder event E is, by (10.86), independent of the choice of subsequence. Therefore, the weak limit in (10.75) exists, and the theorem is proved. □

Proof of Theorem 10.82. (a) Let ρ be as given. We claim that

$$\phi^1_{\Lambda_n,p,q}(0 \leftrightarrow \partial\Lambda_n) \to g_q(\rho) \qquad \text{as } n \to \infty. \tag{10.87}$$

By series/parallel replacement as in the proof of Theorem 10.74,

$$\theta_n(p,q) = \phi^1_{\Lambda_n,p,q}(0 \leftrightarrow \partial\Lambda_n)$$

satisfies

$$\theta_n(p,q) = \theta_1(f^{(n)}_{p,q}(1), q).$$

By (10.83), $\theta_n(p,q) \to \theta_1(\rho,q)$ as $n \to \infty$. It is an easy calculation that $\theta_1(z,q) = g_q(z)$, and (10.87) follows.

The proof of Proposition 5.11 is valid in the current setting, whence

$$\theta(p,q) = \lim_{n\to\infty} \theta_n(p,q) = g_q(\rho), \qquad \text{whenever } q \in [1,\infty).$$

This proves (a) for $q \in [1,\infty)$.

Suppose that $q \in (0,1)$. The situation is now harder since we may not appeal to positive association. Instead, we use the weaker inequalities (5.117)–(5.118) which we summarize as:

$$\phi_{G,p,1} \leq_{\mathrm{st}} \phi_{G,p,q} \leq_{\mathrm{st}} \phi_{G,\pi,1}, \tag{10.88}$$

for any finite graph G, where $\pi = p/[p+q(1-p)]$. By Proposition 4.10(a), corresponding inequalities hold for the weak limits of random-cluster measures.

Let $p \leq \kappa_q$, so that $\rho = 0$. Then $\pi = p/[p+q(1-p)] \leq \frac{1}{2}$, and therefore $\phi^1_\pi(0 \leftrightarrow \infty) = 0$. By (10.88), $\theta(p,q) = \rho = 0$ as claimed.

Let $p > \kappa_q$, so that $\rho > 0$. By Theorem 10.74,

$$\begin{aligned} \theta(p,q) &= \lim_{r\to\infty} \phi^1_{p,q}(0 \leftrightarrow \partial\Lambda_r) \\ &= \lim_{r\to\infty}\lim_{s\to\infty} \phi^1_{\Lambda_s,p,q}(0 \leftrightarrow \partial\Lambda_r). \end{aligned} \tag{10.89}$$

Now,

$$\phi^1_{\Lambda_s,p,q}(0 \leftrightarrow \partial\Lambda_r) \geq \phi^1_{\Lambda_s,p,q}(0 \leftrightarrow \partial\Lambda_s), \qquad r \leq s,$$

and therefore, by (10.87),

$$\theta(p,q) \geq g_q(\rho). \tag{10.90}$$

By (10.87) and (10.89),

$$\begin{aligned} \theta(p,q) - g_q(\rho) &= \lim_{r\to\infty}\lim_{s\to\infty} \phi^1_{\Lambda_s,p,q}(0 \leftrightarrow \partial\Lambda_r,\ 0 \not\leftrightarrow \partial\Lambda_s) \\ &= \lim_{r\to\infty} \phi^1_{r,\infty}(0 \leftrightarrow \partial\Lambda_r,\ 0 \not\leftrightarrow \partial\Lambda_{r+1}), \end{aligned} \tag{10.91}$$

where $\phi^1_{r,\infty}$ is defined after (10.85).

For $\omega \in \Omega$ and $r \geq 0$, let G_r be the set of vertices $x \in \partial\Lambda_r$ such that 0 is joined to x by an open path of the tree, and write $N_r = |G_r|$. We claim that

$$\text{(10.92)} \qquad \text{for } k = 1, 2, \dots, \qquad \phi^1_{p,q}(1 \leq N_r \leq k) \to 0 \qquad \text{as } r \to \infty,$$

and we prove this as follows. Let $k \in \{1, 2, \dots\}$, and define the random sequence $R(0), R(1), R(2), \dots$ by $R(0) = 0$ and

$$R(i+1) = \min\{s > R(i) : 1 \leq N_s \leq k\}, \qquad i \geq 0.$$

The length of the sequence is $I + 1$ where $I = I(\omega) = |\{r \geq 1 : 1 \leq N_r \leq k\}|$, and we prove next that

$$\text{(10.93)} \qquad \phi^1_{p,q}(I < \infty) = 1.$$

Let $i \geq 0$, and suppose we are given that $I(\omega) \geq i$. Conditional on $R(0), R(1), R(2), \dots, R(i)$, and on the states of all edges in $E_{\Lambda_{R(i)}}$, there is a certain (conditional) probability that, for all $x \in G_{R(i)}$, x is incident to no vertex in $\partial\Lambda_{R(i)+1}$. By Theorem 3.1(a), the appropriate (conditional) probability measure is a random-cluster measure on a certain graph obtained from T by the deletion and contraction of edges in $E_{\Lambda_{R(i)}}$. Since $|G_{R(i)}| \leq k$, there are no more than $2k$ edges of T joining $G_{R(i)}$ to $\partial\Lambda_{R(i)+1}$ and, by the second inequality of (10.88),

$$\phi^1_{p,q}(I = i \mid I \geq i) \geq (1-\pi)^{2k}.$$

Therefore,

$$\phi^1_{p,q}(I \geq i+1 \mid I \geq i) \leq 1 - (1-\pi)^{2k}, \qquad i \geq 0,$$

whence

$$\phi^1_{p,q}(I \geq i) \leq \big[1 - (1-\pi)^{2k}\big]^i, \qquad i \geq 0,$$

and, in particular, (10.93) holds. Hence, $M = \sup\{r : 1 \leq N_r \leq k\}$ satisfies $\phi^1_{p,q}(M < \infty) = 1$, implying as required that

$$\text{(10.94)} \qquad \phi^1_{p,q}(1 \leq N_r \leq k) \leq \phi^1_{p,q}(M \geq r) \to 0 \qquad \text{as } r \to \infty.$$

By a similar argument,

$$\begin{aligned}
\phi^1_{r,\infty}(0 \leftrightarrow \partial\Lambda_r,\ 0 \nleftrightarrow \partial\Lambda_{r+1}) &= \sum_{l=1}^{\infty} \phi^1_{r,\infty}(N_r = l,\ 0 \nleftrightarrow \partial\Lambda_{r+1}) \\
&\leq \sum_{l=1}^{\infty} (1-\rho)^{2l} \phi^1_{p,q}(N_r = l) \quad \text{by (10.88)} \\
&\leq \phi^1_{p,q}(1 \leq N_r \leq k) + (1-\rho)^{2k} \\
&\to (1-\rho)^{2k} \qquad \text{as } r \to \infty, \quad \text{by (10.92)} \\
&\to 0 \qquad \text{as } k \to \infty.
\end{aligned}$$

By (10.90) and (10.91), $\theta(p,q) = g_q(\rho)$.
(b) This is an immediate consequence of part (a), Proposition 10.81, and the definition of $p_c(q)$.
(c) Let $q \in (0,\infty)$. We shall show that $\phi^1_{p,q}$ satisfies (10.71) for $e \in E$. As in (10.72), for $e = \langle x, y\rangle \in E$,

$$\text{(10.95)} \qquad \phi^1_{p,q}(J_e \mid \mathcal{T}_e)(\xi) = \lim_{\Lambda\uparrow V}\lim_{\Delta\uparrow V} \phi^1_{\Delta,p,q}(J_e \mid [\xi]_{\Lambda\setminus e}), \qquad \phi^1_{p,q}\text{-a.s.}$$

If $\xi \notin K^1_e$, then $[\xi]_{\Lambda\setminus e} \cap K^1_e = \varnothing$ for large Λ, and thus

$$\phi^1_{\Delta,p,q}(J_e \mid [\xi]_{\Lambda\setminus e}) = \frac{p}{p+q(1-p)} \qquad \text{for large } \Lambda,\ \Delta,$$

by Theorem 3.1. By (10.95),

$$\text{(10.96)} \qquad \phi^1_{p,q}(J_e \mid \mathcal{T}_e) = \frac{p}{p+q(1-p)}, \qquad \phi^1_{p,q}\text{-a.s. on } \Omega \setminus K^1_e.$$

Suppose that $\phi^1_{p,q}(K^1_e) > 0$, let $\xi \in K^1_e$, and take $\Lambda = \Lambda_r$ in the notation of the previous proof. As in that proof, for $e \in E_{\Lambda_r}$,

$$\lim_{\Delta\uparrow V} \phi^1_{\Delta,p,q}(J_e \mid [\xi]_r) = \phi^1_{r,\infty}(J_e \mid [\xi]_r),$$

where $[\xi]_r = [\xi]_{\Lambda_r\setminus e}$. Let $N_r(u)$ be the number of vertices in $\partial\Lambda_r$ joined to u by an open path. As in the previous proof,

$$N_r(x),\ N_r(y) \to \infty \quad \text{as } r \to \infty, \qquad \phi^1_{p,q}\text{-a.s. on } K^1_e,$$

whence, for $\phi^1_{p,q}$-almost-every $\xi \in K^1_e$,

$$\left|\phi^1_{r,\infty}(J_e \mid [\xi]_r) - \phi^1_{r,\infty}(J_e \mid x, y \leftrightarrow \partial\Lambda_{r+1} \text{ off } e)\right| \to 0 \qquad \text{as } r \to \infty.$$

By Theorem 3.1,

$$\phi^1_{r,\infty}(J_e \mid x, y \leftrightarrow \partial\Lambda_{r+1} \text{ off } e) = p,$$

and therefore,

$$\phi^1_{p,q}(J_e \mid \mathcal{T}_e) = p, \qquad \phi^1_{p,q}\text{-a.s. on } K^1_e.$$

When combined with (10.96), this implies (10.71), and the claim follows.
(d) Let $\phi \in \mathcal{R}^1_{p,q}$, where p and q are such that $\theta(p,q) = 0$. By the argument in the proof of part (a), $\phi(0 \leftrightarrow \infty) = 0$, and therefore $\phi(K^1_e) = 0$ for $e \in E$. By (10.71), $\phi(J_e \mid \mathcal{T}_e) = \pi$, ϕ-almost-surely, whence $\phi = \phi_\pi$ as claimed. □

10.11 (Non-)uniqueness of measures on trees

For which p, q is there a *unique* wired random-cluster measure on the binary tree T? We assume for simplicity that $q \in [1, \infty)$. By Theorem 10.82, $\mathcal{R}^1_{p,q} = \{\phi_\pi\}$ whenever p is sufficiently small that $\phi^1_{p,q}(0 \leftrightarrow \infty) = 0$. The last holds if and only if

$$p \begin{cases} \le \kappa_q & \text{for } q \in [1, 2], \\ < \kappa_q & \text{for } q \in (2, \infty), \end{cases}$$

where κ_q is given in (10.78). Larger values of p are considered in the following conjecture.

(10.97) Conjecture [167]. *We have that* $|\mathcal{R}^1_{p,q}| = 1$ *if: either* $q \in [1, 2]$, *or* $q \in (2, \infty)$ *and* $p > q/(q+1)$.

When $q \in (2, \infty)$ and $\kappa_q \le p \le q/(q+1)$, there exists a continuum of wired random-cluster measures, see [167]. These may be cooked up on the basis of the following two facts:

(i) $\phi_\pi(0 \leftrightarrow \infty) = 0$ when $p \le q/(q+1)$,

(ii) $\phi^1_{p,q} \ne \phi_\pi$ when $q \in (2, \infty)$ and $p \ge \kappa_q$,

where $\pi = p/[p + q(1-p)]$. The recipe is as follows. Let x be a vertex of T other than its root. The set $\mathcal{R}_x$ of x-rays constitutes an infinite binary tree denoted by $T_x = (V_x, E_x)$ with root x (the vertex x has degree 2 in T_x). Let e_x denote the unique edge of T with endvertex x and not belonging to T_x, and let $E'_x = E_x \cup \{e_x\}$. Let μ_x be the measure on $(\Omega, \mathcal{F})$ given by:

(a) the states of edges in E'_x are independent of those of edges in $E \setminus E'_x$, and have as law the product measure on $\{0, 1\}^{E_x}$ with density π,

(b) the states of edges in $E \setminus E'_x$ have as law the conditional measure of $\phi^1_{p,q}$ given that e_x is closed.

That $\mu_x \in \mathcal{R}^1_{p,q}$ may be seen in very much the same way as in the proof of Theorem 10.67(c), under the condition that there exist, ϕ_π-almost-surely, no infinite open clusters. Thus, $\mu_x \in \mathcal{R}^1_{p,q}$ if $p \le q/(q+1)$. If, in addition, $q \in (2, \infty)$ and $p \ge \kappa_q$, then $\phi^1_{p,q}(0 \leftrightarrow \infty) > 0$. This implies that

$$\phi^1_{p,q}(x \leftrightarrow \infty \text{ in } T_x \mid e_x \text{ is closed}) > 0,$$

whence $\mu_x \ne \phi^1_{p,q}$. It is not hard to see that $\mu_x \ne \mu_y$ whenever $x \ne y$, subject to the above conditions on p, q. Since V is countably infinite, there exist (at least) countably infinitely many members of $\mathcal{R}^1_{p,q}$.

This conclusion may be strengthened by choosing an infinite sequence $\mathbf{x} = (x_i : i = 1, 2, \dots)$ of vertices such that: for every i, x_i is incident to no $e \in E'_{x_j}$ with $j < i$. One performs a construction similar to the above, but with product measure on each of the sets E_{x_i}, $i = 1, 2, \dots$. This results in a probability measure $\mu_{\mathbf{x}}$ belonging to $\mathcal{R}^1_{p,q}$ and labelled uniquely by the sequence $\mathbf{x}$. There are uncountably many choices for $\mathbf{x}$, and therefore uncountably many distinct members of $\mathcal{R}^1_{p,q}$. For the sake of clarity, we point out that one way to choose a large class of possible $\mathbf{x}$ is to take an infinite directed path Π of T, and to consider the power set of the set of all neighbours of Π that do not belong to Π.

Partial progress towards a verification of Conjecture 10.97 may be found in [196]. A broader class of equivalence relations has been considered in [160].

(10.98) Theorem [160, 196]. *Let $q \in [1, \infty)$ and let $p \geq 2q/(2q+1)$. The set $\mathcal{R}^1_{p,q}$ comprises the singleton $\phi^1_{p,q}$ only.*

The condition of this theorem is not best possible in the case $q = 1$, and therefore is unlikely to be best possible for $q \in (1, \infty)$.

There has been extensive study of the Ising model on a tree. It turns out that there are two critical points on the binary tree T. The first critical point corresponds to the random-cluster transition at the point $p = \kappa_2 = \frac{2}{3}$, and the second arises as follows. Consider the Ising model on T with free boundary conditions. There is a critical value of the inverse-temperature at which the corresponding Gibbs state ceases to be extremal. In the parametrization of this chapter, this critical point is given by $p_{\mathrm{sg}} = 2/(1+\sqrt{2})$, see [49, 188, 189, 250]. This value arises also in the study of a related 'Edwards–Anderson' spin-glass problem on T, see [89] and Section 11.5. It may be seen by a process of spin-flipping that the spin-glass model with ± 1 interactions can be mapped to a ferromagnetic Ising model with boundary conditions taken uniformly and independently from the spin space $\{-1, +1\}$. It turns out that this model has critical value p_{sg} also, and for this reason p_{sg} is commonly referred to as the 'spin-glass critical point'.

In summary, for $p = 1 - e^{-\beta} < \frac{2}{3}$, the Ising model has a unique Gibbs state. For $p \in (\frac{2}{3}, p_{\mathrm{sg}})$, the $+$ Gibbs state differs from the free state, whereas 'typical' boundary conditions (in the sense of boundary conditions chosen randomly according to the free state) result in the free measure. When $p > p_{\mathrm{sg}}$, the free state is no longer an extremal Gibbs state. This double transition is not evident in the analysis of this chapter since it is restricted to boundary conditions of 'unconditioned' random-cluster-type.

Sketch proof of Theorem 10.98. Note first that $p \geq 2q/(2q+1)$ if and only if $\pi = p/[p + q(1-p)]$ satisfies $\pi \geq \frac{2}{3}$. Under this condition we may obtain, by a branching-process argument, the ϕ_π-almost-sure existence in T of a (random) set W of vertices such that: (i) every 0-ray passes through some vertex of W, and (ii) every $w \in W$ is the root of an infinite open sub-tree of T. The argument then continues rather as in the proof of Theorem 5.33(b). The details may be found in [160, 196]. $\square$

10.12 On non-amenable graphs

The properties of interacting systems on trees are often quite different from those of lattice systems, for two reasons. Firstly, trees have a multiplicity of 'infinite ends', and secondly, the surface/volume ratios of boxes are bounded away from 0. The latter property is especially interesting and leads to an important categorization of graphs. Let $G = (V, E)$ be an infinite connected locally finite graph. We call G *amenable* if its 'isoperimetric constant'

$$\chi(G) = \inf\left\{\frac{|\partial W|}{|W|} : W \subseteq V,\ 0 < |W| < \infty\right\} \tag{10.99}$$

satisfies $\chi(G) = 0$. The graph is called *non-amenable* if $\chi(G) > 0$. It is easily seen that the lattices $\mathbb{L}^d$ and the regular m-ary tree T_m satisfy

$$\chi(\mathbb{L}^d) = 0, \qquad \chi(T_m) > 0 \text{ for } m \geq 2,$$

so that lattices are amenable, and regular trees of degree 3 or more are not.

It is convenient to make certain assumptions of homogeneity on the graph $G = (V, E)$. An automorphism[8] of G is a bijection $\gamma : V \to V$ such that $\langle x, y\rangle \in E$ if and only if $\langle \gamma x, \gamma y\rangle \in E$. A subgroup Γ of the automorphism group $\mathrm{Aut}(G)$ is said to *act transitively* on G if, for every pair $x, y \in V$, there exists $\gamma \in \Gamma$ such that $\gamma x = y$. We say that Γ acts *quasi-transitively* if V may be partitioned as the finite union $V = \bigcup_{i=1}^m V_i$ such that, for every $i = 1, 2, \dots, m$ and every pair $x, y \in V_i$, there exists $\gamma \in \Gamma$ such that $\gamma x = y$. The graph G is called *transitive* (respectively, *quasi-transitive*) if $\mathrm{Aut}(G)$ acts transitively (respectively, *quasi-transitively*). Results for transitive graphs are usually provable for quasi-transitive graphs also and, for simplicity, we shall usually assume G to be transitive.

For any graph G, the *stabilizer* $S(x)$ of the vertex x is defined to be the set of automorphisms of G that do not move x,

$$S(x) = \{\gamma \in \mathrm{Aut}(G) : \gamma x = x\}.$$

We write $S(x)y$ for the set of images of $y \in V$ under members of $S(x)$,

$$S(x)y = \{\gamma y : \gamma \in S(x)\},$$

and we call G *unimodular*[9] if $|S(x)y| = |S(y)x|$ whenever x and y belong to the same orbit of $\mathrm{Aut}(G)$.

---

[8]See Section 4.3 for the basic definitions associated with the automorphism group $\mathrm{Aut}(G)$.

[9]The terms 'amenable' and 'unimodular' come from group theory, see [265, 290, 312]. The assumption of unimodularity is equivalent to requiring that the left and right Haar measures on $\mathrm{Aut}(G)$ be the same.

There is a useful class of graphs arising from group theory. Let Γ be a finitely generated group and let S be a symmetric generating set. The associated (*right*) *Cayley graph* is the graph $G = (V, E)$ with $V = \Gamma$ and

$$E = \big\{\langle x, y\rangle : x, y \in \Gamma,\ xg = y \text{ for some } g \in S\big\}.$$

There are many Cayley graphs of interest to probabilists, including the lattices $\mathbb{L}^d$ and the trees T_m. All Cayley graphs are unimodular, see [241, Chapter 7]. One may take Cartesian products of Cayley graphs to obtain further graphs of interest, including the well-known example $\mathbb{L}^d \times T_m$, which has been studied in some depth in the context of percolation, [162].

The graph-property of (non-)amenability first became important in probability through the work of Kesten on random walks, [205, 206]. In [162] it was shown that percolation on the non-amenable graph $\mathbb{L}^d \times T_m$ possesses three phases. Pemantle [267] developed a related theory for the contact model on a tree, while Benjamini and Schramm [32] laid down further challenges for non-amenable graphs. There has been a healthy interest since in stochastic models on non-amenable graphs, and a systematic theory has developed. More recent references include [29, 30, 174, 176, 196, 197, 240, 241, 293].

Let $G = (V, E)$ be an infinite, connected, locally finite, transitive graph, and let $\Omega = \{0, 1\}^E$. As usual, for $F \subseteq E$, we write $\mathcal{F}_F$ for the σ-field generated by the states of edges in F, $\mathcal{T}_F = \mathcal{F}_{E\setminus F}$, and $\mathcal{F}$ for the σ-field generated by the finite-dimensional cylinders. The *tail σ-field* is $\mathcal{T} = \bigcap_F \mathcal{T}_F$ where the intersection is over all finite subsets F of E. A probability measure μ on $(\Omega, \mathcal{F})$ is called *tail-trivial* if $\mu(A) \in \{0, 1\}$ for all $A \in \mathcal{T}$.

The translations of $\mathbb{L}^d$ play a special role in considerations of mixing and ergodicity. For graphs G of the above type, this role is played by automorphism subgroups with infinite orbits. Let Γ be a subgroup of $\mathrm{Aut}(G)$. We say that Γ *has an infinite orbit* if there exists $x \in V$ such that the set $\{\gamma x : \gamma \in \Gamma\}$ has infinite cardinality. It is easy to see that a group Γ of automorphisms has an infinite orbit if and only if every orbit of Γ is infinite.

We turn now to random-cluster measures on the graph $G = (V, E)$. Let $p \in (0, 1)$, and assume for simplicity that $q \in [1, \infty)$. Let $\mathbf{\Lambda} = (\Lambda_n : n = 1, 2, \dots)$ be an increasing sequence of finite sets of vertices such that $\Lambda_n \uparrow V$ as $n \to \infty$. We concentrate as usual on two extremal random-cluster measures given very much as in Section 10.9, and we specify these informally as follows. Let Λ be a finite subset of V, and let $\phi_{\Lambda,p,q}$ be the random-cluster measure on Ω^0_Λ with parameters p, q, as in (4.11) with $\xi = 0$. By stochastic monotonicity, the limit

$$\phi^0_{p,q} = \lim_{n\to\infty} \phi_{\Lambda_n,p,q}$$

exists, and it is called the 'free' random-cluster measure on G. We note as before that the limit measure $\phi^0_{p,q}$ does not depend on the choice of $\mathbf{\Lambda}$, and that $\phi^0_{p,q}$ is automorphism-invariant.

In defining the wired measure, we veer towards the recipe of Section 10.9 rather than the lattice-theoretic (4.11). This amounts in rough terms to the following. Let Λ be a finite subset of V, and identify the set $\partial\Lambda$ as a single vertex. Write $\phi^1_{\Lambda,p,q}$ for the random-cluster measure with parameters p, q on this new graph, and view $\phi^1_{\Lambda,p,q}$ as a measure on the infinite measurable pair $(\Omega, \mathcal{F})$. As above, the limit

$$\phi^1_{p,q} = \lim_{n\to\infty} \phi^1_{\Lambda_n,p,q}$$

exists and does not depend on the choice of $\boldsymbol{\Lambda}$. We call $\phi^1_{p,q}$ the 'wired' random-cluster measure on G, and we note that $\phi^1_{p,q}$ is automorphism-invariant.

As pointed out in [240], the method of proof of Theorem 4.19(d) is valid for general graphs, and implies that the measures $\phi^b_{p,q}$ are tail-trivial. Let Γ be a subgroup of Aut(G) with an infinite orbit. By an adaptation of the proof of Theorem 4.19, the $\phi^b_{p,q}$ are Γ-ergodic. Indeed, the $\phi^b_{p,q}$ satisfy the following form of the mixing property. Since Γ has an infinite orbit, all its orbits are infinite. For $x \in V$ and y lying in the orbit of x under Γ, let $\gamma_{x,y} \in \Gamma$ be an automorphism mapping x to y. For $x \in V$ and $A, B \in \mathcal{F}$,

$$\lim_{\delta(x,y)\to\infty} \phi^b_{p,q}(A \cap \gamma_{x,y}B) = \phi^b_{p,q}(A)\phi^b_{p,q}(B), \qquad b = 0, 1, \tag{10.100}$$

in that, for $\epsilon > 0$, there exists N such that

$$\left|\phi^b_{p,q}(A \cap \gamma_{x,y}B) - \phi^b_{p,q}(A)\phi^b_{p,q}(B)\right| < \epsilon \qquad \text{if} \quad \delta(x,y) \ge N,$$

where $\delta(x, y)$ denotes the length of the shortest path from x to y.

The measures $\phi^b_{p,q}$ satisfy different 'one-point specifications', namely:

$$\begin{aligned} \phi^0_{p,q}(J_e \mid \mathcal{T}_e) &= \pi + (p-\pi)1_{K_e}, \qquad && \phi^0_{p,q}\text{-a.s.},\\ \phi^1_{p,q}(J_e \mid \mathcal{T}_e) &= \pi + (p-\pi)1_{K^1_e}, && \phi^1_{p,q}\text{-a.s.}, \end{aligned}$$

for $e = \langle x, y\rangle$. Here, as in (10.71), J_e is the event that e is open, $\mathcal{T}_e$ is the σ-field generated by states of edges other than e, and $\pi = p/[p+q(1-p)]$. In addition,

$$\begin{aligned} K_e &= \{x \leftrightarrow y \text{ off } e\},\\ K^1_e &= \{x \leftrightarrow y \text{ off } e\} \cup \{x \leftrightarrow \infty,\ y \leftrightarrow \infty\}. \end{aligned}$$

Many questions may be asked about the free and wired measures on a general graph G. We restrict ourselves here to the existence and number I of infinite open clusters. The critical points are defined by

$$p^b_c(q) = \sup\{p : \phi^b_{p,q}(I = 0) = 1\}, \qquad b = 0, 1.$$

By the tail-triviality of the $\phi^b_{p,q}$,

$$\phi^b_{p,q}(I = 0) = \begin{cases} 1 & \text{if } p < p^b_c(q), \\ 0 & \text{if } p > p^b_c(q). \end{cases}$$

We note the elementary inequality $p^1_c(q) \le p^0_c(q)$. It is an open question to decide when strict inequality holds here. As in (5.4), we have that $p^1_c(q) = p^0_c(q)$ for lattices, and the proof of this may be extended to all amenable graphs, [196]. On the other hand, by Theorem 10.82, $p^1_c(q) < p^0_c(q)$ for the regular binary tree T_2 when $q \in (2, \infty)$.

If there exists an infinite open cluster with positive probability, under what further conditions is this cluster almost-surely unique? The property of having a unique infinite cluster is not monotone in the configuration: there exist $\omega_1, \omega_2 \in \Omega$ such that $\omega_1 \le \omega_2$ and $I(\omega_1) = 1$, $I(\omega_2) \ge 2$. Nevertheless, it turns out that, for transitive unimodular graphs, the set of values of p for which $I = 1$ is indeed (almost surely) an interval.

The 'uniqueness critical point' is given by

$$p^b_u(q) = \inf\{p : \phi^b_{p,q}(I = 1) = 1\}, \qquad b = 0, 1.$$

and satisfies

$$p^b_c(q) \le p^b_u(q), \qquad b = 0, 1.$$

Since G is transitive, Aut(G) has an infinite orbit. The event $\{I = 1\}$ is Aut(G)-invariant whence, by the Aut(G)-ergodicity of the $\phi^b_{p,q}$,

$$\phi^b_{p,q}(I = 1) = 0, \qquad p < p^b_u.$$

(10.101) Theorem [240]. *Let G be an infinite connected locally finite graph that is transitive and unimodular, and let $b \in \{0, 1\}$. If $\phi^b_{p,q}(I = 1) = 1$ then $\phi^b_{p',q}(I = 1) = 1$ for $p' \ge p$. In particular,*

$$\phi^b_{p,q}(I = 1) = 1, \qquad p > p^b_u.$$

The proof is based upon the following proposition whose proof is omitted from the current work. A probability measure μ on $(\Omega, \mathcal{F})$ is called *insertion-tolerant* if, for all $e \in E$ and $A \in \mathcal{F}$,

$$\mu(A^e) > 0 \quad \text{whenever} \quad \mu(A) > 0,$$

where A^e is the set of configurations obtained from members of A by declaring e to be an open edge. Insertion-tolerance is a weak form of finite-energy, see (3.4). The symbol 0 denotes an arbitary vertex of G called its 'origin'.

(10.102) Proposition [242]. *Let G be an infinite connected locally finite graph that is transitive and unimodular, and let μ be an* $\mathrm{Aut}(G)$*-ergodic probability measure on $(\Omega, \mathcal{F})$ that is positively associated and insertion-tolerant. Then $\mu(I = 1) = 1$ if and only if*

$$\inf_{x \in V} \mu(0 \leftrightarrow x) > 0.$$

Theorem 10.101 is an immediate consequence, since the $\phi^b_{p,q}(0 \leftrightarrow x)$ are non-decreasing in p.

Suppose that G is unimodular. By Theorem 10.101 and a well known argument from [261], the free and wired random-cluster measures have (each) three phases: for $b = 0, 1$,

$$I = \begin{cases} 0 & \text{if } p < p^b_{\mathrm{c}}(q), \\ \infty & \text{if } p^b_{\mathrm{c}}(q) < p < p^b_{\mathrm{u}}(q), \\ 1 & \text{if } p > p^b_{\mathrm{u}}(q). \end{cases} \qquad \phi^b_{p,q}\text{-a.s.}$$

It is an open problem to obtain necessary and sufficient criteria for the strict inequalities

$$p^1_{\mathrm{c}}(q) < p^1_{\mathrm{u}}(q), \quad p^0_{\mathrm{c}}(q) < p^0_{\mathrm{u}}(q), \tag{10.103}$$

and the reader is referred to [174] for a discussion of this. The Burton–Keane argument, [72, 129], may be adapted to show that equalities hold in (10.103) when G is amenable. On the other hand, the inequalities may be strict, see [174, 240].

It is natural to ask for the value of I when p *equals* one of the critical values p^b_{c}, p^b_{u}. The picture is far from complete, and the reader is referred to [29, 30, 33, 167, 174] and Section 10.11 for the current state of knowledge.

Chapter 11

Graphical Methods for Spin Systems

Summary. Five applications are presented of the random-cluster model to lattice spin-systems, namely the Potts and Ashkin–Teller models, the disordered Potts ferromagnet, the Edwards–Anderson spin-glass model, and the Widom–Rowlinson lattice gas model.

11.1 Random-cluster representations

The interacting systems of lattice statistical mechanics are mostly 'vertex-models' in the sense that the configurations are spin-vectors indexed by the vertices. Such spins may take values in a general state-space, and the nature of the interaction between different vertices is specified within the Hamiltonian. A substantial technology has been developed for such systems. One of the techniques is to seek a transformation to an 'edge-model' that enables the use of geometric arguments in the study of correlations. The standard example of this is the mapping of Section 1.4 linking the Potts model and the random-cluster model. Such arguments are sometimes known as 'graphical methods', and some examples are summarized briefly in this chapter.

No attempt is made in this chapter to be encyclopaedic. Instead, we describe five cases of special interest, namely the Potts and Ashkin–Teller models for a ferromagnet, the disordered Potts model, the Edwards–Anderson model for a spin glass, and the Widom–Rowlinson model for a two-type lattice gas. There is a common theme to these examples. The first step in each case is to find a corresponding model of random-cluster type, with the property that the original spin system may be obtained by assigning spins to its clusters. It turns out that there exists a unique Gibbs state for the original spin system if and only if the new model has (almost surely) only finite clusters. The existence or not of an infinite cluster may be studied either directly, or by comparison with a known system such as a percolation model.

Accounts of the use of graphical methods for these and other classical models may be found, for example, in the work of Alexander [15], Chayes and Machta [93, 94], Graham and Grimmett [142], and in the reviews of Georgii, Häggström, Maes [136], and Häggström [169], as well as in the literature listed later in this chapter. The use of random-cluster methods in quantum spin systems is exemplified in [11, 12, 258].

11.2 The Potts model

The random-cluster model was introduced in part as a means to study the Potts model. No attempt is made here to compress the ensuing theory into a few pages. Instead, we state and prove one theorem concerning a random-cluster analysis of the (non-)uniqueness of Gibbs states for the Potts model.

The Potts model on a finite graph $G = (V, E)$ has an integer number $q \in \{2, 3, \dots\}$ of states and an 'inverse-temperature' $\beta \in (0, \infty)$. We shall consider the case of zero external-field, and we recall the notation of Section 1.3. We write $\Sigma = \{1, 2, \dots, q\}^V$ for the configuration space. For $e = \langle x, y \rangle \in E$ and $\sigma = (\sigma_x : x \in V) \in \Sigma$, let $\delta_e(\sigma)$ be defined by the Kronecker delta

$$\delta_e(\sigma) = \delta_{\sigma_x, \sigma_y} = \begin{cases} 1 & \text{if } \sigma_x = \sigma_y, \\ 0 & \text{otherwise.} \end{cases}$$

The Potts probability measure is defined as

$$\pi_{G,\beta,q}(\sigma) = \frac{1}{Z_{\mathrm{P}}} e^{-\beta H(\sigma)}, \qquad \sigma \in \Sigma,$$

where the Hamiltonian $H(\sigma)$ is given by

$$H(\sigma) = -\sum_{e=\langle x,y \rangle \in E} \delta_e(\sigma),$$

and $Z_{\mathrm{P}} = Z_{\mathrm{P}}(G, \beta, q)$ is the appropriate normalizing constant.

Consider now the lattice $\mathbb{L}^d$ with $d \ge 2$. The spin space is the set $\Sigma = \{1, 2, \dots, q\}^{\mathbb{Z}^d}$, and the appropriate σ-field $\mathcal{G}$ is that generated by the finite-dimensional cylinders of Σ. Let Λ be a finite box of $\mathbb{L}^d$, which we consider as a graph with edge-set $\mathbb{E}_\Lambda$. For $\tau \in \Sigma$, let Σ_Λ^τ be the subset of Σ containing all configurations that agree with τ off $\Lambda \setminus \partial\Lambda$. The Potts measure on Λ 'with boundary condition τ' is the probability measure on $(\Sigma, \mathcal{G})$ satisfying

$$\pi_{\Lambda,\beta,q}^\tau(\sigma) = \begin{cases} c_\Lambda^\tau \pi_{\Lambda,\beta,q}(\sigma_\Lambda) & \text{if } \sigma \in \Sigma_\Lambda^\tau, \\ 0 & \text{otherwise,} \end{cases} \tag{11.1}$$

where σ_Λ is the partial vector $(\sigma_x : x \in \Lambda)$ comprising spins in Λ, and c_Λ^τ is the normalizing constant. Of particular interest are the boundary conditions $\tau \equiv i$ for given $i \in \{1, 2, \dots, q\}$, in which case we write $\pi_{\Lambda,\beta,q}^i$. The symbol $\mathcal{U}_\Lambda$ will be used to denote the σ-field generated by the spins $(\sigma_y : y \notin \Lambda \setminus \partial\Lambda)$.

(11.2) Definition. A probability measure π on $(\Sigma, \mathcal{F})$ is called a *Gibbs state* of the q-state Potts model with inverse-temperature β if it satisfies the DLR condition:

$$\text{for all } A \in \mathcal{F} \text{ and boxes } \Lambda, \quad \pi(A \mid \mathcal{U}_\Lambda)(\tau) = \pi^{\tau}_{\Lambda,\beta,q}(A) \quad \text{for } \pi\text{-a.e. } \tau.$$

The principal question concerning Gibbs states is the following. For which values of the inverse-temperature β does there exist a unique (respectively, a multiplicity of) Gibbs states? It turns out that there is a unique Gibbs state if and only if the corresponding wired random-cluster model possesses (almost surely) no infinite cluster. Prior to the formal statement of this claim, which will be given in a form borrowed from [136, Thm 6.10], the reader is reminded of the weak limits[1]

$$\pi^0_{\beta,q} = \lim_{\Lambda \uparrow \mathbb{Z}^d} \pi_{\Lambda,\beta,q}, \quad \pi^1_{\beta,q} = \lim_{\Lambda \uparrow \mathbb{Z}^d} \pi^1_{\Lambda,\beta,q},$$

given in Theorem 4.91 and the remark immediately following. The measure $\pi^0_{\beta,q}$ is called the 'free' Potts measure.

(11.3) Theorem. *Let $\beta \in (0, \infty)$, $q \in \{2, 3, \dots\}$, and let $p = 1 - e^{-\beta}$.*

(a) *The measures $\pi^0_{\beta,q}$, $\pi^1_{\beta,q}$ are translation-invariant Gibbs states.*

(b) [8] *The following statements are equivalent:*

  (i) *there exists a unique Gibbs state,*

  (ii) $\pi^1_{\beta,q}(\sigma_0 = 1) = q^{-1}$,

  (iii) *the wired random-cluster measure $\phi^1_{p,q}$ satisfies $\phi^1_{p,q}(0 \leftrightarrow \infty) = 0$.*

We have highlighted the Potts measure $\pi^1_{\beta,q}$ with boundary condition 1. One may construct further measures $\pi^i_{\beta,q}$ with boundary condition $i \in \{2, 3, \dots, q\}$. Such measures differ from $\pi^1_{\beta,q}$ only through a re-labelling of the spin values $1, 2, \dots, q$.

The main claim of the theorem is that there exists a unique Gibbs state if and only if $\phi^1_{p,q}(0 \leftrightarrow \infty) = 0$. When $\phi^1_{p,q}(0 \leftrightarrow \infty) > 0$, there exists more than one Gibbs state, but how many? It is easily seen from the theorem that the measures $\pi^i_{\beta,q}$, $i \in \{1, 2, \dots, q\}$, are then distinct Gibbs states, but do there exist further states? The set of Gibbs states for given β, q is convex, and thus we are asking about the number of extremal Gibbs states. There are three situations to consider. The parameters p and β are related throughout by $p = 1 - e^{-\beta}$.

1. *Two dimensions* ($d = 2$). It is believed that the $\pi^i_{\beta,q}$ are the unique extremal Gibbs states whenever $p > p_c(q)$. At the point of a discontinuous phase transition (see Conjecture 6.32), the set of extremal Gibbs states is believed to be $\pi^i_{p,q}$ for $i \in \{0, 1, 2, \dots, q\}$.

[1] There is a technical detail here in that $\pi_{\Lambda,\beta,q}$ is defined on Λ rather than on $\mathbb{Z}^d$, but we overlook this.

2. *Supercritical phase* ($d \geq 3$). Suppose $p > p_c(q)$. It is believed that the $\pi^i_{\beta,q}$ are the unique extremal translation-invariant Gibbs states. On the other hand, there exist non-translation-invariant Gibbs states (see Theorem 7.89) when β (and hence p) is sufficiently large.
3. *Critical case* ($d \geq 3$). Let $p = p_c(q)$. By Theorem 11.3, there exists a unique Gibbs state if the phase transition is continuous in the sense that $\phi^1_{p_c,q}(0 \leftrightarrow \infty) = 0$. When q is sufficiently large, the transition is discontinuous and there exist exactly $q+1$ translation-invariant extremal Gibbs states $\pi^i_{\beta,q}$, $i \in \{0, 1, 2, \dots, q\}$, [251]. There is in addition an infinity of non-translation-invariant extremal Gibbs states, [85, 254].

To each vertex of a q-state Potts model is allocated one of the states $1, 2, \dots, q$. The so-called 'Potts lattice gas' has an augmented state space $0, 1, 2, \dots, q$, where the vertices labelled 0 are considered as 'empty'. The Potts lattice gas may be studied via the so-called 'asymmetric random cluster model', see [15]. A similar augmentation of the state space was introduced for the Ising model by Blume and Capel in a study of first-order phase transitions, [50, 79]. This gives rise to a 'Blume–Capel–Potts model' which may be studied via a random-cluster representation, see [142].

Proof of Theorem 11.3. (a) The existence of the measures is proved in Theorem 4.91 and the comments immediately thereafter. Their translation-invariance follows from the translation-invariance of $\phi^b_{p,q}$ for $b = 0, 1$, see Theorem 4.19(b), on following the recipes of Section 4.6.

We prove next that $\pi^1_{\beta,q}$ is a Gibbs state, and the same proof is valid for $\pi^0_{\beta,q}$. For boxes Λ, Δ satisfying $\Lambda \subseteq \Delta$, let $\mathcal{V}_{\Delta\setminus\Lambda}$ denote the σ-field generated by the states of vertices in $\Delta \setminus (\Lambda \setminus \partial\Lambda)$. Let $A \in \mathcal{G}$. By the martingale convergence theorem (see [164, Ex. 12.3.9]),

$$\pi^1_{\beta,q}(A \mid \mathcal{U}_\Lambda) = \lim_{\Delta \uparrow \mathbb{Z}^d} \pi^1_{\beta,q}(A \mid \mathcal{V}_{\Delta\setminus\Lambda}), \qquad \pi^1_{\beta,q}\text{-a.s.}$$

By weak convergence, Theorem 4.91,

$$\pi^1_{\beta,q}(A \mid \mathcal{V}_{\Delta\setminus\Lambda}) = \lim_{\Delta' \uparrow \mathbb{Z}^d} \pi^1_{\Delta',\beta,q}(A \mid \mathcal{V}_{\Delta\setminus\Lambda}),$$

and it is a simple calculation based on the definition of the finite-volume Potts measures that

$$\pi^1_{\Delta',\beta,q}(A \mid \mathcal{V}_{\Delta\setminus\Lambda})(\tau) = \pi^\tau_{\Lambda,\beta,q}(A).$$

Combining the last three equations, we find as required that

$$\pi^1_{\beta,q}(A \mid \mathcal{U}_\Lambda)(\tau) = \pi^\tau_{\Lambda,\beta,q}(A), \qquad \pi^1_{\beta,q}\text{-a.s.}$$

(b) We prove first that (i) implies (ii). Assume that (i) holds, so that, in particular, $\pi^1_{\beta,q} = \pi^i_{\beta,q}$ for $i = 2, 3, \dots, q$. Then,

$$\begin{aligned}\pi^1_{\beta,q}(\sigma_0 = 1) &= \pi^j_{\beta,q}(\sigma_0 = j)\\ &= \pi^1_{\beta,q}(\sigma_0 = j), \qquad j = 1, 2, \dots, q.\end{aligned}$$

However,

$$\sum_{j=1}^{q} \pi_{\beta,q}^{1}(\sigma_0 = j) = 1,$$

and (ii) follows.

By Theorem 1.16 applied to Λ with the wired boundary condition,

$$\pi_{\Lambda,\beta,q}^{1}(\sigma_0 = 1) - \frac{1}{q} = (1 - q^{-1})\phi_{\Lambda,p,q}^{1}(0 \leftrightarrow \infty).$$

Let $\Lambda \uparrow \mathbb{Z}^d$ and deduce by Theorems 4.91 and 5.11 that

$$\pi_{\beta,q}^{1}(\sigma_0 = 1) - \frac{1}{q} = (1 - q^{-1})\phi_{p,q}^{1}(0 \leftrightarrow \infty).$$

Therefore, (ii) and (iii) are equivalent.

Finally, we prove that (iii) implies (i). Assume (iii), and let π be a Gibbs state for the Potts model with parameters β, q. Let $A \in \mathcal{G}$ be a cylinder event, and let $\pi_{\beta,q}^{0}$ denote the Potts measure on $\mathbb{L}^d$ with the free boundary condition. We claim that

$$\pi(A) = \pi_{\beta,q}^{0}(A), \tag{11.4}$$

which implies (i) since the cylinder events generate $\mathcal{G}$. Let $\epsilon > 0$. We shall prove that

$$\left|\pi_{\Lambda,\beta,q}^{\tau}(A) - \pi_{\beta,q}^{0}(A)\right| < \epsilon, \qquad \text{for some box } \Lambda \text{ and all } \tau \in \Sigma. \tag{11.5}$$

Equation (11.4) follows by (π-)averaging over τ and appealing to Definition 11.2.

We concentrate for the moment on the measure $\pi_{\Lambda,\beta,q}^{\tau}$. We may couple this measure with a certain random-cluster-type measure in the same manner as described in Section 1.4 for the free measures. For $\omega \in \Omega_\Lambda = \{0,1\}^{\mathbb{E}_\Lambda}$, let $k^\tau(\omega)$ be the number of open clusters in the graph obtained from $(\Lambda, \mathbb{E}_\Lambda)$ by identifying each of the sets $V_i = \{x \in \partial\Lambda : \tau_x = i\}$, $i = 1, 2, \dots, q$, as a single vertex. Let $\overline{\phi}$ be the random-cluster measure on Ω_Λ with the usual cluster-count $k(\omega)$ replaced by $k^\tau(\omega)$. Finally, let $\phi_{\Lambda,p,q}^{\tau}$ denote[2] $\overline{\phi}$ conditioned on the event

$$D^\tau = \bigl\{\omega \in \Omega : V_i \not\leftrightarrow V_j \text{ in } \Lambda, \text{ for all distinct } i, j \in \{1, 2, \dots, q\}\bigr\}. \tag{11.6}$$

It is left as an exercise to prove that $\pi_{\Lambda,\beta,q}^{\tau}$ is the law of the spin-vector obtained as follows. If $x \notin \Lambda \setminus \partial\Lambda$, assign spin τ_x to x. For vertices in $\Lambda \setminus \partial\Lambda$, first sample $\omega \in \Omega_\Lambda$ according to $\phi_{\Lambda,p,q}^{\tau}$, and then assign uniformly distributed random spins

[2] Since $k^\tau(\omega)$ differs from $k^1(\omega)$ by a constant (depending on τ), we could take $\phi_{\Lambda,p,q}^{\tau}$ to be the wired measure $\phi_{\Lambda,p,q}^{1}$ conditioned on D^τ.

to each open cluster of ω subject to the constraint: if $x \leftrightarrow V_i$, then x is assigned spin i.

By positive association,

$$\phi^{\tau}_{\Lambda,p,q} \leq_{\text{st}} \phi^{1}_{\Lambda,p,q} \tag{11.7}$$

where the latter measure is to be interpreted as its projection onto $\{0,1\}^{\mathbb{E}_\Lambda}$.

We return to the cylinder event A, and we let $\epsilon > 0$. By the remark after Theorem 4.91, $\pi^0_{\Lambda,\beta,q} \Rightarrow \pi^0_{\beta,q}$ as $\Lambda \uparrow \mathbb{Z}^d$, and thus we may find a box B such that

$$\left|\pi^0_{\gamma,\beta,q}(A) - \pi^0_{\beta,q}(A)\right| < \epsilon \qquad \text{for all } \gamma \supseteq B. \tag{11.8}$$

Let Δ' be a box sufficiently large that: $B \subseteq \Delta'$, and A is defined in terms of the vertex-spins within Δ'. By (iii), we may choose a box Θ such that $\Delta' \subseteq \Theta$ and

$$\phi^1_{p,q}(\Delta' \leftrightarrow \partial\Theta) < \epsilon. \tag{11.9}$$

Since $\phi^1_{\Lambda,p,q} \Rightarrow \phi^1_{p,q}$ as $\Lambda \uparrow \mathbb{Z}^d$, we may find a box Λ such that $\Theta \subseteq \Lambda$ and

$$\phi^1_{\Lambda,p,q}(\Delta' \leftrightarrow \partial\Theta) < 2\epsilon. \tag{11.10}$$

Let $\tau \in \Sigma$. By (11.7) and (11.10),

$$\phi^{\tau}_{\Lambda,p,q}(\Delta' \leftrightarrow \partial\Theta) < 2\epsilon, \qquad \tau \in \Sigma. \tag{11.11}$$

On the event $\{\omega \in \Omega : \Delta' \not\leftrightarrow \partial\Theta\}$, there exists a connected subgraph Γ of $\Theta \setminus \partial\Theta$, containing Δ' and with closed external edge-boundary $\Delta_{\mathrm{e}}\Theta$. Let Γ be the maximal graph with this property, and let $\mathcal{H}$ be the set of all possible outcomes of Γ. For $\gamma \in \mathcal{H}$, the event $\{\Gamma = \gamma\}$ is measurable on the σ-field generated by the states of edges not belonging to γ. (There is a similar step in the proof of Proposition 5.30.) The marginal measure on γ of $\phi^{\tau}_{\Lambda,p,q}(\cdot \mid \Gamma = \gamma)$ is therefore the free measure $\phi^0_{\gamma,p,q}$ and hence, by coupling,

$$\left|\pi^{\tau}_{\Lambda,\beta,q}(A) - \sum_{\gamma\in\mathcal{H}} \pi^0_{\gamma,\beta,q}(A)\phi^{\tau}_{\Lambda,p,q}(\Gamma = \gamma)\right| \leq \phi^{\tau}_{\Lambda,p,q}(\Delta' \leftrightarrow \partial\Delta).$$

By (11.8) and (11.11),

$$\left|\pi^{\tau}_{\Lambda,\beta,q}(A) - \pi^0_{\beta,q}(A)\right| < 5\epsilon,$$

whence (11.5) holds with an adjusted value of ϵ, and (i) is proved. □

11.3 The Ashkin–Teller model

Each vertex may be in either of two states of the Ising model. The Potts model was proposed in 1952, and allows a general number q of local states. Nearly ten years earlier, Ashkin and Teller [21] proposed a 4-state model which, with hindsight, may be viewed as an interpolation between the Ising model and the 4-state Potts model. Their model amounts to the following.

Let $G = (V, E)$ be a finite graph. The set of local spin-values is taken to be $\{A, B, C, D\}$, so that the configuration space is $\Sigma = \{A, B, C, D\}^V$. Let J_1, J_2 be edge-interactions satisfying $0 \le J_1 \le J_2$, and let $\beta \in (0, \infty)$. The spins at the endvertices x and y of the edge $e = \langle x, y \rangle$ interact according to a function δ given as follows:

$$\delta(A, B) = \delta(C, D) = J_1,$$
$$\delta(A, C) = \delta(A, D) = \delta(B, C) = \delta(B, D) = J_2.$$

There is symmetry within the pair $\{A, B\}$ and within the pair $\{C, D\}$, but asymmetry between the pairs. The Ashkin–Teller measure on G is the probability measure given by

$$\alpha_{G,\beta}(\sigma) = \frac{1}{Z_{\mathrm{AT}}} e^{-\beta H(\sigma)}, \qquad \sigma \in \Sigma,$$

where Z_{AT} is the appropriate normalizing constant and

$$H(\sigma) = \sum_{\substack{e=\langle x,y \rangle: \\ \sigma_x \neq \sigma_y}} \delta(\sigma_x, \sigma_y), \qquad \sigma \in \Sigma.$$

Neighbouring pairs prefer to have the same spin, failing which they prefer to have spins in one of the sets $\{A, B\}$, $\{C, D\}$, and failing that either of the spins in the other pair. When $J_1 = 0$, the Ashkin–Teller model is equivalent to the Ising model. When $J_1 = J_2$, it is equivalent to the 4-state Potts model.

Consider the lattice $\mathbb{L}^d$ with $d \ge 2$. The spin space is $\Sigma = \{A, B, C, D\}^{\mathbb{Z}^d}$, and $\mathcal{G}$ denotes the σ-field of Σ generated by the cylinder events. In order to define Ashkin–Teller measures on the infinite lattice, we follow the standard recipe outlined in the last section around Definition 11.2. For $\tau \in \Sigma$ and a box Λ, one may define an Ashkin–Teller measure $\alpha^\tau_{\Lambda,\beta}$ on Λ with boundary condition τ. A probability measure α on $(\Sigma, \mathcal{G})$ is called a Gibbs state of the Ashkin–Teller model with inverse-temperature β if, for any box Λ, the conditional measure on Λ, given the configuration τ off $\Lambda \setminus \partial\Lambda$, is $\alpha^\tau_{\Lambda,\beta}$.

For what values of β does there exist a unique Gibbs state?

(11.12) Theorem [271]. *Consider the Ashkin–Teller model on $\mathbb{L}^d$ with $d \geq 2$ and $0 < J_1 < J_2$. There exist β_1, β_2 satisfying $0 < \beta_1 \leq \beta_2 < \infty$ such that the following hold.*

(a) *There is a unique Gibbs state if $\beta < \beta_1$.*

(b) *If $\beta \in (\beta_1, \beta_2)$, there is a multiplicity of Gibbs states each of which is invariant under the re-labellings $A \leftrightarrow B$ and $C \leftrightarrow D$.*

(c) *If $\beta > \beta_2$ then, for each $s \in \{A, B, C, D\}$, there exists a Gibbs state in which the local state s dominates. That is, for each s there exists a Gibbs state α such that*

$$\alpha(\sigma_x = s) > \tfrac{1}{4}, \quad \alpha(\sigma_x = t) < \tfrac{1}{4}, \qquad x \in \mathbb{Z}^d,\ t \in \{A, B, C, D\} \setminus \{s\}.$$

Furthermore, $\beta_1 < \beta_2$ if J_2/J_1 is sufficiently large.

It is an open question to decide whether $\beta_1 < \beta_2$ whenever $J_1 < J_2$. Perhaps the answer depends on the choice of lattice.

Theorem 11.12 may be found in [271], and it is proved here via a random-cluster representation following the treatment in [169]. Further results for the Ashkin–Teller model and its random-cluster representation may be found in [93, 273, 289, 321].

The relevant graphical method makes uses the following edge-model. Let $G = (V, E)$ be a finite graph as before, and take as configuration space the set $\Omega = \{0, 1, 2\}^E$. For $\omega \in \Omega$ and $i \in \{0, 1, 2\}$, we write $\eta_i(\omega)$ for the set of edges e with $\omega(e) = i$. Let $\mathbf{p} = (p_0, p_1, p_2)$ be a vector of non-negative reals with sum 1. The *Ashkin–Teller random-cluster measure* on G is the probability measure $\phi_{G,\mathbf{p}}$ on Ω given by

$$\phi_{G,\mathbf{p}}(\omega) = \frac{1}{Z_{\mathrm{ATRC}}} p_0^{|\eta_0|} p_1^{|\eta_1|} p_2^{|\eta_2|} 2^{k(\eta_1 \cup \eta_2) + k(\eta_2)}, \qquad \omega \in \Omega,$$

where $\eta_i = \eta_i(\omega)$ and Z_{ATRC} is the appropriate normalizing constant.

Suppose that $\beta \in (0, \infty)$, $0 < J_1 \leq J_2$, and let $\mathbf{p} = (p_1, p_1, p_2)$ satisfy

$$(11.13) \qquad p_0 = e^{-\beta J_2}, \quad p_1 = e^{-\beta J_1} - e^{-\beta J_2}, \quad p_2 = 1 - e^{-\beta J_1}.$$

We describe next how to couple $\alpha_{G,\beta}$ and $\phi_{G,\mathbf{p}}$. Let ω have law $\phi_{G,\mathbf{p}}$. For each cluster C_{12} of the graph $(V, \eta_1(\omega) \cup \eta_2(\omega))$, we flip a fair coin to determine whether the spins in C_{12} are drawn from the pair $\{A, B\}$ or from the pair $\{C, D\}$. Having done this for each C_{12}, we consider the clusters of the graph $(V, \eta_2(\omega))$. For each such cluster C_2, we flip a fair coin to determine which of the two possibilities will be allocated to the vertices of C_2. Thus, for example, if $C_2 \subseteq C_{12}$ and vertices in C_{12} are to receive spins from the pair $\{A, B\}$, then either every vertex in C_2 receives spin A, or every vertex receives spin B, each such possibility having (conditional) probability $\frac{1}{2}$. This recipe results in a random spin-vector $\sigma \in \{A, B, C, D\}^V$, and it is left as an exercise to check that σ has law $\alpha_{G,\beta}$.

The key question in deciding the multiplicity of Gibbs states is whether a weak limit of the $\phi_{\Lambda,\mathbf{p}}$ may possess an infinite cluster of edges each of which has either state 1 or state 2 (respectively, each of which has state 2). We begin the proof of Theorem 11.12 with a lemma. The configuration space $\Omega = \{0, 1, 2\}^E$ may be viewed as a partially ordered set. For probability measures μ_1, μ_2 on Ω, we write $\mu_1 \le_{\mathrm{st}} \mu_2$ if $\mu_1(f) \le \mu_2(f)$ for all non-decreasing functions $f : \Omega \to \mathbb{R}$. See Section 2.1.

(11.14) Lemma. *Suppose* $0 < J_1 \le J_2$. *Let* $\beta \in (0,\infty)$, *and let* $\mathbf{p} = \mathbf{p}(\beta)$ *satisfy* (11.13). *The probability measures* $\Phi_\beta = \phi_{G,\mathbf{p}(\beta)}$ *satisfy*

$$\Phi_{\beta_1} \le_{\mathrm{st}} \Phi_{\beta_2}, \qquad 0 < \beta \le \beta_2 < \infty. \tag{11.15}$$

Proof. Each Φ_β is a probability measure on the partially ordered set Ω. By Theorems 2.1 and 2.6[3], inequality (11.15) holds if, for $v = 1, 2$ and every $e \in E$,

$$\pi_{\beta,e}(v,\xi) = \Phi_\beta\bigl(\omega(e) \ge v \,\big|\, \omega(f) = \xi(f) \text{ for all } f \in E \setminus \{e\}\bigr)$$

is increasing (that is, non-decreasing) in $\beta \in (0,\infty)$ and $\xi \in \Omega$.

For $e \in E$ and $\xi \in \Omega$, let $\kappa_2(e,\xi)$ (respectively, $\kappa_{12}(e,\xi)$) be the number of clusters of the graph $(V, \eta_2(\xi) \setminus \{e\})$ (respectively, $(V, \eta_1(\xi) \cup \eta_2(\xi) \setminus \{e\})$) that intersect the endvertices of e. It is an easy calculation that

$$\pi_{\beta,e}(v,\xi) = \begin{cases} \dfrac{p_2}{\gamma_0 p_0 + \gamma_1 p_1 + p_2} & \text{if } v = 2, \\[2ex] 1 - \dfrac{\gamma_0 p_0}{\gamma_0 p_0 + \gamma_1 p_1 + p_2} & \text{if } v = 1, \end{cases} \tag{11.16}$$

where

$$\gamma_0 = 2^{\kappa_{12}(e,\xi)+\kappa_2(e,\xi)-2}, \qquad \gamma_1 = 2^{\kappa_{12}(e,\xi)-1}. \tag{11.17}$$

Note that

$$\gamma_0 \ge \gamma_1 \ge 1, \tag{11.18}$$

and, in addition, γ_0, γ_1, γ_0/γ_1, and $\gamma_0 - \gamma_1$ are decreasing functions of ξ.

Now,

$$\frac{\gamma_0 p_0 + \gamma_1 p_1 + p_2}{p_2} = 1 + (\gamma_0 - \gamma_1)\frac{p_0}{p_2} + \gamma_1\left(\frac{1}{p_2} - 1\right), \tag{11.19}$$

$$\frac{\gamma_0 p_0 + \gamma_1 p_1 + p_2}{\gamma_0 p_0} = 1 + \frac{\gamma_1}{\gamma_0}\cdot\frac{p_1}{p_0} + \frac{1}{\gamma_0}\cdot\frac{p_2}{p_0}. \tag{11.20}$$

[3] These were proved for the case $\Omega = \{0, 1\}^E$, but similar results are valid in the more general setting when $\Omega = T^E$ and T is a finite subset of $\mathbb{R}$. See, for example, [136, Section 4].

It is easily checked from (11.13) that p_0, p_0/p_1, and $1/p_2$ are decreasing in β. By (11.18), (11.19) is decreasing in β. By the remark after (11.18), (11.19) is decreasing in ξ, and therefore $\pi_{\beta,e}(2,\xi)$ is increasing in β and ξ as required.

Similarly, (11.20) is increasing in β and ξ, and therefore so is $\pi_{\beta,e}(1,\xi)$ in (11.16). We conclude that each $\pi_{\beta,e}(v,\xi)$ is increasing in β and ξ, and (11.15) follows. □

Sketch proof of Theorem 11.12. We follow [169]. For $\omega \in \{0,1,2\}^E$, a *cluster of type* 1/2 (respectively, type 2) is a cluster formed by the edges e with $\omega(e) \in \{1,2\}$ (respectively, $\omega(e) = 2$). As in the Potts case of the previous section, there is a unique Gibbs state if and only if every weak limit, as $\Lambda \uparrow \mathbb{Z}^d$, of $\phi_{\Lambda,\mathbf{p}}$ possesses (almost surely) no infinite cluster of type 1/2. By Lemma 11.14, the last statement about the $\phi_{\Lambda,\mathbf{p}}$ is a decreasing property of β: if it holds when $\beta = \beta'$ then it holds for $\beta \le \beta'$. Therefore, there exists a critical value β_1 such that there exists a unique (respectively, multiplicity) of Gibbs states when $\beta < \beta_1$ (respectively, $\beta > \beta_1$).

By (11.16)–(11.17),

$$\pi_{\beta,e}(1,\xi) \le 2p_1 + p_2 = p^*(\beta), \qquad \xi \in \Omega,\ e \in E,$$

where

$$p^*(\beta) = 2(e^{-\beta J_1} - e^{-\beta J_2}) + 1 - e^{-\beta J_1}.$$

By Theorems 2.1 and 2.6, the law of the set of edges of type 1/2 in G is dominated by a product measure with density $p^*(\beta)$. When $p^*(\beta) < p_c^{\text{bond}}(\mathbb{L}^d)$, no weak limit of $\phi_{\Lambda,\mathbf{p}}$ may possess an infinite cluster of type 1/2. Here, $p_c^{\text{bond}}(\mathbb{L}^d)$ denotes the critical probability of bond percolation on $\mathbb{L}^d$. We deduce that $\beta_1 > 0$.

The same argument may be applied to the existence (or not) of an infinite cluster of type 2. Once again, there exists a critical value β_2 marking the onset of the existence of such a cluster, and it is elementary that $\beta_1 \le \beta_2$. By (11.16)–(11.17),

$$\pi_{\beta,e}(2,\xi) \ge \tfrac{1}{4}p_2, \qquad \xi \in \Omega,\ e \in E,$$

implying as above that, when β is large, every weak limit of $\phi_{\Lambda,\mathbf{p}}$ possesses (almost surely) an infinite cluster of type 2. Therefore, $\beta_2 < \infty$. Statement (c) is easily seen to follow and, in addition, statement (b) when $\beta_1 < \beta_2$.

By (11.16),

$$\pi_{\beta,e}(1,\xi) \ge 1 - 4p_0 = 1 - 4e^{-\beta J_2},$$
$$\pi_{\beta,e}(2,\xi) \le p_2 = 1 - e^{-\beta J_1}.$$

Suppose there exists a non-empty interval I of values of β such that

$$1 - e^{-\beta J_1} < p_c^{\text{bond}}(\mathbb{L}^d) < 1 - 4e^{-\beta J_2}. \tag{11.21}$$

If $\beta \in I$, the edges of type 1/2 dominate a supercritical product measure, and those of type 2 are dominated by a subcritical product measure. Therefore, $\beta_1 \le \beta \le \beta_2$, and hence I is a sub-interval of $[\beta_1, \beta_2]$, implying that $\beta_1 < \beta_2$. We may indeed find such an interval I if J_2/J_1 is sufficiently large. □

11.4 The disordered Potts ferromagnet

All our models have been assumed so far to be homogeneous in the sense that their edge-parameters have been assumed equal. In a 'disordered' system, one begins instead with a general family of edge-parameters indexed by the edge-set E. It is potentially a major complication that the ensuing measures may not be automorphism-invariant, and one may not apply techniques such as the ergodic theorem. A degree of statistical homogeneity may be re-introduced by assuming that the edge-parameters are chosen according to some given translation-invariant random field. We restrict ourselves for simplicity here to the situation in which this random field is a product measure with a given marginal distribution.

The disordered Potts model on a finite graph $G = (V, E)$ is given as follows. One begins with a family $\mathbf{J} = (J_e : e \in E)$ of 'random interactions'[4]. These are independent, identically distributed random variables taking values in the half-open interval $[0, \infty)$ according to a given law ν. Let $\beta \in (0, \infty)$ and $q \in \{2, 3, \dots\}$. The corresponding Potts (random) measure on the configuration space $\Sigma = \{1, 2, \dots, q\}^V$ is

$$\pi_{\mathbf{J},q}(\sigma) = \frac{1}{Z_{\mathbf{J}}} e^{-\beta H(\sigma)}, \qquad \sigma \in \Sigma, \tag{11.22}$$

where $Z_{\mathbf{J}}$ is the appropriate (random) normalizing constant and

$$H(\sigma) = -\sum_{e=\langle x,y\rangle \in E} J_e \delta_e(\sigma), \qquad \delta_e(\sigma) = \delta_{\sigma_x,\sigma_y}.$$

Such a model is ferromagnetic in that the J_e are *non-negative* random variables. The non-ferromagnetic case is much harder, and the reader is referred to Section 11.5 for some partial results of random-cluster type.

The 'disordered random-cluster model' is defined similarly on $G = (V, E)$. Let $q \in (0, \infty)$, and let $\mathbf{p} = (p_e : e \in E)$ be a family of independent, identically distributed random variables chosen from the interval $[0, 1]$. The corresponding random-cluster (random) measure $\phi_{\mathbf{p},q}$ on $\Omega = \{0, 1\}^E$ is given as usual by

$$\phi_{\mathbf{p},q}(\omega) = \frac{1}{Z_{\mathbf{p}}} \left\{ \prod_{e \in E} p_e^{\omega(e)} (1 - p_e)^{1-\omega(e)} \right\} q^{k(\omega)}, \qquad \omega \in \Omega, \tag{11.23}$$

where $Z_{\mathbf{p}}$ is the appropriate (random) normalizing constant.

With q, β, and the J_e as above, let

$$p_e = 1 - e^{-\beta J_e}, \qquad e \in E. \tag{11.24}$$

The measures $\phi_{\mathbf{p},q}$ and $\pi_{\mathbf{J},q}$ may be coupled as in Section 1.4. As in Theorem 1.16,

$$\pi_{\mathbf{J},q}(\sigma_x = \sigma_y) - \frac{1}{q} = (1 - q^{-1}) \phi_{\mathbf{p},q}(x \leftrightarrow y), \qquad x, y \in V. \tag{11.25}$$

[4]Disordered systems were introduced in [143], and early papers include [132, 133].

Consider the lattice $\mathbb{L}^d$ with $d \geq 2$. In developing the theory of disordered random-cluster measures on $\mathbb{L}^d$, one needs to take care to avoid the use of spatial homogeneity. It turns out that quite a lot of the theory of Chapters 1–4 remains valid in this setting, including the comparison inequalities. When working on a finite box Λ of the lattice $\mathbb{L}^d$ with $q \in [1, \infty)$, one may therefore pass to the infinite-volume limit as $\Lambda \uparrow \mathbb{Z}^d$, as in Section 4.3. Without more ado, we shall use the notation introduced earlier, including that of the infinite-volume random-cluster measures $\phi^0_{\mathbf{p},q}$, $\phi^1_{\mathbf{p},q}$.

The disordered Potts model has a random set of Gibbs states, and we seek a condition under which this set comprises (almost surely) a singleton only. As in the previous sections, for given $\beta\mathbf{J}$, there is a unique Gibbs state if and only if the corresponding wired random-cluster measure possesses no infinite cluster.

Let $I = \{\omega \in \Omega : \omega$ possesses an infinite cluster$\}$ and consider the probability $\phi^1_{\mathbf{p},q}(I)$, viewed as a function of β. By the comparison inequalities, $\phi^1_{\mathbf{p},q}(I)$ is non-decreasing in β, and we define the critical point $\beta_c(\mathbf{J})$ by

$$\beta_c(\mathbf{J}) = \sup\{\beta > 0 : \phi^1_{\mathbf{p},q}(I) = 0\},$$

noting that $\beta_c(\mathbf{J})$ is a random variable. The random variable $\phi^1_{\mathbf{p},q}(I)$ is invariant under lattice-translations, and the invariant σ-field of the p_e is trivial, whence $\phi^1_{\mathbf{p},q}(I) \in \{0, 1\}$. Therefore, there exists a constant $\beta_c \in [0, \infty]$ such that

$$\mathbb{P}(\beta_c(\mathbf{J}) = \beta_c) = 1, \qquad \phi^1_{\mathbf{p},q}(I) = \begin{cases} 0 & \text{if } \beta < \beta_c, \\ 1 & \text{if } \beta > \beta_c, \end{cases}$$

where $\mathbb{P}$ denotes the product measure with marginals ν on the space $[0, \infty)^{\mathbb{E}^d}$.

A pivotal role is played by the atom of ν at 0,

$$\nu(0) = \mathbb{P}(J_e = 0).$$

By (11.24), $\mathbb{P}(J_e = 0) = \mathbb{P}(p_e = 0)$. By the comparison inequality (3.22), $\phi^1_{\mathbf{p},q}(I) = 0$, ($\mathbb{P}$-almost-surely), if $1 - \nu(0) < p_c^{\mathrm{bond}}(\mathbb{L}^d)$. Therefore,

$$\text{(11.26)} \qquad \beta_c = \infty \quad \text{if} \quad \nu(0) > 1 - p_c^{\mathrm{bond}}(\mathbb{L}^d).$$

The situation is more interesting when $\nu(0) < 1 - p_c^{\mathrm{bond}}(\mathbb{L}^d)$.

(11.27) Theorem [7]. *Let $d \geq 2$, and consider the disordered Potts model on $\mathbb{L}^d$ with edge-interaction law ν.*

(a) *If $\nu(0) > 1 - p_c^{\mathrm{bond}}(\mathbb{L}^d)$, then $\beta_c = \infty$.*

(b) *If $\nu(0) < 1 - p_c^{\mathrm{bond}}(\mathbb{L}^d)$, there exists $\beta_c = \beta_c(\nu) \in (0, \infty)$ such that: there exists ($\mathbb{P}$-almost-surely) a unique Gibbs state if $\beta < \beta_c$, and ($\mathbb{P}$-almost-surely) a multiplicity of Gibbs states if $\beta > \beta_c$.*

The literature on disordered Potts models is substantial, see for example [7, 155] and the bibliographies of [136, Section 9], [169, 259, 260]. Lower and upper bounds for β_c may be found at (11.28)–(11.29).

Proof. Part (a) was proved at (11.26). Suppose that $\nu(0) < 1 - p_c^{\text{bond}}(\mathbb{L}^d)$. By the earlier remarks, it suffices to prove that $0 < \beta_c < \infty$. By the comparison inequality (3.22),

$$\phi^1_{\mathbf{p},q}(I) \le \phi_{\mathbf{p}}(I)$$

where $\phi_{\mathbf{p}}$ is the product measure on Ω in which edge e is open with probability p_e. Therefore,

$$\mathbb{P}[\phi^1_{\mathbf{p},q}(I)] \le \mathbb{P}[\phi_{\mathbf{p}}(I)] = \phi_{\mathbb{P}(\mathbf{p})}(I),$$

since the average of a product measure is a product measure. Now,

$$\mathbb{P}(p_e) = \mathbb{P}(1 - e^{-\beta J_e}) \to 0 \qquad \text{as } \beta \downarrow 0,$$

by monotone convergence. Therefore,

$$\beta_c \ge \sup\{\beta > 0 : \mathbb{P}(1 - e^{-\beta J_e}) < p_c^{\text{bond}}(\mathbb{L}^d)\} > 0. \tag{11.28}$$

We turn to the upper bound for β_c. By the other comparison inequality (3.23),

$$\phi^1_{\mathbf{p},q}(I) \ge \phi_{\mathbf{p}'}(I)$$

where $\phi_{\mathbf{p}'}$ is the product measure on Ω in which edge e is open with probability

$$p'_e = \frac{p_e}{p_e + q(1 - p_e)} = \frac{1 - e^{-\beta J_e}}{1 + (q-1)e^{-\beta J_e}}.$$

By monotone convergence,

$$\mathbb{P}(p'_e) \to 1 - \nu(0) \qquad \text{as } \beta \to \infty,$$

and $1 - \nu(0) > p_c^{\text{bond}}(\mathbb{L}^d)$, by assumption. Arguing as above,

$$\beta_c \le \inf\left\{\beta > 0 : \mathbb{P}\left(\frac{1 - e^{-\beta J_e}}{1 + (q-1)e^{-\beta J_e}}\right) > p_c^{\text{bond}}(\mathbb{L}^d)\right\} < \infty, \tag{11.29}$$

and the theorem is proved. □

11.5 The Edwards–Anderson spin-glass model

The Ising/Potts models with positive edge-interactions J_e are termed 'ferromagnetic': like spins attract one another, unlike spins repel. The corresponding edge-variables $p_e = 1 - e^{-\beta J_e}$ satisfy $p_e \in [0, 1)$, and the random-cluster model is a satisfactory tool for the analysis of the correlation structure. Conversely, when the J_e can be of either sign, the model is non-ferromagnetic, and the analysis is relatively difficult and incomplete[5]. The random-cluster model plays a role in this situation also, as described in this section in the context of an Ising model with real-valued edge-interactions.

In the last section, the J_e were allowed to be random variables taking values in the half-line $[0, \infty)$. A model which is especially interesting and relatively poorly understood is the so-called 'Edwards–Anderson spin-glass model', [109], in which the J_e are independent, identically distributed random variables taking values in $\mathbb{R}$ with a symmetric distribution (that is, J_e and $-J_e$ have the same law). Two natural distributions for the J_e are the normal distribution, and the symmetric distribution on the two-point space $\{-1, 1\}$. There are several beautiful open problems concerning the Edwards–Anderson model. We refer the reader to [260] for an account of the theory, and to [262, 263] for recent results and speculations.

Let $G = (V, E)$ be a finite graph, and write $\Sigma = \{-1, 1\}^V$ and $\Omega = \{0, 1\}^E$ for the associated vertex- and edge-configuration spaces[6]. Let $\mathbf{J} = (J_e : e \in E)$ be a given vector of reals, which may be negative or positive. We shall be interested in the Ising[7] measure $\pi_{\beta\mathbf{J}} = \pi_{G,\beta\mathbf{J}}$ given by

$$\pi_{\beta\mathbf{J}}(\sigma) = \frac{1}{Z_\mathrm{I}} e^{-\beta H(\sigma)}, \qquad \sigma \in \Sigma, \tag{11.30}$$

$$H(\sigma) = - \sum_{e=\langle x,y\rangle \in E} \tfrac{1}{2} J_e \sigma_x \sigma_y. \tag{11.31}$$

The inverse-temperature $\beta \in (0, \infty)$ is regarded as the parameter to be varied.

When $J_e > 0$ (respectively, $J_e < 0$), the spins at the endvertices of the edge e prefer to be equal (respectively, opposite). The usual stochastic orderings of the measures are invalid when some of the J_e are negative, and the consequent theory is substantially less developed than that of the ferromagnetic case. This notwithstanding, the measure $\pi_{\beta\mathbf{J}}$ may be coupled as follows with a random-cluster-type measure on Ω with edge-parameters $(p_e : e \in E)$ given by

$$p_e = 1 - e^{-\beta |J_e|}, \qquad e \in E. \tag{11.32}$$

Let P be the product measure on $\Sigma \times \Omega$ given by

$$P = \left\{ \prod_{x \in V} \phi^x \right\} \times \left\{ \prod_{e \in E} \phi^e \right\},$$

[5] See Kasteleyn's remark about the anti-ferromagnet in Paragraph 12 of the Appendix.

[6] We take the vertex-spins to be -1 and 1 in order to highlight a symmetry.

[7] The term 'Ising' is normally used in the ferromagnetic case only, but we choose to retain it in this disordered model.

where

$$\phi^x(\sigma_x = -1) = \phi^x(\sigma_x = 1) = \tfrac{1}{2}, \qquad x \in V,$$
$$\phi^e(\omega(e) = 1) = 1 - \phi^e(\omega(e) = 0) = p_e, \qquad e \in E.$$

Let $W = W_G \subseteq \Sigma \times \Omega$ be the (non-empty) event

(11.33) $$W = \big\{(\sigma, \omega) : J_e \sigma_x \sigma_y > 0 \text{ for all } e = \langle x, y \rangle \text{ with } \omega(e) = 1\big\}.$$

That is, W is the set of pairs $(\sigma, \omega) \in \Sigma \times \Omega$ such that the spins at the endvertices of every open edge e have the same sign (respectively, opposite signs) if $J_e > 0$ (respectively, $J_e < 0$). We now define the probability measure μ to be P conditioned on W,

(11.34) $$\mu(\sigma, \omega) = \frac{1}{P(W)} \mu(\sigma, \omega) 1_W(\sigma, \omega), \qquad (\sigma, \omega) \in \Sigma \times \Omega.$$

Let $U = U_G \subseteq \Omega$ be the event

(11.35) $$U = \big\{\omega \in \Omega : \text{there exists } \sigma \in \Sigma \text{ with } (\sigma, \omega) \in W\big\}.$$

A configuration $\omega \in \Omega$ is called *frustrated* if $\omega \notin U$. It is left as an exercise[8] to show that the marginal measure on Σ of μ is the Ising measure (11.30), and the marginal measure on Ω is the random-cluster measure with parameters $\mathbf{p} = (p_e : e \in E)$, $q = 2$, *conditioned on the event* U. We write this as

(11.36) $$\mu(\sigma) = \pi_{\beta \mathbf{J}}(\sigma), \qquad \sigma \in \Sigma,$$
$$\mu(\omega) = \overline{\phi}_{\mathbf{p}}(\omega), \qquad \omega \in \Omega,$$

where

(11.37) $$\overline{\phi}_{\mathbf{p}}(\omega) = \overline{\phi}_{G,\mathbf{p}}(\omega) = \frac{1}{Z} \phi_{\mathbf{p},2}(\omega) 1_U(\omega), \quad Z = \sum_{\omega \in U} \phi_{\mathbf{p},2}(\omega).$$

The conditional measure on Σ of μ is determined as follows, the derivation is omitted. First, we sample $\omega \in \Omega$ according to the relevant marginal $\overline{\phi}_{\mathbf{p}}$. Given ω, the (conditional) law of the random spin σ has as support the set

$$S(\omega) = \big\{\sigma \in \Sigma : (\sigma, \omega) \in W\big\},$$

which is non-empty since $\omega \in U$ (μ-almost-surely). Let C be an open cluster of ω, and let x, y be distinct vertices in C. Let ρ be an open path from x to y. Since every edge e of ρ is open, it must be the case that $p_e > 0$, and therefore $J_e \neq 0$.

[8] This coupling may be found in [129] and the present account draws on [259, 260]. The first use of a random-cluster representation in this context appears to be in [202].

Let $\sigma \in S(\omega)$. By (11.33), $\sigma_y = \eta_{x,y}\sigma_x$ where $\eta_{x,y}$ is the product of the signs of the J_e for $e \in \rho$. Thus, the relative signs of the spins on C are determined by knowledge of ω. Since there are two possible choices for the spin at any given x, there are two choices for the spin-configuration on C, and we choose between these according to the flip of a fair coin. In summary, we assign spins randomly to V in such a way that: the spins within a cluster satisfy $\sigma_y = \eta_{x,y}\sigma_x$ as above, and the spins of different clusters are independent.

Let $\omega \in \Omega$. We extend the definition of $\eta_{x,y}$ by setting $\eta_{x,y} = 0$ if $x \nleftrightarrow y$, and we arrive at a proposition which may be viewed as a generalization of Theorem 1.16 to situations in which $q = 2$ and the J_e may be of either sign.

(11.38) Proposition [259]. *For any finite graph $G = (V, E)$,*

$$\pi_{\beta\mathbf{J}}(\sigma_x\sigma_y) = \overline{\phi}_{\mathbf{p}}(\eta_{x,y}), \qquad x, y \in V.$$

When $J_e \geq 0$ for all $e \in E$, then $\eta_{x,y} = 1_{\{x\leftrightarrow y\}}$, and the conclusion of Theorem 1.16 is retrieved.

We pass now to the infinite-volume limit. Let $d \geq 2$, let Λ be a finite box of $\mathbb{L}^d$, and write $\Omega_\Lambda = \{0,1\}^{\mathbb{E}_\Lambda}$. For $\tau \in \Sigma$, let Σ_Λ^τ be given as in Section 11.2. We may construct a measure μ_Λ^τ on $\Sigma_\Lambda^\tau \times \Omega_\Lambda$ by adapting the definition of μ given above. The reference product measure P is given similarly but subject to $\sigma_x = \tau_x$ for $x \in \partial\Lambda$, and μ_Λ^τ is obtained by conditioning P on the event $W = W_\Lambda$. The marginal of μ_Λ^τ on Σ_Λ^τ is an Ising measure with boundary condition τ.

A *(EA-)Gibbs state* for the Edwards–Anderson model on $\mathbb{L}^d$ is defined to be a probability measure π on $\Sigma = \{-1,1\}^{\mathbb{Z}^d}$ satisfying the DLR condition as in Definition 11.2. The principal problem is to determine, for a given vector $\mathbf{J} = (J_e : e \in \mathbb{E}^d)$, the set of values of β for which there exists a unique Gibbs state. Only a limited amount is known about this problem. One of the main difficulties is that correlations are not generally monotonic in β, and thus we know no satisfactory definition of a critical value of β. Nevertheless, for given $\mathbf{J}$ we may define

(11.39)
$$\beta_c(\mathbf{J}) = \sup\big\{\beta : \text{there is a unique Gibbs state at inverse-temperature } \beta\big\}.$$

The following is proved as an application of the random-cluster method.

(11.40) Theorem [259]. *Consider the Ising model on $\mathbb{L}^d$ with real-valued edge-interactions $\mathbf{J} = (J_e : e \in \mathbb{E}^d)$ and inverse-temperature β. We have that $\beta_c(\mathbf{J}) \geq \beta_c(|\mathbf{J}|)$, where the latter is the critical inverse-temperature for the ferromagnetic Ising model with edge-interactions $|\mathbf{J}| = (|J_e| : e \in \mathbb{E}^d)$.*

It is an important open problem to decide whether or not there is non-uniqueness of Gibbs states on $\mathbb{L}^d$ for large β, [260]. There has been a considerable amount of discussion of and speculation around this question, for an account of which the reader is referred to the work of Newman and Stein [262, 263].

We consider briefly the special case in which the J_e have the symmetric distribution on the two-point space $\{-1, 1\}$. The quantity $\beta_c(\mathbf{J})$ is a translation-invariant function of a family of independent random variables. Therefore, there exists a real number β_c^{EA} such that $P(\beta_c(\mathbf{J}) = \beta_c^{\text{EA}}) = 1$. The theorem implies the uniqueness of Gibbs states for every possible value of the vector $\mathbf{J}$, whenever $0 < \beta < \beta_c(1)$ with $\beta_c(1)$ the critical inverse-temperature for the ferromagnetic Ising model with constant edge-interaction 1. The weak inequality $\beta_c^{\text{EA}} \ge \beta_c(1)$ may be strengthened to strict inequality for this case, [100].

Proof of Theorem 11.40. We begin with a discussion of boundary conditions. Let $\mathbf{J} = (J_e : e \in \mathbb{E}^d)$ be given, and $\beta \in (0, \infty)$. For $\tau \in \Sigma$ and a box Λ, write $\pi^\tau_{\Lambda,\beta\mathbf{J}}$ for the corresponding Ising measure on Λ with boundary condition τ, as in (11.1). Let A be a cylinder event of Σ, and suppose β is such that,

$$\text{(11.41)} \qquad \text{for all } \tau, \tau' \in \Sigma, \qquad \pi^\tau_{\Lambda,\beta\mathbf{J}}(A) - \pi^{\tau'}_{\Lambda,\beta\mathbf{J}}(A) \to 0 \qquad \text{as } \Lambda \uparrow \mathbb{Z}^d.$$

Let π, π' be Gibbs states at inverse-temperature β. We may sample τ according to π, and τ' according to π', thereby obtaining from (11.41) and the definition of a Gibbs state (as in, for example, Definition 11.2), that $\pi(A) = \pi'(A)$. Since the cylinder events generate the requisite σ-field of Σ, we deduce that $\pi = \pi'$. It will therefore suffice to prove (11.41) under the assumption that $\beta < \beta_c(|\mathbf{J}|)$, and this will be achieved via a transformation to the random-cluster model.

We construct next the random-cluster measure on Λ corresponding to the Ising measure $\pi^\tau_{\Lambda,\beta\mathbf{J}}$, and we remind the reader of the ferromagnetic case around (11.6). Let $\Omega_\Lambda = \{0, 1\}^{\mathbb{E}_\Lambda}$ and

$$\text{(11.42)} \qquad U^\tau_\Lambda = \big\{\omega \in \Omega_\Lambda : \text{there exists } \sigma \in \Sigma^\tau_\Lambda \text{ such that } (\sigma, \omega) \in W^\tau_\Lambda\big\},$$

where $W^\tau_\Lambda \subseteq \Sigma^\tau_\Lambda \times \Omega_\Lambda$ is given by

$$\text{(11.43)} \quad W^\tau_\Lambda = \big\{(\sigma, \omega) : J_e\sigma_x\sigma_y > 0 \text{ for all } e = \langle x, y\rangle \in \mathbb{E}_\Lambda \text{ with } \omega(e) = 1\big\}.$$

Let $\mathbf{p} = (p_e : e \in \mathbb{E}^d)$ satisfy (11.32). As in Section 11.2 (see the footnote on page 324), we let $\phi^\tau_{\Lambda,\mathbf{p}}$ be the wired random-cluster measure $\phi^1_{\Lambda,\mathbf{p},2}$ conditioned on the event U^τ_Λ.

The event U^τ_Λ is a decreasing subset of Ω_Λ, so that, by positive association,

$$\text{(11.44)} \qquad \phi^\tau_{\Lambda,\mathbf{p}} \le_{\text{st}} \phi^1_{\Lambda,\mathbf{p},2}.$$

There is a close link between stochastic inequalities and couplings. For $\omega \in \Omega_\Lambda$, let $S(\omega) = \{x \in \Lambda : x \leftrightarrow \partial\Lambda\}$, and $G = \Lambda \setminus S(\omega)$. We claim that there exists a probability measure κ on $\Omega_\Lambda \times \Omega_\Lambda$ such that:

(i) the first marginal is $\phi^\tau_{\Lambda,\mathbf{p}}$, and the second marginal is $\phi^1_{\Lambda,\mathbf{p},2}$,

(ii) the support of κ is the set of pairs (ω_0, ω_1) satisfying $\omega_0 \le \omega_1$,

(iii) for any suitable g, conditional on the event $\{G(\omega_1) = g\}$, the marginal law of $\{\omega_0(e) : e \in \mathbb{E}_g\}$ is the free measure $\overline{\phi}_{g,\mathbf{p}}$.

The full proof of this step is omitted, and the reader is referred to [259] and to the closely related proof of Proposition 5.30. The idea is to sample the states $\omega_0(e), \omega_1(e)$ of edges recursively, beginning with edges e incident to $\partial\Lambda$. At each stage, one checks the stochastic domination (conditional on the past history of the construction) that is necessary to continue the pointwise ordering.

Let Δ, Λ be boxes such that: A is defined in terms of the spins within Δ, and $\Delta \subseteq \Lambda$. Let S, G, and κ be given as above. If $\omega_1 \in \{\Delta \nleftrightarrow \partial\Lambda\}$, then $G(\omega_1) \supseteq \Delta$, and we write $\mathcal{H}$ for the set of possible values of G on this event. Using the coupling of the Ising and random-cluster measures, together with the remarks above, it follows by conditioning on the event $\{\Delta \nleftrightarrow \partial\Lambda\}$ that

$$\pi^{\tau}_{\Lambda,\beta\mathbf{J}}(A) = \sum_{g\in\mathcal{H}} \phi^1_{\Lambda,\mathbf{p},2}(G = g)\pi_{g,\beta\mathbf{J}}(A) + \phi^1_{\Lambda,\mathbf{p},2}(\Delta \leftrightarrow \partial\Lambda)m^{\tau}_{\Lambda},$$

for some m^{τ}_{Λ} satisfying $0 \le m^{\tau}_{\Lambda} \le 1$. Similarly,

$$\pi^{\tau'}_{\Lambda,\beta\mathbf{J}}(A) = \sum_{g\in\mathcal{H}} \phi^1_{\Lambda,\mathbf{p},2}(G = g)\pi_{g,\beta\mathbf{J}}(A) + \phi^1_{\Lambda,\mathbf{p},2}(\Delta \leftrightarrow \partial\Lambda)m^{\tau'}_{\Lambda}.$$

By subtraction,

$$\big|\pi^{\tau}_{\Lambda,\beta\mathbf{J}}(A) - \pi^{\tau'}_{\Lambda,\beta\mathbf{J}}(A)\big| \le \phi^1_{\Lambda,\mathbf{p},2}(\Delta \leftrightarrow \partial\Lambda). \tag{11.45}$$

For $\beta < \beta_c(|\mathbf{J}|)$, the right side of (11.45) approaches 0 as $\Lambda \uparrow \mathbb{Z}^d$, and (11.41) follows as required. □

11.6 The Widom–Rowlinson lattice gas

Particles of two types, type 1 and type 2 say, are distributed randomly within a bounded measurable subset Λ of $\mathbb{R}^d$ in such a way that no 1-particle is within unit distance of any 2-particle. A simple probabilistic model for this physical model is the following, termed the Widom–Rowlinson model after the authors of the paper [319] on the liquid/vapour transition. Let $\lambda \in (0, \infty)$. Let Π_1 and Π_2 be independent subsets of Λ chosen as spatial Poisson processes[9] with intensity λ. Let D_Λ be the event

$$D_\Lambda = \big\{|x - y| > 1 \text{ for all } x \in \Pi_1,\ y \in \Pi_2\big\},$$

and let $\mu_{\Lambda,\lambda}$ be the law of the pair (Π_1, Π_2) conditioned on the event D_Λ. This measure is well defined since $\mathbb{P}(D_\Lambda) > 0$ for bounded Λ.

[9]See [164, Section 6.13] for an introduction to the theory of spatial Poisson processes.

The definition of the Widom–Rowlinson measure $\mu_{\Lambda,\lambda}$ may be extended to the whole of $\mathbb{R}^d$ in the usual manner, following. A probability measure on pairs of countable subsets of $\mathbb{R}^d$ is called a *(WR-)Gibbs state* if, conditional on the configuration off any bounded measurable set Λ, the configuration within Λ is that of two independent Poisson processes on Λ conditional on no 1-particle in $\mathbb{R}^d$ being within unit distance of any 2-particle.

How many Gibbs states exist for a given value of λ? The following theorem may be proved using random-cluster methods in the continuum.

(11.46) Theorem [285]. *Consider the Widom–Rowlinson model on $\mathbb{R}^d$ with $d \geq 2$. There exist constants λ_1, λ_2 satisfying $0 < \lambda_1 \leq \lambda_2 < \infty$ such that: there is a unique Gibbs state when $\lambda < \lambda_1$, and there exist multiple Gibbs states when $\lambda > \lambda_2$.*

It is an open problem to show the existence of a single critical value marking the onset of multiple Gibbs states. In advance of the proof, which is sketched at the end of the section, we turn to a lattice version of this model introduced in [232].

Let $G = (V, E)$ be a finite graph. To each vertex we allocate a 'type' from the 'type-space' $\{0, 1, 2\}$, and we write $\Sigma_V = \{0, 1, 2\}^V$ for the ensuing spin space. For $\sigma \in \Sigma$, let $z(\sigma)$ be the number of vertices x with $\sigma_x = 0$. Let $\lambda \in (0, \infty)$, and consider the probability measure on Σ_V given by

$$\mu_{G,\lambda}(\sigma) = \begin{cases} \dfrac{1}{Z_{\mathrm{WR}}} \lambda^{-z(\sigma)} & \text{if } \sigma \in D, \\ 0 & \text{otherwise,} \end{cases}$$

where D is the event that, for all $x, y \in V$, $x \nsim y$ whenever $\sigma_x = 1$ and $\sigma_y = 2$, and Z_{WR} is the appropriate normalizing constant.

Consider now the infinite lattice $\mathbb{L}^d$ where $d \geq 2$, and let $\Sigma = \{0, 1, 2\}^{\mathbb{Z}^d}$, endowed with the usual σ-field $\mathcal{G}$. We may define a Gibbs state in the manner given above: a probability measure μ on $(\Sigma, \mathcal{G})$ is called a lattice (WR-)Gibbs state if it satisfies the appropriate DLR condition.

(11.47) Theorem [232]. *Consider the lattice Widom–Rowlinson model on $\mathbb{L}^d$ with $d \geq 2$. There exist constants λ_1, λ_2 satisfying $0 < \lambda_1 \leq \lambda_2 < \infty$ such that: there is a unique Gibbs state when $\lambda < \lambda_1$, and there exist multiple Gibbs states when $\lambda > \lambda_2$.*

It is an open problem to show the existence of a single critical value of λ. Proofs of such facts hinge usually on monotonicity, but such monotonicity is not generally valid for this model, see [69]. Progress has been made for certain lattices, [171], but the case of $\mathbb{L}^d$ remains unsolved.

The main ingredient in the proof of the latter theorem is a certain 'site-random-cluster measure', given as follows for the finite graph G. The configuration space is $\Omega_V = \{0, 1\}^V$. For $\omega \in \Omega_V$, Let $k(\omega)$ be the number of components in the graph obtained from G by deleting every vertex x with $\omega(x) = 0$. The *site-random-cluster measure* $\psi_{G,p,q}$ is given by

$$(11.48)\quad \psi_{G,p,q}(\omega) = \frac{1}{Z_{\mathrm{SRC}}}\left\{\prod_{x\in V} p^{\omega(x)}(1-p)^{1-\omega(x)}\right\} q^{k(\omega)}, \qquad \omega \in \Omega_V,$$

where $p \in [0, 1]$, $q \in (0, \infty)$, and Z_{SRC} is the appropriate normalizing constant. This measure reduces when $q = 1$ to the product measure on Ω_V otherwise known as site percolation.

At first sight, one might guess that the theory of such measures may be developed in much the same manner as that of the usual random-cluster model, but this is false. The problem is that, even for $q \in [1, \infty)$, the measures $\psi_{G,p,q}$ lack the stochastic monotonicity which has proved so useful in the other case. Specifically, the function k does not satisfy inequality (3.11).

Proof of Theorem 11.47. We follow [136], see also [86]. Let $G = (V, E)$ be a finite graph, and let $q = 2$, $\lambda \in (0, \infty)$, and $p = \lambda/(1 + \lambda)$. We show first how to couple $\mu_{G,\lambda}$ and $\psi_{G,p,q}$. Let ω be sampled from Ω_V according to $\psi_{G,p,q}$. If $\omega(x) = 0$, we set $\sigma_x = 0$. To each vertex y with $\omega(y) = 1$, we allocate a type from the set $\{1, 2\}$, each value having probability $\frac{1}{2}$, and we do this by allocating a given type to each given cluster of ω, these types being constant within clusters, and independent between clusters. The outcome is a spin vector σ taking values in Σ_V, and it is left as an exercise to check that σ has law $\mu_{G,\lambda}$.

Next, we compare $\psi_{G,p,q}$ with a product measure on Ω_V. It is immediate from (11.48) that, for $\xi \in \Omega$ and $x \in V$,

$$\psi_{G,p,q}\big(\omega(x) = 1 \,\big|\, \omega(y) = \xi(y) \text{ for all } y \notin V \setminus \{x\}\big) = \frac{pq}{pq + (1-p)q^{\kappa(x,\xi)}},$$

where $\kappa(x, \xi)$ is the number of open clusters of ξ_x that contain neighbours of x. [Here, ξ_x denotes the configuration obtained from ξ by setting the state of x to 0.] If the maximum degree of vertices in G is Δ, then $0 \le \kappa(x, \xi) \le \Delta$, and

$$p_1 \le \psi_{G,p,q}\big(\omega(x) = 1 \,\big|\, \omega(y) = \xi(y) \text{ for all } y \notin V \setminus \{x\}\big) \le p_2,$$

where

$$p_1 = \frac{pq}{pq + (1-p)^{\Delta}}, \qquad p_2 = \frac{pq}{pq + 1 - p}.$$

By Theorems 2.1 and 2.6,

$$(11.49)\qquad \phi_{G,p_1} \le_{\mathrm{st}} \psi_{G,p,q} \le_{\mathrm{st}} \phi_{G,p_2}$$

where $\phi_{G,r}$ is product measure on Ω_V with density r.

Consider now a finite box Λ of $\mathbb{L}^d$, with $\Delta = 2d$. It may be seen as in the case of the Potts model of Section 11.2 that there is a multiplicity of WR-Gibbs states if and only if the $\psi_{\Lambda,p,q}$ have a weak limit (as $\Lambda \uparrow \mathbb{Z}^d$) that possesses an infinite cluster with strictly positive probability. By (11.49), this cannot occur if $p_2 < p_c^{\text{site}}(\mathbb{L}^d)$, but this does indeed occur if $p_1 > p_c^{\text{site}}(\mathbb{L}^d)$. Here, $p_c^{\text{site}}(\mathbb{L}^d)$ denotes the critical probability of site percolation on $\mathbb{L}^d$, see [154]. □

Sketch proof of Theorem 11.46. The full proof is not included here, and interested readers are referred to [136, Thm 10.2] for further details[10]. Rather as in the previous proof, we relate the Widom–Rowlinson model to a type of 'continuum site-random-cluster measure'. Let Λ be a bounded measurable subset of $\mathbb{R}^d$. For any countable subset Π of Λ, let $N(\Pi)$ be the union of the closed $\frac{1}{2}$-neighbourhoods of the points in Π, and let $k(\Pi)$ be the number of (topologically) connected components of $N(\Pi)$. Consider now the probability measure $\overline{\pi}_{\Lambda,\lambda}$ on the family of countable subsets of Λ given by

$$\overline{\pi}_{\Lambda,\lambda}(d\Pi) = \frac{1}{Z_\Lambda} 2^{k(\Pi)} \pi_{\Lambda,\lambda}(d\Pi)$$

where $\pi_{\Lambda,\lambda}$ is the law of a Poisson process on Λ with intensity λ, and Z_Λ is a normalizing constant.

It is not hard to verify the following coupling. Let Π be a random countable subset of Λ with law $\overline{\pi}_{\Lambda,\lambda}$. To each point $x \in \Pi$ we allocate either type 1 or type 2, each possibility having probability $\frac{1}{2}$. This is done simultaneously for all $x \in \Pi$ by allocating a random type to each component of $N(\Pi)$, this type being constant within components, and independent between components. The outcome is a configuration (Π_1, Π_2) of two sets of points labelled 1 and 2, respectively, and it may be checked that the law of (Π_1, Π_2) is $\mu_{\Lambda,\lambda}$.

One uses arguments of stochastic domination next, but in the continuum. The methods of Section 2.1 may be adapted to the continuum to obtain a criterion under which $\overline{\pi}_{\Lambda,\lambda}$ may be compared to some $\pi_{\Lambda,\lambda'}$. It turns out that there exists $\alpha = \alpha(d) \in (0, \infty)$ such that

$$\pi_{\Lambda,\alpha\lambda} \leq_{\text{st}} \overline{\pi}_{\Lambda,\lambda} \leq_{\text{st}} \pi_{\Lambda,2\lambda} \qquad \text{for bounded measurable } \Lambda. \tag{11.50}$$

Let π_λ be the law of a Poisson process on $\mathbb{R}^d$ with intensity λ. It is a central fact of continuum percolation, see [154, Section 12.10] and [253], that there exists $\lambda_c \in (0, \infty)$ such that the percolation probability

$$\rho(\lambda) = \pi_\lambda\big(N(\Pi) \text{ possesses an unbounded component}\big) \tag{11.51}$$

satisfies

$$\rho(\lambda) = \begin{cases} 0 & \text{if } \lambda < \lambda_c, \\ 1 & \text{if } \lambda > \lambda_c. \end{cases}$$

It may be seen as in Section 11.2 that there exists a multiplicity of WR-Gibbs states if and only the $\overline{\pi}_{\Lambda,\lambda}$ have a weak limit (as $\Lambda \uparrow \mathbb{R}^d$) that allocates strictly positive probability to the occurrence of an unbounded component. By (11.50)–(11.51), this cannot occur when $\lambda < \lambda_1 = \frac{1}{2}\lambda_c$, but does indeed occur when $\lambda > \lambda_2 = \lambda_c/\alpha$. □

[10] The proof utilizes arguments of [86, 138].

Appendix

The Origins of FK(G)

The basic theory of the random-cluster model was presented in a series of papers by Kees [Cees] Fortuin and Piet Kasteleyn around 1970, and in the 1971 doctoral thesis of Fortuin. This early work contains several of the principal ingredients of Chapters 2 and 3 of the current book. The impact of the approach within the physics community was attenuated at the time by the combinatorial style and the level of abstraction of these papers.

The random-cluster model has had substantial impact on the study of Ising and Potts models. It has, in addition, led to the celebrated FKG inequality, [124], of which the history is as follows[1]. Following a suggestion of Kasteleyn, Fortuin proved an extension of Harris's positive-correlation inequality, [181, Lemma 4.1], to the random-cluster model, [122]. Kasteleyn spoke of related work during a lecture at the IHES in 1970, with Jean Ginibre in the audience. Ginibre realized subsequently that the inequality could be set in the general context of a probability measure μ on the power set of a finite set, subject to the condition $\mu(X \cup Y)\mu(X \cap Y) \geq \mu(X)\mu(Y)$, and he proceeded to write the first draft of the ensuing publication. Meanwhile, Fortuin met Ginibre at the 1970 Les Houches Summer School on 'Statistical mechanics and quantum field theory'.

In a reply dated 23 September 1970 to Ginibre's first draft, Kasteleyn made a number of suggestions, including to extend the domain of the main theorem to a finite distributive lattice, thereby generalizing the result to include both a totally ordered finite set and the power set of a finite set. He proposed the use of the standard result that any finite distributive lattice is lattice-isomorphic to a sublattice of the power set of some finite set. The article was re-drafted accordingly. The two Dutch co-authors later "thought it worthwhile to develop a self-supporting *lattice-theoretic proof*" of the principal proposition[2]. Ginibre placed his own name third in the list of authors, and the subsequent paper, [124], was published in the

[1] I am indebted to Cees Fortuin and Jean Ginibre for their recollections of the events leading to the formulation and proof of the FKG inequality, and to Frank den Hollander for passing on material from Piet Kasteleyn's papers.

[2] The quotation is taken from the notes written by Kasteleyn on Ginibre's second draft.

Communications in Mathematical Physics in 1971.

Ginibre recollects chatting with Kasteleyn at an AMS meeting during Spring 1971, and summarizing the situation as follows: "You had the proof and the conclusion of the theorem, and I provided the assumption". This first proof used induction, the coupling proof of Chapter 2 was found later by Holley, [185].

In response to an enquiry concerning the discovery of random-cluster measures, Piet Kasteleyn kindly contributed the following material, quoted from two letters to the present author dated November 1992.

First letter from Piet Kasteleyn to GRG, dated 11 November 1992.

You asked me about the origin of Kees Fortuin's and my ideas on the random-cluster model. I have excavated my recollections and here, and in the subsequent pages, is what came up.

When in the late '60s Fortuin came from the Technical University of Delft to Leiden for a PhD study, I had for some time been intrigued by a similarity between a number of very elementary facts concerning three different models defined on finite graphs. I was at that time actively interested in graph theory and I had begun to toy with a few ideas in order to find out if there was more behind this similarity than sheer accidents or trivialities.

I told Fortuin about the data that had struck me and proposed him to look closer at them and at related problems. So we began to cooperate and first attacked the case of finite graphs. When this led to success, we turned to infinite graphs. Since Kees was very good, he mastered in a short time the necessary mathematics and began to work more and more independently. It became a good piece of PhD research, of which the results were set down in his thesis and (identically) in our and his papers on the percolation model and the random-cluster model. The details you find on the following pages. As you will see, the first few steps were all extremely simple. Therefore you may find my account unnecessarily detailed. I found it fun, however, to go through this history once again.

Let $G = (V, E)$ be a finite connected graph with vertex set V and edge set E (multiple edges allowed).

A. Consider a function $R : E \to [0, \infty)$. Then (G, R) may be considered as representing an *electric network* consisting of 'branches' (resistors) and 'nodes' (for brevity identify edges with branches and nodes with vertices), where $R(e)$ is the resistance of the branch e.

(i) Suppose E contains two edges e_1 and e_2, placed 'in series' (i.e. having end points (u, v) and (v, w), respectively, with $u \neq w$):

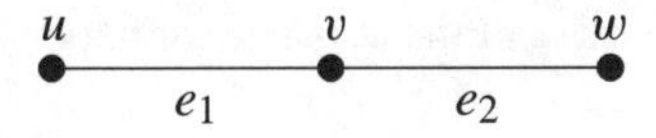

Write $R(e_1) = R_1$, $R(e_2) = R_2$. We can replace e_1 and e_2 by a *single* new edge e between u and w without affecting the electric currents and potentials that arise in the rest of the network when a potential difference is imposed on an arbitrary pair of nodes ($\neq v$), *provided* we attribute to e a resistance $R(e) = R = R_1 + R_2$.

(ii) Suppose next that E contains two edges e_1 and e_2 which are 'parallel' (i.e. which have the same pair of end points):

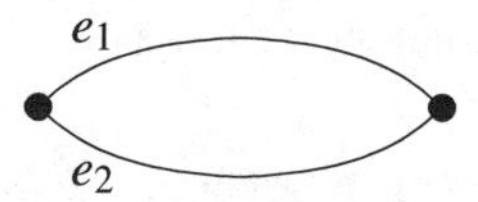

We can now replace e_1 and e_2 by a single edge e without affecting the rest of the network, as before, provided we choose $R(e) = R$ according to the following 'substitution rule':

$$R = \frac{R_1 R_2}{R_1 + R_2}.$$

B. Consider a function $J : E \to [0, \infty)$. Then (G, J) may be considered as representing an *Ising model* consisting of spins taking values ± 1, having ferromagnetic interactions described by the Hamiltonian

$$H(\sigma) = -\sum_{e \in E} J(e)\sigma(e),$$

where $\sigma : V \to \{-1, 1\}$ and $\sigma(e) = \sigma(u)\sigma(v)$ where u and v are the end points of e. The probability of the spin state σ is

$$\pi(\sigma) = Z^{-1} e^{-\beta H(\sigma)}$$

with Z the normalizing factor (partition function).

(i) Let e_1 and e_2 be two edges in series with $J_{1,2} = J(e_{1,2})$, as before. We can replace e_1 and e_2 by a single edge e without affecting the probabilities of the spin states on $V \setminus \{v\}$ *provided* we choose

$$J(e) = J = (2\beta)^{-1} \log \frac{1 + e^{2\beta(J_1 + J_2)}}{e^{2\beta J_1} + e^{2\beta J_2}}.$$

This can be seen as the result of 'summing out' the variable $\sigma(v)$.

(ii) Let e_1 and e_2 be parallel edges. We can replace them by a single edge e without affecting the rest of the system provided we choose $J(e) = J = J_1 + J_2$. Trivial: the two similar terms in H are combined to one.

C. Consider $p : E \to [0, 1]$. Then (G, p) may be considered as representing a *percolation model* with the usual interpretation of $p(e)$.

(i) As before, with suitable translations of concepts. The substitution rule is now $p = p_1 p_2$.

(ii) As before. The substitution rule is $p = p_1 + p_2 - p_1 p_2$.

So far for the facts. *Question*: do they reflect some relation between the three systems? To answer this question we began with the following elementary steps (in which order, I do not remember exactly; the one I give here will not be far from the actual one).

1. To bring case B somewhat more in line with A and C, go over from $J(e)$ to $w(e) = \exp(-2\beta J(e))$. Then the substitution rules are

$$\text{(i)} \quad w = \frac{w_1 + w_2}{1 + w_1 w_2}; \qquad \text{(ii)} \quad w = w_1 w_2.$$

2. Introduce new variables, viz.
in case A: $R^* = R^{-1}$ (conductivity);
in case B: $w^* = (1-w)/(1+w) = \tanh(\beta J)$;
in case C: $p^* = 1 - p$. This reduces the substitution rules to:

	(i)	(ii)
A	$R = R_1 + R_2$	$R^* = R_1^* + R_2^*$
B	$w = \dfrac{w_1 + w_2}{1 + w_1 w_2}$	$w^* = \dfrac{w_1^* + w_2^*}{1 + w_1^* w_2^*}$
C	$p = p_1 p_2$	$p^* = p_1^* p_2^*$

3. (Sideline) Note that if G is planar and G^* is its dual, then the situation •—•—• in G corresponds to (two vertices joined by two parallel edges) in G^* and conversely. So the starred variables can be considered as 'dual' to the original ones (note that $R^{**} = R$ etc).

4. Consider now a q-state Potts model with $\sigma : V \to \{1, 2, \ldots, q\}$ and

$$H = 2 \sum_{e \in E} J_e (1 - \delta_\sigma(e))$$

with $J(e) \geq 0$ for all e; the factor 2 is inserted for the sake of comparison and a constant term is omitted. Define $w(e)$ as for the Ising model. For case (i), a simple calculation (summation over $\sigma(v)$) shows that the substitution rule is now

$$w = \frac{w_1 + w_2 + (q-2) w_1 w_2}{1 + (q-1) w_1 w_2}.$$

That for case (ii) is, as in the Ising model, $w = w_1 w_2$. It takes the same form as for case (i), with w^* instead of w, if we define

$$w^* = \frac{1 - w}{1 + (q - 1)w}.$$

5. Now note that by substituting $q = 1$ in the last few formulae we get $w = w_1 + w_2 - w_1 w_2$, or $1 - w = (1 - w_1)(1 - w_2)$ and $w^* = 1 - w$. Hence, if we write $1 - w = p$, we recover the rules for C. So in this very special sense, the percolation model behaves, just like the Ising model, as a special case of the Potts model.

6. At first sight the electric network does not seem to fit into the Potts model. It does so, however, if we take an appropriate formal limit. Define for the Potts model $S = q^{-\frac{1}{2}}(1 - w)$. In the limit $q \to 0$ (which at this stage is still meaningless, since q is an integer), the substitution rules reduce to

$$\text{(i)} \quad S = \frac{S_1 S_2}{S_1 + S_2}; \qquad \text{(ii)} \quad S = S_1 + S_2$$

and the duality rule is $S^* = 1/S$. In other words, we recover the rules for the network, with $S = R^{-1}$ $(= R^*)$.

7. So far we had got only a first indication about a relationship between the systems A–C and the Potts model. We then wanted to turn to more general situations. We had observed that (for arbitrary G) certain characteristic quantities in A–C, such as

(A) the total current flowing through an electric network when a unit potential difference is imposed on two arbitrary vertices x and y,

(B) the two-spin correlation $E[\sigma(x)\sigma(y)]$ of the Ising model,

(C) the pair connectivity $E[I(x \leftrightarrow y)]$ of the percolation model

can all be written in the form $P(X)/Q(X)$ (where X stands for $S = R^{-1}$, $p = 1 - w$, and p, respectively), with P and Q polynomials in the edge variables $X(e)$ that are linear in each variable separately. For the electric network, P and Q are homogeneous in all variables $S(e)$ together (of degree $|V| - 2$ and $|V| - 1$, respectively); Q is the generating function of spanning trees of G, and P is the generating function of spanning forests which consist of two trees, with x and y in different trees. For the Ising model, Q is the partition function (which has also a graph-theoretical interpretation, viz. in terms of cycles). For the percolation model, $Q = 1$ identically.

8. From the linearity in the $X(e)$ it follows that $P(X) = P(X, G)$ satisfies, for each edge e, the recursion relation

$$P(X, G) = P_1 + X(e) P_2,$$

with P_1 and P_2 polynomials in all $X(f)$ with $f \neq e$. This relation holds also for the Potts model with arbitrary q, again with $p = 1 - w$. A similar relation holds for Q.

9. Now if in the Potts model we have $p(e) = 0$, i.e. $w(e) = 1$, for some edge e, this means that $J(e) = 0$; this is equivalent to having an interaction graph G with the edge e *deleted*. Similarly, $p(e) = 1$, $w(e) = 0$, means that the interaction is infinitely strong; this is equivalent to having an interaction graph with the edge e *contracted* (i.e. deleted and its end points identified). If $D_e G$ and $C_e G$, respectively, are the graphs thus obtained from G, then obviously,

$$P(p, D_e G) = P_1,$$
$$P(p, C_e G) = P_1 + P_2.$$

Hence, we can write

$$\begin{aligned} P(p, G) &= P(p, D_e G) + p(e)\{P(p, C_e G) - P(p, D_e G)\} \\ &= [1 - p(e)]P(p, D_e G) + p(e)P(p, C_e G). \end{aligned}$$

10. Iteration of the last step leads directly to the expansion of P and Q in the variables $p(e)$ and $1 - p(e)$. Since $0 \leq p(e) \leq 1$, we could interpret the $p(e)$ as probabilities and the entire system as an example (the first one we knew) of 'weighted' (we would now say 'dependent') percolation. The generalization to arbitrary positive q (and even to complex q!) was now obvious. Then the limit $q \to 0$, as described above, could be taken correctly, and what came out was the electric network with all its properties.

11. It is the system obtained in this way which — for lack of a more instructive name — we called the random-cluster model. Most people have just called it the Potts model, and of course, it is closely related to the generalized spin model bearing this name. (We referred to the latter model as the Ashkin–Teller–Potts model, because Ashkin and Teller were really the first to consider generalizations of the Ising model to more than two spin states; one of these was the 4-state Potts model.) Fortuin and I preferred, however, to distinguish between the two systems, because they are different in principle. It is only in the paper by Edwards and Sokal that the relation between the two was fully established for integral values of q. It is now obvious that to every function $f(\sigma)$ of the spin state in the Potts model there corresponds a function $F(\omega)$ of the edge state in the random-cluster model such that the expectation of F with respect to the random-cluster measure equals the expectation of f with respect to the Potts measure, and conversely. F and f are transformed into each other via kernels which are nothing but the conditional probabilities of Edwards and Sokal. The relations which Fortuin and I found between spin correlations in the Ising model and certain connectivity probabilities in the random-cluster model with $q = 2$ were special cases.

12. After thus having introduced the random-cluster model for finite graphs, we were prepared to tackle infinite graphs. Fortuin wanted absolutely to treat these in as general a setting as possible and not restrict himself from the outset to regular lattices, as I had suggested to him. This admittedly makes his papers less accessible, but, in my opinion, also richer than they would have been if he had followed my suggestion. But this is a matter of taste.

So far about history. Looking at the subsequent developments, I am somewhat surprised by the fact that (to my knowledge) no one has given any attention to the domain of q between 0 and 1, not even to the limit in which one recovers the electric network, where life becomes much simpler. Of course, the FKG inequality does not hold in this domain, but does that imply that nothing of interest can be done? I admit that Fortuin was the first to restrict himself to the region where FKG holds, but that was because the time for his PhD research was limited! In fact, if I remember correctly, some mathematician once published a paper in which a graph-theoretical interpretation was given to the random-cluster model (probably under the name of dichromatic polynomial) for $q = -1$ (or $q = -2$, I am not sure).

Then there is the 'antiferromagnetic' Potts model, where $J(e)$ is allowed to be negative. If it is, $p(e)$ is also negative, so that a standard probability interpretation is impossible. This case has not been investigated either, as far as I know. Still, it is of interest, if only because, for integral positive q, the limit where all $J(e)$ become infinitely negative leads one into the theory of vertex colourings with q colours! In this connection I may point out that for two-dimensional regular lattices the value $q = 4$ plays a very special role in the random-cluster model: for $q < 4$ the phase transition is 'of second order' (i.e. the percolation probability is continuous), for $q > 4$ it is 'of first order'. So it may be that there is more to be said about the four-colour problem than we know at present!

Second letter from Piet Kasteleyn to GRG, dated 17 November 1992.

... I have been a bit too hasty in my conclusion about the connection between functions $f(\sigma)$ in the Potts model (PM) and functions $F(\omega)$ in the random-cluster model (RCM). What I wrote about the 'transformation' from f to F and vice versa may be formally true, but it is trivial. What is not trivial, is the question whether to each f there corresponds an F depending *only* on the edge configuration ω, and *not* (parametrically) on $p = (p_e : e \in E)$, and vice versa. I do not remember having seen this question discussed in the literature. In one direction there is no problem. If for given $f(\sigma)$ we define

$$f(\sigma) \mapsto F(\omega) = \sum_{\sigma} f(\sigma)\mu(\sigma \mid \omega),$$

this $F(\omega)$ satisfies the requirement I mentioned, because $\mu(\sigma \mid \omega)$ does not depend on p. However, the map

$$F(\omega) \mapsto f(\sigma) = \sum_{\omega} F(\omega)\mu(\omega \mid \sigma)$$

does *not* satisfy the requirement, $\mu(\omega \mid \sigma)$ depends on p_e explicitly.

To analyse this point we can proceed as follows. Using your notation we have

$$\pi(\sigma) = \pi_{G,p,q}(\sigma) = Z^{-1} \prod_{e \in E} \big((1-p_e) + p_e \delta_\sigma(e)\big),$$

$$\phi(\omega) = \phi_{G,p,q}(\omega) = Z^{-1} \bigg(\prod_{e \in E} \big((1-p_e)(1-\omega(e)) + p_e \omega(e)\big) \bigg) q^{k(\omega)}.$$

The expectation w.r.t. π of a function $f(\sigma) = f(\sigma, G)$ can be written as

$$\begin{aligned} E_\pi f &= \sum_\sigma f(\sigma) Z^{-1} \prod_{e \in E} \big(1 + p_e\{\delta_\sigma(e) - 1\}\big) \\ &= Z^{-1} \sum_\sigma f(\sigma) \sum_{D \subset E} \bigg(\prod_{e \in D} p_e \bigg) \bigg(\prod_{e \in D} \{\delta_\sigma(e) - 1\} \bigg). \end{aligned}$$

The expectation w.r.t. ϕ of a function $F(\omega) = F(\omega, G)$ can be written as

$$\begin{aligned} E_\phi F &= \sum_\omega F(\omega) Z^{-1} \bigg(\prod_e \big(\{1 - \omega(e)\} + p_e\{2\omega(e) - 1\}\big) \bigg) q^{k(\omega)} \\ &= Z^{-1} \sum_\omega q^{k(\omega)} F(\omega) \\ &\qquad \times \sum_{D \subset E} \bigg(\prod_{e \in D} p_e \bigg) \bigg(\prod_{e \in D} \{2\omega(e) - 1\} \bigg) \bigg(\prod_{e \in E \setminus D} \{1 - \omega(e)\} \bigg). \end{aligned}$$

In order that $E_\pi f = E_\phi F$ identically in p we must have

$$\begin{aligned} \forall D \subset E : \sum_\sigma f(\sigma) \bigg(\prod_{e \in D} \{\delta_\sigma(e) - 1\} \bigg) \\ = \sum_\omega q^{k(\omega)} F(\omega) \bigg(\prod_{e \in D} \{2\omega(e) - 1\} \bigg) \bigg(\prod_{e \in E \setminus D} \{1 - \omega(e)\} \bigg). \quad (*) \end{aligned}$$

It follows from what I remarked on $\mu(\sigma \mid \omega)$ that for given $f(\sigma)$ there is a solution $F(\omega)$ of this equation. (It is readily verified.) *Question*: is there a solution $f(\sigma)$ for given $F(\omega)$? The answer is *not in general*. If, e.g., G contains the subgraph K_r (the complete graph on r vertices, having $\frac{1}{2}r(r-1)$ edges), and we choose $D =$ the edge-set of this subgraph, then there is *no* σ such that $\prod_{e \in D} \{\delta_\sigma(e) - 1\} \neq 0$ if $r > q$. The reason is that it is not possible to have different spin values for *every* pair of adjacent vertices in D if you have only q different values at your disposal. Hence the l.h.s. of $(*)$ equals 0 for this D, so that $F(\omega)$ has to satisfy the condition

$$\sum_\omega q^{k(\omega)} F(\omega) \bigg(\prod_{e \in D} \{2\omega(e) - 1\} \bigg) \bigg(\prod_{e \in E \setminus D} \{1 - \omega(e)\} \bigg) = 0. \qquad (**)$$

This can be rewritten, if we denote an edge configuration not by ω, but by the set of open edges. Let us denote this by C (your $\eta(\omega)$). Then

$$2\omega(e) - 1 = \begin{cases} 1 & \text{if } e \in C, \\ -1 & \text{if } e \in E \setminus C, \end{cases} \qquad 1 - \omega(e) = \begin{cases} 0 & \text{if } e \in C, \\ 1 & \text{if } e \in E \setminus C. \end{cases}$$

Hence, the l.h.s. of the condition $(**)$ reduces to

$$0 = \sum_{C \subset E} q^{k(C)} F(C)(-1)^{|D \cap (E \setminus C)|} I_{\{C \subset D\}} = \sum_{C \subset D} q^{k(C)} F(C)(-1)^{|D \setminus C|}$$

where $\sum_{C \subset D} = \sum_{C:\, C \subset D}$ So the condition reads:

$$\sum_{C \subset D} (-1)^{|D \setminus C|} q^{k(C)} F(C) = 0.$$

For $q = 2$ (Ising model), the existence of a triangle in G already causes a relation (it cannot accommodate 3 unequal pairs of spins). This is, e.g., satisfied by $F(C) = I_{\{u \leftrightarrow v\}}$, but not by $F(C) = I_{\{u \leftrightarrow v \leftrightarrow w\}}$, where u, v, w are vertices.

You may be amused to see what happens in the case $q = 1$!

My conclusion is that the PM is 'included' (in the spirit of this analysis) in the RCM, but that generically the RCM is 'richer': there are questions one can ask in the RCM which have no counterpart in the PM. In addition, of course, the RCM makes also sense for $q \notin \mathbb{N}$, but the PM — as far as we know (Fortuin and I tried hard!) — not.

Postscript by Cees Fortuin, 11 September 2003.

I remember especially the first time he [Piet] told me about his ideas (end of 1966 when I still was doing my military service and he already had invited me to work with him): we were sitting next each other at the table in front of the window, which he used for working sessions, and he explained his ideas (the ABC of the letter) while I was listening. My first work was then 'putting the electrical current of a network into the scheme'. The actual formulating of the model took some time: I guess that it was end 1968/begin 1969 before on my blackboard the formula with 2^n appeared (the reformulation of the Ising model); I then went to his office and said something like: "I have found what we sought" (but half and half expecting he would say that he already knew!). We walked back to my office where he overlooked the blackboard and remarked that this was a special moment(!).

List of Notation

Graphs and sets:

$G = (V, E)$	15	Graph with vertex-set V and edge-set E
E_W	16	Set of edges having both endvertices in W
V_E	174	Set of vertices incident to edges in the set E
$\mathrm{Aut}(G)$	74	Automorphism group of G
G_{d}	133	Dual graph of the planar graph G
$\langle x, y \rangle$	15	Edge joining vertices x and y
$x \sim y$	15	x is adjacent to y
$\mathbb{R}$	18	The real line $(-\infty, \infty)$
$\mathbb{Z}$	17	The set $\{\dots, -2, -1, 0, 1, 2, \dots\}$ of integers
$\mathbb{Z}_+$	18	The set $\{0, 1, 2, \dots\}$ of non-negative integers
$\mathbb{N}$	18	The natural numbers $\mathbb{Z}_+ \setminus \{0\}$
$\mathbb{L}^d$	18	The d-dimensional (hyper)cubic lattice
$\mathbb{E}^d$	18	The set of edges of $\mathbb{L}^d$
$\mathbb{E}_V$	18	Subset of edges having both endvertices in V
$\mathbb{T}$	159	The triangular lattice
$\mathbb{H}$	159	The hexagonal lattice
	169	The set of plaquettes of $\mathbb{L}^3$
$\mathbb{U}$	164	The upper half-plane
x_i	17	The ith component of the vertex $x \in \mathbb{Z}^d$
$\Lambda_{a,b}$	18	The box with vertex-set $\prod_{i=1}^d [a_i, b_i]$
Λ_n	18	The box with vertex-set $[-n, n]^d$
$S(L, n)$	124	The box $[0, L-1] \times [-n, n]^{d-1}$
$\deg(u)$	58	The degree of the vertex u
$\deg(W)$	43	The maximal degree of a spanning set W of vertices
∂A	17	The surface of the set A of vertices
$\Delta_{\mathrm{e}} W$	17	The edge boundary of W
$\Delta_{\mathrm{e},\delta} C$	170	Edge-set given in terms of a surface δ of plaquettes
$\partial_{\mathrm{e}} D$	174	The 1-edge-boundary of the edge-set D
$\partial_{\mathrm{ext}} F$	147	Set of vertices of V_F in infinite paths of the complement
$\Delta_{\mathrm{ext}} D$	174	The external edge-boundary of the edge-set D
$\Delta_{\mathrm{int}} D$	174	The internal edge-boundary of the edge-set D
$\Delta_{\mathrm{v},\delta} C$	170	Subset of C given in terms of a surface δ of plaquettes

$\partial^+\Lambda$, $\partial^-\Lambda$	197	Upper and lower boundaries of Λ
$\mathrm{rad}(D)$	110	Radius of a subgraph D of $\mathbb{L}^d$ containing 0
$\delta(x, y)$	18	Number of edges in the shortest path from x to y
$\lvert x \rvert$	18	$\delta(0, x)$
$\lVert x \rVert$	18	$\max\{\lvert x_i \rvert : 1 \le i \le d\}$
$h(e)$	169	The plaquette intersecting the edge $e \in \mathbb{E}^3$
$[H]$	170	Subset of $\mathbb{R}^3$ lying in some plaquette of H
$E(H)$	169	Set of edges corresponding to the set H of plaquettes
$\overline{\delta}$	170	The closure or extended interface of a set δ of plaquettes
$\mathcal{D}_{L,M}$	201	The set of interfaces
δ_0	201	The regular interface
$\overset{s}{\sim}$	169	s-connectedness for plaquettes
$\lVert h_1, h_2 \rVert$	169	The L^∞ distance between the centres of plaquettes h_1, h_2
$\mathrm{ins}(T)$	169	Union of the bounded connected components of $\mathbb{R}^d \setminus T$
$\mathrm{out}(T)$	169	Union of the unbounded connected components of $\mathbb{R}^d \setminus T$
$\lvert A \rvert$	17	Cardinality of A, or number of vertices of A
$A \mathbin{\triangle} B$	60	Symmetric difference of A and B

Probability notation:

$\mu(X)$	18	Expectation of the random variable X under the measure μ
p, q	4	Edge and cluster-weighting parameters
$\phi_{G,p,q}$, $\phi_{p,q}$	4	Random-cluster measure on G with parameters p, q
$\phi_{G,p}$, ϕ_p	4	Product measure with density p on edges of G
$Z_G(p, q)$	4	Random-cluster partition function
$\lambda_{\beta,h}$	7	Ising probability measure
$\pi_{\beta,h}$	7	Potts probability measure
$\phi^\xi_{\Lambda,p,q}$	38	Random-cluster measure on Λ with boundary condition ξ
$\phi^b_{p,q}$	75	Random-cluster measure on $\mathbb{L}^d$ with boundary condition b
$\mathcal{W}_{p,q}$	72	Set of limit-random-cluster measures
$\mathcal{R}_{p,q}$	78	Set of DLR-random-cluster measures
UCS	13	Uniform connected graph
UST	13	Uniform spanning tree
USF	13	Uniform (spanning) forest
1_A	15	Indicator function of an event A
$\mathrm{cov}_{p,q}$	41	Covariance corresponding to $\phi_{p,q}$
cov_p	33	Covariance corresponding to μ_p
$\mathrm{var}_{p,q}$	56	Variance corresponding to $\phi_{p,q}$
ω	15	Typical realization of open and closed edges
Ω	15	The space $\{0, 1\}^E$ of configurations
$\mathcal{F}$	16	The σ-field of Ω generated by the cylinders
$\mathcal{F}_\Lambda$	16	The σ-field generated by states of edges in $\mathbb{E}_\Lambda$
$\mathcal{T}_\Lambda$	16	The σ-field generated by states of edges in $\mathbb{E}^d \setminus \mathbb{E}_\Lambda$
$\mathcal{T}$	16	The tail σ-field
Ω^ξ_F	27	Set of configurations that agree with ξ off F

$\omega_e,\ \omega^e$	16	Configuration ω with e declared closed/open
$\omega_1 \vee \omega_2$	20	Maximum configuration of ω_1 and ω_2
$\omega_1 \wedge \omega_2$	20	Minimum configuration of ω_1 and ω_2
$H(\omega_1, \omega_2)$	16	Hamming distance between ω_1 and ω_2
$\eta(\omega)$	16	The set of edges that are open in ω
$k(\omega)$	17	Number of open components in ω
$I(\omega)$	79	Number of infinite open clusters in ω
$\overline{A}, A^c$	16	Complement of event A
$A \Box B$	64	Event that A and B occur 'disjointly'
$B(X)$	241	Space of bounded measurable functions from X to $\mathbb{R}$
$C(X)$	233	Space of continuous functions from X to $\mathbb{R}$
D_X	82	Discontinuity set of the random variable X
$I_A(e)$	30	Influence of the edge e on the event A
$\leq_{st}$	19	Stochastic domination inequality
$\Rightarrow$	69	Weak convergence

Random-cluster notation:

C_x	17	Open cluster at x
C	18	Open cluster C_0 at 0
$p_c(q)$	99	Critical value of p under $\phi_{p,q}$
$p_{sd}(q)$	135	The self-dual point of the random-cluster model on $\mathbb{L}^2$
$\widehat{p}_c(q)$	124	Critical point defined via slab connections
$\widetilde{p}_c(q)$	113	Critical point for polynomial/exponential decay
$\overline{p}(q)$	197	Critical point for the roughening transition
p_c^{bond}	329	Critical probability of bond percolation
p_c^{site}	340	Critical probability of site percolation
$p_{tc}(q)$	114	Critical point for the time-constant
$p_g(q)$	115	Critical point for exponential decay of connectivity
$\eta(\mu)$	114	The time-constant associated with the measure μ
$\theta^b(p, q)$	98	Percolation probability under $\phi_{p,q}^b$
$\xi(p, q)$	115	Correlation length
$\delta_e(\sigma)$	7	Indicator function that the endvertices of e have equal spin
$\{A \leftrightarrow B\}$	17	Event that there exist $a \in A$ and $b \in B$ such that $a \leftrightarrow b$
$\{A \not\leftrightarrow B\}$	17	Complement of the event $\{A \leftrightarrow B\}$
J_e	15	Event that e is open; also the indicator function of this event
K_e	37	Event that endvertices of e are joined by an open path not using e

Finally:

$a \vee b$	18	Maximum of a and b
$a \wedge b$	18	Minimum of a and b
$\lfloor c \rfloor$	18	Least integer not less than c
$\lceil c \rceil$	18	Greatest integer not greater than c
$\delta_{u,v}$	7	The Kronecker delta
λ_s	171	s-dimensional Lebesgue measure

References

Abraham, D. B., Newman, C. M.

1. *The wetting transition in a random surface model*, Journal of Statistical Physics 63 (1991), 1097–1111.

Ahlswede, R., Daykin, D. E.

2. *An inequality for the weights of two families of sets, their unions and intersections*, Zeitschrift für Wahrscheinlichkeitstheorie und Verwandte Gebiete 43 (1978), 183–185.

Aizenman, M.

3. *Geometric analysis of ϕ^4 fields and Ising models*, Communications in Mathematical Physics 86 (1982), 1–48.

Aizenman, M., Barsky, D. J.

4. *Sharpness of the phase transition in percolation models*, Communications in Mathematical Physics 108 (1987), 489–526.

Aizenman, M., Barsky, D. J., Fernández, R.

5. *The phase transition in a general class of Ising-type models is sharp*, Journal of Statistical Physics 47 (1987), 343–374.

Aizenman, M., Chayes, J. T., Chayes, L., Fröhlich, J., Russo, L.

6. *On a sharp transition from area law to perimeter law in a system of random surfaces*, Communications in Mathematical Physics 92 (1983), 19–69.

Aizenman, M., Chayes, J. T., Chayes, L., Newman, C. M.

7. *The phase boundary in dilute and random Ising and Potts ferromagnets*, Journal of Physics A: Mathematical and General 20 (1987), L313–L318.
8. *Discontinuity of the magnetization in one-dimensional $1/|x-y|^2$ Ising and Potts models*, Journal of Statistical Physics 50 (1988), 1–40.

Aizenman, M., Fernández, R.

9. *On the critical behavior of the magnetization in high-dimensional Ising models*, Journal of Statistical Physics 44 (1986), 393–454.

Aizenman, M., Grimmett, G. R.

10. *Strict monotonicity for critical points in percolation and ferromagnetic models*, Journal of Statistical Physics 63 (1991), 817–835.

Aizenman, M., Klein, A., Newman, C. M.

11. *Percolation methods for disordered quantum Ising models*, Phase Transitions: Mathematics, Physics, Biology, ... (R. Kotecký, ed.), 1992, pp. 129–137.

Aizenman, M., Nachtergaele, B.

12. *Geometric aspects of quantum spin systems*, Communications in Mathematical Physics 164 (1994), 17–63.

Alexander, K.

13. *Simultaneous uniqueness of infinite clusters in stationary random labeled graphs*, Communications in Mathematical Physics 168 (1995), 39–55.
14. *Weak mixing in lattice models*, Probability Theory and Related Fields 110 (1998), 441–471.
15. *The asymmetric random cluster model and comparison of Ising and Potts models*, Probability Theory and Related Fields 120 (2001), 395–444.
16. *Power-law corrections to exponential decay of connectivities and correlations in lattice models*, Annals of Probability 29 (2001), 92–122.
17. *Cube-root boundary fluctuations for droplets in random cluster models*, Communications in Mathematical Physics 224 (2001), 733–781.
18. *The single-droplet theorem for random-cluster models*, In and Out of Equilibrium (V. Sidoravicius, ed.), Birkhäuser, Boston, 2002, pp. 47–73.
19. *Mixing properties and exponential decay for lattice systems in finite volumes*, Annals of Probability 32, 441–487.

Alexander, K., Chayes, L.

20. *Non-perturbative criteria for Gibbsian uniqueness*, Communications in Mathematical Physics 189 (1997), 447–464.

Ashkin, J., Teller, E.

21. *Statistics of two-dimensional lattices with four components*, The Physical Review 64 (1943), 178–184.

Barlow, R. N., Proschan, F.

22. *Mathematical Theory of Reliability*, Wiley, New York, 1965.

Barsky, D. J., Aizenman, M.

23. *Percolation critical exponents under the triangle condition*, Annals of Probability 19 (1991), 1520–1536.

Barsky, D. J., Grimmett, G. R., Newman, C. M.

24. *Percolation in half spaces: equality of critical probabilities and continuity of the percolation probability*, Probability Theory and Related Fields 90 (1991), 111–148.

Batty, C. J. K., Bollmann, H. W.

25. *Generalized Holley–Preston inequalities on measure spaces and their products*, Zeitschrift für Wahrscheinlichkeitstheorie und Verwandte Gebiete 53 (1980), 157–174.

Baxter, R. J.

26. *Exactly Solved Models in Statistical Mechanics*, Academic Press, London, 1982.

Baxter, R. J., Kelland, S. B., Wu, F. Y.

27. *Equivalence of the Potts model or Whitney polynomial with an ice-type model*, Journal of Physics A: Mathematical and General 9 (1976), 397–406.

Beijeren, H. van

28. *Interface sharpness in the Ising system*, Communications in Mathematical Physics 40 (1975), 1–6.

Benjamini, I., Lyons, R., Peres, Y., Schramm, O.

29. *Group-invariant percolation on graphs*, Geometric and Functional Analysis 9 (1999), 29–66.

30. *Critical percolation on any nonamenable group has no infinite clusters*, Annals of Probability 27 (1999), 1347–1356.
31. *Uniform spanning forests*, Annals of Probability 29 (2001), 1–65.

Benjamini, I., Schramm, O.

32. *Percolation beyond $\mathbb{Z}^d$, many questions and a few answers*, Electronic Communications in Probability 1 (1996), 71–82.
33. *Percolation in the hyperbolic plane*, Journal of the American Mathematical Society 14 (2001), 487–507.

Ben-Or, M., Linial, N.

34. *Collective coin flipping*, Randomness and Computation, Academic Press, New York, 1990, pp. 91–115.

Berg, J. van den, Häggström, O., Kahn, J.

35. *Some conditional correlation inequalities for percolation and related processes*, Random Structures and Algorithms (2006) (to appear).

Berg, J. van den, Keane, M.

36. *On the continuity of the percolation probability function*, Particle Systems, Random Media and Large Deviations (R. T. Durrett, ed.), Contemporary Mathematics Series, vol. 26, American Mathematical Society, Providence, RI, 1985, pp. 61–65.

Berg, J. van den, Kesten, H.

37. *Inequalities with applications to percolation and reliability*, Journal of Applied Probability 22 (1985), 556–569.

Berg, J. van den, Maes, C.

38. *Disagreement percolation in the study of Markov fields*, Annals of Probability 22 (1994), 749–763.

Bezuidenhout, C. E., Grimmett, G. R., Kesten, H.

39. *Strict inequality for critical values of Potts models and random-cluster processes*, Communications in Mathematical Physics 158 (1993), 1–16.

Biggs, N. L.

40. *Algebraic Graph Theory*, Cambridge University Press, Cambridge, 1984.
41. *Interaction Models*, London Mathematical Society Lecture Notes, vol. 30, Cambridge University Press, Cambridge, 1977.

Billingsley, P.

42. *Convergence of Probability Measures*, Wiley, New York, 1968.

Biskup, M.

43. *Reflection positivity of the random-cluster measure invalidated for non-integer q*, Journal of Statistical Physics 92 (1998), 369–375.

Biskup, M., Borgs, C., Chayes, J. T., Kotecký, R.

44. *Gibbs states of graphical representations of the Potts model with external fields. Probabilistic techniques in equilibrium and nonequilibrium statistical physics*, Journal of Mathematical Physics 41 (2000), 1170–1210.

Biskup, M., Chayes, L.

45. *Rigorous analysis of discontinuous phase transitions via mean-field bounds*, Communications in Mathematical Physics 238 (2003), 53–93.

Biskup, M., Chayes, L., Crawford, N.

46. *Mean-field driven first-order phase transitions*, Journal of Statistical Physics (2005) (to appear).

Biskup, M., Chayes, L., Kotecký, R.
47. *On the continuity of the magnetization and the energy density for Potts models on two-dimensional graphs* (1999).

Bleher, P. M., Ruiz, J., Schonmann, R. H., Shlosman, S., Zagrebnov, V. A.
48. *Rigidity of the critical phases on a Cayley tree*, Moscow Mathematical Journal 1 (2001), 345–363, 470.

Bleher, P. M., Ruiz, J., Zagrebnov, V. A.
49. *On the purity of the limiting Gibbs state for the Ising model on the Bethe lattice*, Journal of Statistical Physics 79 (1995), 473–482.

Blume, M.
50. *Theory of the first-order magnetic phase change in UO_2*, The Physical Review 141 (1966), 517–524.

Blumenthal, R. M., Getoor, R. K.
51. *Markov Processes and Potential Theory*, Academic Press, New York, 1968.

Bodineau, T.
52. *The Wulff construction in three and more dimensions*, Communications in Mathematical Physics 207 (1999), 197–229.
53. *Slab percolation for the Ising model*, Probability Theory and Related Fields 132 (2005), 83–118.
54. *Translation invariant Gibbs states for the Ising model*, Probability Theory and Related Fields (2006) (to appear).

Bodineau, T., Ioffe, D., Velenik, Y.
55. *Rigorous probabilistic analysis of equilibrium crystal shapes*, Journal of Mathematical Physics 41 (2000), 1033–1098.

Boel, R. J., Kasteleyn, P. W.
56. *Correlation-function identities and inequalities for Ising models with pair interactions*, Communications in Mathematical Physics 61 (1978), 191–208.
57. *On a class of inequalities and identities for spin correlation functions of general Ising models*, Physics Letters A 70 (1979), 220–222.

Boivin, D.
58. *First passage percolation: the stationary case*, Probability Theory and Related Fields 86 (1990), 491–499.

Bollobás, B.
59. *Graph Theory, An Introductory Course*, Springer, Berlin, 1979.
60. *The evolution of random graphs*, Transactions of the American Mathematical Society 286 (1984), 257–274.
61. *Random Graphs*, Academic Press, London, 1985.

Bollobás, B., Grimmett, G. R., Janson, S.
62. *The random-cluster process on the complete graph*, Probability Theory and Related Fields 104 (1996), 283–317.

Borgs, C., Chayes, J. T.
63. *The covariance matrix of the Potts model: A random-cluster analysis*, Journal of Statistical Physics 82 (1996), 1235–1297.

Borgs, C., Chayes, J. T., Frieze, A. M., Kim, J. H., Tetali, E., Vigoda, E., Vu, V. V.
64. *Torpid mixing of some MCMC algorithms in statistical physics*, Proceedings of the 40th IEEE Symposium on the Foundations of Computer Science (1999), 218–229.

Borgs, C., Kotecký, R., Medved', I.

65. *Finite-size effects for the Potts model with weak boundary conditions*, Journal of Statistical Physics 109 (2002), 67–131.

Borgs, C., Kotecký, R., Miracle-Solé, S.

66. *Finite-size scaling for the Potts models*, Journal of Statistical Physics 62 (1991), 529–552.

Bourgain, J., Kahn, J., Kalai, G., Katznelson, Y., Linial, N.

67. *The influence of variables in product spaces*, Israel Journal of Mathematics 77 (1992), 55–64.

Bricmont, J., Kuroda, K., Lebowitz, J. L.

68. *First order phase transitions in lattice and continuous systems: Extension of Pirogov–Sinai theory*, Communications in Mathematical Physics 101 (1985), 501–538.

Brightwell, G. R., Häggström, O., Winkler, P.

69. *Nonmonotonic behavior in hard-core and Widom–Rowlinson models*, Journal of Statistical Physics 94 (1999), 415–435.

Broadbent, S. R., Hammersley, J. M.

70. *Percolation processes I. Crystals and mazes*, Proceedings of the Cambridge Philosophical Society 53 (1957), 629–641.

Brooks, R. L., Smith, C. A. B., Stone, A. H., Tutte, W. T.

71. *The dissection of rectangles into squares*, Duke Mathematical Journal 7 (1940), 312–340.

Burton, R. M., Keane, M.

72. *Density and uniqueness in percolation*, Communications in Mathematical Physics 121 (1989), 501–505.
73. *Topological and metric properties of infinite clusters in stationary two-dimensional site percolation*, Israel Journal of Mathematics 76 (1991), 299–316.

Burton, R. M., Pemantle, R.

74. *Local characteristics, entropy and limit theorems for spanning trees and domino tilings via transfer-impedances*, Annals of Probability 21 (1993), 1329–1371.

Camia, F., Newman, C. M.

75. *Continuum nonsimple loops and 2D critical percolation*, Journal of Statistical Physics 116 (2004), 157–173.
76. *The full scaling limit of two-dimensional critical percolation* (2005).

Campanino, M., Chayes, J. T., Chayes, L.

77. *Gaussian fluctuations of connectivities in the subcritical regime of percolation*, Probability Theory and Related Fields 88 (1991), 269–341.

Campanino, M., Ioffe, D., Velenik, Y.

78. *Ornstein–Zernike theory for the finite range Ising models above T_c*, Probability Theory and Related Fields 125 (2002), 305–349.

Capel, H. W.

79. *On the possibility of first-order transitions in Ising systems of triplet ions with zero-field splitting*, Physica 32 (1966), 966–988; 33 (1967), 295–331; 37 (1967), 423–441.

Cardy, J.

80. *Critical percolation in finite geometries*, Journal of Physics A: Mathematical and General 25 (1992), L201.

Cerf, R.

81. *The Wulff crystal in Ising and percolation models*, Ecole d'Eté de Probabilités de Saint Flour XXXIV–2004 (J. Picard, ed.), Lecture Notes in Mathematics, vol. 1878, Springer, Berlin, 2006.

Cerf, R., Kenyon, R.

82. *The low-temperature expansion of the Wulff crystal in the 3D Ising model*, Communications in Mathematical Physics 222 (2001), 147–179.

Cerf, R., Pisztora, Á.

83. *On the Wulff crystal in the Ising model*, Annals of Probability 28 (2000), 947–1017.
84. *Phase coexistence in Ising, Potts and percolation models*, Annales de l'Institut Henri Poincaré, Probabilités et Statistiques 37 (2001), 643–724.

Černý, J., Kotecký, R.

85. *Interfaces for random cluster models*, Journal of Statistical Physics 111 (2003), 73–106.

Chayes, J. T., Chayes, L., Kotecký, R.

86. *The analysis of the Widom–Rowlinson model by stochastic geometric methods*, Communications in Mathematical Physics 172 (1995), 551–569.

Chayes, J. T., Chayes, L., Newman, C. M.

87. *Bernoulli percolation above threshold: an invasion percolation analysis*, Annals of Probability 15 (1987), 1272–1287.

Chayes, J. T., Chayes, L., Schonmann, R. H.

88. *Exponential decay of connectivities in the two-dimensional Ising model*, Journal of Statistical Physics 49 (1987), 433–445.

Chayes, J. T., Chayes, L., Sethna, J. P., Thouless, D. J.

89. *A mean-field spin glass with short-range interactions*, Communications in Mathematical Physics 106 (1986), 41–89.

Chayes, L.

90. *The density of Peierls contours in $d = 2$ and the height of the wedding cake*, Journal of Physics A: Mathematical and General 26 (1993), L481–L488.

Chayes, L., Kotecký, R.

91. *Intermediate phase for a classical continuum model*, Physical Review B 54 (1996), 9221–9224.

Chayes, L., Lei, H. K.

92. *Random cluster models on the triangular lattice* (2005).

Chayes, L., Machta, J.

93. *Graphical representations and cluster algorithms, Part I: discrete spin systems*, Physica A 239 (1997), 542–601.
94. *Graphical representations and cluster algorithms, II*, Physica A 254 (1998), 477–516.

Couronné, O.

95. *Poisson approximation for large clusters in the supercritical FK model* (2004).

Couronné, O., Messikh, R. J.

96. *Surface order large deviations for 2d FK-percolation and Potts models*, Stochastic Processes and their Applications 113 (2004), 81–99.

Cox, J. T., Gandolfi, A., Griffin, P., Kesten, H.

97. *Greedy lattice animals I: Upper bounds*, Advances in Applied Probability 3 (1993), 1151–1169.

Curie, P.
98. *Propriétés magnétiques des corps à diverses températures*, Thesis, Annales de Chimie et de Physique, séries 7 (1895), 289–405.

Dembo, A., Zeitouni, O.
99. *Large Deviations Techniques and Applications*, 2nd edition, Springer, New York, 1998.

De Santis, E.
100. *Strict inequality for phase transition between ferromagnetic and frustrated systems*, Electronic Journal of Probability 6 (2001), Paper 6.

Deuschel, J.-D., Pisztora, Á.
101. *Surface order large deviations for high-density percolation*, Probability Theory and Related Fields 104 (1996), 467–482.

Dobrushin, R. L.
102. *Gibbsian random fields for lattice systems and pairwise interactions*, Functional Analysis and its Applications 2 (1968), 292–301.
103. *Gibbs state describing coexistence of phases for a three–dimensional Ising model*, Theory of Probability and its Applications 18 (1972), 582–600.

Dobrushin, R. L., Kotecký, R., Shlosman, S.
104. *Wulff Construction. A Global Shape from Local Interaction*, Translations of Mathematical Monographs, vol. 104, American Mathematical Society, Providence, RI, 1992.

Domb, C.
105. *On Hammersley's method for one-dimensional covering problems*, Disorder in Physical Systems (G. R. Grimmett, D. J. A. Welsh, eds.), Oxford University Press, Oxford, 1990, pp. 33–53.

Doyle, P. G., Snell, J. L.
106. *Random Walks and Electric Networks*, Carus Mathematical Monographs, vol. 22, Mathematical Association of America, Washington, DC, 1984.

Dudley, R. M.
107. *Real Analysis and Probability*, Wadsworth and Brooks/Cole, Belmont, California, 1989.

Edwards, R. G., Sokal, A. D.
108. *Generalization of the Fortuin–Kasteleyn–Swendsen–Wang representation and Monte Carlo algorithm*, The Physical Review D 38 (1988), 2009–2012.

Edwards, S. F., Anderson, P. W.
109. *Theory of spin glasses*, Journal of Physics F: Metal Physics 5 (1975), 965–974.

Enter, A. van, Fernández, R., Schonmann, R. H., Shlosman, S.
110. *Complete analyticity of the $2D$ Potts model above the critical temperature*, Communications in Mathematical Physics 189 (1997), 373–393.

Enter, A. van, Fernández, R., Sokal, A.
111. *Regularity properties and pathologies of position-space renormalization-group transformations: scope and limitations of Gibbsian theory*, Journal of Statistical Physics 72 (1993), 879–1167.

Essam, J. W., Tsallis, C.
112. *The Potts model and flows: I. The pair correlation function*, Journal of Physics A: Mathematical and General 19 (1986), 409–422.

Ethier, S. N., Kurtz, T. G.
113. *Markov Processes*, Wiley, New York, 1986.

Evans, W., Kenyon, C., Peres, Y., Schulman, L. J.
114. *Broadcasting on trees and the Ising model*, Annals of Applied Probability 10 (2000), 410–433.

Falconer, K. J.
115. *The Geometry of Fractal Sets*, Cambridge University Press, Cambridge, 1985.

Feder, T., Mihail, M.
116. *Balanced matroids*, Proceedings of the 24th ACM Symposium on the Theory of Computing (1992), 26–38.

Fernández, R., Ferrari, P. A., Garcia, N. L.
117. *Loss network representation for Peierls contours*, Annals of Probability 29 (2001), 902–937.

Fernández, R., Fröhlich, J., Sokal, A. D.
118. *Random Walks, Critical Phenomena, and Triviality in Quantum Field Theory*, Springer, Berlin, 1992.

Fontes, L., Newman, C. M.
119. *First passage percolation for random colorings of* $\mathbb{Z}^d$, Annals of Applied Probability 3 (1993), 746–762; *Erratum* 4, 254.

Fortuin, C. M.
120. *On the Random-Cluster Model*, Doctoral thesis, University of Leiden, 1971.
121. *On the random-cluster model. II. The percolation model*, Physica 58 (1972), 393–418.
122. *On the random-cluster model. III. The simple random-cluster process*, Physica 59 (1972), 545–570.

Fortuin, C. M., Kasteleyn, P. W.
123. *On the random-cluster model. I. Introduction and relation to other models*, Physica 57 (1972), 536–564.

Fortuin, C. M., Kasteleyn, P. W., Ginibre, J.
124. *Correlation inequalities on some partially ordered sets*, Communications in Mathematical Physics 22 (1971), 89–103.

Friedgut, E.
125. *Influences in product spaces: KKL and BKKKL revisited*, Combinatorics, Probability, Computing 13 (2004), 17–29.

Friedgut, E., Kalai, G.
126. *Every monotone graph property has a sharp threshold*, Proceedings of the American Mathematical Society 124 (1996), 2993–3002.

Gallavotti, G.
127. *The phase separation line in the two-dimensional Ising model*, Communications in Mathematical Physics 27 (1972), 103–136.

Gallavotti, G., Miracle-Solé, S.
128. *Equilibrium states of the Ising model in the two-phase region*, Physical Review B 5 (1972), 2555–2559.

Gandolfi, A., Keane, M., Newman, C. M.
129. *Uniqueness of the infinite component in a random graph with applications to percolation and spin glasses*, Probability Theory and Related Fields 92 (1992), 511–527.

Gandolfi, A., Keane, M., Russo, L.
130. *On the uniqueness of the infinite occupied cluster in dependent two-dimensional site percolation*, Annals of Probability 16 (1988), 1147–1157.

Garet, O.
131. *Central limit theorems for the Potts model*, Mathematical Physics Electronic Journal 11 (2005), Paper 4.

Georgii, H.-O.
132. *Spontaneous magnetization of randomly dilute ferromagnets*, Journal of Statistical Physics 25 (1981), 369–396.
133. *On the ferromagnetic and the percolative region of random spin systems*, Advances in Applied Probability 16 (1984), 732–765.
134. *Gibbs Measures and Phase Transitions*, Walter de Gruyter, Berlin, 1988.

Georgii, H.-O., Häggström, O.
135. *Phase transition in continuum Potts models*, Communications in Mathematical Physics 181 (1996), 507–528.

Georgii, H.-O., Häggström, O., Maes, C.
136. *The random geometry of equilibrium phases*, Phase Transitions and Critical Phenomena (C. Domb, J. L. Lebowitz, eds.), vol. 18, Academic Press, London, 2000, pp. 1–142,.

Georgii, H.-O., Higuchi, Y.
137. *Percolation and number of phases in the two-dimensional Ising model*, Journal of Mathematical Physics 41 (2000), 1153–1169.

Giacomin, G., Lebowitz, J. L., Maes, C.
138. *Agreement percolation and phase coexistence in some Gibbs systems*, Journal of Statistical Physics 80 (1995), 1379–1403.

Gielis, G., Grimmett, G. R.
139. *Rigidity of the interface in percolation and random-cluster models*, Journal of Statistical Physics 109 (2002), 1–37.

Gobron, T., Merola, I.
140. *Phase transition induced by increasing the range in Potts model* (2005).

Graham, B. T., Grimmett, G. R.
141. *Influence and sharp-threshold theorems for monotonic measures*, Annals of Probability (2005) (to appear).
142. *Random-cluster representation of the Blume–Capel model* (2005).

Griffiths, R. B., Lebowitz, J. L.
143. *Random spin systems: some rigorous results*, Journal of Mathematical Physics 9 (1968), 1284–1292.

Grimmett, G. R.
144. *A theorem about random fields*, Bulletin of the London Mathematical Society 5 (1973), 81–84.
145. *The rank functions of large random lattices*, Journal of London Mathematical Society 18 (1978), 567–575.
146. *Unpublished* (1991).
147. *Differential inequalities for Potts and random-cluster processes*, Cellular Automata and Cooperative Systems (N. Boccara, E. Goles, S. Martínez, P. Picco, eds.), Kluwer, Dordrecht, 1993, pp. 227–236.
148. *Potts models and random-cluster processes with many-body interactions*, Journal of Statistical Physics 75 (1994), 67–121.

149. *The random-cluster model*, Probability, Statistics and Optimisation (F. P. Kelly, ed.), Wiley, Chichester, 1994, pp. 49–63.
150. *Percolative problems*, Probability and Phase Transition (G. R. Grimmett, ed.), Kluwer, Dordrecht, 1994, pp. 69–86.
151. *Comparison and disjoint-occurrence inequalities for random-cluster models*, Journal of Statistical Physics 78 (1995), 1311–1324.
152. *The stochastic random-cluster process and the uniqueness of random-cluster measures*, Annals of Probability 23 (1995), 1461–1510.
153. *Percolation and disordered systems*, Ecole d'Eté de Probabilités de Saint Flour XXVI–1996 (P. Bernard, ed.), Lecture Notes in Mathematics, vol. 1665, Springer, Berlin, 1997, pp. 153–300.
154. *Percolation*, 2nd edition, Springer, Berlin, 1999.
155. *Inequalities and entanglements for percolation and random-cluster models*, Perplexing Problems in Probability; Festschrift in Honor of Harry Kesten (M. Bramson, R. Durrett, eds.), Birkhäuser, Boston, 1999, pp. 91–105.
156. *The random-cluster model*, Probability on Discrete Structures (H. Kesten, ed.), Encyclopaedia of Mathematical Sciences, vol. 110, Springer, Berlin, 2003, pp. 73–123.
157. *Flows and ferromagnets*, Combinatorics, Complexity, and Chance (G. R. Grimmett, C. J. H. McDiarmid, eds.), Oxford University Press, Oxford, 2006 (to appear).
158. *Uniqueness and multiplicity of infinite clusters*, Dynamics and Stochastics (F. den Hollander, D. Denteneer, E. Verbitskiy, eds.), Institute of Mathematical Statistics, 2006, pp. 24–36.

Grimmett, G. R., Holroyd, A. E.
159. *Entanglement in percolation*, Proceedings of the London Mathematical Society 81 (2000), 485–512.

Grimmett, G. R., Janson, S.
160. *Branching processes, and random-cluster measures on trees*, Journal of the European Mathematical Society 7 (2005), 253–281.

Grimmett, G. R., Marstrand, J. M.
161. *The supercritical phase of percolation is well behaved*, Proceedings of the Royal Society (London), Series A 430 (1990), 439–457.

Grimmett, G. R., Newman, C. M.
162. *Percolation in $\infty + 1$ dimensions*, Disorder in Physical Systems (G. R. Grimmett, D. J. A. Welsh, eds.), Oxford University Press, Oxford, 1990, pp. 219–240.

Grimmett, G. R., Piza, M. S. T.
163. *Decay of correlations in subcritical Potts and random-cluster models*, Communications in Mathematical Physics 189 (1997), 465–480.

Grimmett, G. R., Stirzaker, D. R.
164. *Probability and Random Processes*, 3rd edition, Oxford University Press, Oxford, 2001.

Grimmett, G. R., Winkler, S. N.
165. *Negative association in uniform forests and connected graphs*, Random Structures and Algorithms 24 (2004), 444–460.

Häggström, O.
166. *Random-cluster measures and uniform spanning trees*, Stochastic Processes and their Applications 59 (1995), 267–275.

167. *The random-cluster model on a homogeneous tree*, Probability Theory and Related Fields 104 (1996), 231–253.
168. *Almost sure quasilocality fails for the random-cluster model on a tree*, Journal of Statistical Physics 84 (1996), 1351–1361.
169. *Random-cluster representations in the study of phase transitions*, Markov Processes and Related Fields 4 (1998), 275–321.
170. *Positive correlations in the fuzzy Potts model*, Annals of Applied Probability 9 (1999), 1149–1159.
171. *A monotonicity result for hard-core and Widom–Rowlinson models on certain d-dimensional lattices*, Electronic Communications in Probability 7 (2002), 67–78.
172. *Is the fuzzy Potts model Gibbsian*, Annales de l'Institut Henri Poincaré, Probabilités et Statistiques 39 (2002), 891–917.
173. *Finite Markov Chains and Algorithmic Applications*, Cambridge University Press, Cambridge, 2002.

Häggström, O., Jonasson, J., Lyons, R.
174. *Explicit isoperimetric constants and phase transitions in the random-cluster model*, Annals of Probability 30 (2002), 443–473.
175. *Coupling and Bernoullicity in random-cluster and Potts models*, Bernoulli 8 (2002), 275–294.

Häggström, O., Peres, Y.
176. *Monotonicity and uniqueness for percolation on Cayley graphs: all infinite clusters are born simultaneously*, Probability Theory and Related Fields 113 (1999), 273–285.

Hammersley, J. M.
177. *Percolation processes. Lower bounds for the critical probability*, Annals of Mathematical Statistics 28 (1957), 790–795.
178. *A Monte Carlo solution of percolation in a cubic lattice*, Methods in Computational Physics, Volume 1 (B. Alder, S. Fernbach, M. Rotenberg, eds.), Academic Press, London, 1963, pp. 281–298.

Hara, T., Slade, G.
179. *Mean-field critical behaviour for percolation in high dimensions*, Communications in Mathematical Physics 128 (1990), 333–391.
180. *The scaling limit of the incipient infinite cluster in high-dimensional percolation. II. Integrated super-Brownian excursion*, Journal of Mathematical Physics 41 (2000), 1244–1293.

Harris, T. E.
181. *A lower bound for the critical probability in a certain percolation process*, Proceedings of the Cambridge Philosophical Society 56 (1960), 13–20.

Higuchi, Y.
182. *A sharp transition for the two-dimensional Ising percolation*, Probability Theory and Related Fields 97 (1993), 489–514.

Hintermann, D., Kunz, H., Wu, F. Y.
183. *Exact results for the Potts model in two dimensions*, Journal of Statistical Physics 19 (1978), 623–632.

Hollander, W. Th. F. den, Keane, M.
184. *Inequalities of FKG type*, Physica 138A (1986), 167–182.

Holley, R.

185. *Remarks on the FKG inequalities*, Communications in Mathematical Physics 36 (1974), 227–231.

Holley, R., Stroock, D.

186. *A martingale approach to infinite systems of interacting particles*, Annals of Probability 4 (1976), 195–228.

Hryniv, O.

187. *On local behaviour of the phase separation line in the 2D Ising model*, Probability Theory and Related Fields 110 (1998), 91–107.

Ioffe, D.

188. *A note on the extremality of the disordered state for the Ising model on the Bethe lattice*, Letters in Mathematical Physics 37 (1996), 137–143.
189. *A note on the extremality of the disordered state for the Ising model on the Bethe lattice*, Trees (B. Chauvin, S. Cohen, A. Roualt, eds.), Birkhäuser, Basel, 1996, pp. 3–14.

Ising, E.

190. *Beitrag zur Theorie des Ferromagnetismus*, Zeitschrift für Physik 31 (1925), 253–258.

Jagers, P.

191. *Branching Processes with Biological Applications*, Wiley, Chichester, 1975.

Janson, S.

192. *Multicyclic components in a random graph process*, Random Structures and Algorithms 4 (1993), 71–84.

Janson, S., Knuth, D. E., Łuczak, T., Pittel, B.

193. *The birth of the giant component*, Random Structures and Algorithms 4 (1993), 233–358.

Janson, S., Łuczak, T., Ruciński, A.

194. *Random Graphs*, Wiley, New York, 2000.

Jerrum, M.

195. *Mathematical foundations of the Markov chain Monte Carlo method*, Probabilistic Methods for Algorithmic Discrete Mathematics (M. Habib, C. McDiarmid, J. Ramirez-Alfonsin, B. Reed, eds.), Springer, Berlin, 1998, pp. 116–165.

Jonasson, J.

196. *The random cluster model on a general graph and a phase transition characterization of nonamenability*, Stochastic Processes and their Applications 79 (1999), 335–354.

Jonasson, J., Steif, J.

197. *Amenability and phase transition in the Ising model*, Journal of Theoretical Probability 12 (1999), 549–559.

Jössang, P., Jössang, A.

198. *Monsieur C. S. M. Pouillet, de l'Académie, qui découvrit le point "de Curie" en ... 1832*, `http://www.tribunes.com/tribune/art97/jos2f.htm`, Science Tribune (1997).

Kahn, J.

199. *A normal law for matchings*, Combinatorica 20 (2000), 339–391.

Kahn, J., Kalai, G., Linial, N.

200. *The influence of variables on Boolean functions*, Proceedings of 29th Symposium on the Foundations of Computer Science, Computer Science Press, 1988, pp. 68–80.

Kalai, G., Safra, S.

201. *Threshold phenomena and influence*, Computational Complexity and Statistical Physics (A. G. Percus, G. Istrate, C. Moore, eds.), Oxford University Press, New York, 2006.

Kasai, J., Okiji, A.

202. *Percolation problem describing $\pm J$ Ising spin glass system*, Progress in Theoretical Physics 79 (1988), 1080–1094.

Kasteleyn, P. W., Fortuin, C. M.

203. *Phase transitions in lattice systems with random local properties*, Journal of the Physical Society of Japan 26, (1969), 11–14, Supplement.

Kemperman, J. H. B.

204. *On the FKG-inequality for measures on a partially ordered set*, Indagationes Mathematicae 39 (1977), 313–331.

Kesten, H.

205. *Symmetric random walks on groups*, Transactions of the American Mathematical Society 92 (1959), 336–354.
206. *Full Banach mean values on countable groups*, Mathematica Scandinavica 7 (1959), 146–156.
207. *The critical probability of bond percolation on the square lattice equals $\frac{1}{2}$*, Communications in Mathematical Physics 74 (1980), 41–59.
208. *On the time constant and path length of first-passage percolation*, Advances in Applied Probability 12 (1980), 848–863.
209. *Analyticity properties and power law estimates in percolation theory*, Journal of Statistical Physics 25 (1981), 717–756.
210. *Percolation Theory for Mathematicians*, Birkhäuser, Boston, 1982.
211. *Aspects of first-passage percolation*, Ecole d'Eté de Probabilités de Saint Flour XIV-1984 (P. L. Hennequin, ed.), Lecture Notes in Mathematics, vol. 1180, Springer, Berlin, 1986, pp. 125–264.

Kesten, H., Schonmann, R. H.

212. *Behavior in large dimensions of the Potts and Heisenberg models*, Reviews in Mathematical Physics 1 (1990), 147–182.

Kihara, T., Midzuno, Y., Shizume, J.

213. *Statistics of two-dimensional lattices with many components*, Journal of the Physical Society of Japan 9 (1954), 681–687.

Kim, D., Joseph, R. I.

214. *Exact transition temperatures for the Potts model with q states per site for the triangular and honeycomb lattices*, Journal of Physics C: Solid State Physics 7 (1974), L167–L169.

Kirchhoff, G.

215. *Über die Auflösung der Gleichungen, auf welche man bei der Untersuchung der linearen Verteilung galvanischer Strome gefuhrt wird*, Annalen der Physik und Chemie 72 (1847), 497–508.

Kotecký, R.

216. *Geometric representations of lattice models and large volume asymptotics*, Probability and Phase Transition (G. R. Grimmett, ed.), Kluwer, Dordrecht, pp. 153–176.
217. *Phase transitions: on a crossroads of probability and analysis*, Highlights of Mathematical Physics (A. Fokas, J. Halliwell, T. Kibble, B. Zegarlinski, eds.), American Mathematical Society, Providence, RI, 2002, pp. 191–207.

Kotecký, R., Laanait, L., Messager, A., Ruiz, J.

218. *The q-state Potts model in the standard Pirogov–Sinai theory: surface tension and Wilson loops*, Journal of Statistical Physics 58 (1990), 199–248.

Kotecký, R., Preiss, D.

219. *Cluster expansion for abstract polymer models*, Communications in Mathematical Physics 103 (1986), 491–498.

Kotecký, R., Shlosman, S.

220. *First order phase transitions in large entropy lattice systems*, Communications in Mathematical Physics 83 (1982), 493–515.

Kramers, H. A., Wannier, G. H.

221. *Statistics of the two-dimensional ferromagnet, I, II*, The Physical Review 60 (1941), 252–276.

Krengel, U.

222. *Ergodic Theorems*, Walter de Gruyter, Berlin, 1985.

Kuratowski, K.

223. *Topology*, Volume 2, Academic Press, New York, 1968.

Laanait, L., Messager, A., Miracle-Solé, S., Ruiz, J., Shlosman, S.

224. *Interfaces in the Potts model I: Pirogov–Sinai theory of the Fortuin–Kasteleyn representation*, Communications in Mathematical Physics 140 (1991), 81–91.

Laanait, L., Messager, A., Ruiz, J.

225. *Phase coexistence and surface tensions for the Potts model*, Communications in Mathematical Physics 105 (1986), 527–545.

Lanford, O. E., Ruelle, D.

226. *Observables at infinity and states with short range correlations in statistical mechanics*, Communications in Mathematical Physics 13 (1969), 194–215.

Langlands, R., Pouliot, P., Saint-Aubin, Y.

227. *Conformal invariance in two-dimensional percolation*, Bulletin of the American Mathematical Society 30 (1994), 1–61.

Lawler, G. F., Schramm, O., Werner, W.

228. *The dimension of the planar Brownian frontier is* 4/3, Mathematics Research Letters 8 (2001), 401–411.
229. *Values of Brownian intersection exponents III: Two-sided exponents*, Annales de l'Institut Henri Poincaré, Probabilités et Statistiques 38 (2002), 109–123.
230. *One-arm exponent for critical 2D percolation*, Electronic Journal of Probability 7 (2002), Paper 2.
231. *Conformal invariance of planar loop-erased random walks and uniform spanning trees*, Annals of Probability 32 (2004), 939–995.

Lebowitz, J. L., Gallavotti, G.

232. *Phase transtions in binary lattice gases*, Journal of Mathematical Physics 12 (1971), 1129–1133.

Lebowitz, J. L., Martin-Löf, A.

233. *On the uniqueness of the equilibrium state for Ising spin systems*, Communications in Mathematical Physics 25 (1972), 276–282.

Lieb, E. H.

234. *A refinement of Simon's correlation inequality*, Communications in Mathematical Physics 77 (1980), 127–135.

Liggett, T. M.

235. *Interacting Particle Systems*, Springer, Berlin, 1985.

Liggett, T. M., Steif, J.

236. *Stochastic domination: the contact process, Ising models, and FKG measures*, Annales de l'Institut Henri Poincaré, Probabilités et Statistiques 42 (2006), 223–243.

Lindvall, T.

237. *Lectures on the Coupling Method*, Wiley, New York, 1992.

Luczak, M., Łuczak, T.

238. *The phase transition in the cluster-scaled model of a random graph*, Random Structures and Algorithms 28 (2006), 215–246.

Łuczak, T., Pittel, B., Wierman, J. C.

239. *The structure of a random graph at the point of the phase transition*, Transactions of the American Mathematical Society 341 (1994), 721–748.

Lyons, R.

240. *Phase transitions on nonamenable graphs*, Journal of Mathematical Physics 41 (2001), 1099–1126.

Lyons, R., Peres, Y.

241. *Probability on Trees and Networks*, `http://mypage.iu.edu/~rdlyons/prbtree/prbtree.html`, Cambridge University Press, 2007 (to appear).

Lyons, R., Schramm, O.

242. *Indistinguishability of percolation clusters*, Annals of Probability 27 (1999), 1809–1836.

MacKay, D. J. C.

243. *Information Theory, Inference, and Learning Algorithms*, Cambridge University Press, Cambridge, 2003.

Madras, N., Slade, G.

244. *The Self-Avoiding Walk*, Birkhäuser, Boston, 1993.

Maes, C., Vande Velde, K.

245. *The fuzzy Potts model*, Journal of Physics A: Mathematical and General 28 (1995), 4261–4271.

Magalhães, A. C. N. de, Essam, J. W.

246. *The Potts model and flows. II. Many-spin correlation function*, Journal of Physics A: Mathematical and General 19 (1986), 1655–1679.

247. *The Potts model and flows. III. Standard and subgraph break-collapse methods*, Journal of Physics A: Mathematical and General 21 (1988), 473–500.

Martin, J.

248. *Reconstruction thresholds on regular trees*, Discrete Mathematics and Theoretical Computer Science, Proceedings AC (C. Banderier, C. Krattenthaler, eds.), 2003, pp. 191–204.

Martinelli, F.

249. *Lectures on Glauber dynamics for discrete spin models*, Ecole d'Eté de Probabilités de Saint Flour XXVII–1997 (P. Bernard, ed.), Lecture Notes in Mathematics, vol. 1717, Springer, Berlin, 1999, pp. 93–191.

Martinelli, F., Sinclair, A., Weitz, D.

250. *Glauber dynamics on trees: Boundary conditions and mixing time*, Communications in Mathematical Physics 250 (2004), 310–334.

Martirosian, D. H.

251. *Translation invariant Gibbs states in the q-state Potts model*, Communications in Mathematical Physics 105 (1986), 281–290.

McCoy, B. M., Wu, T. T.

252. *The two-dimensional Ising model*, Harvard University Press, Cambridge, MA, 1973.

Meester, R., Roy, R.

253. *Continuum Percolation*, Cambridge University Press, Cambridge, 1996.

Messager, A., Miracle-Solé, S., Ruiz, J., Shlosman, S.

254. *Interfaces in the Potts model. II. Antonov's rule and rigidity of the order disorder interface*, Communications in Mathematical Physics 140 (1991), 275–290.

Moran, P. A. P.

255. *An Introduction to Probability Theory*, Clarendon Press, Oxford, 1968.

Mossel, E.

256. *Survey: Information flow on trees*, Graphs, Morphisms and Statistical Physics (J. Nešetřil, P. Winkler, eds.), American Mathematical Society, DIMACS, 2004, pp. 155–170.

Mossel, E., O'Donnell, R., Regev, O., Steif, J., Sudakov, B.

257. *Non-interactive correlation distillation, inhomogeneous Markov chains, and the reverse Bonami–Beckner inequality*, Israel Journal of Mathematics (2006).

Nachtergaele, B.

258. *A stochastic geometric approach to quantum spin systems*, Probability and Phase Transition (G. R. Grimmett, ed.), Kluwer, Dordrecht, 1994, pp. 237–246.

Newman, C. M.

259. *Disordered Ising systems and random cluster representations*, Probability and Phase Transition (G. R. Grimmett, ed.), Kluwer, Dordrecht, 1994, pp. 247–260.
260. *Topics in Disordered Systems*, Birkhäuser, Boston, 1997.

Newman, C. M., Schulman, L. S.

261. *Infinite clusters in percolation models*, Journal of Statistical Physics 26 (1981), 613–628.

Newman, C. M., Stein, D. L.

262. *Short-range spin glasses: results and speculations*, Spin Glass Theory (E. Bolthausen, A. Bovier, eds.), Springer, Berlin, 2006 (to appear).
263. *Short-range spin glasses: selected open problems*, Mathematical Statistical Physics (A. Bovier, F. Dunlop, F. den Hollander, A. van Enter, eds.), Proceedings of 2005 Les Houches Summer School LXXXIII, Elsevier, 2006 (to appear).

Onsager, L.

264. *Crystal statistics, I. A two-dimensional model with an order–disorder transition*, The Physical Review 65 (1944), 117–149.

Paterson, A. L. T.

265. *Amenability*, American Mathematical Society, Providence, RI, 1988.

Peierls, R.

266. *On Ising's model of ferromagnetism*, Proceedings of the Cambridge Philosophical Society 36 (1936), 477–481.

Pemantle, R.

267. *The contact process on trees*, Annals of Probability 20 (1992), 2089–2016.
268. *Towards a theory of negative dependence*, Journal of Mathematical Physics 41 (2000), 1371–1390.

Petritis, D.

269. *Equilibrium statistical mechanics of frustrated spin glasses: a survey of mathematical results*, Annales de l'Institut Henri Poincaré, Physique Théorique 84 (1996), 255–288.

Pfister, C.-E.

270. *Translation invariant equilibrium states of ferromagnetic abelian lattice systems*, Communications in Mathematical Physics 86 (1982), 375–390.
271. *Phase transitions in the Ashkin–Teller model*, Journal of Statistical Physics 29 (1982), 113–116.

Pfister, C.-E., Vande Velde, K.

272. *Almost sure quasilocality in the random cluster model*, Journal of Statistical Physics 79 (1995), 765–774.

Pfister, C.-E., Velenik, Y.

273. *Random-cluster representation for the Ashkin–Teller model*, Journal of Statistical Physics 88 (1997), 1295–1331.

Pirogov, S. A., Sinaĭ, Ya. G.

274. *Phase diagrams of classical lattice systems*, Theoretical and Mathematical Physics 25 (1975), 1185–1192.
275. *Phase diagrams of classical lattice systems, continuation*, Theoretical and Mathematical Physics 26 (1976), 39–49.

Pisztora, Á.

276. *Surface order large deviations for Ising, Potts and percolation models*, Probability Theory and Related Fields 104 (1996), 427–466.

Pittel, B.

277. *On tree census and the giant component of sparse random graphs*, Random Structures and Algorithms 1 (1990), 311–342.

Potts, R. B.

278. *Some generalized order–disorder transformations*, Proceedings of the Cambridge Philosophical Society 48 (1952), 106–109.

Preston, C. J.

279. *Gibbs States on Countable Sets*, Cambridge University Press, Cambridge, 1974.
280. *A generalization of the FKG inequalities*, Communications in Mathematical Physics 36 (1974), 233–241.

Procacci, A., Scoppola, B.

281. *Convergent expansion for random cluster model with $q > 0$ on infinite graphs* (2005).

Propp, J. G., Wilson, D. B.

282. *Exact sampling with coupled Markov chains and applications to statistical mechanics*, Random Structures and Algorithms 9 (1996), 223–252.

Reimer, D.

283. *Proof of the van den Berg–Kesten conjecture*, Combinatorics, Probability, Computing 9 (2000), 27–32.

Rohde, S., Schramm, O.

284. *Basic properties of SLE*, Annals of Mathematics 161 (2005), 879–920.

Ruelle, D.

285. *Existence of a phase transition in a continuous classical system*, Physical Review Letters 26 (1971), 303–304.

Russo, L.

286. *A note on percolation*, Zeitschrift für Wahrscheinlichkeitstheorie und Verwandte Gebiete 43 (1978), 39–48.

287. *On the critical percolation probabilities*, Zeitschrift für Wahrscheinlichkeitstheorie und Verwandte Gebiete 56 (1981), 229–237.

Sakai, A.

288. *Lace expansion for the Ising model* (2005).

Salas, J., Sokal, A. D.

289. *Dynamic critical behavior of a Swendsen–Wang-type algorithm for the Ashkin–Teller model*, Journal of Statistical Physics 85 (1996), 297–361.

Schlicting, G.

290. *Polynomidentitäten und Permutationsdarstellungen lokalkompakter Gruppen*, Inventiones Mathematicae 55 (1979), 97–106.

Schneider, R.

291. *Convex Bodies: The Brunn–Minkowski Theory*, Cambridge University Press, Cambridge, 1993.

Schonmann, R. H.

292. *Metastability and the Ising model*, Documenta Mathematica, Extra volume (G. Fischer, U. Rehmann, eds.), Proceedings of the International Congress of Mathematicians, Berlin 1998, vol. III, pp. 173–181.

293. *Multiplicity of phase transitions and mean-field criticality on highly non-amenable graphs*, Communications in Mathematical Physics 219 (2001), 271–322.

Schramm, O.

294. *Scaling limits of loop-erased walks and uniform spanning trees*, Israel Journal of Mathematics 118 (2000), 221–288.

295. *Conformally invariant scaling limits*, Proceedings of the International Congress of Mathematicians, Madrid, 2006 (to appear).

Schramm, O., Sheffield, S.

296. *Harmonic explorer and its convergence to* SLE_4, Annals of Probability 33 (2005), 2127–2148.

297. *Contour lines of the 2D Gaussian free field* (2006) (to appear).

Seppäläinen, T.

298. *Entropy for translation-invariant random-cluster measures*, Annals of Probability 26 (1998), 1139–1178.

Sharpe, M.

299. *General Theory of Markov Processes*, Academic Press, San Diego, 1988.

Simon, B.

300. *Correlation inequalities and the decay of correlations in ferromagnets*, Communications in Mathematical Physics 77 (1980), 111–126.

Sinaĭ, Y. G.

301. *Theory of Phase Transitions: Rigorous Results*, Pergamon Press, Oxford, 1982.

Slade, G.

302. *Bounds on the self-avoiding walk connective constant*, Journal of Fourier Analysis and its Applications, Special Issue: Proceedings of the Conference in Honor of Jean-Pierre Kahane, 1993 (1995), 525–533.
303. *The lace expansion and its applications*, Ecole d'Eté de Probabilités de Saint Flour (J. Picard, ed.), Lecture Notes in Mathematics, vol. 1879, Springer, Berlin, 2006.

Smirnov, S.

304. *Critical percolation in the plane: conformal invariance, Cardy's formula, scaling limits*, Comptes Rendus des Séances de l'Académie des Sciences. Série I. Mathématique 333 (2001), 239–244.
305. *Critical percolation in the plane. I. Conformal invariance and Cardy's formula. II. Continuum scaling limit* (2001).
306. *In preparation* (2006).

Smirnov, S., Werner, W.

307. *Critical exponents for two-dimensional percolation*, Mathematics Research Letters 8 (2001), 729–744.

Sokal, A.

308. *The multivariate Tutte polynomial (alias Potts model) for graphs and matroids* (2005).

Stigler, S. M.

309. *Stigler's law of eponymy*, Transactions of the New York Academy of Sciences 39 (1980), 147–157; Reprinted in *Statistics on the Table*, by Stigler, S. M. (1999), Harvard University Press, Cambridge, MA.

Swendsen, R. H., Wang, J. S.

310. *Nonuniversal critical dynamics in Monte Carlo simulations*, Physical Review Letters 58 (1987), 86–88.

Sykes, M. E., Essam, J. W.

311. *Exact critical percolation probabilities for site and bond problems in two dimensions*, Journal of Mathematical Physics 5 (1964), 1117–1127.

Trofimov, V. I.

312. *Automorphism groups of graphs as topological groups*, Mathematical Notes 38 (1985), 717–720.

Tutte, W. T.

313. *Graph Theory*, first published in 1984, Cambridge University Press, Cambridge, 2001.

Welsh, D. J. A.

314. *Percolation in the random-cluster process*, Journal of Physics A: Mathematical and General 26 (1993), 2471–2483.

Welsh, D. J. A., Merino, C.

315. *The Potts model and the Tutte polynomial*, Journal of Mathematical Physics 41 (2000), 1127–1152.

Werner, W.

316. *Random planar curves and Schramm–Loewner evolutions*, Ecole d'Eté de Probabilités de Saint Flour XXXII–2002 (J. Picard, ed.), Lecture Notes in Mathematics, vol. 1840, Springer, Berlin, 2004, pp. 107–195.

Whittle, P. W.

317. *Systems in Stochastic Equilibrium*, Wiley, Chichester, 1986.
318. *Polymer models and generalized Potts–Kasteleyn models*, Journal of Statistical Physics 75 (1994), 1063–1092.

Widom, B., Rowlinson, J. S.

319. *New model for the study of liquid–vapor phase transition*, Journal of Chemical Physics 52 (1970), 1670–1684.

Wilson, R. J.

320. *Introduction to Graph Theory*, Longman, London, 1979.

Wiseman, S., Domany, E.

321. *Cluster method for the Ashkin–Teller model*, Physical Review E 48 (1993), 4080–4090.

Wolff, U.

322. *Collective Monte Carlo updating for spin systems*, Physical Review Letters 62 (1989), 361–364.

Wood, De Volson

323. *Problem 5*, American Mathematical Monthly 1 (1894), 99, 211–212.

Wu, F. Y.

324. *The Potts model*, Reviews in Modern Physics 54 (1982), 235–268.

Wulff, G.

325. *Zur Frage der Geschwindigkeit des Wachsturms und der Auflösung der Krystallflächen*, Zeitschrift für Krystallographie und Mineralogie 34 (1901), 449–530.

Zahradník, M.

326. *An alternate version of Pirogov–Sinai theory*, Communications in Mathematical Physics 93, 559–581.

Subject Index

Grundlehren der mathematischen Wissenschaften

A Series of Comprehensive Studies in Mathematics

A Selection

246. Naimark/Stern: Theory of Group Representations
247. Suzuki: Group Theory I
248. Suzuki: Group Theory II
249. Chung: Lectures from Markov Processes to Brownian Motion
250. Arnold: Geometrical Methods in the Theory of Ordinary Differential Equations
251. Chow/Hale: Methods of Bifurcation Theory
252. Aubin: Nonlinear Analysis on Manifolds. Monge-Ampère Equations
253. Dwork: Lectures on ρ-adic Differential Equations
254. Freitag: Siegelsche Modulfunktionen
255. Lang: Complex Multiplication
256. Hörmander: The Analysis of Linear Partial Differential Operators I
257. Hörmander: The Analysis of Linear Partial Differential Operators II
258. Smoller: Shock Waves and Reaction-Diffusion Equations
259. Duren: Univalent Functions
260. Freidlin/Wentzell: Random Perturbations of Dynamical Systems
261. Bosch/Güntzer/Remmert: Non Archimedian Analysis – A System Approach to Rigid Analytic Geometry
262. Doob: Classical Potential Theory and Its Probabilistic Counterpart
263. Krasnosel'skiĭ/Zabreĭko: Geometrical Methods of Nonlinear Analysis
264. Aubin/Cellina: Differential Inclusions
265. Grauert/Remmert: Coherent Analytic Sheaves
266. de Rham: Differentiable Manifolds
267. Arbarello/Cornalba/Griffiths/Harris: Geometry of Algebraic Curves, Vol. I
268. Arbarello/Cornalba/Griffiths/Harris: Geometry of Algebraic Curves, Vol. II
269. Schapira: Microdifferential Systems in the Complex Domain
270. Scharlau: Quadratic and Hermitian Forms
271. Ellis: Entropy, Large Deviations, and Statistical Mechanics
272. Elliott: Arithmetic Functions and Integer Products
273. Nikol'skiĭ: Treatise on the shift Operator
274. Hörmander: The Analysis of Linear Partial Differential Operators III
275. Hörmander: The Analysis of Linear Partial Differential Operators IV
276. Liggett: Interacting Particle Systems
277. Fulton/Lang: Riemann-Roch Algebra
278. Barr/Wells: Toposes, Triples and Theories
279. Bishop/Bridges: Constructive Analysis
280. Neukirch: Class Field Theory
281. Chandrasekharan: Elliptic Functions
282. Lelong/Gruman: Entire Functions of Several Complex Variables
283. Kodaira: Complex Manifolds and Deformation of Complex Structures
284. Finn: Equilibrium Capillary Surfaces
285. Burago/Zalgaller: Geometric Inequalities
286. Andrianaov: Quadratic Forms and Hecke Operators
287. Maskit: Kleinian Groups
288. Jacod/Shiryaev: Limit Theorems for Stochastic Processes

289. Manin: Gauge Field Theory and Complex Geometry
290. Conway/Sloane: Sphere Packings, Lattices and Groups
291. Hahn/O'Meara: The Classical Groups and K-Theory
292. Kashiwara/Schapira: Sheaves on Manifolds
293. Revuz/Yor: Continuous Martingales and Brownian Motion
294. Knus: Quadratic and Hermitian Forms over Rings
295. Dierkes/Hildebrandt/Küster/Wohlrab: Minimal Surfaces I
296. Dierkes/Hildebrandt/Küster/Wohlrab: Minimal Surfaces II
297. Pastur/Figotin: Spectra of Random and Almost-Periodic Operators
298. Berline/Getzler/Vergne: Heat Kernels and Dirac Operators
299. Pommerenke: Boundary Behaviour of Conformal Maps
300. Orlik/Terao: Arrangements of Hyperplanes
301. Loday: Cyclic Homology
302. Lange/Birkenhake: Complex Abelian Varieties
303. DeVore/Lorentz: Constructive Approximation
304. Lorentz/v. Golitschek/Makovoz: Construcitve Approximation. Advanced Problems
305. Hiriart-Urruty/Lemaréchal: Convex Analysis and Minimization Algorithms I. Fundamentals
306. Hiriart-Urruty/Lemaréchal: Convex Analysis and Minimization Algorithms II. Advanced Theory and Bundle Methods
307. Schwarz: Quantum Field Theory and Topology
308. Schwarz: Topology for Physicists
309. Adem/Milgram: Cohomology of Finite Groups
310. Giaquinta/Hildebrandt: Calculus of Variations I: The Lagrangian Formalism
311. Giaquinta/Hildebrandt: Calculus of Variations II: The Hamiltonian Formalism
312. Chung/Zhao: From Brownian Motion to Schrödinger's Equation
313. Malliavin: Stochastic Analysis
314. Adams/Hedberg: Function spaces and Potential Theory
315. Bürgisser/Clausen/Shokrollahi: Algebraic Complexity Theory
316. Saff/Totik: Logarithmic Potentials with External Fields
317. Rockafellar/Wets: Variational Analysis
318. Kobayashi: Hyperbolic Complex Spaces
319. Bridson/Haefliger: Metric Spaces of Non-Positive Curvature
320. Kipnis/Landim: Scaling Limits of Interacting Particle Systems
321. Grimmett: Percolation
322. Neukirch: Algebraic Number Theory
323. Neukirch/Schmidt/Wingberg: Cohomology of Number Fields
324. Liggett: Stochastic Interacting Systems: Contact, Voter and Exclusion Processes
325. Dafermos: Hyperbolic Conservation Laws in Continuum Physics
326. Waldschmidt: Diophantine Approximation on Linear Algebraic Groups
327. Martinet: Perfect Lattices in Euclidean Spaces
328. Van der Put/Singer: Galois Theory of Linear Differential Equations
329. Korevaar: Tauberian Theory. A Century of Developments
330. Mordukhovich: Variational Analysis and Generalized Differentiation I: Basic Theory
331. Mordukhovich: Variational Analysis and Generalized Differentiation II: Applications
332. Kashiwara/Schapira: Categories and Sheaves. An Introduction to Ind-Objects and Derived Categories
333. Grimmett: The Random-Cluster Model
334. Sernesi: Deformations of Algebraic Schemes
335. Bushnell/Henniart: The Local Langlands Conjecture for GL(2)

Printing: Krips bv, Meppel
Binding: Stürtz, Würzburg